Cell Fusion

Cell Fusion

Edited by
Arthur E. Sowers

American National Red Cross
Rockville, Maryland

Plenum Press • New York and London

Library of Congress Cataloging in Publication Data

Cell fusion.

Includes bibliographies and index.
1. Cell hybridization. I. Sowers, Arthur E.
QH451.C444 1987 574.87′6 86-30392
ISBN 0-306-42351-0

A Division of Plenum Publishing Corporation
233 Spring Street, New York, N.Y. 10013

Printed in the United States of America

Contributors

Patricia A. Baldwin, Cancer Research Institute, School of Medicine, University of California at San Francisco, San Francisco, California 94143. *Present address:* California Biotechnology, Inc., Mountain View, California 94043

George W. Bates, Department of Biological Science and Institute of Molecular Biophysics, Florida State University, Tallahassee, Florida 32306

Joe Bentz, Departments of Pharmacy and Pharmaceutical Chemistry, School of Pharmacy, University of California at San Francisco, San Francisco, California 94143

L. T. Boni, Department of Physiology and Biophysics, Harvard Medical School, Boston, Massachusetts 02115

Wendy F. Boss, Botany Department, North Carolina State University, Raleigh, North Carolina 27695

R. Malcolm Brown, Jr., Department of Botany, University of Texas at Austin, Austin, Texas 78712

E. L. Chang, Naval Research Laboratory, Bio/Molecular Engineering Branch, Code 6190, Washington, D.C. 20375

Purnell W. Choppin, Virology Laboratory, Rockefeller University, New York, New York 10021. *Present address:* Howard Hughes Medical Institute, Bethesda, Maryland 20814

Jerome Connor, Department of Biochemistry, University of Tennessee, Knoxville, Tennessee 37996. *Present Address:* Department of Cell Biology, M. D. Anderson Tumor Institute, Houston, Texas 77030

Mary K. Conrad, National Institute on Drug Abuse, Addiction Research Center, Baltimore, Maryland 21224

Christine B. Couch, Department of Biochemistry and Neurology, Program in Neuroscience, Baylor College of Medicine, Houston, Texas 77030

Carl E. Creutz, Department of Pharmacology, Programs in Biophysics, Neuroscience, and Cell and Molecular Biology, and Center for Diabetes and Cancer Research and Training, University of Virginia, Charlottesville, Virginia 22908

Richard L. Davidson, Center for Genetics, University of Illinois, Chicago, Illinois 60612

Nejat Düzgüneş, Cancer Research Institute, School of Medicine, and Department of Pharmaceutical Chemistry, School of Pharmacy, University of California at San Francisco, San Francisco, California 94143

Helen C. Hamman, Department of Pharmacology, Programs in Biophysics, Neuroscience and Cell and Molecular Biology, and Center for Diabetes and Cancer Research and Training, University of Virginia, Charlottesville, Virginia 22908

Clare A. Hasenkampf, Department of Biological Science and Institute of Molecular Biophysics, Florida State University, Tallahassee, Florida 32306. *Present address:* Department of Biology, University of California at San Diego, La Jolla, California 92093

Robert M. Hochmuth, Department of Biomedical Engineering, Duke University, Durham, North Carolina 27706. *Present address:* Department of Mechanical Engineering, Duke University, Durham, North Carolina 27706

Keelung Hong, Cancer Research Institute, School of Medicine, University of California at San Francisco, San Francisco, California 94143

Leaf Huang, Department of Biochemistry, University of Tennessee, Knoxville, Tennessee 37996

S. W. Hui, Department of Biophysics, Roswell Park Memorial Institute, Buffalo, New York 14263

Veena Kapoor, American National Red Cross, Jerome H. Holland Laboratory for the Biomedical Sciences, Rockville, Maryland 20855

Seong Yun Lee, Department of Botany, University of Texas at Austin, Austin, Texas 78712

Michael R. Lieber, Laboratory of Pathology, National Institutes of Health, Bethesda, Maryland 20205. *Present address:* Laboratory of Molecular Biology, NIDDK, National Institutes of Health, Bethesda, Maryland 20892

John W. Littlefield, Department of Physiology, Johns Hopkins University School of Medicine, Baltimore, Maryland 21205

Mathew M. S. Lo, National Institute on Drug Abuse, Addiction Research Center, Baltimore, Maryland 21224

William H. Martin, Department of Pharmacology, Programs in Biophysics, Neuroscience and Cell and Molecular Biology, and Center for Diabetes and Cancer Research, University of Virginia, Charlottesville, Virginia 22908

Paul R. Meers, Cancer Research Institute, School of Medicine, University of California at San Francisco, San Francisco, California 94143

Donna M. Miles, Department of Biomedical Engineering, Duke University, Durham, North Carolina 27706

Dorothy I. Mundy, Department of Biochemistry and Neurology, Program in Neuroscience, Baylor College of Medicine, Houston, Texas 77030

Lawrence J. Nea, Department of Biological Science and Institute of Molecular Biophysics, Florida State University, Tallahassee, Florida 32306

Shlomo Nir, Seagram Centre for Soil and Water Sciences, Hebrew University of Jerusalem, Rehovot 76-100, Israel

Shinpei Ohki, Department of Biophysical Sciences, School of Medicine, State University of New York at Buffalo, Buffalo, New York 14214

Demetrios Papahadjopoulos, Cancer Research Institute and Department of Pharmacology, School of Medicine, University of California at San Francisco, San Francisco, California 94143

Helmut Plattner, Faculty of Biology, University of Konstanz, D-7750 Konstanz, Federal Republic of Germany

David S. Roos, Department of Biology, Stanford University, Stanford, California 94305

James A. Saunders, Germplasm Quality and Enhancement Laboratory, USDA Plant Genetics and Germplasm Institute, Beltsville, Maryland 20705

Einhard Schierenberg, Max-Planck-Institut für Experimentelle Medizin, Abteilung Chemie, D-3400 Göttingen, Federal Republic of Germany

Richard Schlegel, Laboratory of Tumor Virus Biology, National Cancer Institute, Bethesda, Maryland 20892

Robert A. Schlegel, Department of Molecular and Cell Biology, Pennsylvania State University, University Park, Pennsylvania 16802

Dave P. Siegel, Miami Valley Laboratories, Proctor and Gamble Company, Cincinnati, Ohio 45247

Solomon H. Snyder, Departments of Neuroscience, Pharmacology and Experimental Therapeutics, and Psychiatry and Behavioral Sciences, Johns Hopkins University School of Medicine, Baltimore, Maryland 21205

Arthur E. Sowers, American National Red Cross, Jerome H. Holland Laboratory for the Biomedical Sciences, Rockville, Maryland 20855

Patricia G. Spear, Department of Molecular Genetics and Cell Biology and the Committee on Virology, University of Chicago, Chicago, Illinois 60637

Warren J. Strittmatter, Department of Biochemistry and Neurology, Program in Neuroscience, Baylor College of Medicine, Houston, Texas 77030

Francis C. Szoka, Departments of Pharmacy and Pharmaceutical Chemistry, School of Pharmacy, University of California at San Francisco, San Francisco, California 94143

Tian Y. Tsong, Department of Biological Chemistry, Johns Hopkins University School of Medicine, Baltimore, Maryland 21205

Woodring E. Wright, Department of Cell Biology, University of Texas Southwestern Medical School, Dallas, Texas 75235

William J. Zaks, Department of Pharmacology, Programs in Biophysics, Neuroscience and Cell and Molecular Biology, and Center for Diabetes and Cancer Research and Training, University of Virginia, Charlottesville, Virginia 22908

Preface

Over the last decade the volume *Membrane Fusion,* edited by Poste and Nicholson, has probably served as one of the major sources of review information on fusion in membrane systems. Since its publication much new information has been collected. New methods of inducing fusion have been invented or discovered, and new applications for fusion have been found. The need for an up-to-date monograph that covers and integrates these subjects, reviews established material, and rationalizes and integrates the old and the new is thus obvious. This book is the product of efforts to meet this need.

Most of the current work in the field of membrane fusion takes place within the context of intact or modified cells. Hence this book emphasizes the plasma membrane. Each chapter is either a review, a report, or a short historical overview, depending, respectively, on whether the subject is large in scope and has a long history, or the subject is in such an early stage of development that most of what is known is still in the hands of a relatively small number of investigators and is best covered in report form.

Since all fusion research involves either the goal of learning more about how fusion occurs or the goal of achieving a delivery function, this book covers both fundamental and applied topics. It is expected that workers concerned with chapters dealing primarily with one of these two goals might, sooner or later, have a look at the other chapters for new ideas. This book is geared toward recognition of current trends and rapidly developing ideas and techniques. Also, it recognizes fusion as it relates to today's biotechnology scene. In this regard plant systems have often been overlooked as systems for study and subjects for chapters in cell fusion books despite the fact that plants not only are essential for the existence of all animal life but, as the reader will appreciate, also have some novel characteristics of significance for fusion research.

This book is divided into four parts of roughly equal importance. The first two parts are devoted to fusion mechanisms and mechanism-related phenomenology in cell membranes and artificial membranes, respectively. In Part I, Fusion in Plasma Membranes, the authors and topics were selected to illustrate the successful application of some novel experimental approach or unique biological characteristics to study fusion in some plasma membrane. The material covered was selected to emphasize depth rather than breadth.

In Part II, Fusion in Model Membranes, we devote attention to studies which involve artificial membranes, i.e., vesicles or liposomes. Although these membranes are rarely reconstituted with more than one or two of the many lipid components found in natural membranes or with the lipid asymmetry and integral proteins known to exist in natural membranes, they nevertheless are almost the only practical way to have a simple and clean experimental system to simplify the complex fusion-related interactions at molecular levels and analyze the effects of lipid structure and chemical factors on membrane fusion events. Indeed much, if not most, of what we know about fusion in membranes has come from work with vesicle systems. The chapter by Siegel and the chapter by Szoka are complementary reviews. The first author emphasizes a more conceptual perspective while the second emphasizes a more empirical viewpoint. A newcomer to the field of membrane fusion would do well to read these two pace-setting chapters before any of the others. The next two chapters summarize the effects of chemical conditions on vesicle fusion and survey protein modulation of vesicle fusion, while the remaining chapters specialize in the major topics of acid-induced fusion, polyethylene-glycol-induced fusion, and the development of a formalization of membrane fusion based on physical principles. The last chapter deals with a newly recognized experimental variable—hydrostatic pressure as a probe of fusion mechanisms. A book on fusion would not be complete without a section on fusion in model membranes, and a reader should at least develop an appreciation for what has been learned from them.

The next two parts are devoted to those researchers primarily interested in achieving a goal via a process in which fusion is but one step. Part III, New Membrane Fusion Methods, reviews electrofusion and describes the laser-induced fusion method. Both of these fusion methods are new, having been discovered or developed in the last half decade, and depend on physical effects rather than changes in chemical conditions brought about by added substances. Laser-induced fusion is significant because it is probably the only way to selectively fuse two cells in a tissue mass at one specific point in time without disturbing the surrounding cells. Electrofusion can potentially induce high yields with high time resolution and is attracting much interest as an efficient and convenient fusion method. Our own experimental investigations on the mechanisms of electrofusion are beginning to reveal that electrofusion may be more than just another fusion method. Rather it may reveal additional new knowledge relevant not only to fusion phenomena but also to membrane structure and properties.

In Part IV, Applications of Membrane Fusion, we not only give examples of how electrofusion can be used to obtain better monoclonal antibodies, improve plant genetic engineering procedures, and make a unique measurement of membrane elasticity properties but also present an overview of how fusion was exploited in somatic cell genetics studies, a description of the red cell ghost microinjection technique, a review of chemically induced plant protoplast fusion, and finally a review of novel methods of selecting desired fusion products out of the many that are produced.

Finally, I would like to thank all of the authors for their fine contributions and acknowledge the American Red Cross, National Headquarters Sector, for support during the editing of this book.

The work on this volume was completed after the American National Red Cross moved its research activities from Bethesda to the Jerome H. Holland Laboratory for the Biomedical Sciences in Rockville, Maryland.

Arthur E. Sowers

Rockville, Maryland

Contents

I

Fusion in Cell Membranes

Chapter 1

Virus-Induced Cell Fusion

Patricia G. Spear

1. INTRODUCTION

1.1. Purpose

Attention is focused in this review on viral proteins that are known to mediate or influence virus-induced cell fusion. Because these proteins also mediate or influence entry of virus into cells, the virion–cell fusion required for this entry is discussed as well. The emphasis is on new information that has emerged since publication of an earlier review on these subjects (White *et al.,* 1983). It was not the intention to provide comprehensive treatment of these subjects.

1.2. Viruses That Induce Cell Fusion

Representatives of many families of animal viruses have been shown to induce cell fusion under appropriate conditions (Table I). These viruses have both similarities and differences that are relevant to an analysis of virus-induced cell fusion. The similarities include the following: All animal viruses capable of inducing cell fusion have one structural compo-

PATRICIA G. SPEAR • Department of Molecular Genetics and Cell Biology and the Committee on Virology, University of Chicago, Chicago, Illinois 60637.

Table I. Some Fusion-Inducing Viruses and Selected Properties

Virus family	Representative member	Genome		Nucleocapsid		Site of envelopment	Reference[b]
		Nucleic acid[a]	Size (kb or kbp)	Symmetry	Site of assembly		
Rhabdovirus	Vesicular stomatitis (VSV)	ssRNA(−)	12	Helical	Cytoplasm	Plasma membrane	Wagner (1975)
Paramyxovirus	Sendai	ssRNA(−)	16–18	Helical	Cytoplasm	Plasma membrane	Choppin and Compans (1975)
Orthomyxovirus	Influenza	ssRNA(−)	13.5–14	Helical	Cytoplasm	Plasma membrane	Lamb and Choppin (1983)
Bunyavirus	LaCross	ssRNA(−)	14–17	Helical	Cytoplasm	Golgi complex	Bishop and Shope (1979)
Togavirus	Semliki Forest (SFV)	ssRNA(+)	12	Icosahedral	Cytoplasm	Plasma membrane	Garoff *et al.* (1982)
Coronavirus	Infectious bronchitis (IBV)	ssRNA(+)	18	Helical	Cytoplasm	Endoplasmic reticulum	Siddell *et al.* (1983)
Retrovirus	Mouse mammary tumor (MMTV)	ssRNA(+)	8–10[c]	Icosahedral	Cytoplasm	Plasma membrane	Varmus and Swanstrom (1984)
Herpesvirus	Herpes simplex (HSV)	dsDNA	140–250	Icosahedral	Nucleus	Inner nuclear membrane	Spear and Roizman (1980)
Poxvirus	Vaccinia	dsDNA	200–330	Complex	Cytoplasm	Golgi complex[d]	Moss (1974)

[a]ss, single-stranded; ds, double-stranded; (+) or (−) denotes whether genomic RNA has the same polarity as mRNA (+) or opposite polarity (−).
[b]These reviews summarize properties characteristic of the virus families. Replication strategies of most of the RNA viruses listed are also described and compared in Strauss and Strauss (1983). Some of the information is documented in papers cited in the text.
[c]Two identical RNAs of this size are present in each virion.
[d]Poxviruses are assembled in the cytoplasm to form a particle of complex architecture that can be infectious in cell culture without acquiring an envelope derived from cell membrane. The form of the virus that is infectious in nature probably has an envelope, however, and this envelope is derived from membrane of the Golgi complex (Payne and Kristenson, 1979; Hiller and Weber, 1985).

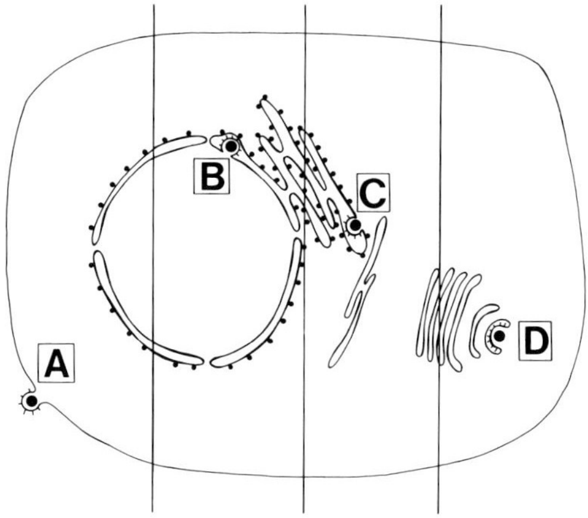

Figure 1. Envelopment of nucleocapsids by modified patches of cell membrane during final stages of virion morphogenesis. Members of different virus families characteristically acquire their envelopes by budding through the plasma membrane (A), the inner nuclear membrane (B), the membrane of the endoplasmic reticulum (C), or the Golgi apparatus (D). Table I lists the sites of envelopment for different virus families.

nent in common—namely, a lipid-containing membrane (envelope) that is the outer coat of the virus particle (virion) containing viral glycoproteins essential for infectivity (ability of the virus to attach to and invade a cell so as to initiate viral gene expression). In all cases this envelope is derived, during virion morphogenesis in the virus-producing cell, from a patch of cell membrane that has been modified by the incorporation of viral-specified glycoproteins and by the exclusion of most or all cell proteins. The viral nucleocapsid (genome plus proteins arranged in a regular helical or icosahedral structure) buds through this modified patch of cell membrane, and the neck of the bud seals up to yield the virion with its intact envelope (Fig. 1). Once released from the virus-producing cell, the virion may infect another cell. Infectivity depends on ability of the viral envelope glycoprotein(s) to mediate attachment of the virion to a cell surface and to induce fusion of the virion envelope with a cell membrane (Fig. 2). This fusion introduces the nucleocapsid into the cytoplasm of the cell so that viral gene expression can be initiated.

Differences among the viruses to be discussed include differences in genetic complexity, in the intracellular site at which envelopment of nucleocapsids characteristically occurs during virion morphogenesis (Table I and Fig. 1) and in the site at which membrane fusion occurs to initiate a new infectious cycle (Fig. 2). To infect a cell and initiate a new infectious cycle, some enveloped viruses probably fuse with the cell at the plasma membrane, some may fuse with the membrane of an endosome following receptor-mediated endocytosis, and some may do either depending on circumstances.

Virus-induced cell fusion can be mediated either by the virion itself or by viral proteins incorporated into the plasma membrane of the infec-

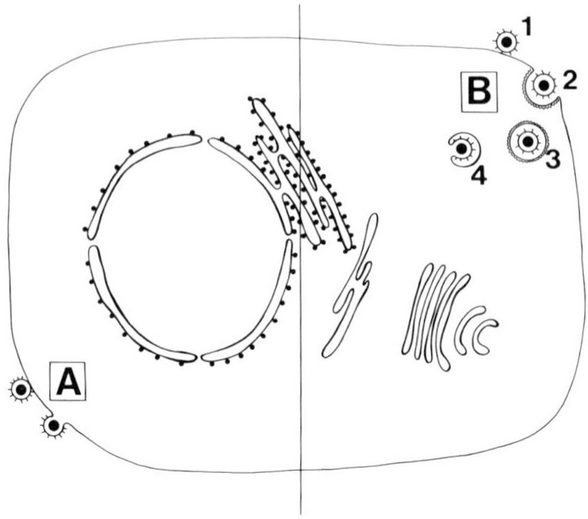

Figure 2. Pathways of viral entry during initiation of infection. The virion–cell fusion required to release the nucleocapsid into the cytoplasm may occur at the plasma membrane (A) or with the membrane of an endosome (B). Stages of the endocytic pathway (B) are thought to be similar to those defined for receptor-mediated endocytosis of other ligands (Goldstein *et al.*, 1985). They include attachment of virus to the cell surface, lateral movement to a clathrin-coated pit, ingestion by the cell in a coated vesicle, transition of the vesicle from coated to uncoated (endosome), and fusion of the virion with the membrane of the endosome.

ted cell. These operationally distinct types of virus-induced cell fusion have been called fusion from without (FFWO) in the first instance and fusion from within (FFWI) in the second (Kohn, 1965; Bratt and Gallaher, 1969). For FFWO the cells must be exposed to large numbers of virions, but the virions need not necessarily be capable of initiating or sustaining viral gene expression. FFWI can be induced by virus at low multiplicities of infection and requires viral gene expression. A number of reviews have been published on the subject of FFWI and FFWO (Roizman, 1962; Poste, 1972; Hosaku and Shimuzu, 1977; Knutton, 1978; Poste and Pasternak, 1978). In this review the emphasis is on FFWI, except where noted otherwise.

Events that have been monitored as indicative of virus-induced membrane fusion include hemolysis as well as FFWO and FFWI. Fusion of virions with red cells can result in hemolysis, either as a concomitant of the fusion reaction or due to incorporation of damaged viral envelopes into the red cell membrane.

The viral glycoproteins responsible for mediating virion-cell fusion to initiate infection also mediate cell fusion. A virus may be able to fuse with a cell to initiate infection, however, and yet apparently not induce cell fusion, either FFWO or FFWI. Whether fusion is observed in cultured cells depends on culture conditions and on both viral and cell determinants. For example, different strains of the same virus may differ in their ability to induce cell fusion. Moreover, a single virus strain may fuse some cells but not others.

In general, failure to detect cell fusion after infection with a virus capable of inducing membrane fusion could be due to one of several circumstances, including the following: (1) failure of the viral fusion protein(s) to be incorporated at sufficiently high concentration into the plasma membrane (this could be compatible with viral replication especially if envelopment occurs at another membrane); (2) inappropriate conditions at the cell surface for activation of the viral fusion protein(s); (3) the presence of viral or cell components in the plasma membrane that inhibit or modulate activity of the viral fusion protein(s); (4) the absence of appropriate receptors in the plasma membrane of uninfected cells—receptors that could be required, not for attachment of virus, but for the membrane fusion process (this could be compatible with viral infectivity provided the virus can attach to the cell surface and then fuse with the membrane of an endosome); or (5) failure to recognize that membrane fusion has occurred because two cells that are perhaps joined by cytoplasmic bridges retain the shape of two adherent spheres (or discoids) instead of assuming the shape of a single sphere.

1.3. Significance of Virus-Induced Cell Fusion

Cell fusion is not an invariant consequence of infection with viruses that have the potential to induce membrane fusion. In fact, for some virus families (orthomyxoviruses, bunyaviruses, togaviruses, retroviruses), infection rarely if ever results in cell fusion under physiological conditions. For other virus families, the extent of cell fusion may be highly strain dependent. Matters seem to have been arranged so that expression of overt membrane-fusing activity is often limited to the virion in its interactions with uninfected cells but is sometimes also observed to occur at infected cell surfaces and to result in cell fusion.

Virus-induced cell fusion can be of conseqence in pathogenicity, as fusion affords a means of spreading infection that does not depend on production or survival of infectious virus. For example, antibodies to the receptor-binding glycoprotein of a paramyxovirus can neutralize viral infectivity but do not prevent spread of infection in cell culture by cell fusion, whereas antibodies to the fusion protein prevent the fusion (Merz *et al.,* 1980). In addition, inactivated vaccines for several paramyxoviruses have proved deficient in inducing antibodies to the fusion protein, but not the receptor-binding protein, and have also proved ineffective (Norrby *et al.,* 1975; Norrby and Penttinen, 1978).

For defining mechanisms by which fusion of biological membranes can be induced and regulated, virus-infected cells provide an attractive experimental system. At least some of the components that either induce or regulate the fusion are viral proteins, and the genes for these proteins are readily altered to provide the mutants that can aid in defining structure–

function relationships. Moreover, there are a variety of ways in which membrane fusion can be assayed in virus-infected cells.

2. INFECTIVITY OF ENVELOPED ANIMAL VIRUSES

2.1. Pathway of Entry into Cells

Defining the site(s) at which a virion can fuse with a cell membrane to initiate infection is relevant to defining the conditions required for cell–cell fusion. For example, if a virion cannot fuse with the plasma membrane but only with the membrane of an endosome, the viral protein(s) responsible for inducing membrane fusion may not be active at the cell surface and therefore may not induce cell fusion, either FFWO or FFWI. As the following illustrates, it can be difficult to establish, however, which cell membranes are targets for virion fusion and which are not.

Viruses such as Semliki Forest virus (SFV, a togavirus), influenza virus (an orthomyxovirus), and vesicular stomatitis virus (VSV, a rhabdovirus) can initiate infection via an endocytic pathway of entry (Marsh, 1984). It has been suggested that the low pH of endosomes may be required to trigger fusion activity of these viruses based principally on the following observations: (1) initiation of infection by these viruses is blocked by agents (ammonium ions, chloroquine, amantadine) that raise the pH of endosomes, and (2) these viruses do not induce FFWO at neutral pH but can do so if the infected cells are exposed briefly to medium of low pH (reviewed by White *et al.,* 1983; Marsh, 1984).

Although these viruses clearly can infect by an endocytic pathway (and this may be the major pathway), the possibility that they can also fuse with the plasma membrane has not been ruled out, despite the apparent requirement for low pH to trigger fusion. There are several relevant arguments. First, both influenza virus and VSV can exhibit membrane-fusing activity at neutral pH, as will be seen.

Second, the findings that FFWO could be induced by SFV, VSV, and influenza virus at low pH but not at neutral pH (White *et al.,* 1981) cannot be taken as evidence that these viruses are unable to fuse at the cell surface at neutral pH. Fusion of virus with the plasma membrane may not necessarily lead to FFWO, particularly if conditions are not appropriate for a virion to fuse with two cells simultaneously, thereby forming a cytoplasmic bridge, or if the bridge cannot enlarge so that fusion can be recognized to have occurred.

Third, if a virus can infect either by fusion at the cell surface or after endocytosis, certain conditions may favor use of one pathway over another. In general, experiments providing evidence for an endocytic pathway have been done under conditions that may favor this pathway over alternatives. Specifically, cells have been exposed to virus in the cold, washed, and then warmed to 37°C. This kind of temperature shift has been shown to stimulate a burst of endocytosis (Anderson *et al.,* 1977; Marsh and Helenius, 1980).

The possibility exists, therefore, that pH dependence of fusion activity will not provide the full explanation for inability of SFV, VSV, and influenza virus to induce FFWO at neutral pH.

Factors that determine efficiency and pathways of viral entry include the following: Adsorption of an enveloped virus to a cell surface will be followed by elution, fusion, or endocytosis of the virus. If a virion fails to fuse with the plasma membrane and also fails to fuse after endocytosis, it will probably be delivered to a lysosome for destruction. Lysosomal destruction of the virus and differences in the fraction of input virus destroyed by different cells undoubtedly accounts in part for the relatively high particle : plaque-forming unit ratio in most preparations of animal viruses and for differences in this ratio when titrations of plaque-forming units are done on different cell types. The probability that virus attached to a cell surface will fuse with the plasma membrane depends on at least two factors: rate of endocytosis and rate of fusion under conditions that prevail at the cell surface. Endocytosis of different viruses by the same cell type clearly occurs at different rates (Marsh, 1984). Adequate methods for directly measuring the rate of viral fusion at the cell surface have not yet been described. The probability that ingested virus can fuse with the membrane of an endosome probably depends on transit time in prelysosomal vesicles and on the rate of fusion under the conditions prevailing in endocytic vesicles. Too little is known about the range of conditions and factors that influence fusion to permit conclusions that a particular virus can fuse only with either the plasma membrane or the membrane of an endosome.

2.2. Roles of the Viral Glycoproteins

Table II lists the envelope glycoproteins of selected viruses and briefly describes their organization in virions and proposed roles in infectivity. Two activities known to be required for infectivity and expected to be mediated by one or another of the envelope glycoproteins are (1) binding to a cell surface receptor, and (2) induction of fusion between the virion envelope and a cell membrane.

For many of these viruses, the envelope glycoproteins form uniform structures or spikes (on virions but not necessarily in cell membranes) that protrude from the envelope; they can be recognized by electron microscopy to have a characteristic size and shape. In several instances only one

Table II. Envelope Glycoproteins of Selected Viruses

Virus	Glycoprotein			Spike			References[b]
	Precursor[a]	Mature forms	Approx. M_r ($\times 10^{-3}$)	Subunit associations	Membrane-spanning domains	Activities	
Vesicular stomatitis (VSV)	—	G	62	?	Near C-terminus	Receptor binding; fusion	Rose and Gallione (1981)
Sendai	—	HN	67	?	Near N-terminus	Receptor binding; neuraminidase	Blumberg *et al.* (1985*b*)
	F_0	$F_2 + F_1$	12; 53	$[F_2 + F_1]_?$	Near C-terminus of F_1	Fusion	Blumberg *et al.* (1985*a*)
Influenza	—	NA	60	$[NA]_4$	Near N-terminus	Neuraminidase	Varghese *et al.* (1983)
	HA	HA1 + HA2	44; 30	$[HA1 + HA2]_3$	Near C-terminus of HA2	Receptor binding; fusion	Wilson *et al.* (1981)
Semliki Forest (SFV)	—	E1	51	$[E1 + E2 + E3]_1$	Near C-termini of E1 and E2	Receptor binding; fusion	Garoff *et al.* (1980)
	p62	E3 + E2	11; 52				
Herpes simplex (HSV)	—	gB	120	$[gB]_2$	Near C-terminus	?	See legend to Fig. 5
	—	gC	130/75[c]	?	Near C-terminus	C3b binding	
	—	gD	60	?	Near C-terminus	Receptor binding?	
	—	gE	75	?	Near C-terminus	Fc binding	
	—	gG	?/125[c]	?	?	?	
	—	gH	110	?	?	?	

[a]Only relatively stable precursors detectable in infected cells are listed here.
[b]Most of the publications listed describe primary sequence or three-dimensional structure of the indicated glycoproteins and contain citations to publications providing other information about the proteins.
[c]The number before the slash is for HSV type 1 and the number after the slash for HSV type 2.

kind of spike is present. This spike may be composed of a single or multiple species of glycopolypeptide and has all activities required for infectivity.

In other instances at least two different kinds of spikes are present, and they have different activities. For example, spikes containing the cleaved form of hemagglutinin (HA) (HA1 + HA2) in influenza virus mediate adsorption of virus to sialic acid moieties on cell surfaces as well as fusion of the viral envelope with a cell membrane. The other glycoprotein of influenza virus, neuraminidase (NA), forms separate spikes. For the evolutionary related Sendai virus (a paramyxovirus), these three activities are differently distributed between two kinds of spikes. That is, spikes composed of hemagglutinin-neuraminidase (HN) mediate adsorption to sialic acid moieties and display neuraminidase activity, whereas spikes composed of the cleaved form of the fusion protein F_0 (F_2 + F_1) induce fusion between the virion envelope and a cell membrane. Neuraminidase activity is not found in all enveloped viruses. Although the role of neuraminidase in orthomyxo- and paramyxovirus replication is not fully understood, its expression and presence in the virion envelope is undoubtedly related to the fact that the cell receptor for these viruses is sialic acid. It has been suggested that the binding of virus to sialic acid on certain cell surface components may not lead to fusion and that one role of neuraminidase could be to release virus from these dead-end receptors in order to enable subsequent binding to productive receptors (Haywood, 1974).

In contrast to the RNA viruses, herpesvirus specify the synthesis of more than six different envelope glycoproteins. Their organization in virions and precise roles in infectivity are not yet well defined. Two of the herpes simplex virus (HSV) glycoproteins have unexpected activities (Fc-binding and C3b-binding) that may be irrelevant to infectivity (Spear, 1984).

3. VIRUS-INDUCED CELL FUSION

3.1. Stages in the Process of Cell Fusion

In discussing virus-induced cell fusion, Knutton (1978) proposed that three stages may be required (Fig. 3):

Stage 1 Adhesion between two cells

Stage 2 Membrane fusion to create a cytoplasmic bridge

Stage 3 Enlargement of the bridge to yield what would be recognized as a polykaryocyte—multiple nuclei occupying a single cytoplasm.

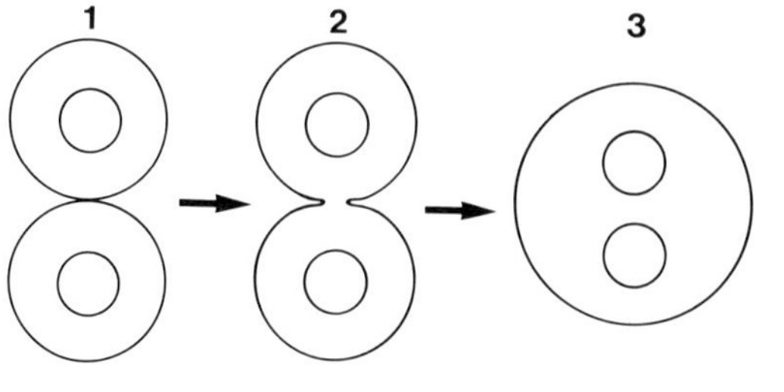

Figure 3. Stages in the process of virus-induced cell fusion as proposed by Knutton (1978). Stage 1 is adhesion between two cells, stage 2 is membrane fusion to form a cytoplasmic bridge, and stage 3 is enlargement of the cytoplasmic bridge with rearrangement and merger of intracellular contents.

Because fusion of cells is usually monitored visually by light microscope, only after the third stage would it be evident that fusion has occurred even though fusion of membranes actually occurs in the second stage.

By analogy with virus attachment and fusion to initiate infection, the first stage of cell fusion could be mediated by the viral receptor-binding glycoprotein and the second stage by the viral fusion glycoprotein. The third stage of cell fusion may not have a counterpart in infectivity unless some specific condition is required to enlarge the initial virion–cytoplasmic bridge so as to permit entry of the nucleocapsid into the cytoplasm.

A change in permeability leading to cell swelling appears to be one aspect of stage 3 (Knutton, 1978; Patel and Pasternak, 1985). For example, preparations of Sendai virus that are most efficient at inducing FFWO have damaged envelopes and apparently, on fusion with cells, introduce lesions into the cell surface that result in increased permeability. Cell lysis rather than fusion can occur unless the medium contains factors, such as calcium, required for repair of the lesions (Okada and Murayama, 1966). Another aspect of stage 3 may be changes in the cell cytoskeleton that could facilitate rearrangement of cell contents.

3.2. Viral Proteins That Induce or Influence Cell Fusion

For the viruses listed in Table II, attention is focused here on the conditions required for cell fusion, the identities of the viral glycoproteins for inducing membrane fusion, structural features of these glycoproteins thought to be important for fusion activity, and the identities and properties of other viral proteins that may influence fusion activity.

The most compelling evidence that a viral glycoprotein has fusion-inducing activity has come from demonstration that the product of a single molecularly cloned viral gene can induce cell fusion. Findings that antibodies of a given specificity can block virus-induced cell fusion do not necessarily identify the fusion protein, despite published claims. For example, antibodies specific for either HN (Miura *et al.,* 1982) or F (Merz *et al.,* 1980) of a paramyxovirus can block fusion, and yet the cleaved form of F alone is sufficient to induce membrane fusion, as discussed in section 3.2.2a. Experiments using liposomes containing purified viral glycoproteins (virosomes) to assess fusion activity were summarized by White *et al.* (1983) and have given results consistent for the most part with the conclusions presented below.

Attempts to relate structure and function of viral fusion proteins are complicated by lack of knowledge as to general features of the mechanism by which these proteins induce membrane fusion. Attention has focused on hydrophobic domains (other than the membrane-spanning regions) of the fusion proteins in part because these domains seem to be more highly conserved than others among related viruses and because, in some instances, these hydrophobic domains are at the new N-termini generated by cleavages required for fusion activity.

Barriers to the close approach of two lipid bilayers required for membrane fusion include the physical bulk of membrane proteins and charge repulsion. There are several ideas about the role of viral fusion proteins in overcoming these barriers. A hydrophobic region of the fusion protein may interact with lipids or proteins in the target membrane, either to draw the target membrane and viral membrane into closer proximity or to destabilize the lipid bilayer of the target membrane for initiation of fusion. Alternatively, the fusion protein may promote aggregation or patching of the viral (and perhaps cell) membrane proteins so that lipid bilayers devoid of protein can approach closely enough to fuse (White *et al.,* 1983). Enhanced curvature of either the viral or target membrane may also be necessary to overcome charge repulsion (Haywood and Boyer, 1981).

If fusion activity requires insertion of a hydrophobic N-terminus of the fusion protein into the lipid bilayer of the target membrane, then one might expect hydrophobicity, but not necessarily amino acid sequence, to be conserved at the N-terminus, as is the case for signal sequences and membrane-spanning domains. The conservation of sequence, within a virus family, at the relevant N-termini of fusion proteins (Fig. 4) suggests that there are requirements for interaction of these domains with cell or viral components other than lipid bilayers, in addition to any other requirements. If members of the different virus families listed in Figure 4 cause fusion by similar mechanisms, common features of the fusion process must not depend on relatedness of amino acid sequence at the N-termini of the proteins shown.

Although it can be shown that individual viral glycoproteins are both necessary and sufficient to induce cell fusion under a particular set of conditions, there is increasing evidence that other viral proteins influence fusion-inducing activity. These other viral proteins may be required for

```
A. Rhabdoviruses
                 ┌- I E - - - - Q H T T - D - - R - - H E - N - -                                [1]
                 │- I E - - - - Q H ? - - ? - - ? -                                              [2]
VSV G:           │- - S - - - - Q S - - - - - - - - - - S - Y -                                  [3]
                 │- - - - - - - Y - - - - - - - - - - - - - - - -                                [4]
                 └K F T I V F P H N Q K G N W K N V P S N Y H Y C P S S S D L N W H N D     L I G   [5]
                  : :   :                 :   :                   : :                       :   :
Rabies G:         K F P I Y T I L D K L G P W S P I D I H H L S C P N N L V V E D E G C T N L S G   [6]

B. Paramyxoviruses

Sendai F1:        F F G A V I G T I A L G V A T A A Q I T A G I A L A E A R E A K R D I A L I K E   [7]
                  :   :   : : :     : : : : : : : : :   : :     : :     :   :       :     :
SV5 F1:           F A G V V I G L A A L G V A T A A Q V T A A V A L V K A N E N A A A I L N L K N   [8]
                  : : : : :       : : : : : : : : : :   : :     : :         :   : :   : :
Measles F1:       F A G V V L A G A A L G V A T A A Q I T A G I A L ? Q S M L N S Q A I D N L T A   [9]
                  :   :       :   : : : : : : : : : : :
NDV F1:           F I G A I I G G V A L G V A T A A Q I T                                          [10]
                  :   :     :
RS F1:            F L G F L L G V G S A I A S G V A V S K V L H L E G E V N K I K S A L L S T N K   [11]

C. Orthomyxoviruses
                 ┌- - - - - - - - - - - - - - - - L V - - - - - - - - - - A Q - E - T - - - Y - - -   [12]
                 │- - - - - - - - - - - G - - Q - - V - - - - - - Y H - S - D Q - S - Y - - - K E - -   [13]
Influenza A HA2: │- - - - - - - - - - - G - - T - - - - - - - - - Y H - - - E Q - S - Y - - - Q - - -   [14]
                 │- I - - - - - - - - - - - - - - - - - - - - - - - - - - - - - - - - - - - - - - -   [15]
                 └G L F G A I A G F I E N G W E G M I D G W Y G F R H Q N S E G T G Q A A D L K S T   [16]
                  :   : : : : : : :   :   : : : : : :   : :   :             :       : : : : : : :
Influenza B HA2:  G F F G A I A G F L E G G W E G M I A G W H G Y T S H G A H G V A V A A D L K S T   [17]
                      : :                                 :       :                 :
Influenza C HA2:    I F G I D D L I I G L L F V A I V E T G I G G Y L L G S R K E S G G G V T K E S   [18]
```

Figure 4. Comparisons of N-terminal amino acid sequences for the fusion proteins or fusion protein subunits indicated. For different strains of the same virus, the dashes indicate identity with the reference sequence. For different viruses within a single family the double dots indicate identity between sequences. The viruses and strains for which data are presented and citations are as follows. *Rhabdoviruses.* VSV G: strains [1] New Jersey Ogden, [2] New Jersey Concan, [3] Cocal, [4] Indiana Toronto (Kotwal *et al.,* 1983), [5] Indiana San Juan (Rose and Gallione, 1981), [6] rabies G (Rose *et al.,* 1982). *Paramyxoviruses.* Primary data for measles F_1 and citations of papers providing information for the other viruses are in Varsanyi *et al.* (1985). *Orthomyxoviruses.* Influenza A HA2: strains [12] fowl 1979, [13] A/Japan/305/57, [14] A/WSN/33 and A/PR/8/34, [15] A/Victoria/3/75, A/England/321/77 and A/Bangkok/1/79, [16] A/NT/60/68/29C, A/Aichi/2/68, A/Mem/102/72 and A/duck/Ukraine/63 (from the protein sequence database of the National Biomedical Research Foundation, Georgetown University Medical Center); influenza B HA2: strains [17] B/Lee/40, B/HK/8/73 and B/Md/59 (Krystal *et al.,* 1983, 1984); [18] influenza C HA2 (Pfeifer and Compans, 1984).

cell fusion under certain conditions. Alternatively, they may modulate or inhibit cell-fusing activity.

It should be kept in mind that, in most of the studies done to date, attempts have not been made to differentiate between the postulated stages 2 and 3 of the cell fusion process (Fig. 3). Usually cell fusion is judged to have occurred only if the final product, the polykaryocyte or greatly enlarged red cell, is observed. The possibility exists, however, that viral proteins may induce cytoplasmic bridges as depicted for stage 2 without inducing the progression to stage 3. Special techniques would be required to detect these bridges. The possibility also exists that formation of the cytoplasmic bridges is reversible if transition to stage 3 does not occur. The transition from stage 1 to stage 2 is topologically equivalent to fusion of a virion with the cell at the cell surface. The reverse reaction (stage 2 to stage 1) is topologically equivalent to the purse–string closure and annular fusion that occur to separate a budding virion from the cell surface.

3.2.1. Rhabdoviruses

FFWO has not been reported for VSV except after exposure of infected cells to low pH (White *et al.,* 1983). Certain strains of VSV induce the FFWI of certain cell types at neutral pH, whereas other VSV strains do not (Takehara, 1975; Nishiyama *et al.,* 1976; Chany-Fournier *et al.,* 1977). The genetic differences responsible for the phenotypic differences are for the most part unknown. Strains of VSV that do not ordinarily cause FFWI can be induced to do so under the following conditions: (1) exposure to low pH of cells infected with VSV or cells expressing only the VSV G protein (Florkiewicz and Rose, 1984; Riedel *et al.,* 1984); (2) infection of cells at nonpermissive temperature with the temperature-sensitive mutant *ts*G31, which has a lesion in the gene for M protein (Hughes *et al.,* 1979*b;* Handa *et al.,* 1982; Storey and Kang, 1985); and (3) inhibition of protein synthesis, especially relatively late in the infectious cycle, in cells infected with wild-type virus (Storey and Kang, 1985). Under conditions (2) and (3), the fusion occurs at neutral pH and may occur only in certain cell types and not others.

3.2.1a. VSV G Protein. Expression of G protein alone (of the five VSV proteins) is sufficient to induce cell fusion provided the cells are exposed to low pH (Florkiewicz and Rose, 1984; Riedel *et al.,* 1984). Cells that expressed a truncated form of G that lacked the membrane-spanning domain and was secreted into the medium failed to fuse, indicating that fusion activity probably depended on anchorage of G to the cell surface (Florkiewicz and Rose, 1984).

In contrast to some other viruses, there is no evidence that proteolytic cleavage of G protein (other than that which removes the signal sequence) is required to activate fusion activity. The N-terminus of G (Fig. 4) may be important for fusion activity, based on findings that a synthetic peptide (24 amino acids) similar in sequence to the N-terminus has hemolytic activity at pH 5.0 (Schlegel and Wade, 1984; see Chapter 2, this volume). VSV virions and purified G protein in liposomes also have hemolytic activity at pH 5.0 (Bailey *et al.*, 1981, 1984). Reducing the size of the synthetic peptide from 25 amino acids to 6 retained hemolytic activity and abolished the low pH dependence of this activity. The basic amino acid at the N-terminus of the peptide (Lys-Phe-Thr-Ile-Val-Phe) may be essential for activity. Activity was retained when Lys was substituted with Arg but not with Glu (Schlegel and Wade, 1985).

In native G protein at neutral pH, the N-terminus appears not to be accessible to antibodies (Schlegel and Wade, 1984) or to protease (Capone *et al.*, 1982). If indeed fusion activity of G depends on interaction of the N-terminus with the target membrane, activation of fusion may require a change in conformation of G. Exposure to low pH may be one way to induce the postulated change in conformation, but presumably it is not the only way.

3.2.1b. VSV M Protein and Nucleocapsids. Although low pH appears to be a requirement for induction of cell fusion in cells expressing only G protein, cell fusion (FFWI) can occur at neutral pH in infected cells expressing other viral proteins. Mutation in the M protein or partial inhibition of viral protein synthesis, as well as low pH, can cause a nonfusing VSV strain to induce cell fusion.

Why should mutation in M protein (as in *ts*G31) permit cell fusion to occur at neutral pH? M protein lines the undersurface of the virion envelope. Interactions of M protein with both the nucleocapsid and G protein are required for virion morphogenesis (Knipe *et al.*, 1977). Apparently the mutation in *ts*G31 prevents budding of nucleocapsids (Hughes *et al.*, 1979*a*) but may not prevent all interactions of M (possibly aberrant) with either G protein or the nucleocapsid. The effects of protein synthesis inhibitors could be to create an imbalance of G protein, M protein, and nucleocapsid available for interaction.

A hypothesis consistent with the results summarized above proposes that interactions of M and G without nucleocapsid trigger fusion activity at neutral pH, whereas adherence of nucleocapsids to the complex blocks fusion activity and initiates budding for virion morphogenesis. A prediction of this hypothesis is that cells expressing only M and G should fuse at neutral pH.

3.2.2. Paramyxoviruses

Paramyxoviruses can induce both FFWO and FFWI at neutral pH (Choppin and Compans, 1975; Hosaka and Shimizu, 1977). No enhancement of fusion activity by exposure to low pH has been reported. On the contrary, it was shown that transient exposure of Sendai virus to pH 9.0 enhanced fusion activity, as assessed by hemolysis and FFWO, and caused an irreversible change in the conformation of F protein (Hsu *et al.*, 1982). Cell fusion is not an invariant consequence of infection with paramyxoviruses, for reasons that are not understood.

3.2.2a. F Protein. A large body of evidence indicates that the F protein of paramyxoviruses displays fusion activity (Choppin and Compans, 1975; White *et al.,* 1983). In contrast to two other well-characterized fusion proteins (VSV G and influenza virus HA), which exhibit receptor-binding as well as fusion activity, F protein apparently has only fusion activity. It has been reported (Peterhans *et al.,* 1983), however, that F can bind to cells independently of HN, the recognized receptor-binding protein. Cells infected with a simian virus (SV40) vector carrying and expressing the F gene only of SV5 (a paramyxovirus) were shown to fuse at neutral pH. The active cleaved form of F was produced by the cells (Paterson *et al.,* 1985).

Fusion activity of the F protein depends on a specific proteolytic cleavage yielding the subunits F_2 and F_1, which remain covalently joined by disulfide linkage (White *et al.,* 1983). The F_1 subunit contains the membrane-spanning domain of F (C-terminus). The N-terminus of F_1, generated by the cleavage required for fusion activity, is hydrophobic and more conserved among different paramyxoviruses (Fig. 4) than are other domains. The sequences shown in Figure 4 are for viruses from four animal species including human.

The possibility that the N-terminus of F_1 may interact directly with the target membrane was suggested by results obtained with synthetic peptides. It was found that peptides similar in sequence to the N-terminus could inhibit viral infectivity, FFWI, and hemolysis without blocking virus adsorption and that these peptides interacted with the cell, not the virus (Richardson *et al.,* 1980; Richardson and Choppin, 1983). There are, however, aspects of the results that are difficult to reconcile with the straightforward interpretation that the peptides act by competing with the F_1 polypeptide for specific sites on the target cell membrane. First, the most effective inhibition was obtained with the peptide Z-D-Phe-L-Phe-Gly (where Z is a carbobenzoxy group), and optimal inhibitory activity depended on presence of both the Z group and the D form of Phe in position 1. Second, much greater inhibition was obtained with measles virus than with Sendai virus, even though the sequence of the inhibitory peptide matches better that of Sendai virus at the N-terminus of F_1 (Fig. 4).

3.2.2b. HN Protein. Although expression of F alone (from at least one paramyxovirus) can induce cell fusion, HN influences cell fusion in ways that are not completely understood. The effect of HN on cell fusion appears to differ for different paramyxoviruses.

Several lines of evidence suggest that HN of Sendai virus may be required for fusion activity under certain circumstances and that this requirement is not solely for the receptor-binding or neuraminidase activity of HN. First, an anti-HN monoclonal antibody that did not inactivate hemagglutination or neuraminidase activity was shown to block Sendai virus-induced cell fusion and hemolysis as well as fusion with cells of virosomes (assessed by delivery of toxin from the virosomes to the cells) (Miura *et al.,* 1982). Second, using this latter assay, fusion activity was found to depend sharply on the ratio of F to HN in the virosomes (optimal being 2), suggesting that HN was not merely serving a receptor-binding function (Nakanishi *et al.,* 1982). Third, under conditions such that the requirement for the receptor-binding activity of HN was bypassed, fusion activity depended on the presence of HN nonetheless. Specifically, Sendai virus induced the fusion and hemolysis of neuraminidase-treated red cells provided that antibodies reactive with the virus were coupled to the red cells. Virosomes could also induce hemolysis of red cells treated in this way, but only if both F and HN were present (Nussbaum *et al.,* 1984). (In an earlier study, virosomes containing only F were shown to lyse red cells provided that wheat germ agglutinin was used to mediate attachment of the virosomes to the red cells (Hsu *et al.,* 1979). The reasons for the apparent discrepancy in these results are unknown.)

Although all aspects of the role of HN in Sendai virus-induced cell fusion are not understood, there are some suggestive observations. For example, it was recently reported that virosomes fused in a nonleaky fashion with liposomes composed of phosphatidylcholine and cholesterol but lysed liposomes also containing sialoglycolipids and sialoglycoproteins (Citovsky and Loyter, 1985). Presence of both F and HN in the virosomes was required for both activities (fusion and lysis). In this same study it was also reported that virosomes could fuse with neuraminidase- and pronase-treated red cell vesicles, provided that the medium was hypotonic. Citovsky and Loyter (1985) suggested that normal biological membranes containing the usual complement of proteins and glycolipids may not have areas of protein-free bilayer accessible to the virus for fusion. Interactions of the virus with sialic acid-bearing receptors could somehow induce changes in membrane permeability, leading to cell swelling and membrane stretching, which could be a prerequisite to fusion of the virus with a biological membrane.

Another role of HN in the fusion induced by Sendai virus has been suggested. Haywood and Boyer (1981) examined by electron microscopy the interaction of virus with liposomes containing sialoglycolipids. Within 1 min. at 37°C, many virions appeared to be partially or almost entirely engulfed by a liposome. Fusion occurred at the leading edge of the developing pit, resulting in a small bridge between interiors of the virion and liposome. Because at this stage most of the virion surface remained in close contact with the surface of the pit in the liposome, probably due to the receptor-binding activity of HN, it was proposed that destruction of the receptors by neuraminidase would be required to enable enlargement of the bridge and complete merging of virion and liposome contents.

If HN is required for fusion activity, as these results suggest, then what is the explanation for cell fusion induced by F alone? At this stage it seems necessary to conclude either that other interpretations will be found for the results discussed above or that certain cell surface components can substitute for HN under some conditions. It should be noted that most of the experiments suggesting a role for HN were done with liposomes or red cells, whereas the cells fused after expression of the cloned F gene were monkey fibroblasts.

With mumps virus it appears that HN in the cell surface may inhibit cell fusion, at least in part because of its neuraminidase activity. Strains of mumps virus differ in their ability to induce cell fusion (FFWI), and there is an inverse correlation between cell fusing activity and level of neuraminidase activity associated with HN (Merz and Wolinsky, 1981). Moreover, treatment of infected cells with chymotrypsin cleaves HN, but not F, and induces fusion of cells infected with nonfusing virus strains, whereas exogenous neuraminidase inhibits the fusion of cells infected with fusing strains (Merz and Wolinsky, 1983). Similar but not identical results have been reported for bovine parainfluenza 3 (Shibuta *et al.*, 1983).

The apparently opposite effects of HN on cell fusion induced by different paramyxoviruses have not yet been explained.

3.2.3. Orthomyxoviruses

Influenza viruses induce neither FFWO or FFWI unless the infected cells are exposed to low pH (White *et al.*, 1983).

3.2.3a. Influenza HA Protein. Transformed cells expressing HA of influenza virus fused after addition of trypsin to the medium and exposure of the cells to medium of low pH (White *et al.*, 1982). The addition of trypsin was required to convert the HA made by the cells to the active cleaved form. This study also showed that a secreted form of HA could not induce cell fusion.

HA of influenza virus is the best characterized of the viral fusion proteins in terms of primary and tertiary structure. The use of bromelain to separate the HA spikes (consisting of the cleaved form of HA) from their membrane-spanning domains permitted crystallization and determination of structure at 3-Å-resolution (Wilson *et al.,* 1981). Each monomer of the trimer has been described as a long fibrous stem topped by a globular domain. The globular domain is composed entirely of HA1 and forms the receptor-binding site for sialic acid located about 135 Å from the viral membrane. The fibrous stem is composed of both HA1 and HA2, with the N-terminus of HA2 located about 35 Å from the membrane and probably buried in the interior of the trimeric stem. The C-terminus of HA2 (in the intact spike) has the membrane-spanning domain for anchorage to the viral envelope or cell membrane.

Cleavage to yield HA1 plus HA2 is essential for fusion activity (Lamb and Choppin, 1983). The N-terminus of HA2 is hydrophobic and is more highly conserved than other parts of HA, even between strains of influenza A and B (Fig. 4). Influenza C seems to have diverged considerably from A and B. Consistent with similiarities between orthomyxoviruses and paramyxoviruses that suggest a common evolutionary origin, similarities between amino acid sequences at the N-terminus of HA2 and the N-terminus of F_1 have been noted (Blumberg *et al.,* 1985*a*).

Opinions differ as to whether exposure of HA1–HA2 to low pH is required for fusion activity. There is general agreement that influenza virus does not induce FFWO, FFWI, or hemolysis at neutral pH but that it can be induced to do so at low pH (<6.0) (White *et al.,* 1983). Although it has been concluded from such results that influenza virus has no fusion activity at neutral pH, it can be argued that FFWO, FFWI, and hemolysis do not necessarily follow virion–cell fusion. Therefore, failure to observe these phenomena may not imply failure of virion–cell fusion to occur.

A more telling argument is the evidence obtained, using different kinds of assays, that influenza virus can exhibit fusion activity at neutral pH. Electron microscopic examination showed that influenza virus fused at neutral pH with liposomes containing sialic acid-bearing glycolipids (Haywood and Boyer, 1985) and with red cell membranes (Fidgen and Tisdale, 1981) and that liposomes containing influenza virus glycoproteins fused with cells (Huang *et al.,* 1980). Also, hemolysis and fusion of red cells could be induced by influenza virus at neutral pH provided the red cells were first pretreated with a solution of low pH (Huang *et al.,* 1985). The latter result indicates that fusion occurs because the target membrane, not the virus, is altered by low pH. On the other hand, when virus and red cells both are exposed to low pH, the pH optimum for hemolysis depends on proteins of the virus. Different strains of influenza virus exhibit different pH optima (Huang *et al.,* 1981) and, in particular, amantadine-resistant mutants of influenza virus induce hemolysis at higher pH than do the parental strains (Daniels *et al.,* 1985).

What are the effects of low pH on both the target membrane and the virus or viral proteins? Even transient exposure of the target membrane to low pH may induce irreversible changes in organization and conformation of membrane components or alter permeability. Possibly changes in membrane permeability before or during exposure to virus can permit cell–virion–cell fusion (stage 2) to progress to visually evident cell fusion (stage 3).

Influenza virions aggregate when exposed to low pH in the absence of cells and are rapidly inactivated with respect to infectivity and fusion. If attachment of virions to cells is permitted prior to exposure to low pH, virion–cell fusion (and FFWO) can occur, but initiation of infection does not result (Matlin *et al.,* 1981). Exposure of purified HA or virions to low pH causes a change in conformation of HA as assessed by altered sensitivity to protease and altered reactivity with monoclonal antibodies (Skehel *et al.,* 1982; Daniels *et al.,* 1983*a*). Moreover, in mutants that induced hemolysis at higher pH in comparison with parental strains, the altered conformation of HA detectable by susceptibility to proteolytic cleavage also occurred at higher pH (Daniels *et al.,* 1985). Altered conformation of purified HA (bromelain-released spike) after exposure to low pH is also evident from its enhanced binding of nonionic detergent and ability to bind to liposomes lacking sialic acid receptors (Skehel *et al.,* 1982); the hydrophobic domain exposed as a result of exposure to low pH appears to be the N-terminus of HA2 (Daniels *et al.,* 1983*b*). Virions exposed to low pH acquire the ability to bind irreversibly at 0°C to liposomes bearing sialic acid receptors, such that elution cannot be achieved with competitive inhibitors for the viral receptor-binding site (Haywood and Boyer, 1985).

Several points emerge from a consideration of the above observations:

1. Fusion of the virion with a liposome or red cell apparently does not require exposure of either to low pH.
2. Exposure of virion–cell complexes to low pH probably enhances the adherence of virion to cell and may accelerate the rate of virion–cell fusion. This statement is based in part on findings that both virions and HA (bromelain-released spike), after exposure to low pH, adhere tightly to membranes even if sialic acid-bearing receptors are absent.

3. If the low pH-induced change in conformation of HA is obligatory for fusion activity, this change should be inducible by conditions other than low pH and should be demonstrable when fusion is induced at neutral pH.
4. The induction of cell–cell fusion (FFWO or FFWI), in contrast to virion–cell fusion, seems to be dependent on exposure of the cells to low pH. It remains to be determined exactly how the effects of low pH on the target membrane and on HA contribute to the induction of cell fusion.

3.2.3b. NA, M_1, and M_2 Proteins. The suggestion that influenza NA has a role in fusion activity comes from several observations. First, virosomes composed of cleaved HA and NA were shown by electron microscopy to fuse with cells at neutral pH. If HA was not cleaved, NA was omitted, or the virosomes treated with anti-NA antibodies, this fusion did not occur. Soluble neuraminidase of influenza virus or *Vibrio cholerae* could substitute for virosome-bound NA, permitting virosomes containing only cleaved HA to fuse (Huang *et al.,* 1980). Second, both fusion and lysis of red cells induced by influenza virus at pH 5.5 could be blocked by an antineuraminidase antiserum and restored by addition of soluble *V. cholerae* neuraminidase (Huang *et al.,* 1985).

Whereas HN of paramyxoviruses has both receptor-binding and neuraminidase activities, and possibly other activities that influence fusion, only neuraminidase activity has been associated with influenza NA. It is this activity that appears to be required for the kinds of fusion measured by the assays described above. Why was there no apparent requirement for neuraminidase activity in fusion of the cells induced by expression of the cloned HA gene? There is not yet an answer to this question, although it has been suggested that a cell neuraminidase could have substituted for viral NA (Huang *et al.,* 1985).

Results implicating M_1 or M_2 in fusion activity were obtained in investigations of amantadine-resistant viral mutants. M_1 and M_2 are encoded by overlapping reading frames on the same genome segment (here designated M), with M_2 being translated from a spliced messenger RNA (mRNA). M_1 of influenza virus, similarly to M of rhabdoviruses, lines the inner surface of the virion envelope and probably interacts both with the nucleocapsid and internal domains of the viral glycoproteins (Lamb and Choppin, 1983). M_2 is an integral membrane protein expressed on the surface of infected cells but not present in virions (Lamb *et al.,* 1985). Amantadine is a drug that interferes with influenza virus replication and is known to increase the pH of intracellular vesicles. Amantadine-resistant mutants have been shown to induce hemolysis with pH optima about 0.1–0.6 higher than that for the parental viruses (Daniels *et al.,* 1985). Naturally occurring strains of influenza differ in their sensitivity to amantadine (Appleyard, 1977) and in their pH optima for hemolysis (Huang *et al.,* 1981), but whether these two types of genetic variability are always correlated is unknown.

What viral gene or genes determine amantadine sensitivity? Because the eight genome segments (each containing one or two genes) of influenza virus vary in size among different strains, it is often possible to correlate segregation of defined phenotypes among recombinant viruses with segregation of particular genome segments. There are limitations to genetic analysis of this kind, in part because two different strains of influenza virus may have diverged sufficiently such that recombinants formed between the two may exhibit altered phenotypes not expressed by either parental virus.

Nevertheless, depending on assays used and on the particular parental strains of influenza virus used to produce recombinants, it has been shown that sensitivity or resistance to amantadine segregated with the HA genome segment (Scholtissek and Faulkner, 1979), with the M genome segment (Lubeck *et al.,* 1978; Hay *et al.,* 1979), or was not assignable to a single genome segment (Appleyard, 1977; Lubeck *et al.,* 1978; Scholtissek and Faulkner, 1979).

More recently, a number of amantadine-resistant mutants were compared with their respective amantadine-sensitive parental strains with respect to pH optima for hemolysis, pH threshold for changes in conformation of HA as assessed by sensitivity to protease, and nucleotide sequences of the HA gene (Daniels *et al.,* 1985). Although all the mutants discussed had amino acid substitutions in HA, genetic tests were not performed to determine whether these substitutions were responsible for all the mutant phenotypes. Undoubtedly, the amino acid substitutions detected account for the altered pH dependence of conformational changes in HA, as these tests were done with bromelain-released soluble HA spikes. Daniels and co-workers suggested that some of the amino acid substitutions could destabilize either the pH 7.0 conformation that holds the N-terminus of HA2 in a hydrophobic crevice or the subunit interactions necessary for trimer stability. The other mutant phenotypes (amantadine resistance, altered pH optima for hemolysis), however, could depend on mutations other than, or in addition to, those detected in the HA gene.

Despite the complexities of the genetic analyses, they strongly suggest that properties of M_1 or M_2, as well as HA, can in part determine amantadine resistance. If amantadine resistance invariably correlates with altered pH optima for virion-induced hemolysis, these results also suggest that M_1 (but not M_2, as it is apparently not present in virions) influences the fusion activity measured by this assay. Because M_1 probably interacts directly with HA, at least in virions, it seems reasonable that M_1 could influence conformation and function of HA. The possibility also exists that M_2, the function of which is unknown, could in part determine whether fusion activity is expressed at the surfaces of infected cells.

3.2.4. Togaviruses

Similarly to orthomyxoviruses, togaviruses usually do not induce FFWO or FFWI except after exposure of infected cells to low pH (White *et al.,* 1981).

It is experimentally difficult to assign functions to the individual viral glycoproteins because the capsid protein and three glycoproteins are all derived by proteolytic cleavage of a single translation product. Cells microinjected with a cloned cDNA containing the coding region for this translation product of SFV were able to express all the normal cleavage product, or all but E1, depending on how the cDNA was cloned. The cells expressing E1 + E2 + E3, but not those expressing only E2 + E3, were induced to fuse by exposure to low pH (Kondor-Koch *et al.,* 1983). These results indicate that E1 is required for fusion activity but they do not rule out a requirement for E2 as well. E3 is thought not to be required for fusion because it can be removed from some togaviruses without loss of infectivity and because it is not an integral membrane protein (Garoff *et al.,* 1982). It has been reported that monoclonal antibodies to E1 can block fusion activity, as assessed by hemolysis, whereas antibodies to E2 and E3 were without this effect (Chanas *et al.,* 1982). Near, but not at, the N-terminus of E1 is located a hydrophobic segment of about 17 amino acids. This segment is highly conserved in the proteins of Sindbis virus and SFV and could have a role in fusion activity (White *et al.,* 1983).

3.2.5. Herpesviruses

Most wild-type strains of HSV do not cause the fusion of cultured cells despite the fact that polykaryoctyes are a common histological feature of herpetic lesions. Cell fusion (FFWI or FFWO) cannot be induced by exposure of infected cells to low pH. Mutants of HSV can induce cell fusion, however, as a consquence of nonlethal alterations in any one of at least four different genes (Fig. 5).

It is probable that the products of some half-dozen viral genes influence the cell-fusing activity of HSV (Spear, 1984). These products include the genes encoding the fusion protein or proteins as well as the genes that must be mutated to permit cell fusion to occur. The latter are designated Syn genes and may include the gene or genes for fusion proteins.

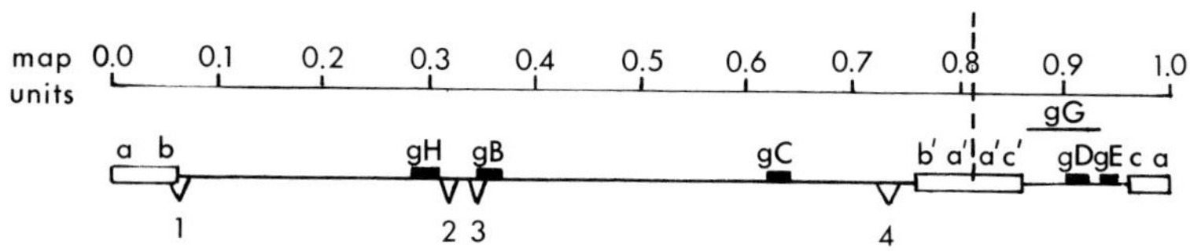

Figure 5. Genome of HSV and locations of genes for glycoproteins (e.g., gB, gC) and of Syn mutations (bracketed regions indicated 1, 2, 3, and 4). The open boxes indicate the positions of inverted repetitive sequences and the filled bars indicate the positions of glycoprotein genes (reviewed in Spear, 1984; see also Buckmaster *et al.*, 1984).

The fusion protein or proteins of HSV have not yet been identified. Temperature-sensitive mutations in the gB gene result in temperature sensitivity of cell fusion (Manservigi *et al.*, 1977), and monoclonal antibodies specific for gD (Noble *et al.*, 1983) or gH (Gompels *et al.*, 1985) can block HSV-induced cell fusion. These three glycoproteins are therefore likely to either induce or influence cell-fusing activity.

Information about the nature of the Syn mutations and the products altered by these mutations is limited. Syn mutations have been mapped on the viral genome by marker transfer or marker rescue experiments. That is, infectious genomic DNA from a wild-type strain is co-transfected into cells with cloned DNA fragments from the mutant, in order to identify which fragments can recombine with the wild-type DNA to yield recombinants with the Syn phenotype (marker transfer). In marker rescue experiments the converse is done.

Most of the Syn mutations mapped appear to be located within the region indicated by the number 4 in Figure 5 (Little and Schaffer, 1981; Bond and Person, 1984; Pogue-Geile *et al.*, 1984). For at least two mutants, nucleotide substitutions have been identified that alter translation in a single open reading frame, the only one likely to be expressed from the region (Debroy *et al.*, 1985; Pogue-Geile and Spear, 1987). Assuming that a protein is translated from this open reading frame, which has not yet been demonstrated, this protein would be very hydrophobic and would probably be membrane-associated and made on membrane-bound ribosomes with a cleavable signal sequence.

Only one Syn mutation has been mapped to region 3 (Fig. 5). This mutation is in a strain designated HSV-1(HFEM)*ts*B5, which also has a ts mutation. Both the ts and Syn mutations, which are segregable by recombination, appear to be in the structural gene for gB. Nucleotide sequence comparisons of *ts*B5 and an unrelated wild-type strain, coupled with the mapping results mentioned above, suggest that the Syn phenotype of

HSV-1(HFEM)*ts*B5 is attributable to an amino acid substitution in the cytoplasmic tail (near the C-terminus) of gB (Bzik *et al.,* 1984).

The Syn mutations of region 2 (Fig. 5) all result from deletions in the thymidine kinase gene (Sanders *et al.,* 1982). Because it seems unlikely that absence of thymidine kinase should result in cell fusion, it is reasonable to speculate that the deletions produced may alter the expression of some other gene, possibly even the adjacent gene for gH.

One mutation, in a strain designated HSV-1(KOS)804, has been mapped to region 1 by marker transfer to the genome of the wild-type parental strain HSV-1(KOS) (Little and Schaffer, 1981). Nothing is known about the nature of the mutation or the product mutated.

In summary, only one Syn mutation (region 3) is in an identified membrane glycoprotein. This glycoprotein, gB, possibly mediates membrane fusion or forms part of an active fusion complex, based on the phenotypes of mutants with ts lesions in this gene. Other Syn mutations (region 4) may be in a previously unrecognized membrane glycoprotein. Syn mutations in regions 1 and 2 must be further characterized to identify the altered products. Identification and characterization of the proteins that must be mutated to permit cell fusion should lead to the formulation of testable hypotheses as to how cell fusion is regulated. It is of interest that the Syn mutation in region 3 appears to alter the cytoplasmic tail of gB. This focuses attention on cytoplasmic and virion proteins that could interact with this domain of gB.

4. CONCLUDING REMARKS

It is evident from the foregoing that the phenomena of virus-induced cell fusion are beginning to be explicable, at least in part, by properties and activities of specific viral proteins. The mechanisms are not yet understood, however. Attention will continue to be focused on the viral fusion proteins. It will be important to define the interactions of these proteins with other viral proteins, cell components, and factors in the medium and to determine how these interactions modulate fusion activity. Better structural definition of the initial, intermediate, and end stages of the virion–cell and cell–cell fusion processes is also needed. Finally, it seems likely that common mechanisms and phenomena may be found in comparing virus-induced cell fusion with other kinds of fusion, such as myoblast fusion. Attempts to define similarities and differences could synergistically advance the studies of both kinds of fusion.

ACKNOWLEDGMENTS. I thank R. A. Lamb and A. M. Haywood for providing me with preprints and other unpublished information. I am grateful to Janice Hoshizaki for unstinting assistance with preparation of this manuscript and literature searches.

REFERENCES

Anderson, R. G. W., Brown, M. S., and Goldstein, J. L., 1977, Role of the coated endocytic vesicle in the uptake of receptor-bound low density lipoprotein in human fibroblasts, *Cell* **10**:351–364.

Appleyard, G., 1977, Amantadine-resistance as a genetic marker for influenza viruses, *J. Gen. Virol.* **36**:249–255.

Bailey, C., Miller, D., and Lenard, J., 1981, Hemolysis of human erythrocytes by vesicular stomatitis virus, *J. Cell Biol.* **91**:111a.

Bailey, C., Miller, D., and Lenard, J., 1984, Effects of DEAE-dextran on infection and hemolysis of VSV. Evidence that non-specific electrostatic interactions mediate effective binding of VSV to cells, *Virology* **133**:111–118.

Bishop, D. H. L., and Shope, R. E., 1979, Bunyaviridae, in: *Comprehensive Virology,* Vol. 14 (H. Fraenkel-Conrat and R. R. Wagner, eds.), Plenum, Press, New York, pp. 1–156.

Blumberg, B., Giorgi, C., Rose, K., and Kolakovsky, D., 1985*a,* Sequence determination of the Sendai virus fusion protein gene, *J. Gen. Virol.* **66**:317–331.

Blumbert, B., Giorgi, C., Roux, L., Dowling, P., Chollet, A., and Kolakofsky, D., 1985*b,* Sequence determination of the Sendai virus HN gene and its comparison to the influenza virus glycoprotein, *Cell* **41**:269–278.

Bond, V. C., and Person, S., 1984, Fine structure physical map locations of alterations that affect cell fusion in herpes simplex virus type 1, *Virology* **132**:368–376.

Bratt, M. A., and Gallaher, W. R., 1969, Preliminary analysis of the requirements for fusion from within and fusion from without by Newcastle disease virus, *Proc. Natl. Acad. Sci. U.S.A.* **64**:536–540.

Buckmaster, E. A., Gompels, U., and Minson, A., 1984, Characterization and physical mapping of an HSV-1 glycoprotein of approximately 115×10^3 molecular weight, *Virology* **139**:408–413.

Bzik, D. J., Fox, B. A., DeLuca, N. A., and Person, S., 1984, Nucleotide sequence of a region of the herpes simplex virus type 1 gB glycoprotein gene: Mutations affecting rate of virus entry and cell fusion, *Virology* **137**:185–190.

Capone, J., Toneguzzo, F., and Ghosh, H., 1982, Synthesis and assembly of membrane glycoproteins: Membrane anchoring COOH terminal domain of vesicular stomatitis virus envelope glycoprotein G contains fatty acids, *J. Biol. Chem.* **257**:16–19.

Chanas, A. C., Gould, E. A., Clegg, J. C. S., and Varma, M. G. R., 1982, Monoclonal antibodies to Sindbis virus glycoprotein E1 can neutralize, enhance infectivity, and independently inhibit haemagglutination or haemolysis, *J. Gen. Virol.* **58**:37–46.

Chany-Fournier, F., Chany, C., and Lafay, F., 1977, Mechanism of polykaryocyte induction by vesicular stomatitis virus in rat XC cells, *J. Gen. Virol.* **34**:305–314.

Choppin, P. W., and Compans, R. W., 1975, Reproduction of paramyxoviruses, in: *Comprehensive Virology,* Vol. 4 (H. Fraenkel-Conrat and R. R. Wagner, eds.), Plenum Press, New York, pp. 95–178.

Citovsky, V., and Loyter, A., 1985, Fusion of Sendai virions or reconstituted Sendai virus envelopes with liposomes or erythrocyte membranes lacking virus receptors, *J. Biol. Chem.* **260:**12072–12077.

Daniels, R. S., Douglas, A. R., Skehel, J. J., and Wiley, D. C., 1983*a,* Analyses of the antigenicity of influenza haemagglutinin at the pH optimum for virus-mediated membrane fusion, *J. Gen. Virol.* **64:**1657–1662.

Daniels, R. S., Douglas, A. R., Skehel, J. J., Waterfield, M. D., Wilson, I. A., and Wiley, D. C., 1983*b,* Studies of the influenza virus haemagglutinin in the pH5 conformation, in: *The Origin of Pandemic Influenza Viruses* (W. G. Laver, ed.), Elsevier, New York, pp. 1–7.

Daniels, R. S., Downie, J. C., Hay, A. J., Knossow, M., Skehel, J. J., Wang, M. L., and Wiley, D. C., 1985, Fusion mutants of the influenza virus hemagglutinin glycoprotein, *Cell* **40:**431–439.

Debroy, C., Pederson, N., and Person, S., 1985, Nucleotide sequence of a herpes simplex virus type 1 gene that causes cell fusion, *Virology* **145:**36–48.

Fidgen, K. J., and Tisdale, M., 1981, An 'on grid' electron microscopic method for studying the interaction and fusion of influenza A virus with human erythrocyte membranes, *J. Virol. Methods* **3:**271–276.

Florkiewicz, R. Z., and Rose, J. K., 1984, A cell line expressing vesicular stomatitis virus glycoprotein fuses at low pH, *Science* **225:**721–723.

Garoff, H., Frischaur, A.-M., Simons, K., Lehrach, H., and Delius, H., 1980, Nucleotide sequence of cDNA coding for Semliki Forest virus membrane glycoprotein, *Nature (Lond.)* **288:**236–241.

Garoff, H., Kondor-Koch, C., and Riedel, H., 1982, Structure and assembly of alphaviruses, *Curr. Topics Microbiol. Immunol.* **99:**1–50.

Goldstein, J. L., Brown, M. S., Anderson, R. G. W., Russell, D. W., and Schneider, W. J., 1985, Receptor-mediated endocytosis: Concepts emerging from the LDL receptor system, *Annu. Rev. Cell Biol.* **1:**1–39.

Gompels, U. A., Richman, D. D., Minson, A. C., and Buckmaster, E. A., 1985, Characterization of glycoprotein H of HSV-1, in: *Tenth International Herpesvirus Workshop,* Ann Arbor, Michigan, p. 135 (abst.).

Handa, K., Chany-Fournier, F., Rousset, S., and Chany, C., 1982, Diffusion of G glycoprotein induced by vesicular stomatitis virus during polykaryocyte formation in cell culture, *Biol. Cell* **44:**261–270.

Hay, A. J., Kennedy, N. C. T., Skehel, J. J., and Appleyard, G., 1979, The matrix protein gene determines amantadine-sensitivity of influenza viruses, *J. Gen. Virol.* **42:**189–191.

Haywood, A. M., 1974, Characteristics of Sendai virus receptors in a model membrane, *J. Mol. Biol.* **83:**427–436.

Haywood, A. M., and Boyer, B. P., 1981, Initiation of fusion and disassembly of Sendai virus membranes into liposomes, *Biochem. Biophys. Acta* **646:**31–35.

Haywood, A. M., and Boyer, B. P., 1985, Fusion of influenza virus membranes with liposomes at pH 7.5, *Proc. Natl. Acad. Sci. U.S.A.* **82:**4611–4615.

Hiller, G., and Weber, K., 1985, Golgi-derived membranes that contain an acylated viral polypeptide are used for vaccinia virus envelopment, *J. Virol.* **55:**651–659.

Hosaka, Y., and Shimizu, K., 1977, Cell fusion by Sendai virus, in: *Virus Infection and the Cell Surface* (G. Poste and G. L. Nicolson, eds.), Elsevier/North-Holland Biomedical Press, Amsterdam, pp. 4–155.

Hsu, M.-C., Scheid, A., and Choppin, P. W., 1979, Reconstitution of membranes with individual paramyxovirus glycoproteins and phospholipid in cholate solution, *Virology* **95:**476–491.

Hsu, M.-C., Scheid, A., and Choppin, P. W., 1982, Enhancement of membrane-fusing activity of Sendai virus by exposure of the virus to basic pH is correlated with a conformational change in the fusion protein, *Proc. Natl. Acad. Sci. U.S.A.* **79**:5862–5866.

Huang, R. T. C., Rott, R., Wahn, K., Klenk, H.-D., and Kohama, T., 1980, The function of the neuraminidase in membrane fusion induced by myxoviruses, *Virology* **107**:313–319.

Huang, R. T. C., Rott, R., and Klenk, H.-D., 1981, Influenza viruses cause hemolysis and fusion of cells, *Virology* **110**:243–247.

Huang, R. T. C., Dietsch, E., and Rott, R., 1985, Further studies on the role of neuraminidase and the mechanism of low pH dependence in influenza virus-induced membrane fusion, *J. Gen. Virol.* **66**:295–301.

Hughes, J. V., Johnson, T. C., Rabinowitz, S. G., and Dal Canto, M. C., 1979*a*, Growth and maturation of a vesicular stomatitis virus temperature-sensitive mutant and its central nervous system isolate, *J. Virol.* **29**:312–321.

Hughes, J. V., Dille, B. J., Thimmig, R. L., Johnson, T. C., Rabinowitz, S. G., and Dal Canto, M. C., 1979*b*, Neuroblastoma cell fusion by a temperature-sensitive mutant of vesicular stomatitis virus, *J. Virol.* **30**:883–890.

Knipe, D. M., Baltimore, D., and Lodish, H., 1977, Separate pathways of maturation of the major structural proteins of vesicular stomatitis virus, *J. Virol.* **21**:1128–1139.

Knutton, S., 1978, The mechanism of virus-induced cell fusion, *Micron* **9**:133–154.

Kohn, A., 1965, Polykaryocytosis induced by Newcastle disease virus in monolayers of animal cells, *Virology* **26**:228–245.

Kondor-Koch, C., Burke, B., and Garoff, H., 1983, Expression of Semliki Forest virus proteins from cloned complementary DNA. I. The fusion activity of the spike glycoprotein, *J. Cell Biol.* **97**:644–651.

Kotwal, G. J., Capone, J., Irving, R. A., Rhee, S. H., Bilan, P., Toneguzzo, F., Hofmann, T., and Ghosh, H. P., 1983, Viral membrane glycoproteins: Comparison of the amino terminal amino acid sequences of the precursor and mature glycoproteins of three serotypes of vesicular stomatitis virus, *Virology* **129**:1–11.

Krystal, M., Young, J. F., Palese, P., Wilson, I. A., Skehel, J. J., and Wiley, D. C., 1983, Sequential mutations in hemagglutinins of influenza B virus isolates: Definition of antigenic domains, *Proc. Natl. Acad. Sci. U.S.A.* **80**:4527–4531.

Krystal, M., Young, J. F., Palese, P., Wilson, I. A., Skehel, J. J., and Wiley, D. C., 1984, (Corrections) Sequential mutations in hemagglutinins of influenza B virus isolates: Definition of antigenic domains, *Proc. Natl. Acad. Sci. U.S.A.* **81**:1261.

Lamb, R. A., and Choppin, P. W., 1983, The gene structure and replication of influenza virus, *Annu. Rev. Biochem.* **52**:467–506.

Lamb, R. A., Zebedee, S. L., and Richardson, C. D., 1985, Influenza virus M_2 protein is an integral membrane protein expressed on the infected-cell surface, *Cell* **40**:627–633.

Little, S. P., and Schaffer, P. A., 1981, Expression of the syncytial (*syn*) phenotype in HSV-1, strain KOS: Genetic and phenotypic studies of mutants in two *syn* loci, *Virology* **112**:686–702.

Lubeck, M. D., Schulman, J. L., and Palese, P., 1978, Susceptibility of influenza A virus to amantadine is influenced by the gene coding for M protein, *J. Virol.* **28**:710–716.

Manservigi, R., Spear, P. G., and Buchan, A., 1977, Cell fusion induced by herpes simplex virus is promoted and suppressed by different viral glycoproteins, *Proc. Natl. Acad. Sci. U.S.A.* **74**:3913–3917.

Marsh, M., 1984, The entry of enveloped viruses into cells by endocytosis, *Biochem. J.* **218**:1–10.

Marsh, M., and Helenius, A., 1980, Adsorptive endocytosis of Semliki Forest virus, *J. Mol. Biol.* **142**:439–454.

Matlin, K. S., Reggio, H., Helenius, A., and Simons, K., 1981, Infectious entry pathway of influenza virus in a canine kidney cell line, *J. Cell Biol.* **91**:601–613.

Merz, D. C., and Wolinsky, J. S., 1981, Biochemical features of mumps virus neuraminidases and their relationship to pathogenicity, *Virology* **114**:218–227.

Merz, D. C., and Wolinsky, J. S., 1983, Conversion of nonfusing mumps virus infections to fusing infections by selective proteolysis of the HN glycoprotein, *Virology* **131**:328–340.

Merz, D. C., Scheid, A., and Choppin, P. W., 1980, Importance of antibodies to the fusion glycoprotein of paramyxoviruses in the prevention of spread of infection, *J. Exp. Med.* **151**:275–288.

Miura, N., Uchida, T., and Okada, Y., 1982, HVJ (Sendai virus)-induced envelope fusion and cell fusion are blocked by monoclonal anti-HN protein antibody that does not inhibit hemagglutination activity of HVJ, *Exp. Cell Res.* **141**:409–420.

Moss, B., 1974, Reproduction of poxviruses, in: *Comprehensive Virology,* Vol. 3, (H. Fraenkel-Conrat and R. R. Wagner, eds.), Plenum Press, New York, pp. 405–474.

Nakanishi, M., Uchida, T., Kim, J., and Okada, Y., 1982, Glycoproteins of Sendai virus (HVJ) have a critical ratio for fusion between virus envelopes and cell membranes, *Exp. Cell Res.* **142**:95–101.

Nishiyama, Y., Ito, Y., Shimokata, K., Kimura, Y., and Nagata, I., 1976, Polykaryocyte formation induced by VSV in mouse L cells, *J. Gen. Virol.* **32**:85–96.

Noble, A. G., Lee, G. T. -Y., Sprague, R., Parish, M. L., and Spear, P. G., 1983, Anti-gD monoclonal antibodies inhibit cell fusion induced by herpes simplex virus type 1, *Virology* **129**:218–224.

Norrby, E., and Penttinen, K., 1978, Differences in antibodies to the surface components of mumps virus after immunization with formalin-inactivated and live virus vaccines, *J. Infect. Dis.* **138**:672–676.

Norrby, E., Enders-Ruckle, G., and ter Meulen, V., 1975, Differences in the appearance of antibodies to structural components of measles virus after immunization with inactivated and live virus, *J. Infect. Dis.* **132**:262–269.

Nussbaum, O., Zakai, N., and Loyter, A., 1984, Membrane-bound antiviral antibodies as receptors for Sendai virions in receptor-depleted erythrocytes, *Virology* **138**:185–197.

Okada, Y., and Murayama, F., 1966, Requirement of calcium ions for the cell fusion reaction of animal cells by HVJ, *Exp. Cell Res.* **44**:527–551.

Patel, K., and Pasternak, C. A., 1985, Permeability changes elicited by influenza and Sendai viruses: Separation of fusion and leakage by pH-jump experiments, *J. Gen. Virol.* **66**:767–775.

Paterson, R. G., Hiebert, S. W., and Lamb, R. A., 1985, Expression at the cell surface of biologically active fusion and hemagglutinin/neuraminidase proteins of the paramyxovirus simian virus 5 from cloned cDNA, *Proc. Natl. Acad. Sci. U.S.A.* **82**:7520–7524.

Payne, L. G., and Kristenson, K., 1979, Mechanism of vaccinia virus release and its specific inhibition by N_1-isonicotinoyl-N_2-3-methyl-4-chlorobenzoylhydrazine, *J. Virol.* **32**:614–622.

Peterhans, E., Baechi, T., and Yewdell, J., 1983, Evidence for different receptor sites in mouse spleen cells for the Sendai viruses hemagglutinin-neuraminidase (HN) and fusion (F) glycoproteins, *Virology* **128**:366–376.

Pfeifer, J. B., and Compans, R. W., 1984, Structure of the influenza C glycoprotein gene as determined from cloned DNA, *Virus Res.* **1**:281–296.

Pogue-Geile, K. L., and Spear, P. G., 1987, The single base pair substitution responsible for the Syn phenotype of herpes simplex virus type 1, strain MP, *Virology,* in press.

Pogue-Geile, K. L., Lee, G. T.-Y., Shapira, S. K., and Spear, P. G., 1984, Fine mapping of mutations in the fusion-inducing MP strain of herpes simplex type 1, *Virology* **136**:100–109.

Poste, G., 1972, Mechanisms of virus-induced cell fusion, *Int. Rev. Cytol.* **33**:157–252.

Poste, G., and Pasternak, C. A., 1978, Virus-induced cell fusion, in: *Membrane Fusion* (G. Poste and G. L. Nicolson, eds.), Elsevier/North- Holland Biomedical Press, Amsterdam, pp. 305–367.

Richardson, C. D., and Choppin, P. W., 1983, Oligopeptides that specifically inhibit membrane fusion by paramyxoviruses: Studies on the site of action, *Virology* **131:**518–532.

Richardson, C. D., Scheid, A., and Choppin, P. W., 1980, Specific inhibition of paramyxovirus and myxovirus replication by oligopeptides with amino acid sequences similar to those at the N-termini of the F_1 or HA_2 viral polypeptides, *Virology* **105:**205–222.

Riedel, H., Kondor-Koch, C., and Garoff, H., 1984, Cell surface expression of fusogenic vesicular stomatitis virus G protein from cloned cDNA, *EMBO J.* **3:**1477–1483.

Roizman, B., 1962, Polykaryocytosis, *Cold Spring Harbor Symp. Quant. Biol.* **27:**327–340.

Rose, J. K., and Gallione, C. J., 1981, Nucleotide sequences of the mRNA's encoding the vesicular stomatitis virus G and M proteins determined from cDNA clones containing the complete coding regions, *J. Virol.* **39:**519–528.

Rose, J. K., Doolittle, R. F., Anilionis, A., Curtis, P. J., and Wunner, W. H., 1982, Homology between the glycoproteins of vesicular stomatitis virus and rabies virus, *J. Virol.* **43:** 361–364.

Sanders, P. G., Wilkie, N. M., and Davison, A. J., 1982, Thymidine kinase deletion mutants of herpes simplex virus type 1, *J. Gen. Virol.* **63:**277–295.

Schlegel, R., and Wade, M., 1984, A synthetic peptide corresponding to the NH_2 terminus of vesicular stomatitis virus glycoprotein is a pH-dependent hemolysin, *J. Biol. Chem.* **259:**4691–4964.

Schlegel, R., and Wade, M., 1985, Biologically active peptides of the vesicular stomatitis virus glycoprotein, *J. Virol.* **53:**319–323.

Scholtissek, C., and Faulkner, G. P., 1979, Amantadine-resistant and sensitive influenza A strains and recombinants, *J. Gen. Virol.* **44:**807–815.

Shibuta, H., Nozawa, A., Shioda, T., and Kanda, T., 1983, Neuraminidase activity and syncytical formation in variants of parainfluenza 3 virus, *Infect. Immun.* **41:**780–788.

Siddell, S., Wege, H., and ter Meulen, V., 1983, The biology of coronaviruses, *J. Gen. Virol.* **64:**761–776.

Skehel, J. J., Bayley, P. M., Brown, E. B., Martin, S. R., Waterfield, M. D., White, J. M., Wilson, I. A., and Wiley, D. C., 1982, Changes in the conformation of influenza virus hemagglutinin at the pH optimum of virus-mediated membrane fusion, *Proc. Natl. Acad. Sci. U.S.A.* **79:**968– 972.

Spear, P. G., 1984, Glycoproteins specified by herpes simplex viruses, in: *The Herpesviruses,* Vol. 3 (B. Roizman, ed.), Plenum Press, New York, pp. 315–356.

Spear, P. G., and Roizman, B., 1980, Herpes simplex viruses, in: *Molecular Biology of Tumor Viruses: DNA Tumor Viruses* (J. Tooze, ed.), Cold Spring Harbor Laboratory, Cold Spring Harbor, New York, pp. 615–745.

Storey, D. G., and Kang, C. Y., 1985, Vesicular stomatitis virus-infected cells fuse when the intracellular pool of functional M protein is reduced in the presence of G protein, *J. Virol.* **53:**374–383.

Strauss, E. G., and Strauss, J. H., 1983, Replication strategies of the single stranded RNA viruses of eukaryotes, *Curr. Top. Microbiol. Immunol.* **105:**1–98.

Takehara, M., 1975, Polykaryocytosis induced by vesicular stomatitis virus infection in BHK-21 cells, *Arch. Virol.* **49:**297–306.

Varghese, J. N., Laver, W. G., and Colemen, P. M., 1983, Structure of the influenza virus glycoprotein antigen neuraminidase at 2.9 Å resolution, *Nature (Lond.)* **303:** 35–40.

Varmus, H., and Swanstrom, R., 1984, Replication of retroviruses, in: *RNA Tumor Viruses,* 2nd ed., Vols. 1 and 2 (R. Weiss, N. Teich, H. Varmus, and J. Coffin, eds.), Cold Spring Harbor

Laboratory, Cold Spring Harbor, New York, pp. 369–512 (Vol. 1) and pp. 75–134 (Vol. 2).

Varsanyi, T. M., Jörnvall, H., and Norrby, E., 1985, Isolation and characterization of the measles F_1 polypeptide: Comparison with other paramyxovirus fusion proteins, *Virology* **147:**110–117.

Wagner, R. R., 1975, Reproduction of rhabdoviruses, in: *Comprehensive Virology,* Vol. 4 (H. Fraenkel-Conrat and R. R. Wagner, eds.), Plenum Press, New York, pp. 1–93.

White, J., Matlin, K., and Helenius, A., 1981, Cell fusion by Semliki Forest, influenza and vesicular stomatitis viruses, *J. Cell Biol.* **89:**674–679.

White, J., Helenius, A., and Gething, M.-J., 1982, Haemagglutinin of influenza virus expressed from a cloned gene promotes membrane fusion, *Nature (Lond.)* **300:**658–659.

White, J., Kielian, M., and Helenius, A., 1983, Membrane fusion proteins of enveloped animal viruses, *Quart. Rev. Biophys.* **16:**151–195.

Wilson, I. A., Skehel, J. J., and Wiley, D. C., 1981, Structure of the haemagglutinin membrane glycoprotein of influenza virus at 3 Å resolution, *Nature (Lond.)* **289:**366–373.

Chapter 2

Probing the Function of Viral Fusion Proteins with Synthetic Peptides

Richard Schlegel

1. INTRODUCTION

Synthetic peptides offer several unique approaches to studying molecular and cellular interactions. Not only do these peptides provide a large quantity of material for experimental analysis, but they also offer a degree of homogeneity that is difficult to obtain through routine biochemical purification. Only during the last several years has the ability to use these peptides for dissecting biochemical interactions begun to be explored. For example, synthetic peptides are now being utilized frequently to generate antisera specific for sequenced genes. This approach has been successfully applied to detecting and characterizing viral and cellular proteins (Neumann *et al.,* 1985; Walter *et al.,* 1981; Walter and Doolittle, 1983) as well as to generate neutralizing antibodies in animals (Chow *et al.,* 1985). In addition to serving as an antigenic stimulus, however, peptides may be used as a biochemical probe to analyze molecular interactions, either as agonists or antagonists. The functional domain of epidermal

RICHARD SCHLEGEL • Laboratory of Tumor Virus Biology, National Cancer Institute, Bethesda, Maryland 20892.

growth factor involved in cell stimulation has been delineated by the methodical use of small peptides to mimic the effect of the intact molecule (Komoriya *et al.*, 1984). Complex interactions such as cell binding to the substratum have been brought to a more elementary level by realizing that the peptide Asp XX Gly serves as an apparent fibronectin attachment site (Pierschbacher and Ruoslahti, 1984).

2. PEPTIDES FOR ANALYZING VIRAL FUSION PROTEINS

While a role for glycoproteins in virus-mediated membrane fusion is certain, the mechanism by which this process is effected remains unclear. One of the best-studied models for viral membrane fusion is that of the paramyxoviruses (e.g., Sendai and measles virus) and orthomyxoviruses (e.g., influenza virus). Sendai virus, for example, encodes a glycoprotein Fo (64,700) that is proteolytically cleaved into two subunits: F_1 (51,500) and F_2 (11,300). Proteolytic cleavage is necessary for activation of the fusogenic properties of the virus and correlates with the generation of a hydrophobic terminus on the F_1 protein (Gething *et al.*, 1978; Scheid and Choppin, 1974; Scheid *et al.*, 1978). It is presumed that this hydrophobic NH_2-terminus of F_1 has a direct role in promoting virus–cell membrane fusion, since a similar hydrophobic terminus is also generated by proteolysis of other paramyxoviruses (NDV, SVS) as well as orthomyxoviruses (influenza). Interestingly, oligopeptides have been used to help define the role of the hydrophobic NH_2-terminus of the F_1 (paramyxovirus) or HA_2 (orthomyxovirus) glycoprotein in membrane fusion. Richardson *et al.* (1980) first described that carbobenzoxy derivatives of the NH_2 terminal amino acids of the F_1 protein were capable of efficiently inhibiting measles plaque formation, cell fusion of CV-1 cells, and hemolysis of African green monkey erythrocytes. These synthetic peptides also inhibited the infectivity of influenza virus which shows strong homology of its HA_2 terminus to the paramyxovirus F_1-amino terminus. Longer peptides (7-amino acids) seemed to be somewhat more active than dipeptides. Somewhat confusing was the observation that the carbobenzoxy derivative of D-Phe-L-Phe-Gly-D-Ala-D-Val-D-Ile-Gly was 10^4-fold more potent than the parent peptide. The etiology of this large increase in biological activity was not delineated. A follow-up study revealed that the inhibitory effects of the NH_2-terminal peptide were not simply due to inhibition of viral binding or uptake (Richardson and Choppin, 1983). Measles virus was still capable of hemagglutinating erythrocytes and binding/entering CV-1 cells in the presence of these peptide inhibitors. Apparently, the cell attachment site for measles virus was not being blocked. The fusion process, therefore, seemed to be the most likely site of inhibition. Additional experiments also indicated that the peptide needed to be bound to the cell surface in order for inhibition to occur and that there were approximately 3×10^6 peptide receptors per cell. It was suggested, therefore, that these sites might be essential for interacting with the F_1 NH_2-terminus and for facilitating membrane fusion. While a defined number of receptors seems inconsistent with experimental findings demonstrating that Sendai virus can fuse with protein-free liposomes (Hsu *et al.,* 1983; Scheid *et al.,* 1978), it is possible that lipid domains function as a component of the interactive site for F_1.

3. SYNTHETIC PEPTIDES OF VSV

Although it is not as well characterized as the HA molecule of influenza, the glycoprotein of VSV (G protein) is easy to purify and has recently been cloned, sequenced, and expressed in mammalian vectors (Florkiewicz and Rose, 1984; Riedel *et al.,* 1984; Rose and Gallione, 1981). As a consequence, we do know a considerable amount about the biological activity of VSV G proteins. This 65,000-M_r protein functions in both the attachment of VSV to the cell surface as well as its fusion with the cell membrane. Thus, purified G protein (incorporated into liposomes) can mediate hemolysis (Bailey *et al.,* 1984) and liposome membrane fusion (Eidelman *et al.,* 1984). Also, cells expressing cloned G protein are fusogenic at low pH (Florkiewicz and Rose, 1984; Riedel *et al.,* 1984). Unlike the fusogenic glycoproteins of ortho- and paramyxoviruses, VSV G protein neither exhibits the proteolytic cleavage necessary for activation of its fusogenic properties nor the presence of a distinct, highly hydrophobic domain which might participate in its interaction with cell membranes. However, the NH_2 terminus of G protein is a relatively conserved domain that contains a small cluster of hydrophobic amino acids and an NH_2-terminal lysine molecule. As G protein appears to mediate fusion best in the presence of acidic phospholipids such as phosphatidylserine (Schlegel *et al.,* 1983), it appeared that the NH_2 terminus might potentially contain both the necessary hydrophobic and charge requirements to interact with this lipid molecule. On the basis of this speculation, several peptides corresponding to this domain of VSV G protein were synthesized and evaluated for either their agonistic or antagonistic properties with respect to G protein-mediated membrane fusion (Schlegel and Wade, 1985). Some peptides were also used to generate antibodies to this domain in order to evaluate the function of the NH_2-terminus.

The consensus sequence for the G protein NH_2 terminus of five strains of VSV is given in Figure 1. Also included in Figure 1 are six synthetic peptides corresponding to this region.

Consensus Sequence (5 strains):

Lys — — Ile Val Phe Pro — — — — Gly Asx Trp Lys — Val Pro — — Tyr — Tyr Cys

KFT (25) (Indiana San Juan):

Lys Phe* Thr Ile* Val* Phe* Pro His Asn Gln Lys Gly Asn Trp Lys Asn Val* Pro Ser Asn Tyr His Tyr Cys* Pro

HNQ (18)

His Asn Gln Lys Gly Asn Trp Lys Asn Val Pro Ser Asn Tyr His Tyr Cys Pro

KFT (6)

Lys Phe Thr Ile Val Phe

EFT (6)

Glu Phe Thr Ile Val Phe

RFT (6)

Arg Phe Thr Ile Val Phe

KFF (6)

Lys Phe Phe Phe Phe Phe

Figure 1. Sequences of the VSV NH_2-terminus and synthesized peptides. The consensus sequence for five strains of VSV (9) and the sequences of six synthetic peptides are shown. Conserved amino acids are underlined; asterisks* (in the KFT(25)) indicate strongly hydrophobic amino acids.

4. NH_2-TERMINAL G PROTEIN PEPTIDES ARE HEMOLYSINS

Since hemolysis is a simple method to follow the fusogenic membrane-destabilizing properties of VSV (Bailey *et al.*, 1984), this assay was used to monitor the activity of the above synthetic peptides. Interestingly, both the KFT(25) and KFT(6) synthetic peptides corresponding to the NH_2 terminus were hemolytic at pH 5.0, similar to intact VSV (Fig. 2). The shorter peptide, in fact, was approximately 10-fold more potent than the larger 25 amino acid peptide. The half-maximal concentration of KFT(25) and KFT(6) was 40 μM and 3 μM, respectively. Variant peptides were also synthesized in an effort to delineate the structural requirements for peptide induced hemolysis. Conservative changes of the NH_2-terminal amino acid from lysine to arginine had only slight effects on peptide hemolytic activity, whereas substitution of lysine with glutamic acid resulted in the total loss of hemolytic function. To determine whether hemolysis was a consequence of a short hydrophobic sequence coupled to a positively charged amino acid, we assayed the activity of KFF(6), which contains an NH_2-terminal lysine followed by five phenylalanine molecules. This peptide exhibited slight hemolysis only at relatively high concentrations ($>$100 μM).

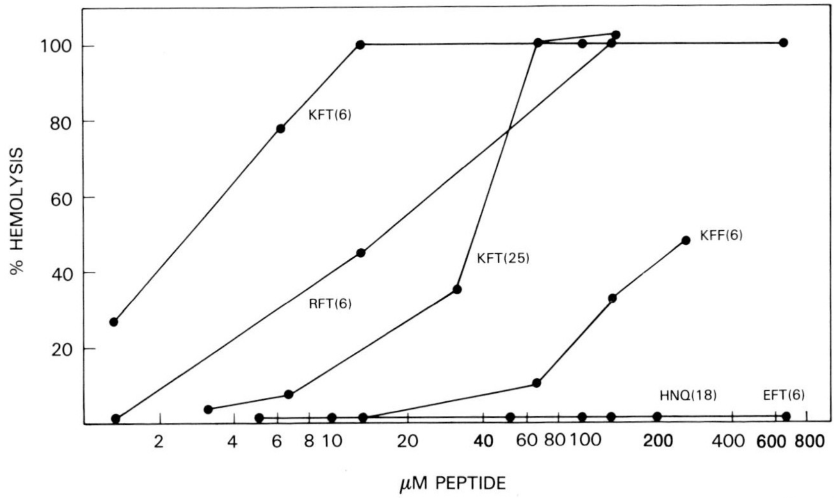

Figure 2. Hemolytic activities of synthetic peptides. Peptide stock solutions were diluted into a suspension of sheep erythrocytes (in pH 5.0 buffer) and the release of hemoglobin measured after 10 min at 37°C. EFT(6) and KFF(6) were dissolved in dimethyl sulfoxide. KFT(6) was the most potent hemolysin (half-maximal hemolysis at 3 μM).

5. SYNTHETIC PEPTIDES AND INTACT VSV HAVE SIMILAR pH/ACTIVITY PROFILES

VSV-mediated hemolysis is dependent upon both pH and temperature (Bailey *et al.,* 1984). It is most rapid at pH 5.0 and decreases with increasing pH up to pH 6.0. Above pH 6.0, hemolysis is very low. Peptide-mediated hemolysis by KFT(25) is analogous. Hemolysis is maximal at pH 5.0 and decreases with increasing pH. The requirement for low pH is VSV-induced hemolysis is believed to be due to conformational changes in G protein, which exposes new regions of the molecule, facilitating membrane fusion. It is perhaps relevant at this point to mention that low pH may be necessary for distinct phases of virus–cell membrane fusion. Changes in viral glycoprotein conformation at low pH have been documented by several techniques and are believed to be important for achieving proper glycoprotein orientation to interact with either the host cell membrane or adjacent glycoprotein molecules. For example, the influenza hemagglutinin molecule undergoes conformational alteration at pH 5.0, which can be monitored by circular dichroism, protease sensitivity, lipid vesicle binding, or detergent binding (Skehel *et al.,* 1982).

Somewhat similar alterations in the VSV G protein have also been noted at acidic pH. Exposure of a cathepsin D fragment of VSV G protein to pH 5.0 results in the increased binding of a flourescent, hydrophobic probe, indicating the uncovering of new hydrophobic domains in the G protein molecule (Crimmins *et al.,* 1983). However, the G protein does not exhibit altered protease sensitivity at pH 5.0. More importantly, whereas the reported conformational changes in influenza hemagglutinin are irreversible, those observed for G protein are reversible. This is an important consideration since the membrane fusion mediated by VSV G protein or intact influenza virus particles is a reversible process (Eidelman *et al.,* 1984; Stegmann *et al.,* 1985). Presumably, the pH-dependent conformational alterations described for influenza may be necessary but not sufficient for membrane fusion. It is also possible that other undetected alterations are occurring in the hemagglutinin (or membrane lipids) at low pH.

Hemolysis mediated by KFT(25) mimics exactly that described above for intact VSV with respect to its pH dependence. Hemolysis induced KFT(6) is weakly pH dependent. It is still unclear as to the mechanism by which pH modulates peptide-induced hemolysis but, due to the small size of these molecules, it appears unlikely that it derives from changes in peptide conformation. In addition, there are no amino acids in these peptides which have pK values which could readily explain the change in hemolytic activity between pH 5.0 and 7.0.

6. PEPTIDE HEMOLYTIC ACTIVITY IS DEPENDENT UPON BUFFER CONDITIONS

Peptide-induced lysis of erythrocytes occurs quickly in buffers made iso-osmotic with small ionic molecules such as NaCl or RbCl. Lysis, however, is delayed when iso-osmotic buffers are made from molecules such as mannitol, sucrose, or raffinose. This inhibition of hemolysis appears to result from a decrease in the rate of hemolysis, as the extent of hemolysis is unaffected. Apparently, the nonionic buffers are inhibiting the process of hemolysis rather than peptide binding. Consistent with this interpretation is the finding that nonionic buffers do not inhibit peptide-induced hemagglutination.

7. HEMOLYTIC PEPTIDES ARE ALSO HEMAGGLUTININS AND CYTOTOXINS

When hemolytic peptides are added to erythrocytes in the presence of sucrose, hemagglutination occurs. Actually, hemagglutination also occurs in NaCl buffers but the event rapidly progresses to hemolysis. Non-hemolytic peptides, such as HNQ(18) and EFT(6), were non-hemagglutinating under all conditions tested. KFT(25) caused hemagglutination at 4 μg/ml in sucrose but required 125 μg/ml in NaCl buffer, suggesting that peptides bind to erythrocytes in nonionic buffers at least as well as in ionic buffers. pH-dependent hemolysis does not result from the mere agglutination of erythrocytes at low pH. For example, polylysine (500 and 1000 μg/ml) causes massive hemagglutination at pH 5.0 but no hemolysis.

Hemolytic peptides were also evaluated for their ability to interact with membranes of nucleated cells. KFT(25) was added to Vero (monkey kidney) cells grown in tissue culture and the pH of the medium adjusted to either 7.4 or 5.5. Membrane permeability to trypan blue was assayed at different times of incubation at 37°C (Fig. 3). Cells exposed to KFT(25) at pH 5.5 showed a dose-dependent increase in permeability to trypan blue, whereas control cells at pH 5.5 did not. In addition, cells exposed to KFT(25) at pH 7.4 also showed no staining. Concomitant with this permeability change was a profound change in cell morphology. The addition of 100 μM KFT(25) at pH 5.5 caused Vero cells to swell and lose their distinct membrane borders. Nuclei became enlarged and cleared and nucleolar structure was lost. No evidence of cell fusion was seen, and electron microscopy confirmed gross vacuolization of the cell cytoplasm, nuclear swelling, and loss of nuclear matrix but retention of the plasma membrane bilayer structure. This finding appears consistent with KFT-(25) causing profound changes in membrane permeability.

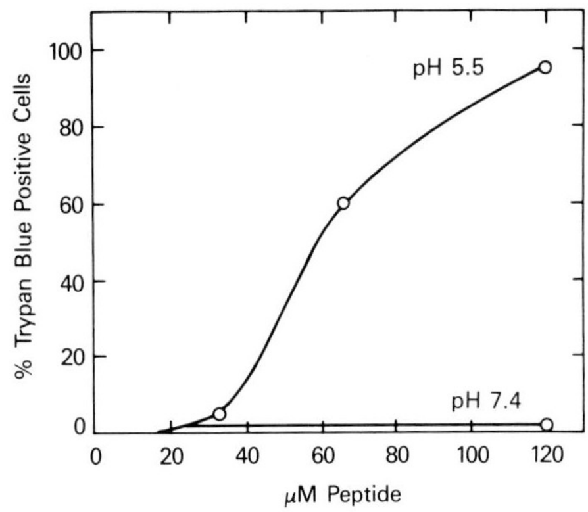

Figure 3. Permeability of peptide-treated cells to trypan blue. 100 μM KFT (25) was added to Vero cells for 20 min at 37°C at the indicated pH values. Cells were then reacted with trypan blue and evaluated for staining.

8. HEMOLYTIC PEPTIDES INDUCE ION PERMEABILITY CHANGES PRIOR TO HEMOLYSIS

The addition of 15 μM KFT(6) to 10^9 sheep erytrocytes labeled with ^{86}RB ions resulted in an immediate and rapid release of intracellular Rb^+ (Fig. 4a). Hemolysis and release of hemoglobin occurred a few minutes later. A greater dissociation of Rb^+ ion and hemoglobin efflux from erythrocytes can be obtained by using 3 μM KFT(6) (Fig. 4b). In this case, approximately 25% of the intracellular Rb^+ ion loss has occurred at 1 min, whereas none of the hemoglobin has been released. It is unlikely that this differential release of ion and protein represents sieving of molecules through a large defect in the cell membrane. Rather, it probably reflects a peptide-induced change in membrane ion permeability with consequent osmotic cell swelling and cell lysis. These changes in ion permeability might be a correlate of membrane destabilization.

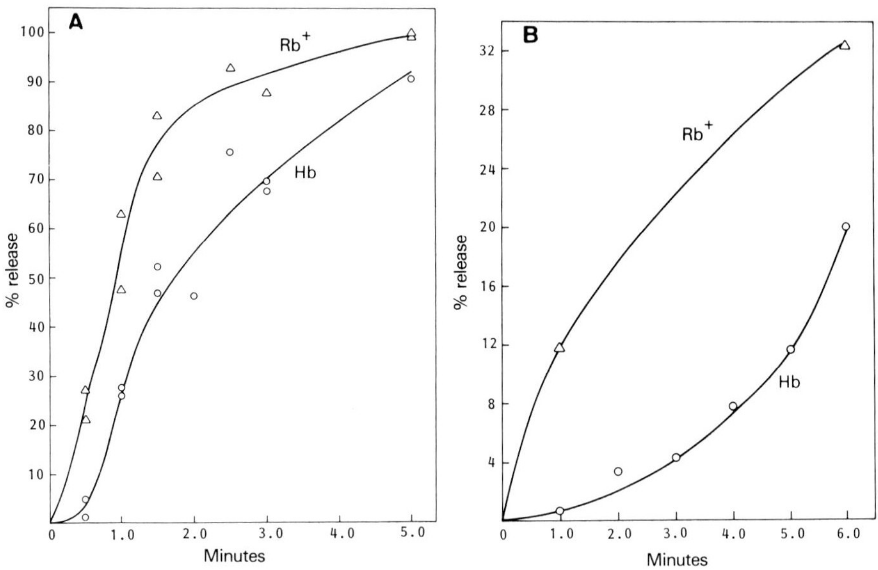

Figure 4. Rb^+ and hemoglobin release from sheep erythrocytes. 10^9 sheep erythrocytes (prelabeled with ^{86}Rb) were incubated with either 15 μM (A) or 3 μM (B) KFT(6) at pH 5.0 for the indicated times. Total hemolysis resulted in the release of 120,000 cpm ^{86}Rb and 0.5 absorption units hemoglobin.

9. IS INCREASED ION PERMEABILITY (AND OSMOTIC SWELLING) ESSENTIAL FOR VSV INFECTIVITY OR FOR VSV-INDUCED CELL FUSION AND HEMOLYSIS?

Although direct studies have not been performed with VSV, analogies to Sendai virus suggest that changes in cell permeability and consequent cell swelling can be dissociated from viral infectivity (Knutton and Pasternak, 1979). Whereas 3-day preparations of Sendai virus induce cell permeability changes and consequent cell swelling, 1-day virus preparations (which have presumably few damaged particles) do not initiate erythrocyte permeability changes or cell swelling despite the fact that they are highly infectious. This would seem to argue that cell permeability changes are not requisite for fusion and subsequent infections. However, erythrocyte swelling and hemolysis need not necessarily correlate with what happens at the virus–host cell membrane. It is clear that whether one observes leaky versus nonleaky fusion of Sendai virus with membranes is critically dependent on the presence of viral receptors. Thus, Sendai virus can effect either lytic or nonlytic fusion with liposomes, depending on the presence of appropriate sialoglycoproteins or sialoglycolipids (Citovsky and Loyter, 1985). Although electron microscopy has illustrated apparent fusion of nonhemolytic Sendai virus with erythrocytes (i.e., the formation of cytoplasmic bridges in the absence of hemolysis), such studies do not offer the quantitation or assurance that membrane fusion has indeed occurred. Although it can be argued that 1-day Sendai virus has fused with erythrocytes, since consequent hypotonic swelling results in observable cell fusion, the process of hypotonic swelling could actually be requisite for the fusion event itself. For example, it has also been postulated that cell swelling and membrane stretching may be necessary to expose the phospholipids of cell membranes in order to permit consequent virus–cell fusion (Blumenthal, 1987). Since fusion of Sendai virus with liposomes correlates with the presence of viral receptors and membrane leakiness, it could be argued that permeability changes are indeed a crucial aspect of virus–cell membrane fusion and that one function of the viral fusion glycoprotein would be to promote membrane permeabilization or destabilization.

It is unclear whether small synthetic peptides can mimic the fusogenic properties of large viral fusion proteins. VSV infection requires the presence of a multifunctional glycoprotein on its surface, which mediates cell attachment as well as subsequent membrane fusion. Presumably, several distinct protein domains are involved in the process of viral penetration, and it might not be possible to effect membrane fusion with a simple synthetic peptide. It is certainly clear that the removal of the hydrophobic transmembrane portion of G protein abolishes its fusion capacity.

While peptides themselves may be insufficient to mediate the entire process of membrane fusion, it might be possible to couple appropriate membrane-destabilizing peptides to hydrophobic molecules (e.g., lipids), which could serve as the equivalent of a transmembrane hydrophobic domain. Alternatively, cross-linking of the hemolytic peptides might permit sufficient interaction between opposing membrane surfaces to effect fusion.

REFERENCES

Bailey, C., Miller, D., and Lenard, J., 1984, Effects of DEAE-dextran on infection and hemolysis by VSV. Evidence that non-specific electrostatic interactions mediate effective binding of VSV to cells, *Virology* **133:**111–118.

Blumenthal, R., 1987, Membrane fusion, *Curr. Top. Membrane Transport* **29:** (in press).

Chow, M., Yabrov, R., Bittle, J., Hogle, J., and Baltimore, D., 1985, Synthetic peptides from four separate regions of the poliovirus type 1 capsid protein VPI induce neutralizing antibodies, *Proc. Natl. Acad. Sci. U.S.A.* **82:**910–914.

Citovsky, V., and Loyter, A., 1985, Fusion of Sendai virions or reconstituted Sendai virus envelopes with liposomes or erythrocyte membranes lacking virus receptors, *J. Biol. Chem.* **260:**12072–12077.

Crimmins, D. L., Mehard, W. B., and Schlesinger, S., 1983, Physical properties of a soluble form of the glycoprotein of vesicular stomatitis virus at neutral and acidic pH, *Biochemistry* **22:**5790–5796.

Eidelman, O., Schlegel, R., Tralka, T., and Blumenthal, R., 1984, pH-dependent fusion induced by vesicular stomatitis virus glycoprotein reconstituted into phospholipid vesicles, *J. Biol. Chem.* **259:**4622–4628.

Florkiewicz, R., and Rose, J., 1984, A cell line expressing vesicular stomatitis virus glycoprotein fuses at low pH, *Science* **225:**721–723.

Gething, M.-J., White, J., and Waterfield, M., 1978, Purification of the fusion protein of Sendai virus: Analysis of the NH_2-terminal sequence generated during precursor activation, *Proc. Natl. Acad. Sci. U.S.A.* **75:**2737–2740.

Hsu, M.-C., Scheid, A., and Choppin, P., 1983, Fusion of Sendai Virus with liposomes: Dependence on the viral fusion protein (F) and the lipid composition of liposomes, *Virology* **126:**361–369.

Knutton, S., and Pasternak, C., 1979, The mechanism of cell–cell fusion, *Trends Biochem. Sci.* **4:**220–223.

Komoriya, A., Hortsch, M., Meyers, C., Smith, M., Kanety, H., and Schlessinger, J., 1984, Biologically active synthetic fragments of epidermal growth factor: Localization of a major receptor-binding region, *Proc. Natl. Acad. Sci. U.S.A.* **81:**1351–1355.

Neumann, D., Gershoni, J., Fridkin, M., and Fuchs, S., 1985, Antibodies to synthetic peptides as probes for the binding site on the α-subunit of the acetylcholine receptor, *Proc. Natl. Acad. Sci. U.S.A.* **82:**3490–3493.

Pierschbacher, M., and Ruoslahti, E., 1984, Cell attachment activity of fibronectin can be duplicated by small synthetic fragments of the molecule, *Nature (Lond.)* **309:**30–33.

Richardson, C., and Choppin, P., 1983, Oligopeptides that specifically inhibit membrane fusion by paramyxovirus: Studies on the site of action, *Virology* **131**:518–532.

Richardson, C., Scheid, A., and Choppin, P., 1980, Specific inhibition of paramyxovirus and myxovirus replication by oligopeptides with amino acid sequences similar to those at the N-termini of the F_1 or HA_2 viral polypeptides, *Virology* **105**:205–222.

Riedel, H., Kondor-Koch, C., and Garoff, H., 1984, Cell surface expression of fusogenic vesicular stomatitis virus G protein from cloned cDNA, *EMBO J.* **3**:1477–1483.

Rose, J., and Gallione, C., 1981, Nucleotide sequences of mRNAs encoding the vesicular stomatitis G and M proteins determined from cDNA clones containing the complete coding regions, *J. Virol.* **39**:519–528.

Scheid, A., and Choppin, P., 1974, Identification of biological activities of paramyxovirus glycoprotein: Activation of cell fusion, hemolysis, and infectivity by proteolytic cleavage of an inactive precursor protein of Sendai virus, *Virology* **57**:475–490.

Scheid, A., Graves, M., Silver, S., and Choppin, P., 1978, Studies on the structure and function of paramyxovirus glycoproteins. In *Negative Strand Viruses and the Host Cell* (B. Mahy and R. Barry, eds.), pp. 183–193, Academic Press, New York.

Schlegel, R., and Wade, M., 1985, Biologically active peptides of the vesicular stomatitis virus glycoprotein, *J. Virol.* **53**:319–323.

Schlegel, R., Tralka, T., Willingham, M., and Pastan, I., 1983, Inhibition of VSV binding and infectivity by phosphatidylserine: is phosphatidylserine a VSV-binding site? *Cell* **32**: 639–646.

Skehel, J. J., Bayley, P. M., Brown, E. B., Martin, S. R., Waterfield, M. D., White, J. N., Wilson, I. A., and Wiley, D. C., 1982, Changes in the conformation of influenza virus hemagglutinin at the pH optimum of virus-mediated membrane fusion, *Proc. Natl. Acad. Sci. U.S.A.* **79**:968–972.

Stegmann, T., Hoekstra, D., Scherphof, G., and Wilshut, J., 1985, Kinetics of pH-dependent fusion between influenza virus and liposomes, *Biochemistry* **24**:3107–3113.

Walter, G., and Doolittle, R., 1983, Antibodies against synthetic peptides, in: *Genetic Engineering: Principles and Methods,* Vol. 5 (J. Setlow and A. Hollsender, eds.), pp. 61–91, Plenum Press, New York.

Walter, G., Hutchinson, M., Hunter, T., and Eckhart, W., 1981, Antibodies specific for the polyoma virus middle-size tumor antigen, *Proc. Natl. Acad. Sci. U.S.A.* **78**:4882–4886.

Chapter 3

The Roles of Ca^{2+}-Dependent Membrane-Binding Proteins in the Regulation and Mechanism of Exocytosis

Carl E. Creutz, William J. Zaks, Helen C. Hamman, and William H. Martin

1. INTRODUCTION

Exocytosis is the process of release of a stored secretory product from a cell by fusion of the membrane of the storage organelle (secretory vesicle) with the surface membrane of the cell. This process appears to have been almost universally adopted as the method for secretion of products that are stored and then suddenly released upon stimulation of the cell. For example, it is the basis for the release of neurotransmitters, such as norepinephrine and acetylcholine (ACH); hormones, such as insulin or other polypeptide hormones; and digestive enzymes, such as amylase and trypsin. Therefore, this is one of the most important membrane fusion events that occurs in mature, differentiated cells. Furthermore, some as-

CARL E. CREUTZ, WILLIAM J. ZAKS, HELEN C. HAMMAN, and WILLIAM H. MARTIN • Department of Pharmacology, Programs in Biophysics, Neuroscience, and Cell and Molecular Biology, and Center for Diabetes and Cancer Research and Training, University of Virginia, Charlottesville, Virginia 22908.

pects of this process may provide a model for other intracellular membrane fusion events that occur during regeneration of the plasma membrane, movement of secretory material between the stacks of the Golgi apparatus, fusion of phagosomes with lysosomes, or, in general, any intracellular transport process that involves membrane fusion.

As is the case for the activation of many cellular processes, the signal for the initiation of exocytosis appears to be an increase in the cytoplasmic concentration of free Ca^{2+}. This has become an important focal point for the biochemical dissection of exoytosis in as much as a great deal of emphasis has been placed on the possible roles of Ca^{2+}-binding proteins in regulating or mediating individual steps in this process. Therefore, we begin this chapter with an outline of the steps in exocytosis and consideration of possible sites of Ca^{2+} action. As this is a new and rapidly advancing area of cell biology, the picture presented in this chapter is rather incomplete and at many points speculative. It is hoped, however, that the presentation will provide a useful organizing framework into which the reader may assimilate future advances in this field.

2. SUBCELLULAR EVENTS IN EXOCYTOSIS

The cytoplasm of a typical secretory cell contains large numbers of secretory vesicles that have a common basic structure consisting of a limiting membrane and a densely packed, usually amorphous core material comprising the secretory product. These vesicles arise from the trans- face of the Golgi apparatus through a poorly understood process of budding from the Golgi stack and subsequently undergo “maturation,” during which the core materials are condensed. The vesicles are then stationed in the cytoplasm, often in discrete locations, or near a particular face of the cell that is adjacent to a blood vessel or secretory duct that is to receive the secretory product. After distribution in this fashion, the vesicles may become essentially immobilized in the cytoplasmic matrix, waiting for the appropriate stimulus to initiate Ca^{2+} fluxes into the cytoplasm. When the cell is activated, through a cell surface receptor, the Ca^{2+} concentration in the cytoplasm increases as a result of entry across the plasma membrane or due to release from intracellular storage sites. The storage sites are presumably components of the endoplasmic reticulum, although mitochondria might also participate in this process. As a result of the increase in Ca^{2+} activity, changes occur in the cytoplasmic matrix. This matrix is composed of a complex arrangement of cytoskeletal and contractile elements including actin, myosin, microtubules, intermediate filaments, and a wide variety of proteins involved in regulating these structures. The changes that occur appear to permit free diffusion or active translocation of vesicles to the plasma membrane for release to occur. Whether free diffusion, or an active, directed motion of the secretory vesicle occurs during exocytosis is not known, but both alternatives seem plausible. In the case of free diffusion, specificity of interaction between vesicle membranes and the plasma membrane may be due to specific recognition proteins in the cytoplasm or on the membranes. A potentially important site of Ca^{2+} action in this part of the exocytotic machinery was identified by Fowler and Pollard (1982), who studied the binding of actin to chromaffin granule membranes. These investigators used a falling ball viscometer to detect the cross-linking of actin filaments by granule membranes *in vitro* to form high-viscosity, gel-like structures. The formation of these gels was completely inhibited by Ca^{2+} at concentrations greater than 1 μM. The gel formed by actin and granule membranes may therefore be representative of the state of chromaffin cell cytoplasm at rest. Activation of the cell and the increase of free Ca^{2+} in the cytoplasm would presumably relax this gel and permit diffusion or translocation of the vesicles.

After translocation occurs, the secretory vesicles form contacts with the plasma membrane, or in the case of *compound exocytosis,* with one another. These contacts are sufficiently intimate that the closely apposed membranes exclude all intervening cytoplasmic material and form a *pentalaminar* junction from the two *trilaminar* membranes (Palade, 1975). This is apparently the initial step in true membrane fusion and must represent a semistable state of the interacting membranes, as these pentalaminar junctions are frequently seen in transmission electron micrographs of stimulated secretory cells. Since typical secretory vesicles, such as chromaffin granules, have a negative surface charge, some mechanism must exist for overcoming the electrostatic barrier, as well as the water of hydration barrier, that would prevent these contacts from forming. As will be seen, formation of membrane contacts may be a primary site of Ca^{2+} action in exocytosis and, in performing this function, Ca^{2+} may act through a number of specific Ca^{2+}-dependent membrane-binding proteins.

After the membrane contacts have formed, the two apposed membranes fuse, thus forming an opening to the outside of the cell through which the condensed secretory product escapes. In the case of compound exocytosis, membranes of secretory vesicles deeper in the cytoplasm fuse with one another so that channels form for escape of secretory product through vesicles that fuse directly with the plasma membrane. These membrane fusion events appear to depend on the destabilization of the pentalaminar junction; *in vitro* models to be described indicate that this may require the action of Ca^{2+}-dependent lipases and osmotic pressure increases in the vesicle core.

Subsequent to the membrane fusion event and release of contents, the secretory vesicle membrane is usually recovered by endocytosis in order to be reutilized. In some cells this recovery may take place at a different location from the site of exocytosis, after incorporation of vesicle membrane into the plasma membrane and lateral flow of membranes to a recovery site (Heuser and Reese, 1973). In other cases, it appears that the membrane is immediately recovered before the basic vesicular shape collapses (Nagasawa and Douglas, 1972). In either case, recovery occurs through the formation of clathrin-coated pits that break off and form coated vesicles. The nucleation of the clathrin coat on the membrane of the expended vesicle may be another site of regulation of the overall secretory process by Ca^{2+}.

There is morphological (Phillips *et al.,* 1983; Dowd *et al.,* 1983) and biochemical (Suchard *et al.,* 1981) evidence that after recovery of vesicle membrane following the secretory event, the membrane is reused to form new secretory vesicles. Presumably this requires transfer of membrane through the Golgi complex, although the precise route taken by recycled membrane is still unclear.

In addition to the specific sites mentioned above, Ca^{2+} may regulate other steps in the overall process of exocytosis. A great deal of attention has been paid to mechanisms that may transduce the Ca^{2+} signal, including binding to calmodulin, regulation of phosphorylation by Ca^{2+}-modulated kinases or phosphatases, and activation of proteases. However, there have been few suggestions concerning what precise mechanical steps are regulated in this way.

3. THE CHROMAFFIN CELL AND MODEL SYSTEMS FOR EXOCYTOSIS

In our laboratory we have focused our studies on the chromaffin cell of the adrenal medulla as a model system. The main attraction of this system is the ease with which large amounts of highly purified secretory vesicles (chromaffin granules) can be isolated from bovine adrenal medullary tissue. These vesicles are representative of many types of secretory organelle in that they contain small molecules such as catecholamines and nucleotides, as well as protein molecules, (e.g., chromogranin A, dopamine-β-hydroxylase, and enkephalin precursors) that are all released by exocytosis. It is also possible to isolate plasma membranes from these cells, as well as various components of the cytoskeleton and the Ca^{2+} dependent membrane-binding proteins, making this secretory cell particularly suitable for the construction of model subcellular systems that permit biochemical analysis of steps in the process of exocytosis.

The construction of models for exocytosis has become potentially more complex as the number of factors thought to be involved in the process has increased. For example, the simplest systems involve studies of isolated chromaffin granules or granules and isolated plasma membranes because it is morphologically obvious that these organelles, at a minimum, are involved. We would now like to suggest, however, that the highly purified chromaffin granule may not be representative of the state of this organelle in a simulated chromaffin cell. Instead, we believe that when the chromaffin cell is stimulated to undergo exocytosis, the chromaffin granules become coated with cytosolic proteins that bind to the vesicle membrane in a Ca^{2+}-dependent manner. In effect, the granule puts on a new suit of clothes before going out. This new surface coating may be the site of interaction with the cytoskeleton, with the plasma membrane, or with clathrin to mediate the various steps in the secretory process. In the sections below, we describe some specific interactions and events mediated by one or more of these proteins in model systems. It is hoped that these models will provide important clues to the molecular events occurring in the stimulated secretory cell.

4. PROTEINS THAT PROMOTE INTERMEMBRANE CONTACT AND FUSION

4.1. Synexin

The first Ca^{2+}-dependent chromaffin granule-binding protein to be described was synexin (Creutz *et al.,* 1978). This protein was encountered as a result of the fortuitous observation that adrenal medullary homogenates scatter more light if the free Ca^{2+} concentration in the homogenate is increased. A deliberate search for factors responsible for this light scattering increase led to the observation that similar Ca^{2+}-dependent increases were seen when a postmicrosomal supernatant from medullary tissue was incubated with isolated chromaffin granules. Observations by phase and electron microscopy indicated that the light-scattering increase was due to aggregation of the chromaffin granules. Using the Ca^{2+}-and temperature-dependent turbidity increase as an assay, the protein synexin (from the Greek *synexis,* meaning "a meeting") was isolated from the supernatant. This protein was found to be not only a membrane-binding protein but also in some sense bivalent, as it bound to two membranes and brought them together.

The mechanism of synexin action in promoting membrane aggregation has turned out to be somewhat more complex than initially expected in that Ca^{2+} appears to act at two different sites with different affinities. The highest affinity Ca^{2+}-binding site appears to regulate the binding of the protein to the membrane (Creutz and Sterner, 1983). This site has an apparent K_d for Ca^{2+} of 4 μM at neutral pH and so is comparable in affinity to sites on other well-characterized Ca^{2+}-dependent regulatory proteins such as calmodulin or troponin C (Potter *et al.,* 1983). The site on the membrane that binds synexin may be an acidic phospholipid, since the protein also binds to liposomes prepared from, for example, phosphatidylserine (Morris and Hughes, 1979) or mixtures of phosphatidic acid and phosphatidylethanolamine (Hong *et al.,* 1982). However, interactions with membrane proteins are also possible and may explain some of the differences seen in aggregation studies with liposomes and with chromaffin granules (Dabrow *et al.,* 1980). The capacity to bind to phosphatidylserine containing membranes is particularly appropriate for a protein that is postulated to bind to cytoplasmic membrane surfaces since, in the case of the secretory vesicle and the plasma membrane, these surfaces are rich in phosphatidylserine.

Although synexin binds to membranes at 1–10 μM Ca^{2+}, membrane–membrane interactions are not promoted until Ca^{2+} is increased above 10 μM (Creutz *et al.,* 1978; Hong *et al.,* 1982). Complete titration of the membrane aggregating activity suggests Ca^{2+} acts at a site with $K_d \sim 150–200$ μM. This second Ca^{2+}-binding site also apparently regulates a self-association reaction that isolated synexin undergoes (Fig. 1). This self-association reaction results in the formation of small polymers of synexin that can be seen in the electron microscope, appearing as 50 × 150 Å rods (Creutz *et al.,* 1979). This has led to the hypothesis that self-association of synexin molecules bound to different membranes may be responsible for bringing two membranes together (Creutz and Sterner, 1983). Some support for this mechanism comes from comparison of the aggregation of chromaffin granules induced by polylysine and by synexin. By virtue of being highly positively charged, polylysine causes aggregation of chromaffin granules. However, one would predict that a single layer of polylysine molecules would act as a bridge between two granules, and that if the polylysine concentration were too high granules would become individually coated with the positively charged protein and therefore repel one another. However, if granules were coated with a layer of synexin molecules these could still interact with one another and bring the granules together. Therefore protein titrations of the granule aggregating activity should show a maximum effect for polylysine, which then would decline at high concentrations, while synexin should show a stable maximum aggregating activity, even at high concentrations. Indeed, as illustrated in Figure 2a, b, these predictions appear to be borne out for granule aggregation by polylysine and by synexin. Curiously, the aggregation by

Figure 1. A hypothesis for the sequence of events that may lead to the formation of synexin induced intermembrane contacts during exocytosis in the chromaffin cell. In the resting cell (top), the Ca^{2+} concentration is less than 1 μM and synexin is freely soluble. In the stimulated cell (middle), as the Ca^{2+} concentration rises to between 1 and 10 μM, synexin attaches to membranes. When the Ca^{2+} concentration further increases above 10 μM Ca^{2+} (bottom), synexin molecules on different membranes self-associate to bring the membranes together. Although isolated synexin in the presence of Ca_o^{2+} self-associates to form 50 × 150 Å rods similar to the structure illustrated at the bottom of the figure, the actual number of synexin molecules that associate to form this structure, the orientation of this structure relative to the membrane, and the number of such structures involved in creating an intermembrane contact are unknown. (From Creutz and Sterner, 1983.)

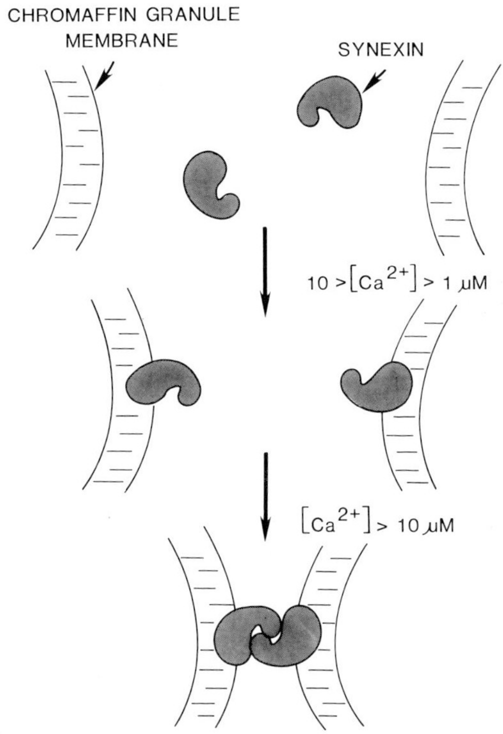

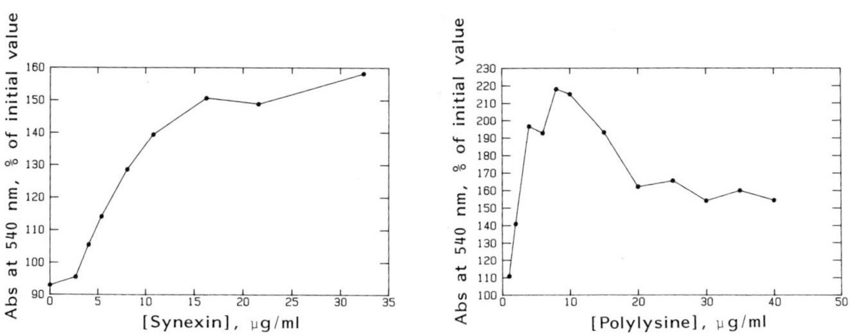

Figure 2. Aggregation of chromaffin granules by synexin (a) or polylysine (b). Dependence on the concentration of the aggregating protein. Granule aggregatoin was measured by a turbidometric assay. Chromaffin granules (50μg/ml) in 0.24 M sucrose, 30 mM KCl, 40 mM Hepes-NaOH, pH 7.0, were incubated with increasing amounts of synexin (a) or polylysine (b) at 25°C. The absorbance change at 540 nm observed 8 min after the addition of 1 mM Ca^{2+} (a) or the indicated concentration of polylysine (b) is plotted as a percentage of the initial absorbance (~0.3).

polylysine is not completely suppressed by high concentrations of polylysine (Fig. 2b). This may indicate that polylysine molecules periodically dissociate from the granule membrane thus briefly exposing the negatively charged membrane surface and permitting granule aggregation to occur.

The affinity of the Ca^{2+}-binding site on synexin that regulates self-association and granule aggregation is so low (Kd ~ 150–200 μM) that it may not be fully occupied even in extensively stimulated cells. We have therefore investigated the possibility that other cofactors may exist in stimulated cells that would enhance the Ca^{2+} sensitivity of this site. Indeed, we have found that cis-unsaturated fatty acids (e.g., arachidonic acid or oleic acid) enhance the polymerization of synexin at low concentrations of Ca^{2+} (Sterner *et al.,* 1985). Oleic acid causes a shift in the Ca^{2+} titration curve for synexin polymerization to the left (Fig. 3), resulting in detectable polymerization below 10 μM Ca^{2+}, where polymerization is not observed in the absence of the free fatty acid. Therefore, the release of cis-unsaturated fatty acids as a result of the action of Ca^{2+}-activated lipases may be necessary to promote the formation of synexin-dependent membrane contacts at physiological concentrations of Ca^{2+}.

Although synexin brings chromaffin granule membranes together, the contacts formed between the membranes are curiously stable, and true fusion of the membranes occurs very slowly or not at all *in vitro.* An additional property of chromaffin granules that we thought might promote fusion is their ability to increase their internal osmotic pressure. This occurs as a result of the action of a membrane ATPase which pumps protons into the vesicle interior thus generating an electrical potential across the vesicle membrane. This in turn will promote the transport of Cl^-, or other permeant anions, into the granule interior. The Cl^- ion itself, and possibly other ions recruited from the core complex, increases the internal osmotic

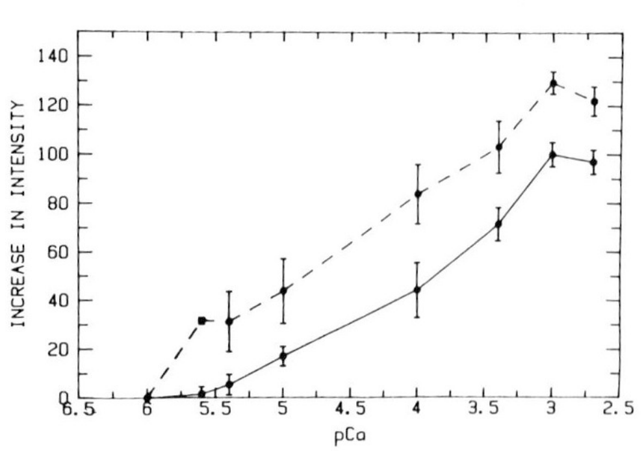

Figure 3. Ca^{2+} dependence of the polymerization of 1 μM synexin in the absence (—) or the presence (---) of 2.3 μM oleic acid. The increase in intensity of 350 nm light scattered at 90°, which reflects synexin polymerization, has been normalized so that 100 equals the increase in light scattered by 1 μM synexin alone 400 sec after the addition of 1 mM Ca^{2+}. pCa, − log [Ca^{2+}]. The data represent the mean ± SD of the means from five independent experiments each conducted in duplicate. (From Sterner *et al.,* 1985.)

pressure of the vesicle. *In vitro,* this process will lead to rupture of the vesicle membrane (Creutz and Pollard, 1980). However, we suspected that possibly rupture would occur at the contact point first, leading to fusion of the granules and consequent relief from the internal osmotic pressure increase. This mechanism would have been consistent with the experimental observations of Finkelstein and his colleagues (Zimmerberg *et al.*, 1980), who found that an osmotic gradient across a liposome membrane would promote its fusion with a planar lipid bilayer. Accordingly, we designed the following experiment to test whether osmotic gradients would promote granule–granule fusion. Chromaffin granule were preincubated with Ca^{2+} and synexin in order to form large granule aggregates. ATP was then added in the presence of 90 mM Cl^- in order to activate chemiosomotic swelling of the aggregated granules. The rather surprising consequence was that the granule membranes consistently broke specifically at regions not involved in forming contacts (Fig. 4). As a consequence, the core materials escaped into the medium, and there was no evidence of membrane fusion. Indeed, it appeared that the pentalaminar contact regions were extraordinarily stable and resistant to breakage. This was certainly troublesome for the hypothesis that these contacts represented the first step in membrane fusion that occurs in exocytosis. However, it

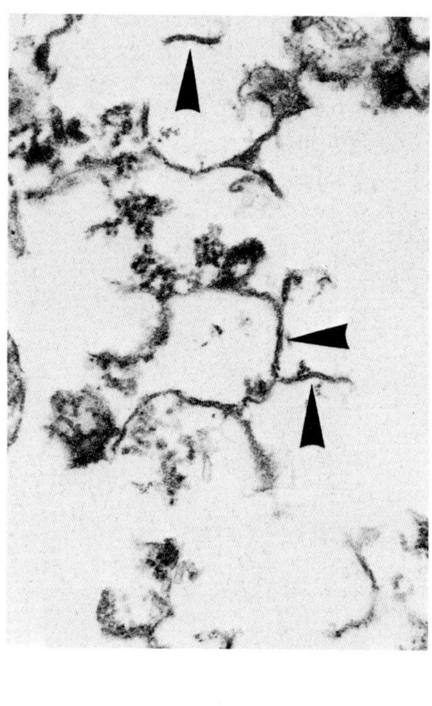

Figure 4. Pentalaminar membrane structures remain intact after chromaffin granules that have been aggregated by synexin are subjected to chemiosmotic lysis. The arrowheads point to some examples of membrane junctions that once formed the boundaries between aggregated chromaffin granules. Photo width: 1.0 μm.

was subsequently found that quite the opposite occurred when free, cis-unsaturated fatty acids were added to the system.

As a result of the accidental observation of granule aggregates undergoing fusion on a contaminated glass microscope slide (Creutz, 1981*b*), conditions were determined under which granule fusion would occur at the point of synexin-induced contact. It was found that an essential cofactor for fusion was a free, cis-unsaturated fatty acid (Fig. 5). Arachidonic acid was most effective at promoting fusion, although other, naturally occurring cis-unsaturated fatty acids, such as linolenic or oleic acids, would also promote fusion. The threshold amount of fatty acid necessary to promote fusion was 4% by mass relative to the lipid content of the granule membranes. Data on the release of arachidonic acid from activated platelets suggest that this concentration might be achieved at least in localized regions of a secretory cell under physiological conditions (Marcus, 1978). It is striking that the fatty acid specifically disrupts the mem-

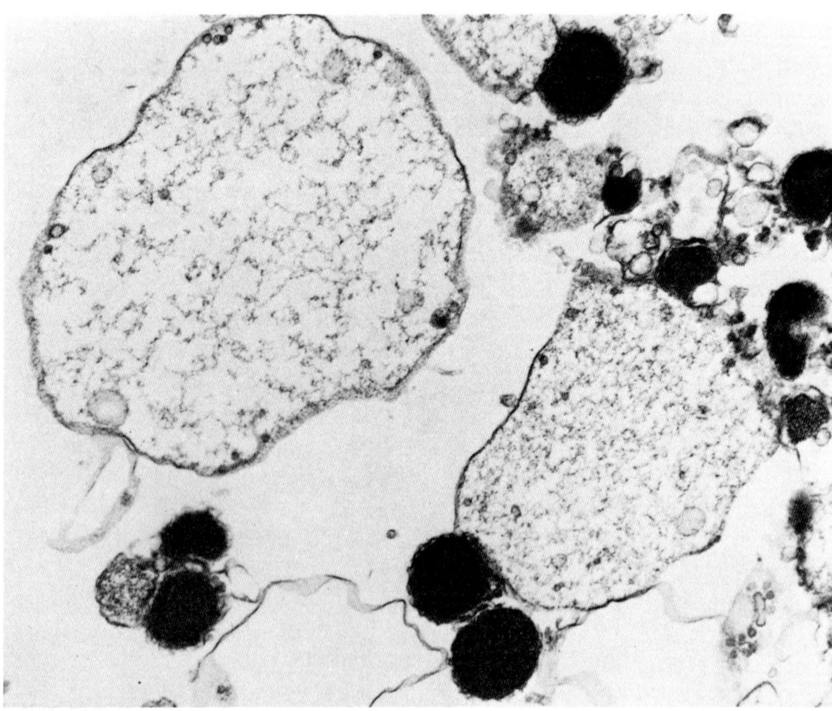

Figure 5. Isolated chromaffin granules fusing with one another in the presence of Ca^{2+}, synexin, and arachidonic acid. This model system appears to recreate membrane contact and fusion events that occur during compound exocytosis. Photo width: 2.7 μm. (From Creutz, 1981*b*.)

brane contact formed by synexin, and does not generally break the membrane of the vesicle at other regions. There are at least two possible reasons for this specific localization of action. First, it may be that the fatty acid promotes the formation of some unusual lipid stucture, such as an inverted micelle that can only form when two or more membranes are in close contact, and this structure then leads to membrane breakage and fusion. Alternatively, it may be that the synexin molecules are located specifically in the contact region between the membranes and that the synexin binds the fatty acid and holds it in such a way as to disrupt membrane structure. Free fatty acids do interact with isolated synexin. Whether a direct interaction also occurs in the presence of a membrane remains to be demonstrated, as well as the possible importance of either of these mechanisms in the membrane fusion event.

Another striking feature of the fusion that occurs between chromaffin granules is that the resultant vesicular structures rapidly swell to contain the maximum volume. If particularly large clumps of chromaffin granules are first formed by synexin action, the fusion can even be observed in the phase microscope because of the large diameter vesicles (>10 μm) that form from the fusion of as many as 1600 chromaffin granules (Fig. 6). The swelling of these vesicles may be osmotic in nature, and indeed can be suppressed by increasing the osmotic pressure of the incubation medium by adding sucrose (Creutz, 1981*b*). One important question that has remained unanswered, however, is whether the swelling of the large vesicles occurs after and as a result of fusion of the granules, or if some initial swelling precedes and is essential for membrane fusion to occur. Recently, we determined that increasing the osmotic pressure of the medium inhibits membrane fusion per se, suggesting that indeed an in-

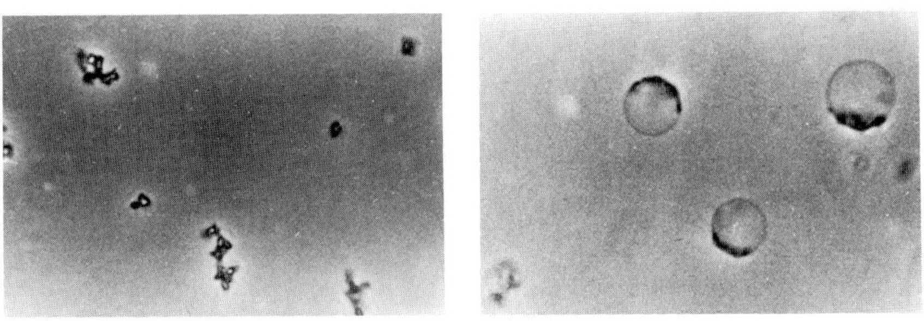

Figure 6. Visualization of chromaffin granule fusion in the phase microscope. (Left) Clumps of granules that have been aggregated by incubation with synexin and Ca^{2+} for 40 min. (Right) Vesicles of various sizes formed after further incubation of the preparation for ~15 min in the presence of 4 μg/ml arachidonic acid. Photo width: 52 μm. (From Creutz, 1981*b*.)

travesicular osmotic pressure increase is essential to drive the membrane fusion process. To make this determination, we applied a novel membrane fusion assay introduced by Hoekstra *et al.* (1984). This assay is based on the increase in fluorescence of octadecyl rhodamine as a consequence of dilution and dequenching of the probe during membrane fusion. Initially the probe is introduced into one population of chromaffin granules at a high, self-quenching concentration. These labeled granules are then mixed with unlabeled granules. Only when fusion is initiated can the lipid probe diffuse into a larger area, reducing its concentration and increasing the fluorescence yield. Using this assay, we found that 200 mM stachyose (an impermeant tetrasacharide) would suppress fusion caused by the synexinlike activity of 32,000-M_r calelectrin, as shown in Figure 7. (The properties of calelectrin, which is a synexinlike protein that was originally isolated from the electric organ of the electric ray Torpedo, are described more fully in Section 4.2.) Similar confirming results were obtained with a fusion assay based on energy transfer between NBD-labeled phosphatidylethanolamine and rhodamine-labeled phosphatidylethanolamine that was orginally applied to chromaffin granule fusion by Morris and Bradley (1984). In addition, we found that membrane fusion could not be induced by calelectrin between granule membrane ghosts (granules freed of secretory content by osmotic lysis and then resealed). Therefore, it appears that some rearrangement of the core materials in the secretory vesicle may be essential for the fusion of granules that can be induced in the *in vitro* model systems. How this change occurs and how it might be regulated in the cell are unknown, but this may be an essential element in the process of membrane fusion in exocytosis.

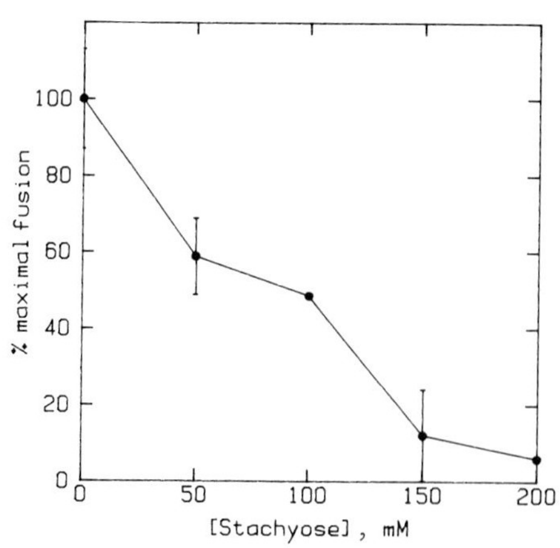

Figure 7. Osmotic suppression of chromaffin granule fusion. Chromaffin granules were labeled with octadecylrhodamine B as described (Hoekstra *et al.*, 1984). Labeled chromaffin granules (1 μg/ml) were mixed with unlabeled chromaffin granules (4 μg/ml) in the presence of 0.5 μg/ml oleic acid and 1.5 μg/ml 32-kDa protein (calelectrin) in buffer containing 0.24 M sucrose, 30 mM KCl, 0.1 mM EGTA, 40 mM Hepes-NaOH, pH 7.0 and increasing amounts of stachyose at 25°C. After recording stable baseline fluorescence (560 excitation, 590 emission), fusion was initiated by addition of 1 mM $CaCl_2$. Fusion was recorded as the absolute change in fluorescence intensity after 10 min and expressed as percentage of the response seen in the absence of added stachyose.

4.2. Additional Proteins with Synexinlike Activity

Recently a group of investigators at the Max Planck Insititute in Göttingen addressed the question of whether synexin could be isolated from a purely cholinergic tissue, the *Torpedo marmorata* electric organ (Walker, 1982; Sudhof *et al.,* 1985). These workers isolated a protein distinct from synexin in that it had a different molecular weight (32,000 for the Torpedo protein, versus 47,000 for synexin) different isoelectric point (5.8 versus 7.0), different peptide map (our unpublished data) and no immunological cross reaction with an antisynexin antiserum (our unpublished data). However, this protein, which they named calelectrin, does aggregate chromaffin granules in a Ca^{2+}-dependent manner (Sudhof *et al.,* 1982), and the granule aggregates are subject to fusion when exposed to arachidonic acid (our unpublished data).

The Göttingen group also prepared an antiserum to *Torpedo* calelectrin and determined that this antiserum cross-reacted with proteins of masses 32,000 and 67,000 daltons in bovine liver and adrenal medulla (Sudhof *et al.,* 1984). The 32,000-dalton protein was subsequently isolated and found to have similar properties to *Torpedo* calelectrin. The 67,000-M_r protein has a very similar amino acid composition to the 32,000-M_r protein and is also activated to aggregate chromaffin granules by Ca^{2+}. However, at intermediate levels of Ca^{2+} ($\sim$10–400μM) the 67,000-M_r protein is a much weaker aggregating agent than synexin or the 32,000-M_r calelectrin. Therefore, in the presence of these other proteins, the 67,000-M_r calelectrin can actually behave as an apparent inhibitor of granule aggregation. In fact, this 67,000-M_r protein is very likely to be identical with the protein "synhibin" described by Pollard and Scott (1982), which at intermediate levels of Ca^{2+} was reported to inhibit the ability of synexin to aggregate chromaffin granules.

The mechanism by which the 67,000-M_r protein inhibits synexin action may be by preventing the binding of synexin to the membrane. This was originally suggested by Pollard and Scott (1982) on the basis of the apparent competitive nature of inhibition; it has been more directly demonstrated in recent studies (Zaks, 1986; Zaks and Creutz, 1986) on the displacement of [^{125}I]labeled synexin from granule membranes as illustrated in Figure 8. Figure 8 also demonstrates that the 67,000-M_r protein is less effective at preventing the binding of synexin to granule membranes at higher Ca^{2+} concentrations (e.g., 630 μM Ca^{2+} versus 40 μM Ca^{2+}). The 67,000-M_r protein also displaces the 32,000-M_r calelectrin from the membrane, indicating that this protein may also bind to the same site as synexin and the 67,000-M_r calelectrin.

It is indeed curious that a single tissue (e.g., the adrenal medulla), would possess two proteins, synexin and 32,000-M_r calelectrin, which have

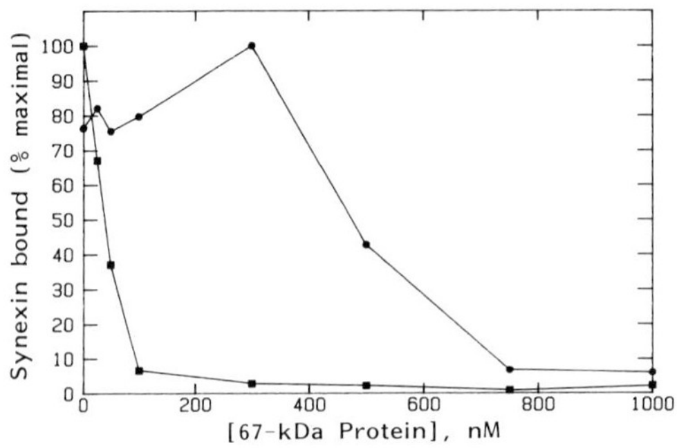

Figure 8. Inhibition of synexin binding to chromaffin granule membranes by the 67-kDa protein. Chromaffin granule membranes (52 μg/ml) in 0.24 M sucrose, 30 mM KCl, 40 mM Hepes-NaOH, pH 7.0 were incubated 15 min at 25°C with 20 nM [^{125}I]synexin (3000 cpm/pmoles) and increasing amounts of 67 kDA protein at pCa 3.2 (●) or 4.4 (■). Samples were centrifuged at 15,000 × *g* for 15 min and the radioactivity in the pellets was measured. Data were calculated as [^{125}I]synexin specifically bound (counts bound in the presence of Ca^{2+} minus counts bound in excess EGTA) and expressed as a percentage of maximum binding observed at each free Ca^{2+} concentration.

such similar Ca^{2+}-dependent membrane-aggregating properties as well as a protein that inhibits their action. Clearly, there is a considerable potential for coordination and regulation of membrane fusion if all these proteins are active in the same part of the chromaffin cell.

There is an additional report of a chromaffin granule aggregating protein obtained from a side fraction of a synexin preparation (Odenwald and Morris, 1983). This protein was termed synexin II, but little has yet been reported in terms of the Ca^{2+}-dependence of action of this protein or its relationship to calelectrin or synexin. With hindsight, we have wondered how these other proteins with synexinlike activity escaped notice when synexin was originally isolated. It is likely that the explanation is that synhibin, or other inhibitory proteins, masked the activity of these granule-aggregating proteins. As synexin was purified away from the inhibitors its increase in activity may have been incorrectly interpreted as recovery of total aggregating activity of the crude postmicrosomal supernatant. Therefore, it remains important to determine how many proteins have synexinlike activity, and the extent to which they are antagonistic or synergistic in their action. We are currently undertaking a systematic approach to this question by studying the properties of proteins that bind to granules in a Ca^{2+}-dependent manner, as described in the following sections.

5. OTHER Ca^{2+}-DEPENDENT SECRETORY VESICLE MEMBRANE-BINDING PROTEINS

The complex sequence of events comprising exocytosis as described in Section 2 may require the involvement of elements that have not been recognized or hypothesized previously, in addition to the cytoskeletal elements and granule-aggregating proteins described above. One approach to search for such factors is to determine what cytosolic components bind to secretory vesicle membrane *in vitro* in the presence of Ca^{2+}. Such an approach has been recently taken using affinity chromatography (Creutz, 1981*a*; Creutz *et al.*, 1983) or a centrifugation protocol (Geisow and Burgoyne, 1982) to isolate such components from adrenal medullary cytosol. This approach is relatively unprejudiced in terms of hypotheses about what proteins one might expect to be involved in exocytosis and depends only on the rationale that regulatory factors in the cytosol must interact physically with the secretory vesicle membrane in a Ca^{2+}-dependent manner. The somewhat surprising result of these studies is that there are a large number ($\geqslant$ 23) of cytosolic proteins that bind to chromaffin granule membranes in the presence of Ca^{2+}. In analogy with the terms *chromogranins,* the proteins in the granule core (Blaschko *et al.*, 1967), and *chromomembrins,* proteins in the granule membrane (Winkler, 1971), we have chosen to refer to these proteins as *chromobindins.* We anticipate that this term may become obsolete as the identities and functions of these proteins become known and more specific names are suggested.

The general procedure used to isolate chromaffin granule membrane binding proteins is by affinity chromatography of adrenal medullary cytosol on a column of granule membranes bound covalently to Sepharose 4B. The basic result of applying this method is illustrated in Figure 9, which represents SDS gels of fractions obtained from the affinity column. Note that a large fraction of the chromobindins have molecular weights in the 30,000–40,000-M_r range. Also note that an additional group of chromobindins is obtained if the affinity chromatography is conducted in the presence of ATP, indicating that some of these proteins are regulated by both Ca^{2+} and ATP.

So far we have been able to identify several of the chromobindins as proteins that have been studied in other contexts. Table 1 is a listing of all the major chromobindins with tentative identification for some of them. The proteins are labeled chromobindins 1–23, corresponding to the labelling of spots that represent these proteins on a two-dimensional gel (Fig. 10). Notice that synexin, calmodulin, protein kinase C, 32,000-M_r calelectrin, and 67,000-M_r calelectrin (synhibin) are all present. Some of these proteins are also likely to be similar to, or identical with the proteins that were described by Moore *et al.* (1984), who have been investigating cytosolic proteins that bind to a fluphenazine–Sepharose affinity matrix in a Ca^{2+}-dependent manner. The 67,000-M_r protein is also likely to be related to a protein that associates with the lymphocyte plasma membrane, as described by Owens and Crumpton (1984).

Some of the chromobindins are phosphoproteins; we are investigating whether their state of phosphorylation may influence their binding to

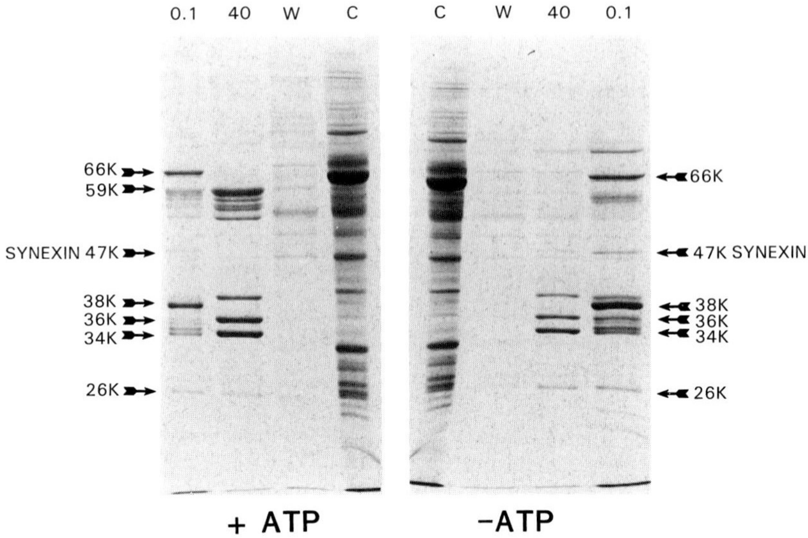

Figure 9. Electrophoretic gels of fractions obtained from the chromaffin granule membrane affinity column. C, adrenal medullary cytosol fraction applied to the column; W, wash fraction obtained immediately before reduction in the calcium concentration; 40, fraction eluting at 40 μM Ca^{2+}; 0.1, fraction eluting at 0.1 μM Ca^{2+}. Interpolated molecular masses of the protein bands are marked in kilodaltons. The position of the 47-kDa synexin band is also marked. +ATP or −ATP, chromatography conducted in the presence or absence of 1 mM ATP. The unmarked band above the 66-kDa band in the 0.1 μM Ca^{2+}, −ATP lane is believed to be protein kinase C. Gels were stained with Coomassie blue. (From Creutz *et al.*, 1983.)

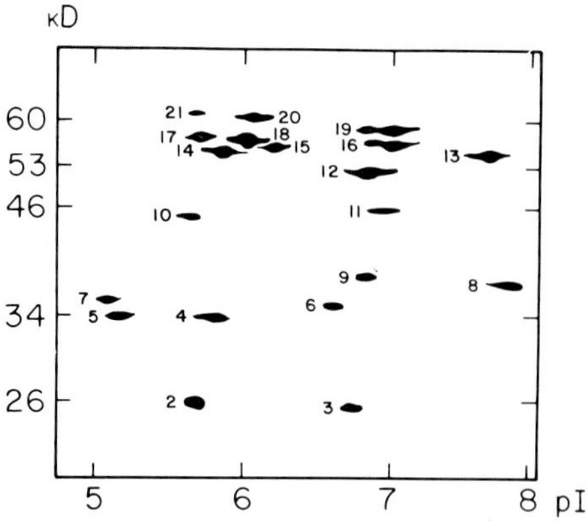

Figure 10. Summary figure illustrating the positions of the chromobindins on two-dimensional gels. The numbers next to the spots correspond to the listing in Table I. (From Creutz *et al.*, 1983.)

Table I. Characteristics of the Chromaffin Granule-Binding Proteins

Chromobindin number[a]	Mass (kd)	pI	Identity
1[b]	19		Calmodulin
2	26	5.7	Unknown
3	26	6.8	Unknown
4	32	5.8	32-kDa Calelectrin
5	32	5.2	Unknown
6	33	6.7	Unknown
7	35	5.1	Unknown
8	36	7.9	p36 (*Src* kinase substrate)
9	37	6.9	Protein kinase C substrate
10	46	5.6	Unknown
11	47	7.0	Synexin
12	53	6.8	Subunit of chromobindin A
13	55	7.8	Subunit of chromobindin A
14	56	5.9	Subunit of chromobindin A
15	56.5	6.2	Subunit of chromobindin A
16	57	7.0	Subunit of chromobindin A
17	57.5	5.7	Unknown
18	58	6.0	Subunit of chromobindin A
19	59	7.0	Subunit of chromobindin A
20[c]	67	6.1	67-kDa Calelectrin (synhibin)
21	68	5.6	Unknown
22[d]	83		Protein kinase C
23[d]			Phosphatidylinositol-specific phospholipase C

[a]The chromobindin numbers correspond to the numbers next to the spots on the two-dimensional gel represented by Figure 10.
[b]Calmodulin is not represented in Figure 10 because of its highly acidic isoelectric point.
[c]This protein was previously incorrectly identified as bovine serum albumin because of its similar electrophoretic parameters (Creutz *et al.*, 1983).
[d]Not represented in Figure 12 because of uncertainty concerning electrophoretic parameters.

the membrane. We have found, for example, that protein kinase C, which can autophosphorylate (Kikkawa *et al.*, 1982), is not obtained from the affinity column if the chromatography is conducted in the presence of ATP (see Fig. 9) (Creutz *et al.*, 1983). This suggests that autophosphorylation of this kinase (or phosphorylation by a membrane kinase) may regulate the interaction of protein kinase C with the granule membrane. We have also found that chromobindin 9 is a substrate for protein kinase C *in vitro* (Summers and Creutz, 1985), and, in addition, becomes phosphorylated in the chromaffin cell when exocytosis is activated (Michener *et al.*, 1986). However, the influence this phosphorylation event has on the binding of this protein to the membrane has not been studied in detail. In addition, we have found that chromobindin 8 is a protein that is referred to as p36 in the oncogene field (C. E. Creutz, S. J. Parsons, and W. H. Martin, unpublished observations). This protein is the primary site of phosphorylation on tyrosine residues in cells that are transformed by Rous sarcoma virus, which codes for the tyrosine kinase $p60^{v\text{-}src}$. This finding is particularly intriguing since we have also found that the normal cellular homologue of this transforming kinase is present in the chromaffin granule membrane (Parsons and Creutz, 1986). The implication seems to be that the *src* oncogene product may play a role in regulating membrane fusion events, perhaps attendant to the release of a transforming growth factor by exocytosis. A fourth phosphoprotein that we have preliminary immunological evidence may be present in the chromobindins is the 35,000-M_r substrate for the EGF receptor-associated tyrosine kinase, as described by Fava and Cohen (1984).

A phospholipid-metabolizing activity is also present in the group of chromobindin proteins. This is a phosphatidylinositol-specific phospholipase C activity that is regulated by Ca^{2+} at a concentration of 600 nM (Creutz *et al.*, 1985). We have not identified which chromobindin has this activity, and it may correspond to a minor protein in the group that has not been otherwise characterized. This activity is potentially important in regulating events during exocytosis as it may lead to the generation of diacylglycerol and inositol phosphates to serve as intracellular signals in the stimulated cell. Furthermore, due to the presence of a Ca^{2+}-independent diacylglycerol lipase in adrenal medullary cytosol (R. Franson, Medical College of Virginia, personal communication) the phospholipase C may be the initial step in a regulated pathway for the release of arachidonic acid to promote membrane fusion.

Seven of the chromobindins are apparently subunits of a large macromolecular complex that we refer to as chromobindin A (Martin and Creutz, 1986). This complex has the particularly intriguing property of being regulated by ATP as well as by Ca^{2+}. Specifically, we have found that this complex binds to the chromaffin granule membrane in the presence of Ca^{2+}. It will then remain bound even if Ca^{2+} is removed by chelation with EGTA. However, if ATP is introduced, in the absence of Ca^{2+}, the complex dissociates from the membrane. This behavior may be analogous to the mechanism by which clathrin is induced to dissociate from coated vesicles by the action of an uncoating ATPase (Schlossman *et al.*, 1984). It is also analogous to the behavior of the contractile apparatus of muscle in that myosin will bind actin if Ca^{2+} is present to cause displacement of the troponin–tropomyosin complex. Actin and myosin will then remain complexed unless ATP is present to dissociate the complex. In the case of chromobindin A, it is not clear where the site of action of ATP is in promoting dissociation, i.e., on the membrane or on the complex itself. However, the isolated complex does contain several bound nucleotides as detected by UV spectroscopy and HPLC.

The functional roles of most of the chromobindins in the secretory process are still unknown. While specific hypotheses have been put forward for the granule aggregating proteins synexin and calelectrin, the greatest uncertainty concerns the function of the 30,000–40,000-M_r phosphoproteins and the ATP-dependent chromobindin A. The phosphoproteins have the potential to participate in a complex regulatory scheme involving specific kinases and phosphatases. The ATP-dependent complex may provide insight into the apparent ATP dependence of exocytosis seen in permeabilized cell models (Knight and Baker, 1982).

6. CONCLUDING REMARKS

6.1. Correlation of *in Vitro* Models with Cell Models

In our studies of exocytosis in the chromaffin cell we have concentrated our efforts on the construction of cell-free models that appear to recreate steps in the overall secretory process. Although this approach has been particularly successful in the development of models for secretory vesicle membrane fusion, we are still faced with the important challenge of demonstrating that these same events occur in the stimulated secretory cell and really do underlie the process of exocytosis. One approach to meeting this challenge is to examine the behavior of chromaffin cells in culture and attempt to establish correlations between the behavior of these cells and the behavior of the model systems. To date, some very successful correlations have been drawn. For example, the secretion of catecholamines from the chromaffin cell, as well as the release of other secretory products such as insulin, parathyroid hormone and serotonin from other cell types, can be suppressed by increasing the osmotic pressure of the incubation medium (Pollard *et al.,* 1977, 1984; Brown *et al.,* 1978; Orci and Malaisse, 1980; Pace and Smith, 1983). This is a direct prediction of the synexin or calelectrin-dependent granule fusion models that involve an osmotic pressure gradient as part of the fusion mechanism. In addition, chromobindin 9, which is phosphorylated by protein kinase C *in vitro* (Summers and Creutz, 1985), undergoes phosphorylation in the intact, stimulated cell, although the kinases responsible for the *in vitro* phosphorylation have yet to be defined (Michener *et al.,* 1986). As another example, synexin-dependent chromaffin granule aggregation can be inhibited by phenothiazines (Creutz *et al.,* 1981), which also inhibit secretion from the chromaffin cell with a similar structure–activity relationship (Pollard *et al.*, 1983).

Another type of correlation may be drawn between the *in vitro* models and the behavior of permeabilized chromaffin cell models. Cells that have been permeabilized by electric discharge (Knight and Baker, 1982) or by detergents (Brooks and Treml, 1983; Wilson and Kirshner, 1983; Dunn and Holz, 1983) will release catecholamines in a manner that is dependent on ATP and on Ca^{2+} in the 1–10-μM range. These are conditions that would regulate the association of many of the chromobindins with membranes. However, the membrane-aggregating activity of synexin and calelectrin would only be slightly activated at these low levels of Ca^{2+}. In this sense, there appears to be a lack of correlation between the *in vitro* fusion models and the permeabilized cell models. Presumably, further advances in two directions will be necessary before the complete blueprint for the machinery of exocytosis will be read: The *in vitro* models will require the addition of appropriate cofactors, and the cell models will need to be further dissected until these two research approaches converge.

6.2. Biotechnology Applications of the Fusion Proteins

We have frequently received inquiries concerning whether synexin and/or calelectrin might be useful for promoting membrane fusion in biotechnological applications. In particular, there is considerable interest in new mechanisms to fuse cells to one another, or to fuse liposomes containing xenobiotics to cells. Unfortunately, these proteins appear to have a high degree of specificity for the cytoplasmic face of biological membranes. This specificity is a result of their requirement for high concentrations of acidic phospholipids in the target membranes, and is appropriate to their postulated function in the cell. We would like to suggest however, that there may be some interesting research applications of these proteins. These proteins are capable of promoting the fusion of large numbers of secretory vesicles to form giant (>10 μm) vacuolelike structures (Fig. 6). Perhaps these giant organelles would be amenable to biophysical studies that would be impossible with the individual secretory vesicles. For example, it may be possible to apply a patch-clamp electrode to such a giant vesicle, or to perform protein or lipid mobility studies using the fluorescence recovery after photobleaching technique. Although extensive *in vitro* fusion has only been demonstrated with chromaffin granules, it may be possible to use these proteins to promote the fusion of other organelles such as lysosomes, mitochondrial inner membranes, other types of secretory vesicles, endocytic vesicles, or inverted plasma membranes, opening new research avenues in cell biology.

ACKNOWLEDGMENTS. Recent studies of chromaffin granule-binding proteins in the authors' laboratory have been supported by research grants (PCM 8206453) from the National Science Foundation, (AM 33151 and CA 40042) from the National Institutes of Health, (#1560) from the Council for Tobacco Research, and (B10-85-017) from the Commonwealth of Virginia Center for Innovative Technology, by training grants from the National Institutes of Health in support of W. J. Zaks and W. H. Martin, and by Established Investigator Award #85-117 from the American Heart Association, in support of C. E. Creutz, with funds contributed in part by the Virginia Affiliate.

REFERENCES

Blaschko, H., Combine, R. S., Schneider, F. H., Silver, M., and Smith, A. D., 1967, Secretion of a chromaffin granule protein, chromogranin, from the adrenal gland after splanchnic stimulation, *Nature (Lond.)* **215**:58–59.

Brooks, J. C., and Treml, S., 1983, Catecholamine secretion by chemically skinned cultured chromaffin cells, *J. Neurochem.* **40**:468–473.

Brown, E. M., Pazoles, C. J., Creutz, C. E., Aurbach, G. D., and Pollard, H. B., 1978, Regulation of parathyroid hormone release from dispersed bovine parathyroid cells by permeant anions, *Proc. Natl. Acad. Sci. U.S.A.* **75**:876–880.

Creutz, C. E., 1981*a*, Secretory vesicle–cytosol interactions in exocytosis, *Biochem. Biophys. Res. Commun.* **103**:1395–1400.

Creutz, C. E., 1981*b*, *Cis*-unsaturated fatty acids induce the fusion of chromaffin granules aggregated by synexin, *J. Cell. Biol.* **91**:247–256.

Creutz, C. E., and Pollard, H. B., 1980, A biophysical model of the chromaffin granule: Accurate description of the kinetics of ATP and Cl^- dependent granule lysis, *Biophys. J.* **31**:255–270.

Creutz, C. E., and Sterner, D. C., 1983, Calcium dependence of the binding of synexin to isolated chromaffin granules, *Biochem. Biophys. Res. Commun.* **114**:355–364.

Creutz, C. E., Pazoles, C. J., and Pollard, H. B., 1978, Identification and purification of an adrenal medullary protein (synexin) that causes calcium dependent aggregation of isolated chromaffin granules, *J. Biol. Chem.* **253**:2858–2866.

Creutz, C. E., Pazoles, C. J., and Pollard, H. B., 1979, Self-association of synexin in the presence of calcium, *J. Biol. Chem.* **254**:553–558.

Creutz, C. E., Scott, J. H., Paxoles, C. J., and Pollard, H. B., 1981, Further characterization of the aggregation and fusion of chromaffin granules as a model for compound exocytosis, *J. Cellular Biochem.* **18**:87–97.

Creutz, C. E., Dowling, L. G., Sando, J. J., Villar-Palasi, C., Whipple, J. H., and Zaks, W. J., 1983, Characterization of the chromobindins: Soluble proteins that bind to the chromaffin granule membrane in the presence of Ca^{2+}, *J. Biol. Chem.* **258**:14664–14674.

Creutz, C. E., Dowling, L. G., Kyger, E. M., and Franson, R. C., 1985, Phosphatidylinositol-specific phospholipase C activity of chromaffin granule-binding proteins, *J. Biol. Chem.* **260**:7171–7173.

Dabrow, M., Zaremba, S., and Hogue-Angeletti, R. A., 1980, Specificity of synexin-induced chromaffin granule aggregation, *Biochem. Biophys. Res. Commun.* **96**:1164–1171.

Dowd, D. J., Edwards, C., Englert, D., Mazurkiewicz, T. E., and Ye, H. Z., 1983, Immunofluorescent evidence for exocytosis and internalization of secretory granule membrane in isolated chromaffin cells, *Neuroscience* **10:**1025–1033.
Dunn, L. A., and Holz, R. W., 1983, Catecholamine secretion from digitonin-treated adrenal medullary chromaffin cells, *J. Biol. Chem.* **258:**4989–4993.
Fava, R. A., and Cohen, S., 1984, Isolation of a calcium-dependent 35-kilodalton substrate for the epidermal growth factor recetor/kinase from A-431 cells, *J. Biol. Chem.* **259:**2636–2645.
Fowler, V. M., and Pollard, H. B., 1982, Chromaffin granule membrane–F-action interactions are calcium sensitive, *Nature (Lond.)* **295:**336–337.
Geisow, M. J., and Burgoyne, R. D., 1982, Calcium-dependent binding of cytosolic proteins by chromaffin granules from adrenal medulla, *J. Neurochem.* **38:**1735–1741.
Heuser, J. E., and Reese, T. S., 1973, Evidence for recycling of synaptic vesicle membrane during transmitter release at the frog neuromuscular junction, *J. Cell. Biol.* **57:**315–344.
Hoekstra, D., de Boer, T., Klappe, K., and Wilschut, J., 1984, Fluorescence method for measuring the kinetics of fusion between biological membranes, *Biochemistry* **23:**5675–5681.
Hong, K., Duzgunes, N., Ekerdt, R., and Papahadjopoulos, D., 1982, Synexin facilitates fusion of specific phospholipid membranes at divalent cation concentrations found intracellularly, *Proc. Natl. Acad. Sci. U.S.A.* **79:**4642–4644.
Kikkawa, O., Takai, Y., Minakuchi, K., Inohara, S., and Nishizuka, Y., 1982, Calcium-activated, phospholipid-dependent protein kinase from rat brain. Subcellular distribution, purification and properties, *J. Biol. Chem.* **257:**13341–13348.
Knight, D. E., and Baker, P. F., 1982, Calcium dependence of catecholamine release from bovine adrenal medullary cells after exposure to intense electric fields, *J. Membr. Biol.* **68:**107–140.
Marcus, A. J., 1978, The role of lipids in platelet function: With particular reference to the arachidonic acid pathway, *J. Lipid Res.* **19:**793–826.
Martin, W. H., and Creutz, C. E., 1986, Chromobindin A: A calcium—and ATP—regulated chromaffin granule membrane-binding protein complex, *J. Biol. Chem.* (in press).
Michener, M. L., Dawson, W. B., and Creutz, C. E., 1986, Phosphorylation of a chromaffin granule-binding protein in stimulated chromaffin cells, *J. Biol. Chem.* **261:**6548–6555.
Moore, P. B., Kraus-Friedman, N., and Dedman, J. R., 1984, Unique calcium-dependent hydrophobic binding proteins: Possible independent mediators of intracellular calcium distinct from calmodulin, *J. Cell Sci.* **72:**121–133.
Morris, S. J., and Bradley, D., 1984, Calcium-promoted fusion of isolated chromaffin granules detected by resonance energy transfer between labeled lipids embedded in the membrane bilayer, *Biochemistry* **23:**4642–4650.
Morris, S. J., and Hughes, J. M. X., 1979, Synexin protein is non-selective in its ability to increase Ca^{2+}-dependent aggregation of biological and artificial membranes, *Biochem. Biophys. Res. Commun.* **91:**345–350.
Nagasawa, J., and Douglas, W. W., 1972, Thorium dioxide uptake into adrenal medullary cells and the problem of recapture of granule membrane following exocytosis, *Brain Res.* **37:**141–145.
Odenwald, W. F., and Morris, S. J., 1983, Identification of a second synexin-like adrenal medullary and liver protein that enhances calcium-induced membrane aggregation, *Biochem. Biophys. Res. Commun.* **112:**147–154.
Orci, L., and Malaisse, W., 1980, Single and chain release of insulin secretory granules is related to anion transport at exocytotic sites, *Diabetes* **29:**943–946.

Owens, R. J., and Crumpton, M. J., 1984, Isolation and characterization of a novel 68,000 M_r Ca^{2+}-binding protein of lymphocyte plasma membrane, *Biochem. J.* **219:**309–316.

Pace, C. S., and Smith, J. S., 1983, The role of chemiosmotic lysis in the exocytotic release of insulin, *Endocrinology* **113:**964–969.

Palade, G., 1975, Intracellular aspects of the process of protein synthesis, *Science* **189:**347–358.

Parsons, S. J., and Creutz, C. E., 1986, $p60^{c\text{-}src}$ activity detected in the chromaffin granule membrane, *Biochem. Biophys. Res. Commun.* **134:**736–742.

Phillips, J. H., Burridge, K., Wilson, S. P., and Kirshner, N., 1983, Visualization of the exocytosis/endocytosis secretory cycle in cultured adrenal chromaffin cells, *J. Cell Biol.* **97:**1906–1917.

Pollard, H. B., and Scott, J. H., 1982, Synhibin: A new calcium-dependent membrane-binding protein that inhibits synexin-induced chromaffin granule aggregation and fusion, *F.E.B.S. Lett.* **150:**201–206.

Pollard, H. B., Tack-Goldman, K. M., Pazoles, C. J., Cruetz, C. E., and Shulman, N. R., 1977, Evidence for control of serotonin secretion from human platelets by hydroxyl ion transport and osmotic lysis, *Proc. Natl. Acad. Sci. U.S.A.* **74:**5295–5299.

Pollard, H. B., Scott, J. H., and Cruetz, C. E., 1983, Inhibition of synexin activity and exocytosis from chromaffin cells by phenothiazine drugs, *Biochem. Biophys. Res. Commun.* **113:**908–915.

Pollard, H. B., Pazoles, C. J., Creutz, C. E., Scott, J. H., Zinder, O., and Hotchkiss, A., 1984, An osmotic mechanism for exocytosis from dissociated chromaffin cells, *J. Biol. Chem.* **259:**1114–1121.

Potter, J. D., Strang-Brown, P., Walker, P. L., and Sida, S., 1983, Ca^{2+} binding to calmodulin, *Meth. Enzymol.* **102:**135–143.

Schlossman, D. M., Schmid, S. L., Braell, W. A., and Rothman, J. E., 1984, An enzyme that removes clathrin coats: Purification of an uncoating ATPase, *J. Cell. Biol.* **99:**723–733.

Sterner, D. C., Zaks, W. J., and Cruetz, C. E., 1985, Stimulation of the polymerization of synexin by *cis*-unsaturated fatty acids, *Biochem. Biophys. Res. Commun.* **132:**505–512.

Suchard, S. J., Corcoran, J. J., Pressman, B. C., and Rubin, R. W., 1981, Evidence for secretory granule membrane recycling in cultured adrenal chromaffin cells, *Cell Biol. Int. Rep.* **5:**953–962.

Sudhof, T. C., Walker, J. H., and Obrocki, T., 1982, Calelectrin self-aggregates and promotes membrane aggregation in the presence of calcium, *EMBO J.* **1:**1167–1170.

Sudhof, T. C., Ebbecke, M., Walker, J. H., Fritsche, U., and Boustead, C., 1984, Isolation of mammalian calelectrins: A new class of ubiquitous Ca^{2+}-regulated proteins, *Biochemistry* **23:**1103–1109.

Sudhof, T. C., Walker, J. H., and Fritsche, U., 1985, Characterization of calelectrin, a Ca^{2+}-binding protein isolated from the electric organ of Torpedo marmorata, *J. Neurochem.* **44:**1302–1307.

Summers, T. A., and Creutz, C. E., 1985, Phosphorylation of a chromaffin granule-binding protein by protein kinase C, *J. Biol. Chem.* **260:**2437–2443.

Walker, J. H., 1982, Isolation from cholinergic synapses of a protein that binds to membranes in a calcium-dependent manner, *J. Neurochem.* **39:**815–823.

Wilson, S. P., and Kirshner, N., 1983, Calcium-evoked secretion from digitonin-permeabilized adrenal medullary chromaffin cells, *J. Biol. Chem.* **258:**4994–5000.

Winkler, H., 1971, The membrane of the chromaffin granule, *Philos. Trans. R. Soc. Ser. B Biol. Sci.* **261:**293–303.

Zaks, W. J., 1986, Ca^{2+}-dependent chromaffin granule aggregating proteins: A comparative study of synexin and the calelectrins, Ph.D. thesis, University of Virginia.

Zaks, W. J., and Cruetz, C. E., 1986, A comparison of synexin and the calelectrins as potentiators of Ca^{2+}-dependent chromaffin granule aggregation and fusion, *J. Biol. Chem.,* (submitted).

Zimmerberg, J., Cohen, F. S., and Finkeletrin, A., 1980, Fusion of phospholipid vesicles with planar phospholipid bilayer membranes: I. Discharge of vesicles contents across the planar membrane, *J. Gen. Physiol.* **75:**241–250.

Chapter 4

Synchronous Exocytosis in *Paramecium* Cells

Helmut Plattner

1. INTRODUCTION

Exocytosis occurs in all eukaryotic cells. It involves the release of cellular biosynthetic products by fusion of the secretory organelle membrane with the plasma membrane. Whereas the general pathway of exocytosis is well established (Palade, 1975), many crucial aspects of the regulation of exocytosis are still poorly understood. This is particularly true for the specificity of membrane–membrane interactions (Palade, 1982) and for the regulation of the actual fusion process (Plattner, 1981; Parsegian *et al.*, 1984; Lucy, 1985). This chapter discusses the results of our work on exocytosis in *Paramecium* as well as the implications of these results for exocytosis in general.

2. THE *PARAMECIUM* CELL AS A MODEL SYSTEM FOR ANALYZING EXOCYTOSIS

The *Paramecium* cell is a ciliated protozoan that represents a well-established model system for studying exocytosis. The most widely

Helmut Plattner • Faculty of Biology, University of Konstanz, D-7750 Konstanz, Federal Republic of Germany.

analyzed species, *P. tetraurelia,* has >1000 secretory organelles (trichocysts) per cell, 95% of which are firmly attached to the cell membrane (Plattner *et al.,* 1984*a,* 1985*b*).

The analysis of trichocyst discharge is facilitated by the fact that in nature the trichocyst docks at precisely predetermined sites in the cell (Jurand and Selman, 1969; Ehret and McArdle, 1974). As shown in preliminary studies by Bachmann *et al.* (1972), Janisch (1972), and Satir *et al.* (1972) and in much more detail by Plattner *et al.* (1973), these sites also display a characteristic appearance in freeze-fracture replicas and thus fusion sites can be pinpointed. Paramecia can also be cultivated in fully synthetic sterile media; their size is convenient for microinjection studies.

Mutations have also been isolated in paramecia that are defective for different stages of secretory activity (Pollack, 1974; Sonneborn, 1974; Beisson *et al.,* 1976; Cohen and Beisson, 1980; Lefort-Tran *et al.,* 1981) and can be an invaluable tool in exocytosis research.

3. TRIGGERING SYNCHRONOUS EXOCYTOSIS

The *in vivo* physiological trigger agent for the trichocyst discharge in *Paramecium* cells is unknown. Although we tested many compounds *in vitro,* only polyamino compounds were found to induce massive exocytosis (Plattner *et al.,* 1984*a*). However, the amino groups must be spaced an average distance of at least 1.0 nm apart (Plattner *et al.,* 1985*b*). Dextran proved the most convenient type of carrier; it also increases the effect of other polyamine agents (Plattner *et al.,* 1985*b*). We now prefer aminoethyldextran (AED) of a molecular weight of 40,000 and with 1 NH_2 per kilodalton.

Polyamines trigger exocytosis in other secretory cells such as mast cells (Ranadive and Cochrane, 1971; Foreman and Lichtenstein, 1980; Lagunoff and Chi, 1980) and adrenal chromaffin cells (Suchard *et al.,* 1980). Many studies with mast cells have used compound 48/80. In paramecia, however, this agent proved less active.

The molecular mechanism of AED is unknown. It induces massive discharge in micromolar concentrations and requires extracellular Ca^{2+} (10 μM). Upon triggering, the concentration of free calcium in the cell increases. This can be monitored by the ciliary beat reversal reaction, which can be considered an endogenous bioassay for cytoplasmic free calcium increase up to micromolar levels (Naitoh and Kaneko, 1972) via voltage-dependent Ca^{2+} channels in the ciliary membrane (Ogura and Takahashi, 1976). The Ca^{2+} influx required for exocytosis occurs via the somatic (rather than the ciliary) cell membrane, since deciliated cells or mutant strains without the functional voltage-dependent Ca^{2+} channels in the cilia—*d4-500r,* a *pawn* mutation (Saimi and Kung, 1980)—perform exocytosis in a normal way (Plattner *et al.,* 1984*a*). It would be interesting to know whether Ca^{2+} influx occurs precisely at those sites of the cell membrane at which trichocysts are docked (in 1–2-μm intervals) and where exocytosis occurs.

The influx of Ca^{2+} is not sufficient, however, to mediate exocytosis. Satir and Oberg (1978) assumed that Ca^{2+} ionophores could bypass Ca^{2+} channels in nondischarge mutations (*nd9-28°C*), but this appears not to be the case (Matt *et al.,* 1980; Gilligan and Satir, 1983).

Triggering can be induced by electric shocks (Wohlfarth-Bottermann, 1953; Yusa, 1963; Hausmann and Allen, 1976; Allen and Fok, 1984) or even by fixatives like picric acid (Jennings, 1906). The electric trigger method is restricted to individual cells in appropriate orientation, requires multiple shocks, and also produces some dead cells (Hausmann and Allen, 1976). With AED, however, cells can be massively and repeatedly triggered in suspension without inducing cell damage (Plattner *et al.,* 1985*b*).

Mg^{2+} (Zieseniss and Plattner, 1985) and polyanionic compounds such as polyglutamic acid or heparin (C. Lumpert and H. Plattner, in preparation) will inhibit AED-induced trichocyst discharge. Polyanions inhibit exocytosis in mast cells (Dufton *et al.,* 1984) also, whereas Mg^{2+} is inhibitory in many systems (Rubin, 1982). Neomycin (10–100 μM) also inhibits AED-mediated discharge (Plattner *et al.,* 1985*b*), possibly because it interferes with the transmembrane Ca^{2+} flux (M. Gustin and T. Hennessey, personal communication), with Ca^{2+}-mediated phase changes in phospholipids (Wang *et al.,* 1984) or with the dephosphorylation of certain phosphoproteins (see Section 14).

AED-induced trichocyst discharge can be localized (Fig. 1) if AED is applied by micropipette (Plattner *et al.,* 1984*a*).

Strains capable of normal exocytosis, such as *K401, 7S* (wild type) contain about 1100–1200 trichocysts per cell; about 90–95% of them can be released by AED (Plattner *et al.,* 1985*b*) (Fig. 2). This corresponds to the percentage docked to the cell membrane while the undocked "free trichocysts" circulate in the cyclosis stream. For unknown reasons, we usually obtain only about a 70% discharge when mass cultures are triggered with AED.

4. TRICHOCYST MUTATIONS

Cohen and Beisson (1980) listed 30 trichocyst mutations in 24 loci of the *Paramecium* genome. Strain *7S* is the wild type from which most other strains are derived; *K401,* a strain we have used frequently, is a mutant with normal exocytosis. Nondischarge mutations contain numerous

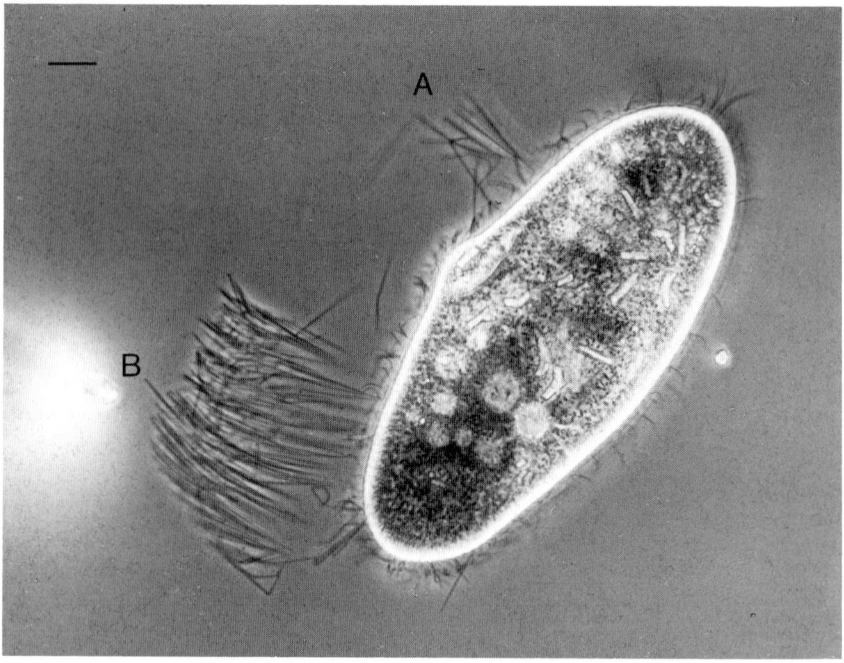

Figure 1. Local exocytosis triggering at sites A and B, where AED was deposited with a micropipette (visible close to B). Trichocysts expand considerably upon discharge. Bar: 10 μm. (From Plattner *et al.*, 1984*a*.)

trichocysts docked to the cell membrane, but they are unable to perform membrane fusion (Beisson *et al.*, 1976; Matt *et al.*, 1980; Lefort-Tran *et al.*, 1981; Pouphile *et al.*, 1986). Strain *ftA* (Pollack, 1974) and strains of the *tam* group (Beisson *et al.*, 1976; Lefort-Tran *et al.*, 1981) keep variable numbers of trichocysts in the cytoplasm. Strain *tl* (*trichless*) does not produce any trichocysts (Pollack, 1974). All these strains show no exocytotic response when AED is added. Another strain, *ptA,* produces pointless trichocysts that can be docked to the cell membrane but not secreted (Pollack, 1974).

The response to AED is not different from that of the fixative picric acid, a routine test traditionally used by geneticists since Jennings in 1906.

5. BIOGENESIS OF EXOCYTOSIS SITES

Analysis of the biogenesis of exocytosis sites is aided by the regularity in the arrangement of exocytosis sites in the protozoa, notably ciliates

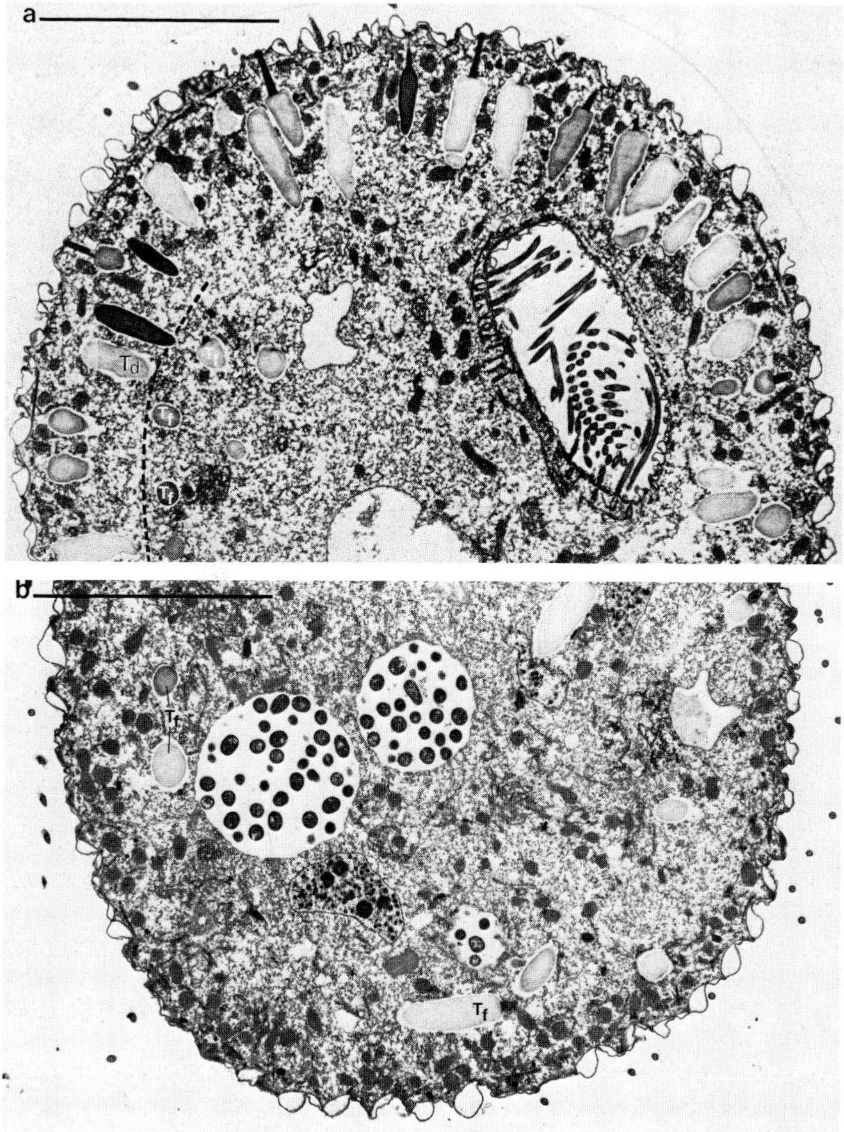

Figure 2. Saponin-permeabilized and tannic acid stained cell (a) before and (b) after AED-induced massive exocytosis. T_d, trichocysts docked to the cell membrane; T_f, free trichocysts in the cytoplasm (not releasable). Bar: 10 µm. (From Pape and Plattner, 1985.)

(Satir *et al.*, 1972, 1973; Plattner *et al.*, 1973, 1984*b*; Bardele, 1977; Hausmann, 1978; Aufderheide *et al.*, 1980; Plattner, 1981) and parasitic sporozoa (see Porchet-Hennere and Nicolas, 1983; Plattner *et al.*, 1984*b*). A similar arrangement is also found in motor end plates of lower vertebrates (Couteaux and Pécot-Dechavassine, 1970; Heuser and Reese, 1981) but less distinctly in other parts of the nervous system.

In paramecia the sites involved in exocytosis are genetically predetermined (Beisson *et al.*, 1976; Cohen and Beisson, 1980; Plattner *et al.*, 1980; Pouphile *et al.*, 1986). Certain ultrastructural elements must be assembled before exocytotic membrane fusion can occur. Two approaches were used to analyze the biogenesis of exoctyosis sites. One was to use different genotypes, the other to analyze structural rearrangements after synchronous exocytosis (Haacke and Plattner, 1984; Pape and Plattner, 1985). The following conclusions were obtained from use of both approaches:

A nonoccupied free docking site displays a parenthesis-shaped (elliptical) aggregate of double membrane-intercalated particles (MIP) rows in the freeze-fractured cell membrane (Fig. 5). These structures are seen to be transformed to a ring (Fig. 3), composed of the same number (69 ± 5) of MIP (Plattner *et al.*, 1985*a*), when a trichocyst is docked underneath (Pape and Plattner, 1985). Concomitantly, triggered cells display many parentheses and only a few rings (Figs. 3, 5). In cells of strains *tl, ftA,* and *tam*

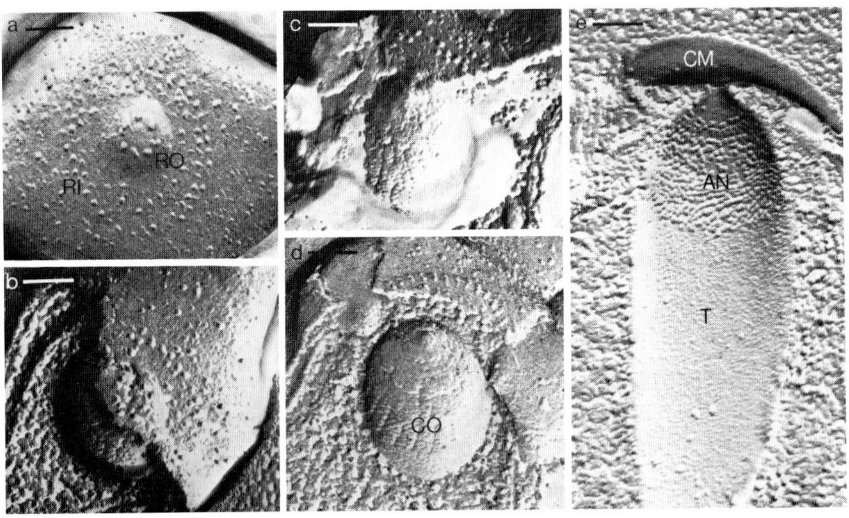

Figure 3. Freeze-fracture of glutaraldehyde-fixed trichocyst attachment site (resting stage of wild-type cells) in top (a) to lateral view (e). AN, annulus; CM, cell membrane; CO, collar; RI, ring; RO, rosette; T, trichocyst. Note the absence of MIP on the top of region of the trichocyst, where it displays a "nipple" closely approaching the cell membrane. Annulus MIP are prominent on PF faces, the collar striations on EF faces. Bars: 0.1 μm. (From K. Olbricht and H. Plattner, unpublished micrographs.)

trichocyst docking sites are present in the parenthesis form (Beisson *et al.*, 1976; Lefort-Tran *et al.*, 1981; Pouphile *et al.*, 1986). When a trichocyst becomes docked to the cell membrane, a parenthesis → ring transformation results. Since rings also occur with *nd* (Beisson *et al.*, 1976) or *ptA* mutations (Pouphile *et al.*, 1986), the transformation and the docking of a trichocyst to the cell membrane appears irrelevant for establishing exocytosis capacity. Figure 4 shows the occurrence of the membrane-associated "connecting materials" between trichocyst and cell membrane in strains capable of exocytosis and its absence from strains that cannot extrude trichocysts.

Membrane fusion capacity is associated with the occurrence of two ultrastructural details: (1) a rosette of MIP, or "fusion rossette" (Fig. 3), at the center of the rings (Beisson *et al.*, 1976; Plattner *et al.*, 1973; Pape and Plattner, 1985; Pouphile *et al.*, 1986); and (2) some ill-defined "connecting materials" (Fig. 4) of amorphous to fibrillar consistency, beneath a rosette (Plattner *et al.*, 1980, 1982; Westphal and Plattner, 1981*a;* Pouphile *et al.*, 1986). Both aspects occur together. Connecting material may account for the assembly of rosettes, or vice versa (Pape and Plattner, 1985). We also found some evidence for a short-lived adaptation stage (<5 min), during which a newly docked trichocyst causes a docking site to mature; it then displays a normal ring structure, but without a rosette inside, which is quite rare in wild-type cells (Pape and Plattner, 1985). This was recognized on replicas after chemical fixation that ties all rosette MIPs to the PF face (Lefort-Tran *et al.*, 1978). The precise origin of the connecting material between trichocyst and cell membrane is still a matter of debate (Pouphile *et al.*, 1986), although it would aid in our understanding how fusion capability develops. On its flanks the tip of a trichocyst is coated by still other protein material ("collar") that, although irrelevant for exocytosis, probably helps restrain the exchange of membrane components during fusion (Hausmann and Allen, 1976; Plattner *et al.*, 1985*a*). Again, results obtained by comparing different genotypic and phenotypic situations led us to conclude that the major part of this collar material is derived from the pluglike structure (Westphal and Plattner, 1981*b;* Plattner *et al.*, 1982), a mass of proteins (Westphal and Plattner, 1981*b*) that accumulates before a trichocyst is docked (see also Beisson *et al.*, 1976). We have therefore postulated that the plug represents a kind of recognition structure during docking (Plattner *et al.*, 1982; Pape and Plattner, 1985; Pouphile *et al.*, 1986). Capillary microinjection of free trichocysts and/or cytoplasmic components from a donor to a recipient strain has also provided important functional clues to trichocyst functioning (Garofalo *et al.*, 1979; Lefort-Tran *et al.*, 1981).

It is noteworthy that a kind of connecting material seems to tie chromaffin granules to the cell membrane in adrenal medullary cells,

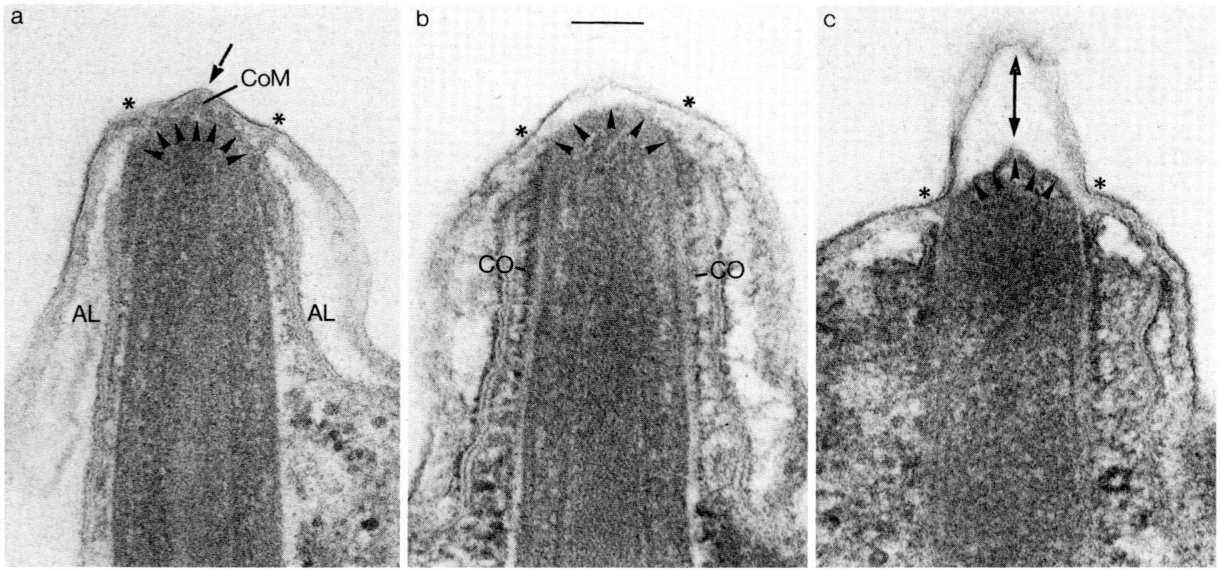

Figure 4. Median sections through trichocyst attachment sites on the cell membrane. (a) Strain *nd9-18°C* (which is indistinguishable from wild-type cells). (b) Strain *nd9-28°C,* incapable of exocytosis; same preparation protocol as in (a). (c) Strain *nd9-28°C* fixed after a hyperosmotic shock. Arrowheads indicate the region engaged in trichocyst membrane—plasma membrane contact. Between the trichocyst tip and cell membrane, some diffuse electron-dense material [connecting material (CoM)] occurs only in strains with dischargeable trichocysts (a), but not in nondischarge mutations (b), where its absence can be documented more clearly after osmotic shock treatment [double arrow in (c)]. In this case, the trichocyst tip is fixed only with its flanks to the surface complex (stars), but not to the cell membrane. Arrows indicate positions of rosette MIP (absent from 28°C cells), stars that of ring MIP (compare Fig. 3). AL, alveolar cavities (a compartment attached to the cell membrane except for special regions, like exocytosis sites). CO, collar (material attached to the trichocyst tip). Bar: 0.1 μm. (From Plattner *et al.*, 1980.)

where it has been partially identified (Pollard *et al.,* 1981) and synaptic vesicles to the active zones of motor end plates (Couteaux and Pécot-Dechavassine, 1970). Similar situations can be found on micrographs from many other cell types (Plattner, 1981). Palade (1982) also considers membrane-associated proteins as possible modulators of membrane–membrane interaction during the secretory process.

6. FREEZE-FRACTURE ELECTRON MICROSCOPIC ANALYSIS OF MEMBRANE FUSION

Figure 5 presents characteristic freeze-fracture micrographs of membranes observed during synchronous exocytosis (Olbricht *et al.,* 1984). These findings are schematically evaluated in Figure 6. The most important events are (1), the exocytotic opening starts as a small hole without diaphragm formation, in the center of a rosette (see also Hausmann and Allen, 1976); (2), while the exocytotic opening expands, these rosette particles become laterally dispersed; (3), the fusogenic zone, or the zone within which the exocytotic opening is formed, is confined by the ring; (4), fusion appears to conserve membrane area, i.e., it does not entail any visible loss of membrane materials; (5), only a temporary integration of the trichocyst membrane occurs into the cell membrane; and (6), resealing during detachment of a trichocyst ghost causes only little exchange of material between both membranes (Plattner *et al.,* 1985*a*). We have determined that the failure rate is only <0.5%. This can be estimated from (1) the rate of integration of characteristic elements of the apical trichocyst membrane in the resealed cell membrane, and (2) from the rather modest integration of glycocalyx elements into the detached trichocyst ghost (Fig. 7).

7. GENERAL ULTRASTRUCTURAL ASPECTS OF EXOCYTOTIC MEMBRANE FUSION

Until recently, quite different theories of the behavior of membranes during exocytotic fusion have been discussed. In Figure 8, the scheme on the left summarizes data combined from ultrathin sectioning (see fusion scheme by Palade, 1975) and freeze fracture (Orci *et al.,* 1981) of different groups. This scheme implies the formation of a large membrane–membrane contact area with the elimination of MIP and also of peripheral proteins between the membranes. [It would allow one to understand membrane fusion on the basis of results obtained by Papahadjopoulos *et al.* (1978) with lipid vesicles; their fusion in response to the addition of

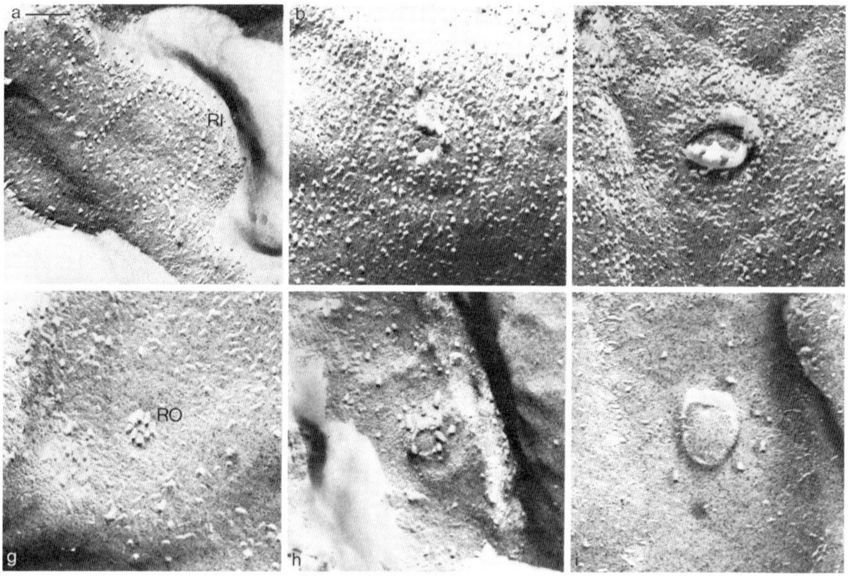

Figure 5. Freeze-fractures of trichocyst exocytosis sites in the cells rapidly frozen (without chemical fixation or antifreeze treatment) during AED triggering. (a–f) PF face, (g–l) EF face. CO, collar (on top of the trichocyst membrane); RI, ring; RO, rosette. Since no chemical fixation was used, in contrast to Figure 3, rosettes are on the EF and rings on the PF face. (For terminology, see Plattner and Zingsheim, 1983.) The following details can be noted. (a,g) Resting stage. A small exocytotic opening is formed within rosette MIP aggregates (b,h). These are laterally dispersed when the exocytotic opening expands (c,i). No diaphragm is formed during fusion, as recognized by the coarse texture of ice within the exocytotic opening in (b,c,h). One can recognize the discharging contents (d,j), completed in (e,k). In this stage, the trichocyst membrane is temporarily integrated into the cell membrane, so that it is difficult to differentiate between ring and annulus MIP (cf. Fig. 3) on the rim of the exocytosis crater. The fused trichocyst membrane is then detached, when rapid and perfect resealing of the exocytotic opening in the cell membrane occurs (f,l); this entails the collapse of a previous ring to a parenthesis stage. There is no evidence for any remarkable exchange of components, visible on freeze-fracture replicas, between the two membranes involved in exocytosis. This series of micrographs is to be compared with Figure 6. Bars: 0.1 μm. (From Olbricht and Plattner, 1984.)

Ca^{2+} was explained by local phase transitions, which are known to segregate MIP from solidified regions (for literature, see Plattner and Bachmann, 1982)]. These phenomena were all artifacts (Chandler and Heuser, 1979, 1980; Plattner, 1981). Antifreezes caused MIP rearrangements and, despite aldehyde fixation and osmiumtetroxide, caused loss of membrane-associated proteins (Plattner, 1981).

In Figure 8 the scheme in the middle represents the mechanism proposed for exocytotic membrane fusion by Pinto da Silva and Nogueira

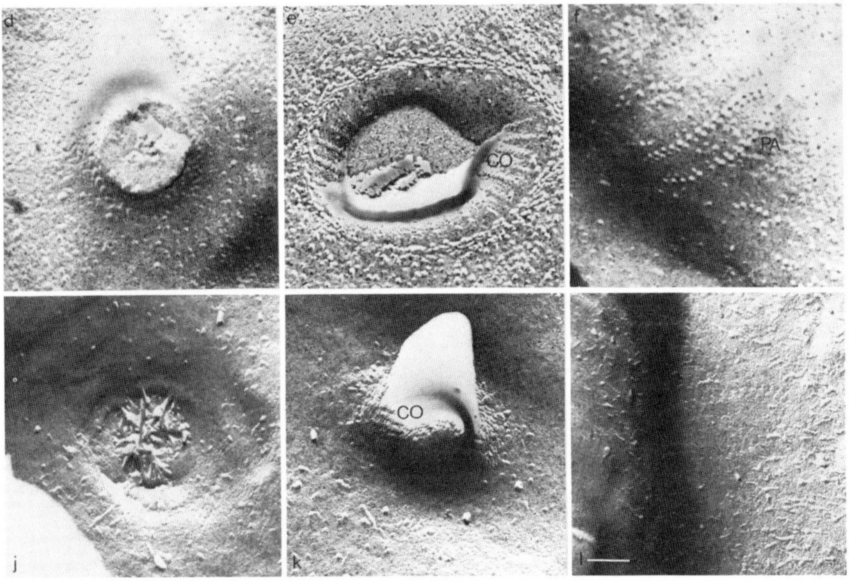

Figure 5. *(continued)*

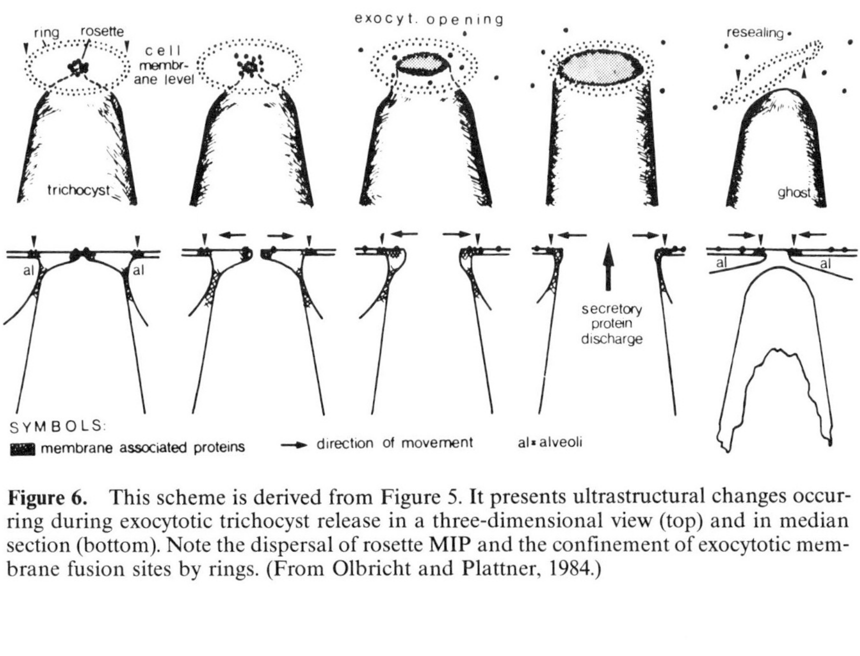

Figure 6. This scheme is derived from Figure 5. It presents ultrastructural changes occurring during exocytotic trichocyst release in a three-dimensional view (top) and in median section (bottom). Note the dispersal of rosette MIP and the confinement of exocytotic membrane fusion sites by rings. (From Olbricht and Plattner, 1984.)

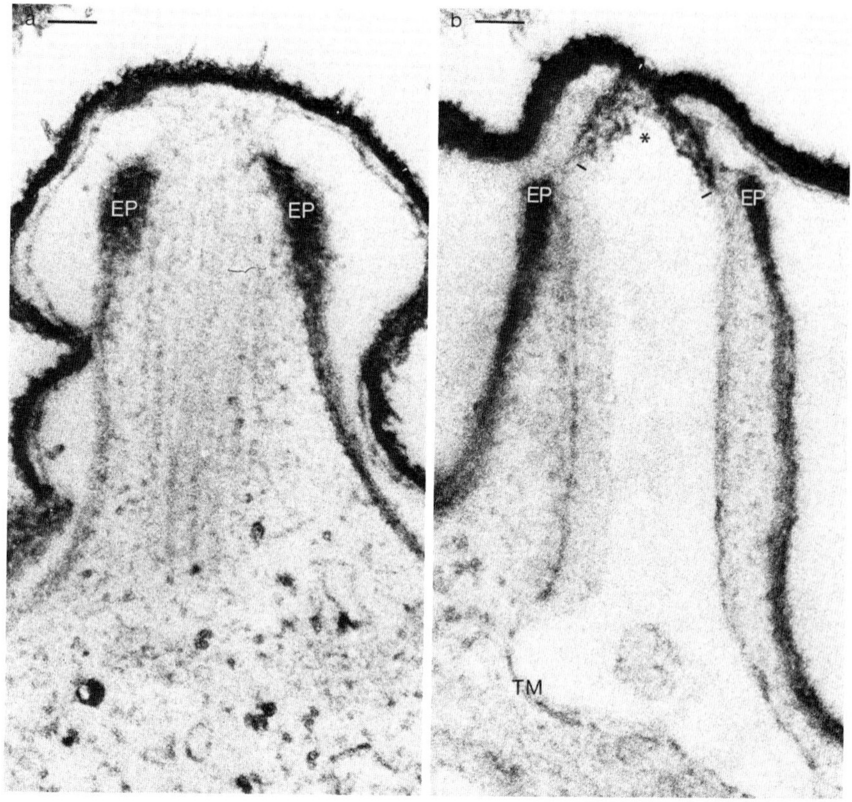

Figure 7. Exocytosis site before (a) and after (b) release of contents. This section was stained with phosphotungstic acid to visualize the rather modest redistribution of glycocalyx elements, which are transferred from the cell membrane onto the uppermost region of a trichocyst membrane (TM) during resealing (asterisk between bars). Only <0.5% of the trichocyst ghost membrane (which is only partly visible here) are thus slightly changed. EP, epiplasmic layer (proteins) outside fusion zone. Bars: 0.1 μm. (From Plattner *et al.*, 1985*a*.)

(1977). This was based on results obtained with *Phythophthora* sporozoa, but Pinto da Silva and Nogueira also referred to some work with *Paramecium* (Beisson *et al.*, 1976) in support of this scheme. However, our results obtained with rapid freezing without chemical pretreatments did not support this assumption (Olbricht *et al.*, 1984). The formation of an inverted lipid micelle (related to hexagonal II phase structures) has also been postulated by Verkleij's group (Verkleij *et al.*, 1984; DeKruijff *et al.*, 1985) as an initiating event for exocytotic membrane fusion. Recently evidence against this hypothesis has been put forward by Papahadjopoulos' group (Düzgünes *et al.*, 1984), whereas Verkleij (1986) interprets some of our own work with adrenal chromaffin cells (Schmidt *et al.*, 1983)

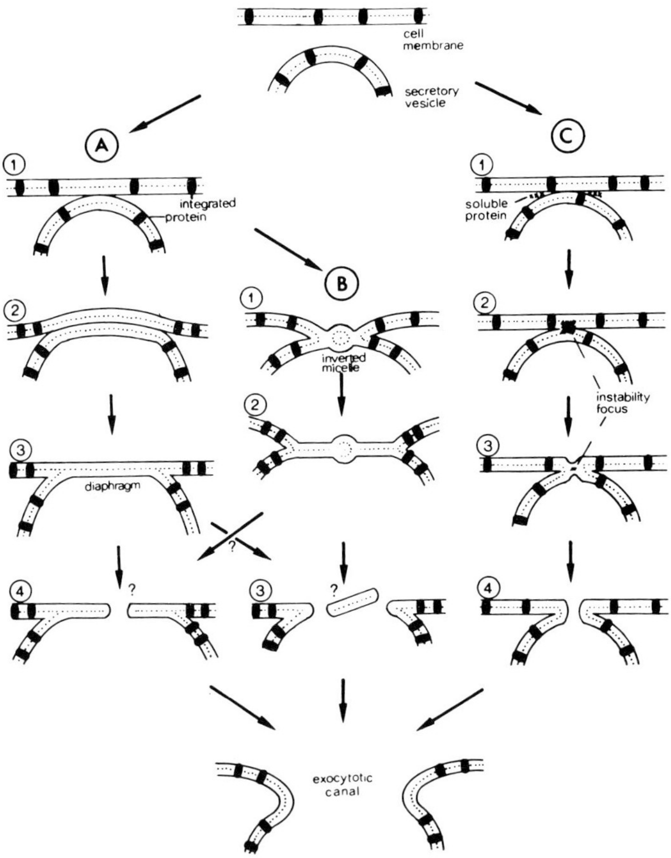

Figure 8. Models for exocytotic membrane fusion according to data in the literature (see text). Black dots represent MIP: membrane-integrated particles = membrane-integrated proteins (cf. Plattner and Zingsheim, 1983). Dotted lines indicate fracture planes in freeze fracturing. Evidence was found by us only for the focal fusion model (c). (From Plattner, 1981.)

in terms of lipid particle formation. This would be difficult to confirm as these particles may represent rather short-lived (msec) structures (Siegel, 1984) this is close to the temporal resolution of fast freezing techniques (Plattner and Bachmann, 1982). The fact that we could not identify lipid particles in paramecia does not preclude that this kind of fusion intermediate might occur.

Heuser and Reese (1981) did not observe long-range MIP rearrangements during neurotransmitter release. Since neurotransmitter vesicles are extremely small, it was important, with the help of advanced cryofixation procedures (Plattner and Bachmann, 1982), to analyze exocytotic systems with larger vesicles. MIP rearrangements were not found to precede exocytosis in: mast cells (Chandler and Heuser, 1980), sea urchin oocytes (Chandler and Heuser, 1979), chromaffin cells isolated from adrenal medullae (Schmidt *et al.*, 1983), and *Paramecium* cells (Olbricht *et al.*, 1984). Also, no diaphragm formation was observed and the exocytotic opening started as a rather small hole which then expanded (see Fig. 8, right). MIP and soluble proteins stay close to the fusion zone, where they may therefore exert some important regulatory functions. The actual fusion process must involve the lipids, but the focal fusion model (Plattner, 1981) requires only a small number of lipids, regardless of whether one would assume the formation of an inverted micelle or of less defined point defects (Hui *et al.*, 1981; Sowers, 1984) for membrane fusion.

8. ELECTROPHYSIOLOGICAL RECORDING OF SYNCHRONOUS EXOCYTOSIS

Most secretory cells require minutes or even hours to release all their secretory material upon stimulation (Rubin, 1982). The time periods required for retrieving the secretory granule membranes after discharge are also frequently quite long. In paramecia, however, the time needed for a secretion event is quite short.

The time course of exocytosis in paramecia is difficult to establish by light or electron microscopic analyses, as the time required for experimental manipulations seems to exceed the time required for exocytosis to take place. With fast freezing techniques we found an increasing number of exocytotic profiles as early as 0.3 sec after adding the trigger agent (Olbricht *et al.*, 1984). When triggering was followed by the fastest chemical fixation we could manage (3 sec) at that time, exocytosis was already completed (Plattner *et al.*, 1984*a*). When the trigger agent, supplemented with a colored marker, was added externally to immobilized cells with a micropipette, no delay could be observed (H. Kersken, personal communication) and the local response was not propagated over the entire surface of the cell (Plattner *et al.*, 1984*a*).

Clearly, electrophysiological recordings were required for determining the time course of exocytosis more precisely. We (Deitmer and Plattner, 1987) have found a transient increase in surface capacitance when individual trichocysts were seen to be discharged. This phenomenon increased in parallel with the amount of trichocysts released when we eventually added the trigger agent AED. We also recorded voltage curves which then exceeded the normal excitation curves. Both individual and cumulative exocytosis events resulted in triangular current curves typical of capacitance changes and we have interpreted these curves as a consequence of the temporary integration of trichocyst membranes into the continuum of the cell membrane. These observations allowed us to estimate (1) the duration of one exocytotic event (50–100 msec), and (2) the degree of synchrony for all events in an individual cell (1 sec).

The time required for actual membrane fusion is unknown but is estimated to be in the μsec to msec range (Siegel, 1984). The discharge of the trichocyst must also be in the millisecond range (Matt *et al.,* 1978). Our electrophysiological and ultrastructure data indicate that membrane resealing is also quite rapid.

Analysis of sections after synchronous exocytosis showed very few exogenous marker molecules of the size of cationic ferritin to be taken up into ghosts (Plattner *et al.,* 1985*a*). Also, little exchange of components occurs between the membranes engaged (Plattner *et al.,* 1985*a*). Therefore, the extremely short time available for exoendocytosis coupling might be at least one factor that maintains membrane specificity during trichocyst discharge. Also, the entire trichocyst membrane is retrieved; both ultrastructural and electrophysiological data agree on these aspects.

Indeed, surface membrane capacitance changes have been observed also with other systems. However, in most cases they could be registered only as cumulative events. Individual capacitance steps were recorded by Neher and Marty (1982) with isolated adrenal medullary cells, but could not be assigned to on–off events on one individual secretory package as with *Paramecium* cells.

9. ROLE OF Ca^{2+} AND OSMOTIC FORCES IN TRICHOCYST DISCHARGE

The osmotic pressure-induced swelling of secretory organelles is sometimes considered a driving force for membrane fusion *in vitro* (Zimmerberg *et al.,* 1980) and *in vivo* (Pollard *et al.,* 1981). However, when we immobilized paramecia and carefully triggered them with AED, no trichocyst swelling could be observed by light microscopy (H. Kersken, personal communication). The same holds for freeze-fracture work. If it happened to occur, trichocyst swelling could either be too small or too fast to be detected by the means currently available. Some arguments against the induction of fusion by osmotic swelling stem from the fact that isolated trichocyst contents without a membrane behave like the organelle during exocytosis *in vivo.* We have shown (Matt *et al.,* 1978; Bilinski *et al.,* 1981*a*) that isolated, membrane-free trichocyst contents can be instantaneously transferred from the condensed state into a decondensed state if $Ca^{2+} > 10$ μM is added. This clearly shows that decondensation of secretory contents does not require a membrane envelope and hence it can take place without osmotic effects. Given the fact that the presence of Ca^{2+} in the outside medium is normally required also for trichocyst membrane fusion (Matt *et al.*, 1980; Plattner *et al.*, 1985*b*) this might therefore guarantee an automatic decondensation of the trichocyst contents *in vivo* upon fusion of the trichocyst membrane with the plasma membrane. Furthermore, no calcium was detected within condensed trichocysts either *in situ* (Schmitz *et al.*, 1985) or after isolation (Bilinski *et al.*, 1981*a*). The decondensation of secretory contents is also in the msec range (Matt *et al.*, 1978).

If the above hypothesis were correct one should be able to induce membrane fusion with manipulation of internal Ca^{2+} pools without inducing the release of secretory contents. This is precisely what we achieved when we triggered fusion by applying an endogenous Ca^{2+}-releasing local anesthetic (Low *et al.*, 1979; Matt *et al.*, 1980) in the presence of EGTA in the culture medium (Matt and Plattner, 1983): Secretory contents remained condensed in the secretory organelles. Normally, membrane fusion would be coupled (via exogenous Ca^{2+}) with the explosive discharge of the secretory contents.

Since secretory organelles in other systems may contain Ca^{2+} (Rubin, 1982), this mechanism in paramecia is certainly not a general feature of exocytosis. In other words, reactions related specifically to the chemical nature of the secretory contents, will evidently be different from system to system.

10. TIME COURSE IN COMPARISON WITH OTHER SYSTEMS

Exocytosis in connection with neurotransmitter release should take place in the minimum time. Although this might be correct, data obtained with the quick freezing methods of Miller and Heuser (1984) for motor end plates in the frog indicate that exocytosis-coupled endocytosis is much more sluggish than that in paramecia. Other fast secretors, such as blood platelets or mast cells require 15–60 sec for full depletion (Painter and Ginsberg, 1984; Ishizuka and Nozawa, 1983). Retrieval times are 2 hr for mast cells (Thilo, 1985) and 40 min for adrenal chromaffin cells (Patzak *et al.*, 1984) to cite just two typical examples.

The very rapid exoendocytosis coupling in our system may account for the evidence of little or no material exchange between the temporarily fusing membranes despite the absence of a clathrin coat from retrieval vesicles (Plattner *et al.*, 1985*a*). One of the prerequisites is the explosive discharge of the contents. As the mechanism does not operate in most other systems, these may require much more time to release their amorphous contents.

The ghosts formed during synchronous trichocyst release are cleared away with half-lives of 5 min or more, depending on the strain (Haacke and Plattner, 1984; Pape and Plattner, 1985). It requires 9–12 hr to regenerate a complete set of dischargeable trichocysts (Pape and Plattner, 1985). Whereas membrane retrieval is the rule (Evered and Collins, 1982), secretory organelle membranes in other systems may only occasionally remain permanently integrated into the cell membrane, e.g., in oocytes during fertilization (Epel and Vacquier, 1978).

11. CURRENT BIOCHEMICAL MODELS FOR MEMBRANE FUSION

The only well known parameter that causes a secretory vesicle to fuse with the cell membrane is the requirement for endogenous free Ca^{2+} (Douglas, 1968; Rubin, 1982) in concentrations of $>10^{-7}$ M (Baker and Knight, 1981; Knight and Baker, 1982). Cytosolic free Ca^{2+} may increase by influx from the extracellular space, as in *Paramecium* (Plattner *et al.*, 1985*b*) or by liberation from intracellular stores (Rubin, 1982). Different hypotheses have been presented in an attempt to understand the molecular basis of membrane fusion and its relation to secretion. Some of these hypotheses assume a direct and others an indirect effect from Ca^{2+} ions (reviewed by Lucy, 1985). In particular, several fusion theories were put forward: (1) reduction of the hydration barrier and surface charge repulsion by Ca^{2+} (Parsegian *et al.*, 1984); (2) Ca^{2+}-mediated effect on phospholipids with local liquid–solid-phase transitions (Papahadjopoulos *et al.*, 1978); (3) formation of inverted lipid micellelike particles as fusion intermediates (DeKruijff *et al.*, 1985; Verkleij *et al.*, 1984; Verkleij, 1986), eventually in conjunction with enzymatic processes; (4) the activation of phosphatidylinositol turnover (Michell *et al.*, 1981; Ishizuka and Nozawa, 1983; Berridge, 1984) with the formation of diacylglycerol (which is known to enhance phospholipase C activity) and of phosphatidic acid (which was assumed to mobilize Ca^{2+} from internal stores); (5) the role played by forces generated by actomyosin (Berl *et al.*, 1973; Poste and Allison, 1973; Allison and Davies, 1974), which has been localized in the surface region of many cells (Stossel *et al.*, 1981); (6) fusogenic proteins previously known only from some virus membranes (White *et al.*, 1983), but now recognized in nonviral systems (Young and Young, 1984; Verkleij *et al.*, 1984; DeKruijff *et al.*, 1985); (7) limited proteolysis is also mentioned in mechanisms of membrane fusion in several different cases (Lucy, 1984, 1985), including exocytosis systems (Jackson *et al.*, 1985; Mundy and Strittmatter, 1985); (8)

protein phosphorylation (e.g., Sieghart *et al.,* 1978; Burgoyne and Geisow, 1982; Nestler *et al.,* 1984; Nishizuka, 1984) and dephosphorylation (Gilligan and Satir, 1982; Spearman *et al.,* 1984; Zieseniss and Plattner, 1985); and (9) removal of adenosinetriphosphate (ATP) from membranes by Ca^{2+}-ATPases (Woodin and Wieneke, 1964; Poste and Allison, 1973). As of this writing, there are no data available regarding phosphatidylinositol turnover during trichocyst exocytosis. Other aspects are discussed in Sections 12–14.

12. IS THERE A ROLE FOR ATP IN MEMBRANE FUSION?

An early observation made with granulocytes by Woodin and Wieneke (1964) initiated many investigations. During stimulated secretory activity, ATP consumption (Johansen, 1980) or respiratory activity (Herzog *et al.,* 1976; Schmidt *et al.,* 1980) has been observed to increase. Alternatively, it has been found that inhibitors of energy metabolism inhibit the secretory response of cells (e.g., Kirshner and Smith, 1966; Jamieson and Palade, 1971). In several permeabilized models, ATP must be added to achieve secretory activity (Baker and Knight, 1981; Knight and Baker, 1982). In 1973, Poste and Allison formulated the theory that hydrolysis of membrane-bound ATP, e.g., by a Ca^{2+}-stimulated ATPase activity, would destabilize membrane structures and that this would cause them to fuse. Although the energy requirement of secretory activity as postulated on a general scale by Poste and Allison (1973) as well as by Viveros (1975) remains unquestioned, much of the data collected on this aspect now appear to confuse the issue because a variety of phenomena interfere with specific tests of certain hypotheses.

Nevertheless, we have also analyzed the potential role of ATP in membrane fusion. Some years ago we had obtained evidence for a Ca^{2+}-dependent ATPase precisely at the trichocyst docking site (Plattner *et al.,* 1977); in a later analysis we found that this was restricted to strains capable of exocytosis (Plattner *et al.,* 1980). Enzymatic measurements with isolated pellicles supported these results (Bilinski *et al.,* 1981*b*) and revealed calmodulin sensitivity of a cortical Ca^{2+}-ATPase (Tiggemann and Plattner, 1982). This seemed to support our hypothesis of a requirement for ATP in exocytosis (Matt *et al.,* 1978). However, the identification of more than one dozen ATPases in paramecia (Andrivon *et al.,* 1983) vastly complicates this hypothesis.

More recently we set up an *in vitro* system that does not involve organelle transport processes. This system involves isolated surface complexes (Vilmart-Seuwen *et al.,* 1986). The system responds to free $Ca^{2+} > 0.1$ μM by exocytotic membrane fusion. ATP at ~1 mM was inhibitory. This could be seen in line with Poste and Allison's (1973) theory that assumes that ATP would have to be removed for membrane fusion to occur. But does the ATP consumption measured correspond precisely to this ATP pool?

It had already been stressed by Knight and Baker (1982) that the precise site of ATP requirement during secretory activity is not known and that endocytosis is coupled to exocytosis in their permeabilized cells. So ATP may rather be required for transcellular transport phenomena (Jamieson and Palade, 1971; Kruse and Bornstein, 1975; Vale *et al.*, 1985).

As we also found a measurable consumption of ATP during trichocyst release (Matt *et al.*, 1978) we have now repeated these measurements with synchronous AED triggering and with the use of a battery of different mutations (Vilmart-Seuwen *et al.*, 1986.) We found that considerable ATP consumption occurs in the same order of magnitude as reported before by Matt *et al.*, (1978) and with a minimal intracellular ATP concentration after 5 sec. This finding implies that ATP concentrations still decrease, when trichocysts are already released. ATP pools increase again before onset of rephosphorylation of the dephosphorylated 65,000-M_r protein shown to be relevant for exocytosis (see Section 14). These data were shown to be statistically significant. In addition, all mutations revealed a similar decay of their ATP pool, which was independent of the actual number of trichocysts discharged.

These findings lead to the conclusion that the decay of the ATP pools measured during exocytosis *in vivo* occurs only after exocytosis. It might be coupled to secondary phenomena, possibly to the elimination of free Ca^{2+} from the cytosol or to the activation of other ATP consuming events. If ATP hydrolysis is then part of a fusogenic mechanism, this fraction of ATP must be small.

From the inhibitory effect of ATP on the release of trichocysts from isolated *Paramecium* cortices, one could derive that ATP is required for maintaining the system in a primed state, as we previously proposed (Plattner *et al.*, 1980). This would also agree with the recent observation by Brooks *et al.* (1984) that thio-ATP permanently blocks exocytosis and with our findings on protein phosphorylation → dephosphorylation phenomena in *Paramecium* (see Section 14).

13. ARE CYTOSKELETAL ELEMENTS INVOLVED?

Since trichocysts (Aufderheide, 1977) as well as other secretory (Lacy *et al.*, 1975; Dustin, 1984) or neurotransmitter vesicles (Vale *et al.*, 1985) are brought to the cell membrane by saltatory movements, microtubules were assumed to be involved (Plattner *et al.*, 1982). However, thin sections never showed any microtubules associated with the trichocyst docking sites, empty or occupied, or even with ghosts that are internalized rapidly after synchronous exocytosis (Plattner *et al.,* 1985*a*). On the other hand, we found microtubules emanating only from nearby ciliary basal bodies; these microtubules were seen to be in contact with trichocyst bodies (Plattner *et al.,* 1982). This could imply that microtubules represent a long-range signal for trichocyst docking. Since these rails do not lead to the terminal, a short-range signal, perhaps the plug, would then be required in addition to these microtubular rails. Also, it would imply that membrane fusion does not require the abutment of a microtubule (Plattner *et al.,* 1982).

Interestingly, quite similar conclusions could be made for filamentous actin. F-actin is absent from the actual trichocyst docking sites, whereas actin bundles encircle these zones at a certain distance (Plattner *et al.,* 1982; Kersken *et al.,* 1986). Membrane fusion, release of secretory contents, and ghost retrieval also appear independent of these microfilaments, as all these functions were unimpaired when all cell motility phenomena were completely inhibited by microinjection of excessively high doses of phalloidin (Kersken *et al.,* 1986). For these studies, we used a fluorescent phalloidin known to bind to, and, thus immobilize, F-actin (Wieland, 1975).

In our system therefore, a mechanism of exocytosis involving a Ca^{2+} activated, ATP-consuming contraction of actomyosin, as postulated for exocytosis in other systems (Allison and Davies, 1974) does not appear very likely to occur. It should also be noted that actin filaments in paramecia are arranged in the wrong direction (Plattner *et al.,* 1982) to generate appropriate forces. From our point of view, microfilaments at best could govern the access of secretory organelles to the cell membrane, as originally proposed by Orci *et al.* (1972). Other possibilities have been also discussed in the literature (reviewed by Plattner, 1981).

14. ROLE OF PROTEINS IN MEMBRANE FUSION

It should be kept in mind that a trichocyst docking site must contain a fusion rosette and connecting material between trichocyst and plasma membrane in order to guarantee exocytosis. Both these structures are sensitive to proteolytic enzymes (Vilmart and Plattner, 1983; Westphal and Plattner, 1981*a*). The absence of these elements in the nondischarge mutations is additional evidence for an absolute requirement for these protein constituents for membrane fusion.

When *Paramecium* cells discharge trichocysts, a phosphoprotein of 65,000 M_r becomes rapidly dephosphorylated. This was first observed in

Satir's laboratory (Gilligan and Satir, 1982), but the cells were triggered with a fixative, picric acid. As cells are killed this way, no dose–response or time-sequence studies were possible.

We have avoided this problem by using AED as a trigger agent (Zieseniss and Plattner, 1985). We allowed the cells to incorporate $^{32}P_i$ before they were triggered and then rapidly (<1 sec) transferred the cells into boiling buffer. We obtained the following results: (1) the 65,000-M_r phosphoprotein is rapidly (<1 sec) dephosphorylated; (2) this effect is strictly dependent on the dose of AED and thus parallels the number of trichocysts discharged; (3) dephosphorylation is reversed within 10–30 sec, depending on the strain; (4) these phenomena do not occur with the mutant strains, which are not capable of membrane fusion (*nd9-28°C*), trichocyst formation (*tl*), or transfer of trichocysts to the cell membrane (*tam, ftA*); (5) what accounts for the capability of performing exocytosis is the capability to dephosphorylate the 65,000-M_r phosphoprotein, but not the original intensity of this band in the resting state; and (6) exocytosis inhibitors, like EGTA, Mg^{2+} or neomycin, also impede the dephosphorylation of the 65,000-M_r phosphoprotein.

The stimulatory effect of AED could reside in the capacity of polyamines to stimulate phosphoprotein phosphatases (Tung *et al.,* 1985) and to depress protein kinase activity (Qi *et al.,* 1983; Auberger *et al.,* 1984), but this is not precisely known.

This suggests a causal connection between the dephosphorylation of the 65,000-M_r phosphoprotein and the exocytosis of trichocysts. It is tempting to speculate about the functional analogues of the ultrastructural defects detected in *nd* mutations. The binding of calmodulin (Klumpp *et al.,* 1983) to the trichocyst attachment sites along the cell membrane has only recently been established by affinity (Fig. 9) and immunocytochemical labeling experiments (Momayezi *et al.,* 1986). In most recent experiments (M. Momayezi and H. Plattner, unpublished observations), trichocyst exocytosis was inhibited by anticalmodulin antibodies. A similar observation has been reported for sea urchin oocytes (Steinhardt and Alderton, 1982) and adrenal chromaffin cells (Kenigsberg and Trifaró, 1985). Calmodulin might confer high Ca^{2+} sensitivity to exocytotic membrane fusion sites. The Ca^{2+}-ATPase and *p*-nitrophenyl phosphatase cytochemically detected at trichocyst extrusion sites (Plattner *et al.,* 1977, 1980) could, in fact, represent the coupled activity of a protein kinase and phosphoprotein phosphatase of the kind described by Zieseniss and Plattner (1985). However, its calmodulin sensitivity also remains to be analyzed.

Further support along these lines comes from function repair studies in which *nd* mutations were microinjected with wild-type cytoplasm, which could then discharge a certain number of trichocysts (Garofalo *et*

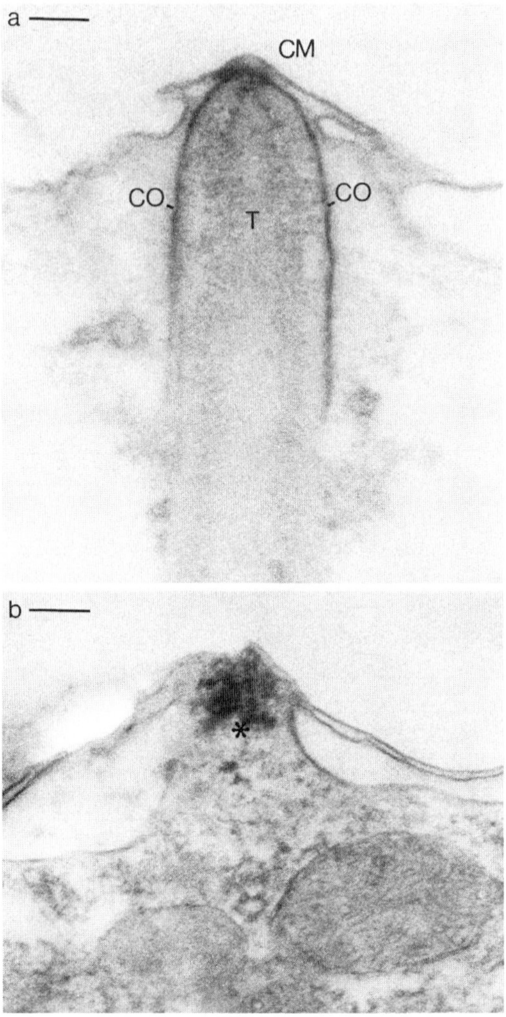

Figure 9. Affinity labeling of a trichocyst docking site by a calmodulin–peroxidase conjugate; wild-type cells permeabilized and incubated in the presence of Ca^{2+}. CM, cell membrane; CO, collar [labeled materials in (a) surrounding the tip of a trichocyst, T]. The asterisk in (b) marks labeled plug material, which is present at free sites beneath the cell membrane before a trichocyst is docked, which then causes the transformation of the plug into the collar. Bars: 0.1 μm. (From R. Tiggemann and H. Plattner, unpublished micrographs.)

al., 1979; Beisson *et al.,* 1980). The chemical nature of the curing component, however, has not yet been identified. Meanwhile, the importance of soluble proteins for exocytosis has been realized also in other systems such as oocytes (Sasaki, 1984).

15. EXOCYTOSIS IN OTHER PROTOZOAN SYSTEMS

Several protozoan parasites, notably sporozoa, possess extrusive organelles that release their contents when infecting a host cell, although the precise causal connection is unknown (Nichols *et al.,* 1983). The docking sites of these extrusomes on the cell membrane also display rosettes of MIP (reviewed by Plattner *et al.,* 1984*b*). It is now interesting to note that biogenic polyamines interfere with the infection by such protozoa and that inhibitors of polyamine biosynthesis reduce infection rates (Cohen, 1982).

NOTE ADDED IN PROOF

Relevant papers published since this chapter was written include those on the role of microfilaments and microtubules (DeLisle and Williams, 1986) and osmotic driving forces (Holz, 1986). Finkelstein and Zimmerberg (1986) and Rand and Parsegian (1986) have cautioned that analysis of exocytosis in *in vitro* systems in terms of osmotic driving forces may not be applicable to *in vivo* systems. Exocytosis in some systems may not involve an increase in free intracellular Ca^{2+} (Gomperts, 1986). The time period required for the recovery of the secretory organelle membrane after its transient integration into the cell membrane may be shorter than previously reported for chromaffin adrenal medullary cells (Geisow and Childs, 1985), and for motor endplates (Torri-Tarelli *et al.,* 1985) and is in better agreement with our findings.

ACKNOWLEDGMENTS. This communication is number IX of a series entitled "Synchronous exocytosis in *Paramecium* cells." I thank Mrs. E. Rehn for correcting the English text and for secretarial help as well as Mrs. D. Bliestle for her assistance with the illustrations. Work presented in this review was supported by SFB-138/85 and SFB-156/B4.

REFERENCES

Allen, R. D., and Fok, A. K., 1984, Membrane behavior of exocytic vesicles. III. Flow of horseradish peroxidase labeled trichocyst membrane remnants in Paramecium, *Eur. J. Cell Biol.* **35**:27–34.

Allison, A. C., and Davies, P., 1974, Interactions of membranes, microfilaments, and microtubules in endocytosis and exocytosis, in: *Cytopharmacology of Secretion* (B. Ceccarelli, F. Clementi, and J. Meldolesi, eds.), pp. 237–248, Raven Press, New York.

Andrivon, C., Brugerolle, G., and Delachambre, D., 1983, A specific Ca^{2+}-ATPase in the ciliary membrane of *Paramecium tetraurelia, Biol. Cell* **47**:351–364.

Auberger, P., Samson, M., LeCam, G., and LeCam, A., 1984, Regulation of protein phosphorylation by polyamines in hepatocytes, *Biochim. Biophys. Acta* **801**:461–469.

Aufderheide, K. J., 1977, Saltatory motility of uninserted trichocysts and mitochondria in *Paramecium tetraurelia, Science* **198**:299–300.

Aufderheide, K. J., Frankel, J., and Williams, N. E., 1980, Formation and positioning of surface-related structures in protozoa, *Microbiol. Rev.* **44**:252–302.

Bachmann, L., Schmitt, W. W., and Plattner, H., 1972, Improved cryofixation: Demonstrated on freeze-etched solutions, cell fractions and unicellular organisms, in:

Proceedings of the Fifth European Conference on Electron Microscopy (V. E. Cosslett, ed.), pp.244–245, The Institute of Physics, Bristol.

Baker, P. F., and Knight, D. E., 1981, Calcium control of exocytosis and endocytosis in bovine adrenal medullary cells, *Philos. Trans. R. Soc. Lond. B* **296:**83–103.

Bardele, C. F., 1977, Membrandifferenzierungen an Anheftungsorten von Extrusomen bei Protozoen, *Verh. Dtsch. Zool. Ges.* **70:**264.

Beisson, J., Lefort-Tran, M., Pouphile, M., Rossignol, M., and Satir, B., 1976, Genetic analysis of membrane differentiation in *Paramecium.* Freeze-fracture study of the trichocyst cycle in wild-type and mutant strains, *J. Cell Biol.* **69:**126–143.

Beisson, J., Cohen, J., Lefort-Tran, M., Pouphile, M., and Rossignol, M., 1980, Control of membrane fusion in exocytosis. Physiological studies on a *Paramecium* mutant blocked in the final step of the trichocyst extrusion process, *J. Cell Biol.* **85:**213–227.

Berl, S., Puszkin, S., and Miklas, W. J., 1973, Actomyosin-like protein in brain, *Science* **179:**441–446.

Berridge, M. J., 1984, Inositol triphosphate and diacylglycerol as second messengers, *Biochem. J.* **220:**345–360.

Bilinski, M., Plattner, H., and Matt, H., 1981*a*, Secretory protein decondensation as a distinct, Ca^{2+}-mediated event during the final steps of exocytosis in *Paramecium* cells, *J. Cell Biol.* **88:**179–188.

Bilinski, M., Plattner, H., and Tiggemann, R., 1981*b,* Isolation of surface membranes from normal and exocytotic mutant strains of Paramecium tetraurelia. Ultrastructural and biochemical characterization, *Eur. J. Cell Biol.* **24:**108–115.

Brooks, J. C., Treml, S., and Brooks, M., 1984, Thiophosphorylation prevents catecholamine secretion by chemically skinned chromaffin cells, *Life Sci.* **35:**569–574.

Burgoyne, R. D., and Geisow, M. J., 1982, Phosphoproteins of the adrenal chromaffin granule membrane, *J. Neurochem.* **39:**1387–1396.

Chandler, D. E., and Heuser, J., 1979, Membrane fusion during secretion. Cortical granule exocytosis in sea urchin eggs as studied by quick-freezing and freeze-fracture, *J. Cell Biol.* **83:**91–108.

Chandler, D. E., and Heuser, J. E., 1980, Arrest of membrane fusion events in mast cells by quick-freezing, *J. Cell Biol.* **86:**666–674.

Cohen, J., and Beisson, J., 1980, Genetic analysis of the relationships between the cell surface and the nuclei in *Paramecium tetraurelia, Genetics* **85:**797–818.

Cohen, S. S., 1982, The polyamines as a growth industry, *Fed. Proc.* **41:**3061–3064.

Couteaux, R., and Pécot-Dechavassine, M., 1970, Vésicules synaptiques et poches au niveau des "zones actives" de la jonction neuromusculaire, *C. R. Acad. Sci. Paris Ser. D* **271:**2346–2349.

Deitmer, J. W., and Plattner, H., 1987, Extremely rapid exo-endocytosis coupling in *Paramecium* cells: An electrophysiological evaluation, manuscript in preparation.

DeKruijff, B., Cullis, P. R., Verkleij, A. J., Hope, M. J., Van Echteld, C. J. A., and Taraschi, T. F., 1985, Lipid polymorphism and membrane function, in: *The Enzymes of Biological Membranes,* Vol. 1 (A. N. Martonosi, ed.), pp. 131–204, Plenum Press, New York.

DeLisle, R. C., and Williams, J. A., 1986, Regulation of membrane fusion in secretory exocytosis, *Ann. Rev. Physiol.* **48:**225–238.

Douglas, W. W., 1968, Stimulus–secretion coupling: The concept and clues from chromaffin and other cells, *Br. J. Pharmacol.* **34:**451–474.

Dufton, M. J., Cherry, R. J., Coleman, J. W., and Stanworth, D. R., 1984, The capacity of basic peptides to trigger exocytosis from mast cells correlates with their capacity to immobilize band 3 proteins in erythrocyte membranes, *Biochem. J.* **223:**67–71.

Dustin, P., 1984, *Microtubules,* 2nd ed., Springer-Verlag, Berlin.

Düzgüneş, N., Paiement, J., Freeman, K. B., Lopez, N. G., Wilschut, J., and Papahadjopoulos, D., 1984, Modulation of membrane fusion by ionotropic and thermotropic phase transitions, *Biochemistry* **23:**3486–3494.

Ehret, C. F., and McArdle, E. W., 1974, The structure of *Paramecium* as viewed from its constituent levels of organization, in: *Paramecium. A Current Survey* (W. J. Van Wagtendonk, ed.), pp. 263–338, Elsevier, Amsterdam.

Epel, D., and Vacquier, V. D., 1978, Membrane fusion events during invertebrate fertilization, in: *Membrane Fusion* (G. Poste and G. L. Nicolson, eds.), pp. 1–63, North-Holland, Amsterdam.

Evered, D., and Collins, G. M., eds., 1982, *Membrane Recycling,* Ciba Foundation Symposium No. 92, Ciba Foundation–Pitman.

Finkelstein, A., and Zimmerberg, J., 1986, Osmotic swelling of vesicles: Its role in the fusion of vesicles with planar phospholipid bilayer membranes and its possible role in exocytosis, *Ann. Rev. Physiol.* **48:**163–174.

Foreman, J. C., and Lichtenstein, L. M., 1980, Induction of histamine secretion by polycations, *Biochim. Biophys. Acta* **629:**587–603.

Garafalo, R. S., Knowles, J. K. C., and Satir, B. H., 1979, Restoration of secretory capacity in a non-discharge *Paramecium* mutant by microinjection of wild type cytoplasmic factors, *J. Cell Biol.* **79:**245a.

Geisow, M. J., and Childs, J., 1985, Cholinergic stimulation of chromaffin cells induces rapid coating of the plasma membrane, *Eur. J. Cell Biol.* **38:**51–56.

Gilligan, D. M., and Satir, B. H., 1982, Protein phosphorylation/dephosphorylation and stimulus–secretion coupling in wild type and mutant *Paramecium, J. Biol. Chem.* **257:**13903–13906.

Gilligan, D. M., and Satir, B. H., 1983, Stimulation and inhibition of secretion in *Paramecium*: Role of divalent cations, *J. Cell Biol.* **97:**224–234.

Gomperts, B. D., 1986, Calcium shares the limelight in stimulus–secretion coupling, *Trends Biochem. Sci.* **11:**290–292.

Haacke, B., and Plattner, H., 1984, Synchronous exocytosis in *Paramecium* cells. III. Rearrangement of membrane-associated structural elements after exocytosis performance, *Exp. Cell Res.* **151:**21–28.

Hausmann, K., 1978, Extrusive organelles in protists, *Int. Rev. Cytol.* **52:**197–276.

Hausmann, K., and Allen, R. D., 1976, Membrane bahavior of exocytic vesicles. II. Fate of the trichocyst membranes in *Paramecium* after induced trichocyst discharge, *J. Cell Biol.* **69:**313–326.

Herzog, V., Sies, H., and Miller, F., 1976, Exocytosis in secretory cells of rat lacrimal gland. Peroxidase release from lobules and isolated cells upon cholinergic stimulation, *J. Cell Biol.* **70:**692–706.

Heuser, J. E., and Reese, T. S., 1981, Structural changes after transmitter release at the frog neuromuscular junction, *J. Cell Biol.* **88:**564–580.

Holz, R. W., 1986, The role of osmotic forces in exocytosis from adrenal chromaffin cells, *Ann. Rev. Physiol.* **48:**175–189.

Hui, S. W., Stewart, T. P., Boni, L. T., and Yeagle, P. L., 1981, Membrane fusion through point defects in bilayers, *Science* **212:**921–923.

Ishizuka, Y., and Nozawa, Y., 1983, Concerted stimulation of PI-turnover, Ca^{2+}-influx and histamine release in antigen-activated rat mast cells, *Biochem. Biophys. Res. Commun.* **117:**710–717.

Jackson, R. C., Ward, K. K., and Haggerty, J. G., 1985, Mild proteolytic digestion restores exocytotic activity to N-ethylmaleimide-inactivated cell surface complex from sea urchin eggs, *J. Cell Biol.* **101:**6–11.

Jamieson, J. D., and Palade, G. E., 1971, Condensing vacuole conversion and zymogen granule discharge in pancreatic exocrine cells: Metabolic studies, *J. Cell Biol.* **48:**503–522.

Janisch, R., 1972, Pellicle of *Paramecium caudatum* as revealed by freeze etching, *J. Protozool.* **19:**470–472.

Jennings, H. S., 1906, *Behavior of the Lower Organisms,* Indiana University Press, Bloomington, Indiana.

Johansen, T., 1980, Adenosine triphosphate levels during histamine release induced by compound 48/80 in rat mast cells in vitro, *Life Sci.* **26:**61–69.

Jurand, A., and Selman, G. G., 1969, *The Anatomy of Paramecium aurelia,* St. Martin's Press, New York.

Kenigsberg, R. L., and Trifaró, J. M., 1985, Microinjection of calmodulin antibodies into cultured chromaffin cells blocks catecholamine release in response to stimulation, *Neuroscience* **14:**335–347.

Kersken, H., Momayezi, M., Braun, C., and Plattner, H., 1986, Filamentous actin in *Paramecium* cells: Functional and structural changes correlated with phalloidin affinity labeling in vivo, *J. Histochem. Cytochem.* **34:**455–465.

Kirshner, N., and Smith, W. J., 1966, Metabolic requirements for secretion from the adrenal medulla, *Science* **154:**422–423.

Klumpp, S., Steiner, A. L., and Schultz, J. E., 1983, Immunocytochemical localization of cyclic GMP, cGMP-dependent protein kinase, calmodulin and calcineurin in Paramecium tetraurelia, *Eur. J. Cell Biol.* **32:**164–170.

Knight, D. E., and Baker, P. F., 1982, Calcium-dependence of catecholamine release from bovine adrenal medullary cells after exposure to intense electric fields, *J. Membr. Biol.* **68:**107–140.

Kruse, N. J., and Bornstein, P., 1975, Metabolic requirements for transcellular movement and secretion of collagen, *J. Biol. Chem.* **250:**4841–4847.

Lacy, P. E., Finke, E. H., and Codilla, R. C., 1975, Cinematography studies on β granule movement in monolayer culture of islet cells, *Lab. Invest.* **33:**570–576.

Lagunoff, D., and Chi, E. Y., 1980, Cell Biology of mast cells and basophils, in: *Cell Biology of Inflammation* (G. Weissmann, ed.), pp. 217–265, Elsevier North-Holland, Amsterdam.

Lefort-Tran, M., Gulik, T., Plattner, H., Beisson, J., and Wiessner, W., 1978, Influence of cryofixation procedures on organization and partition of intramembrane particles, in: *Proceedings of the Ninth International Congress of Electron Microscopy,* Vol. II (J. M. Sturgess, ed.), pp. 146–147, Microscopical Society of Canada, Toronto.

Lefort-Tran, M., Aufderheide, K., Pouphile, M., Rossignol, M., and Beisson, J., 1981, Control of exocytotic processes: Cytological and physiological studies of trichocyst mutants in *Paramecium tetraurelia, J. Cell Biol.* **88:**301–311.

Low, P. S., Lloyd, D. H., Stein, T. M., and Rogers, J. A., 1979, Calcium displacement by local anesthetics. Dependence on pH and anesthetic charge, *J. Biol. Chem.* **254:**4119–4125.

Lucy, J. A., 1984, Do hydrophobic sequences cleaved from cellular polypeptides induce membrane fusion reactions in vivo?, *FEBS Lett.* **166:**223–231.

Lucy, J. A., 1985, Calcium ions, enzymes, and cell fusion, in: *The Enzymes of Biological Membranes,* Vol. 1 (N. Martonosi, ed.), pp. 371–391, Plenum Press, New York.

Matt, H., and Plattner, H., 1983, Decoupling of exocytotic membrane fusion from protein discharge in *Paramecium* cells, *Cell, Biol. Int. Rep.* **7:**1025–1031.

Matt, H., Bilinski, M., and Plattner, H., 1978, Adenosinetriphosphate, calcium and temperature requirements for the final steps of exocytosis in *Paramecium* cells, *J. Cell Sci.* **32:**67–86.

Matt, H., Plattner, H., Reichel, K., Lefort-Tran, M., and Beisson, J., 1980, Genetic dissection of the final exocytosis steps in *Paramecium tetraurelia* cells: Trigger analyses, *J. Cell Sci.* **16:**41–60.

Michell, R. H., Kirk, C. J., Jones, L. M., Downes, C. P., and Creba, J. A., 1981, The stimulation of inositol lipid metabolism that accompanies calcium mobilization in stimulated cells: Defined characteristics and unanswered questions, *Philos. Trans. R. Soc. Lond. B. Biol. Sci.* **296:**123–137.

Miller, T. M., and Heuser, J. E., 1984, Endocytosis of synaptic vesicle membrane at the frog neuromuscular junction, *J. Cell Biol.* **98:**685–698.

Momayezi, M., Kersken, H., Gras, U., Vilmart-Seuwen, J., and Plattner, H., 1986, Calmodulin in *Paramecium tetraurelia:* Localization from the *in vivo* to the ultrastructural level, *J. Histochem. Cytochem.,* in press.

Mundy, D. I., and Strittmatter, W. J., 1985, Requirement for metalloendoprotease in exocytosis: Evidence in mast cells and adrenal chromaffin cells, *Cell* **40:**645–656.

Naitoh, Y., and Kaneko, H., 1972, Reactivated triton-extracted models of Paramecium: Modification of ciliary movement by calcium ions, *Science,* **176:**523–524.

Neher, E., and Marty, A., 1982, Discrete changes of cell membrane capacitance observed under conditions of enhanced secretion in bovine adrenal chromaffin cells, *Proc. Natl. Acad.Sci. U.S.A.* **79:**6712–6719.

Nestler, E. J., Walaas, S. I., and Greengard, P., 1984, Neuronal phosphoproteins: Physiological and clinical implications, *Science* **225:**1357–1364.

Nichols, B. A., Chiappino, M. L., and O'Connor, G. R., 1983, Secretion from the rhoptries of *Toxoplasma gondii* during host-cell invasion, *J. Ultrastruct. Res.* **83:**85–98.

Nishizuka, Y., 1984, Protein kinases in signal transduction, *Trends Biochem. Sci.* **9:**163–166.

Ogura, A., and Takahashi, K., 1976, Artificial deciliation causes loss of calcium-dependent responses in *Paramecium, Nature (Lond.)* **264:**170–172.

Olbricht, K., Plattner, H., and Matt, H., 1984, Synchronous exocytosis in *Paramecium* cells. II. Intramembranous changes analyzed by freeze-fracturing, *Exp. Cell Res.* **151:** 14–20.

Orci, L., Gabbay, K. H., and Malaisse, W. J., 1972, Pancreatic beta-cell web: its possible role in insulin secretion, *Science* **175:**1128–1130.

Orci, L, Amherdt, M., Montesano, R., Vassalli, P., and Perrelet, A., 1981, Topology of morphologically detectable protein and cholesterol in membranes of polypeptide-secreting cells, *Philos. Trans. R. Soc. Lond. B* **296:**47–54.

Painter, R. G., and Ginsberg, M. H., 1984, Centripetal myosin redistribution in thrombin-stimulated platelets. Relationship to platelet factor 4 secretion, *Exp. Cell Res.* **155:**198–212.

Palade, G., 1975, Intracellular aspects of the process of protein synthesis, *Science* **189:**347–358.

Palade, G. E., 1982, Problems in intracellular membrane traffic, in: *Membrane Recycling,* Ciba Foundation Symposium No. 92, pp. 1–14, Pitman, London.

Papahadjopoulos, D., Portis, A., Pangborn, W., and Newton, C., 1978, Fusion of artificial membranes with special emphasis on the role of calcium-induced lipid phase transitions, in: *Transport of Macromolecules in Cellular Systems* (S. C. Silverstein, ed.), pp. 413–430, Dahlem Konferenzen, Berlin.

Pape, R., and Plattner H., 1985, Synchronous exocytosis in Paramecium cells. V. Ultrastructural adaptation phenomena during re-insertion of secretory organelles, *Eur. J. Cell Biol.* **36:**38–47.

Parsegian, V. A., Rand, R. P., and Gingell, D., 1984, Lessons for the study of membrane fusion from membrane interactions in phospholipid systems, in: *Cell Fusion* (D. Evered and J. Whelan, eds.), pp. 9–27, Pitman, London.

Patzak, A., Böck, G., Fischer-Colbrie, R., Schauenstein, K., Schmidt, W., Lingg, G., and Winkler, H., 1984, Exocytotic exposure and retrieval of membrane antigens of chromaffin granules: quantitative evaluation of immunofluorescence on the surface of chromaffin cells, *J. Cell Biol.* **98:**1817–1824.

Pinto da Silva, P., and Nogueira, M. L., 1977, Membrane fusion during secretion. A hypothesis based on electron microscope observation of *Phytophthora palmivora* zoospores during encystment, *J. Cell Biol.* **73:**161–181.

Plattner, H., 1981, Membrane behaviour during exocytosis, *Cell Biol. Int. Rep.* **5:**435–459.

Plattner, H., and Bachmann, L., 1982, Cryofixation: A tool in biological ultrastructural research, *Int. Rev. Cytol.* **79:**237–304.

Plattner, H., and Zingsheim, H. P., 1983, Electron microscopic methods in cellular and molecular biology, in: *Subcellular Biochemistry,* Vol. 9 (D. B. Roodyn, ed.), pp. 1-236, Plenum Press, New York.

Plattner, H., Miller, F., and Bachmann, L., 1973, Membrane specializations in the form of regular membrane-to-membrane attachment sites in *Paramecium.* A correlated freeze-etching and ultrathin-sectioning analysis, *J. Cell Sci.* **13:**687–719.

Plattner, H., Reichel, K., and Matt, H., 1977, Bivalent-cation-stimulated ATPase activity at preformed exocytosis sites in *Paramecium* coincides with membrane-intercalated particle aggregates, *Nature (Lond.)* **267:**702–704.

Plattner, H., and Reichel, K., Matt, H., Beisson, J., Lefort-Tran, M., and Pouphile, M., 1980, Genetic dissection of the final exocytosis steps in *Paramecium tetraurelia* cells: Cytochemical determination of Ca^{2+}-ATPase activity over preformed exocytosis sites, *J. Cell Sci.* **46:**17–40.

Plattner, H., Westphal, C., and Tiggemann, R., 1982, Cytoskeleton-secretory vesicle interactions during the docking of secretory vesicles at the cell membrane in *Paramecium tetraurelia* cells, *J. Cell Biol.* **92:**368– 377.

Plattner, H., Matt, H., Kersken, H., Haacke, B., and Stürzl, R., 1984*a*, Synchronous exocytosis in *Paramecium* cells: A novel approach, *Exp. Cell Res.* **151:**6–13.

Plattner, H., Westphal, C., Tiggemann, R., and Matt, H., 1984*b,* Membrane fusions in protozoa. A survey lecture with special emphasis on exocytosis in ciliates, in: *Progress in Protozoology,* Vol. II (S. Dryl, S. L. Kazubski, L. Kuznicki, and J. Ploszaj, eds.), pp. 267–278, Panstwowe Wydawnictwo Naukowe, Warsaw.

Plattner, H., Pape, R., Haacke, B., Olbricht, K., Westphal, C., and Kersken, H., 1985*a*, Synchronous exocytosis in *Paramecium* cells. VI. Ultrastructural analysis of membrane resealing and retrieval, *J. Cell Sci.,* **77:**1–17.

Plattner, H., Stürzl, R., Matt, H., 1985*b*, Synchronous exocytosis in Paramecium cells. IV. Polyamino compounds as potent trigger agents for repeatable trigger-redocking cycles, *Eur. J. Cell Biol.* **36:**32–37.

Pollack, S., 1974, Mutations affecting the trichocysts in *Paramecium aurelia.* I. Morphology and description of the mutants, *J. Protozool.* **21:**352–362.

Pollard, H. B., Creutz, C. E., Fowler, V., Scott, J., and Pazoles, C. J., 1981, Calcium-dependent regulation of chromaffin granule movement, membrane contact, and fusion during exocytosis, *Cold Spring Harb. Symp.* **46:**819–834.

Porchet-Hennere, E., and Nicolas, G., 1983, Are rhoptries of coccidia really extrusomes?, *J. Ultrastruct. Res.* **84:**194–203.

Poste, G., and Allison, A. C., 1973, Membrane fusion, *Biochim. Biophys. Acta* **300:**421–465.

Pouphile, M., Lefort-Tran, M., Plattner, H., Rossignol, M., and Beisson, J., 1986, Genetic dissection of the morphogenesis of exocytosis sites in *Paramecium, Biol. Cell* **56:**151–162.

Qi, D. F., Schatzmann, R. C., Mazzei, G. J., Turner, R. S., Raynor, R. L., Liao, S., and Kuo, J.

F., 1983, Polyamines inhibit phospholipid-sensitive and calmodulin-sensitive Ca^{2+}-dependent protein kinases, *Biochem. J.* **213:**281–288.

Ranadive, N. S., and Cochrane, C. G., 1971, Mechanism of histamine release from mast cells by cationic protein (band 2) from neutrophil lysosomes, *J. Immunol.* **106:**506–516.

Rand, R. P., and Parsegian, V. A., 1986, Mimicry and mechanism in phospholipid models of membrane fusion, *Ann. Rev. Physiol.* **48:**201–212.

Rubin, R. P., 1982, *Calcium and Cellular Secretion,* Plenum Press, New York.

Saimi, Y., and Kung, C., 1980, A Ca-induced Na-current in *Paramecium, J. Exp. Biol.* **88:**305–325.

Sasaki, H., 1984, Modulation of calcium sensitivity by a specific cortical protein during sea urchin cortical exocytosis, *Dev. Biol.* **101:**125–135.

Satir, B. H., and Oberg, S. G., 1978, *Paramecium* fusion rosettes: Possible function as Ca^{2+} gates, *Science* **199:**536–538.

Satir, B., Schooley, C., and Kung, C., 1972, Internal membrane specializations in *Paramecium aurelia, J. Cell Biol.* **55:**227a.

Satir, B., Schooley, C., and Satir, P., 1973, Membrane fusion in a model system. Mucocyst secretion in *Tetrahymena, J. Cell Biol.* **56:**153–176.

Schmidt, H. W., Herzog, V., and Miller, F., 1980, Oxygen consumption of isolated acini from rat parotid gland, *Eur. J. Cell Biol.* **20:**201–208.

Schmidt, W., Patzak, A., Lingg, G., Winkler, H., and Plattner, H., 1983, Membrane events in adrenal chromaffin cells during exocytosis: A freeze-etching analysis after rapid cryofixation, *Eur. J. Cell Biol.* **32:**31–37.

Schmitz, M., Meyer, R., and Zierold, K., 1985, X-ray microanalysis in cryosections of natively frozen *Paramecium caudatum* with regard to ion distribution in ciliates, *Scann. Electron Microsc.* **1:**433–445.

Siegel, D. P., 1984, Inverted micellar structures in bilayer membranes. Formation rates and half-lives, *Biophys. J.* **45:**399–420.

Sieghart, W., Theoharides, T. C., Alper, S. L., Douglas, W. W., and Greengard, P., 1978, Calcium-dependent protein phosphorylation during secretion by exocytosis in the mast cell, *Nature (Lond.)* **275:**329– 331.

Sonneborn, T. M., 1974, Paramecium tetraurelia, in: *Handbook of Genetics,* Vol. 2 (R. C. Kung, ed.), pp. 469–594, Plenum Press, New York.

Sowers, A. E., 1984, Characterization of electric field-induced fusion in erythrocyte ghost membranes, *J. Cell Biol.* **99:**1989–1996.

Spearman, T. N., Hurley, K. P., Olivas, R., Ulrich, R. G., and Butcher, F. R., 1984, Subcellular location of stimulus-affected endogenous phosphoprotein in the rat parotid gland, *J. Cell Biol.* **99:**1354–1363.

Steinhardt, R. A., and Alderton, J. M., 1982, Calmodulin confers calcium sensitivity on secretory exocytosis, *Nature (Lond.)* **295:**154–155.

Stossel, T. P., Hartwig, J. H., Yin, H. L., Zaner, K. S., and Stendahl, O. I., 1981, Actin gelation and the structure of cortical cytoplasm, *Cold Spring Harbor Symp.* **46:**569–578.

Suchard, S. J., Rubin, R. W., and Pressman, B. C., 1980, Cultured adrenal chromaffin cells as a model system for the study of secretory events, *J. Cell Biol.* **87:**303a.

Thilo, L., 1985, Selective internalization of granule membrane after secretion in mast cells, *Proc. Natl. Acad. Sci. U.S.A.* **82:**1711–1715.

Tiggemann, R., and Plattner, H., 1982, Possible involvement of a calmodulin regulated Ca^{2+}-ATPase in exocytosis performance in *Paramecium tetraurelia* cells, *FEBS Lett.* **148:**226–230.

Torri-Tarelli, F., Grohovaz, F., Fesce, R., Ceccarelli, B., 1985, Temporal coincidence between synaptic vesicle fusion and quantal secretion of acetylcholine, *J. Cell Biol.* **101:**1386–1399.

Tung, H. Y. L., Pelech, S., Fisher, M., Pogson, C. I., and Cohen, P., 1985, The protein phosphatases involved in cellular regulation. Influence of polyamines on the activities of protein phosphatase-1 and protein phosphatase-A2, *Eur. J. Biochem.* **149:**305–313.

Vale, R., Schnapp, B. J., Reese, T. S., and Sheetz, M. P., 1985, Organelle, bead, and microtubule translocations promoted by soluble factors from the squid giant axon, *Cell* **40:**559–569.

Verkleij, A. J., 1986, Role of lipids during fusion of model and biological membranes, in: *Phospholipids in the Nervous System: Biochemical and Molecular Pharmacology* (L. Freysz, L. A. Horrocks, and G. Toffano, eds.), Liviana Press, Padova.

Verkleij, A. J., Leunissen-Bijvelt, J., DeKruijff, B., Hope, M., and Cullis, P. R., 1984, Non-bilayer structures in membrane fusion, in: *Cell Fusion,* Ciba Foundation Symposium No. 103, (D. Evered and J. Whelan, eds.), pp. 45–59, Pitman, London.

Vilmart, J., and Plattner, H., 1983, Membrane-integrated proteins at preformed exocytosis sites, *J. Histochem. Cytochem.* **31:**626–632.

Vilmart-Seuwen, J., Kersken, H., Stürzl, R., and Plattner, H., 1986, ATP keeps exocytosis sites in a primed state but is not required for membrane fusion. An analysis with *Paramecium* cells *in vivo* and *in vitro, J. Cell Biol.,* in press.

Viveros, O. H., 1975, Mechanism of secretion of catecholamines from adrenal medulla, in: *Handbook of Physiology,* Vol. 6 (H. Blaschko, G. Sayers, and A. D. Smith, eds.), pp. 389–426, American Physiological Soc.,Washington, D. C.

Wang, B. M., Weiner, N. D., Ganesan, M. G., and Schach, J., 1984, Interaction of calcium and neomycin with anionic phospholipid–lecithin liposomes. A differential scanning calorimetry study, *Biochem. Pharmacol.* **33:**3787–3791.

Westphal, C., and Plattner, H., 1981*a*, Ultrastructural analysis of the cell membrane-secretory organelle interaction zone in *Paramecium tetraurelia* cells. I. *In situ* characterization by electron "staining" and enzymatic digestion, *Biol. Cell* **42:**125–140.

Westphal, C., and Plattner, H., 1981*b*, Ultrastructural analysis of the cell membrane-secretory organelle interaction zone in *Paramecium tetraurelia* cells. II. Changes involved in the attachment of secretory organelles, *Biol. Cell* **42:**141–146.

White, J., Kielian, M., and Helenius, A., 1983, Membrane fusion proteins of enveloped animal viruses, *Q. Rev. Biophys.* **16:**151–195.

Wieland, T., 1975, Phallotoxins and microfilaments, in: *Molecular Basis of Motility* (L. H. G. Heilmeyer, ed.), pp. 203–214, Springer-Verlag, Berlin.

Wohlfarth-Bottermann, K. E., 1953, Experimentelle und elektronenoptische Untersuchungen zur Funktion der Trichocysten von *Paramecium caudatum, Arch. Protistenk.* **98:**169–226.

Woodin, A. M., and Wieneke, A. A., 1964, The participation of calcium, adenosine triphosphate and adenosine triphosphatase in the extrusion of the granule proteins from polymorphonuclear leucocytes, *Biochem. J.* **90:**498–509.

Young, T. M., and Young, J. D., 1984, Protein-mediated intermembrane contact facilitates fusion of lipid vesicles with planar bilayers, *Biochim. Biophys. Acta* **775:**441–445.

Yusa, A., 1963, An electron microscope study on regeneration of trichocysts in *Paramecium caudatum, J. Protozool.* **10:**253–262.

Zieseniss, E., and Plattner, H., 1985, Synchronous exocytosis in *Paramecium* cells involves very rapid ($\leqslant$1 s), reversible dephosphorylation of a 65-kD phosphoprotein in exocytosis-competent strains, *J. Cell Biol.* **101:**2028–2035.

Zimmerberg, J., Cohen, F. S., and Finkelstein, A., 1980, Micromolar Ca^{2+} stimulates fusion of lipid vesicles with planar bilayers containing a calcium-binding protein, *Science* **210:**906–908.

Chapter 5

Role of Metalloendoprotease in the Fusion of Biological Membranes

Warren J. Strittmatter, Christine B. Couch, and Dorothy I. Mundy

1. INTRODUCTION

Membrane fusion is required for many important cellular functions. Fertilization of the sperm with the egg requires cell–cell fusion followed by the fusion of the two nuclei. During the development of the embryo, muscle precursor cells—myoblasts—fuse to form multinucleated myotubes that differentiate into mature myofibrils. In the mature organism, membrane fusion is required for cell–cell communication, as well as for the proper assembly and insertion of membrane-bound or secreted proteins. Cell–cell communication between neurons at the synapse requires the precisely directed fusion of synaptic vesicles with the presynaptic plasma membrane, releasing neurotransmitter into the synaptic space. Long-distance communication, mediated by hormones, similarly requires exocytotic release by secretory vesicle fusion. Within an individual cell,

WARREN J. STRITTMATTER, CHRISTINE B. COUCH, and DOROTHY I. MUNDY • Department of Biochemistry and Neurology, Program in Neuroscience, Baylor College of Medicine, Houston, Texas 77030.

organelle fusion permits the directed shuttling of proteins from the endoplasmic reticulum to the Golgi apparatus and then to the plasma membrane. All these membrane fusions must be regulated. These fusion events are initiated by specific stimuli, or by developmentally regulated signals. Fusion is also specific for both the site and type of membrane. For example, secretory vesicles fuse with plasma membrane, and not with endoplasmic reticulum membrane; myoblasts fuse with other myoblasts and not with fibroblasts. The specific mechanisms conferring regulation and specificity are only poorly understood. Likewise the mechanism actually causing membrane fusion is not well understood. We have previously reviewed the role of membrane proteins in the regulation of fusion of biological membranes (Strittmatter *et al.,* 1985). Two fundamental roles of proteins in biological membrane fusion are likely: (1) membrane proteins may permit fusion of membranes, thereby determining the specificity of membrane fusion; and (2) membrane protein may actually cause membrane fusion, a role demonstrated for certain viral membrane proteins, reviewed elsewhere in this book.

If membrane proteins play an important role in either regulating or causing membrane fusion, processing of such proteins would be important for their function. For example, as described in Chapter 1, the precise proteolytic processing of the viral coat protein hemagglutinin is required before this protein can mediate the fusion of the virus with its host cell. We are exploring the hypothesis that membrane fusion in vertebrate cells similarly requires a precise proteolytic processing before two membranes can fuse. To examine this hypothesis, we are examining cell–cell fusion, which occurs in muscle development, when myoblasts fuse to form myotubes, and we are also examining organelle fusion, in the exocytotic release of catecholamines by adrenal chromaffin cells. Our observations suggest that protease of the specific class, metal-dependent endoprotease, is required in membrane fusion in these and other cells.

2. METALLOENDOPROTEASES

2.1. Characterization

Metalloendoproteases are proteases requiring metal for catalytic activity and are inhibited by metal chelators such as EDTA and 1,10-phenanthroline, even in the presence of Ca^{2+} (Barrett, 1977). The metal is usually zinc. Inhibition of metalloproteases by chelators can sometimes be reversed by the addition of Zn^{2+} or other metals, but often the apoenzyme is too unstable and is irreversibly denatured (Barrett, 1977). These proteases hydrolyze peptide bonds at the amino group of uncharged aromatic and aliphatic amino acids (Morihara, 1974; Chlebowski and Coleman, 1976). Metalloendoproteases are inhibited by substrate analogues such as CBZ-L-phenylalanine (Morihara, 1974; Burstein *et al.*, 1974) and high-affinity inhibitors (10^{-9}–10^{-7} M) that interact with the active site of metalloendoproteases. These include hydroxamic derivatives, such as 2-(*N*-hydroxycarboxamido)-3-phenylpropanoyl-Ala-Gly-NH_2 (Nishino and Powers, 1979), and *N*-carboxymethyl derivatives, such as Homo-Phe-[N]-l-Phe-β-Ala (Mumford *et al.*, 1982). The natural inhibitor phosphoramidon (rhamnosyl-*p*-Leu-*l*-Trp-OH) also inhibits some metalloendoproteases with high specificity and high affinity (Komiyama *et al.*, 1975). The hydroxamic acid derivative, $ClCH_2CO$(N-OH)-Phe-Ala-Ala-NH_2 irreversibly inhibits by alkylation of the active site (Rasnick and Powers, 1978).

Metalloendoproteases cleave peptide bonds centrally located in proteins, (Morihara, 1974). Metalloendoproteases will also hydrolyze synthetic carbobenzoxy (CBZ)-dipeptide amides but only with large aliphatic amino acids, such as leucine, or noncharged aromatic amino acids, such as phenylalanine, on the amino terminus of the hydrolyzed peptide bond (Morihara, 1974; Chlebowski and Coleman, 1976). These synthetic dipeptides can therefore be used to systematically explore substrate specificity of proteases.

2.2. Methods of Assay

Metalloendoproteases can be assayed using fluorogenic synthetic substrates, which are specific, sensitive, and easily quantitated. These small peptide substrates contain an amino acid sequence that is hydrolyzed by metalloendoproteases. The substrate also contains a fluorescent leaving group that is quenched by energy transfer to either the amino acids or by a special quenching moiety. Two such fluorogenic protease substrates have been particularly useful in the identification and characterization of metalloendoproteases: 2-aminobenzoyl-Ala-Gly-Leu-Ala-4-nitrobenzylamide (AAGLAN) and succinyl-Ala-Ala-Phe-4-aminomethylcoumarin.

The fluorogenic substrate 2-aminobenzoyl-Ala-Gly-Leu-Ala-4-nitrobenzylamide contains the Gly-Leu bond recognized and hydrolyzed by metalloendoproteases (Kam *et al.*, 1979). In the intact substrate, the fluorescent 2-aminobenzoyl is quenched by energy transfer by 4-nitrobenzylamide. After hydrolysis, the fluorescent 2-aminobenzoyl group is no longer quenched, and the amount of fluorescence quantitatively indicates the amount of substrate hydrolyzed, as shown in Figure 1.

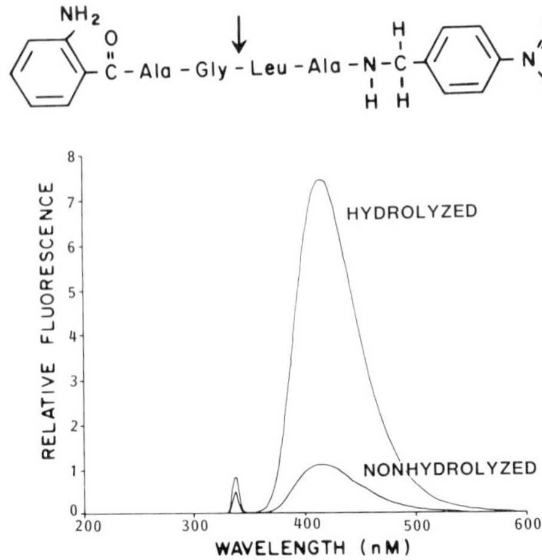

Figure 1. (Top) The metalloendoprotease substrate 2-aminobenzoyl-Ala-Gly-Leu-Ala-4-nitrobenzylamide with the site of hydrolysis indicated by the arrow. (Bottom) Fluorescence spectra of hydrolyzed and nonhydrolyzed substrate.

Succinyl-Ala-Ala-Phe-4-aminomethylcoumarin can also be used to assay metalloendoprotease (Mumford *et al.,* 1980). The fluorescence of the 4-aminomethylcoumarin (AMC) group is quenched by energy transfer to the phenylalanine. Metalloendoproteases will hydrolyze the Ala-Phe bond, leaving Phe-AMC. Exogenous aminopeptidase M then hydrolyzes the Phe-AMC, leaving the free fluoresecent AMC group, as shown in Figure 2.

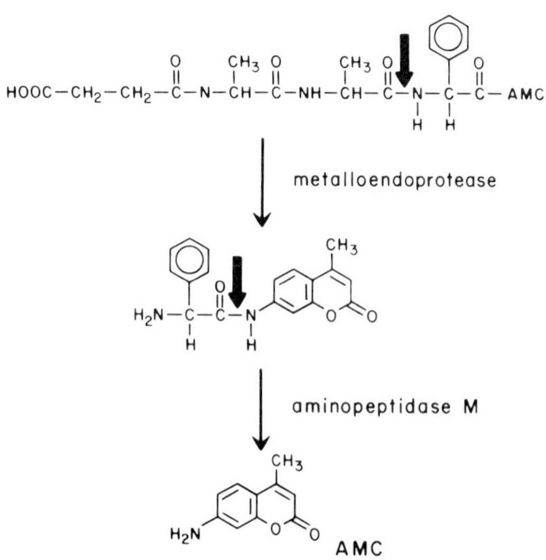

Figure 2. Succinyl-Ala-Ala-Phe-Aminomethylcoumarin hydrolyzed by metalloendoprotease and exogenous aminopeptidase M in a coupled assay.

3. MYOBLAST FUSION

3.1. Regulation and Specificity

During muscle development, mononucleated myoblasts fuse to form multinucleated myotubes. Whereas myoblast fusion requires many months *in vivo,* in tissue culture it occurs in a few days. Myoblast fusion has been studied *in vitro* using primary cultures of embryonic or newborn muscle and with muscle cell lines. Cells proliferate for several days, then withdraw from the cell cycle. Myoblasts become very mobile and aggregate with other myoblasts. Gap junctions form between muscle cells at this stage (Bischoff, 1978; Rash and Fambrough, 1973). Cells fuse together to form large multinucleated myotubes, which then accumulate muscle-specific proteins and develop fibers and contractile ability. Whereas cell fusion is a dramatic event in myogenesis, it is not apparently necessary for the subsequent biochemical differentiation and accumulation of muscle-specific proteins (Bischoff, 1978).

In order to study myoblast fusion, it is important to separate the fusion process from the process by which myoblasts become fusion competent. By switching cells to low (270-μM) calcium about 24 hr after plating (Shainberg *et al.,* 1969), the cells proliferate and withdraw from the cell cycle just as they do in normal media. If kept in low calcium medium, the cells synthesize muscle-specific proteins, indicating that they continue to differentiate, even though they cannot fuse (Bischoff, 1978). When 1.4 mM calcium is added to these cultures, they fuse rapidly (3–24 hr). This fusion does not require further cell division, protein, or RNA synthesis (Bischoff, 1978; Adamo *et al.,* 1976), further indicating the cells are fully fusion competent.

3.2.1. Role of Calcium Myoblast Fusion

The role of calcium in the fusion process is of considerable interest. As with other types of membrane fusion, the divalent cation requirement in myoblast fusion is specific for calcium. Other ions, such as Mg^{2+}, Zn^{2+}, Cu^{2+}, Cd^{2+}, and Li^{+}, cannot substitute for Ca^{2+}, although some fusion does occur with Sr^{2+} (Bischoff, 1978). Przybyski *et al.* (1982) found that cell calcium content measured by atomic absorption goes up about 20% at midfusion and decreases after fusion. David *et al.* (1981) found extracellular calcium insufficient for fusion but that calcium must enter the cells to initiate fusion. This conclusion was based on the following evidence: (1) a measurable net calcium influx occurs just prior (1–2 hr) to fusion, (2) the calcium chelator EGTA blocks fusion even when added after alignment is complete, (3) the calcium ionophore A23186 induces fusion 5–8 hr prior to normal fusion, and (4) the calcium channel blocker D600 inhibits cell fusion, but not alignment.

3.2. Methods Employed in These Studies

3.2.1. Primary Rat Myoblast Culture

Primary cultures of myoblasts are prepared by the techniques described by Yaffe (1973). To prevent myoblast fusion, the medium is switched after 48 hr to calcium-free medium containing dialyzed horse serum. In calcium-deficient medium, the myoblasts proliferate and align but do not fuse. To initiate fusion, $CaCl_2$ is added to a final concentration of 1.4 mM.

3.2.2. Inhibition Studies

Myoblasts grown in calcium-deficient medium for at least 3 days are incubated 2 hr with 0–80 μg/ml (0–0.38 mM) 1,10-phenanthroline (Sigma) in 0.5% ethanol. The cultures are washed in calcium-deficient media with no inhibitor and then switched to media containing 1.4 mM $CaCl_2$. After 24 hr incubation, fusion is assayed as described in the following section. All other inhibitors are preincubated with the cells in the calcium-deficient media for 15 min; then 1.4 mM $CaCl_2$ is added. Blocked amino acids (Sigma) are water soluble. Dipeptide derivatives (Vega Biochemicals) are solubilized in 0.5% DMF. The fluorogenic substrate 2-aminobenzoyl-Ala-Gly-Leu-Ala-4-nitrobenzylamide and the irreversible inhibitor $ClCH_2CO$-(N-OH)-DL-Phe-L-Ala-L-Ala-NH_2 are obtained from Enzyme Systems Products and solubilized in 0.5% dimethylformamide (DMF). The hydroxamic acid derivative 2-(*N*-hydroxycarboxamido)-3-phenylpropanoyl-Ala-Gly-NH_2 (Enzyme Systems Products) is water soluble. Homo-Phe-[*N*]-L-Phe-β-Ala (gift from Richard Mimford, Merck Institute for Therapeutic Research, Rahway, NJ) is made as a 1-mM 4% DMF stock, and the DMF was diluted along with the inhibitor. Phosphoramidon (Transformation Research, Framingham, MA) is water soluble.

3.2.3. Assay of Fusion

Fusion is quantitated 24 hr after the addition of calcium, by photographing each culture plate with phase-contrast microscopy, with a minimum of four frames per culture plate. The negative is enlarged, and the number of myotubes per 0.5 mm^2 (the area of culture dish photographed in one frame) is counted. This method of quantitation compares very favorably with counting the number of nuclei within myotubes and has the advantages of being much easier and not disrupting the cultures, so later time points may also be assayed (Turner *et al.,* 1976).

3.3. Effects of Metalloendoprotease Inhibitors on Myoblast Fusion

To determine whether metalloprotease activity is required for the fusion of myoblasts, metalloprotease inhibitors were added to the medium of calcium-deprived myoblasts just before fusion was triggered by addition of calcium. The metal chelator 1,10-phenanthroline prevented myoblast fusion (Couch and Strittmatter, 1983). Inhibition of fusion by 1,10-phenanthroline was dose dependent, with half-maximal inhibition requiring approximately 20 μg/ml. Complete inhibition of fusion required a 2-hr preincubation with 80 μg/ml 1,10-phenanthroline. Normal fusion occurred if 1,10-phenanthroline was chelated with an equimolar concentration of $ZnCl_2$ before being added to the cultures. Increasing the calcium concentration did not alter inhibition by 1,10-phenanthroline.

Although 1,10-phenanthroline inhibited myoblast fusion by metal chelation, many metalloenzymes and metal binding proteins would be affected. To examine whether a metalloprotease is involved in fusion, other types of metalloprotease inhibitors were used (Couch and Strittmatter, 1983). CBZ-L-phenylalanine, a competitive inhibitor of metalloendoproteases (Burstein *et al.,* 1974), prevented myoblast fusion with half-maximal inhibition of fusion at approximately 1.5 mM, while CBZ-L-tryptophan had little effect at this concentration. The inhibition by CBZ-L-phenylalanine was stereospecific, since CBZ-D-phenylalanine had no effect. Inhibitors of metalloexoproteases, hippuryl-L-arginine and hippuryl-L-phenylalanine (Chlebowski and Colemen, 1977), had no effect on myoblast fusion at 3 mM Captopril, an inhibitor of the metallodipeptidase angiotensin-1-converting enzyme (Ondetti *et al.,* 1977), had no effect on myoblast fusion even at 5 mM.

If a metalloendoprotease is required for myoblast fusion, substrates of known metalloendoproteases should compete with the natural protein substrate and thus prevent fusion. Metalloendoproteases will hydrolyze synthetic CBZ-dipeptideamides only with large aliphatic amino acids, such as leucine, or noncharged aromatic amino acids, such as phenylalanine, on the amino terminus of the hydrolyzed peptide bond. The interaction of dipeptide substrates with this protease class is also altered by the amino acid on the carboxy-terminus of the hydrolyzed peptide bond, with these proteases having a higher affinity for serine and tyrosine than for glycine. Therefore, CBZ-Ser-Leu-amide and CBZ-Tyr-Phe-amide have high affinities for metalloendoproteases, and CBZ-Gly-Leu-amide and CBZ-Gly-Phe-amide have intermediate affinities, whereas CBZ-Gly-Gly-amide is not a substrate. The affinity of a dipeptide substrate varies according to the particular metalloendoprotease and is in the millimolar range. Carbobenzoxydipeptide amide derivatives that interact with metalloendoproteases inhibited myoblast fusion (Couch and Strittmatter, 1983). Cultures incubated in 0.75 mM CBZ-Gly-Phe-amide and calcium demonstrated virtually no fusion after 48 hr, compared with cultures incubated with calcium alone. Inhibition of fusion by these dipeptides was completely reversible by the removal of the dipeptide from the medium. Cell density was not noticeably affected at this concentration of dipeptide. The ability of CBZ-dipeptide amides to inhibit fusion matched their specificity for the metalloendoproteases. In addition, the ability of these dipeptides to inhibit fusion required both the amide- and the carbobenzoxy-blocking groups, which are also required for recognition by the metalloendoproteases. CBZ-Gly-Phe-amide inhibited myoblast fusion, but neither CBZ-Gly-Phe nor Gly-Phe-amide had an effect on fusion. The inhibition by CBZ-Ser-Leu-amide was stereospecific.

The metalloendoprotease inhibitor $ClCH_2CO$-(N-OH)-Phe-Ala-Ala-NH_2 inhibits most metalloendoproteases irreversibly (Rasnick and Powers, 1978). The phenylalanine of the inhibitor occupies the active site of the protease, and the hydroxamic acid functional group complexes the endoprotease metal atom. The protease is then irreversibly inactivated by alkylation. This metalloprotease inhibitor prevented myoblast fusion. Several high-affinity inhibitors of thermolysinlike metalloendoproteases that did not block myoblast fusion were tested. These include 2-(*N*-hydroxycarboxamido)-3-phenylpropanoyl-Ala-Gly-NH_2 (Nishino and Powers, 1979), Homo-Phe-[*N*]-L-Phe-β-Ala (Mumford *et al.,* 1982), and phosphoramidon (Rhamnosyl-P-Leu-L-Trp-OH) (Komiyama *et al.,* 1975).

The fluorogenic substrate, 2-aminobenzoyl-Ala-Gly-Leu-Ala-4-nitrobenzylamide (AAGLAN), may be used to assay hydrolysis quantitatively by metalloendoproteases (Kam *et al.,* 1979). If the fluorescent protease substrate, AAGLAN, is hydrolyzed by the metalloendoprotease involved in myoblast fusion, it should compete with the natural protein substrate for the endoprotease and prevent fusion. This protease substrate inhibited myoblast fusion, with half-maximal inhibition at 500 μM, similar to the concentration of CBZ-Gly-Leu-amide that inhibited fusion. Incubation of AAGLAN with the bacterial metalloendoprotease, thermolysin, hydrolyzes the Gly-Leu peptide bond of this substrate to yield a 2-aminobenzoyl-Ala-Gly fragment and a Leu-Ala-4-nitrobenzylamide fragment. This prehydrolyzed substrate no longer inhibited myoblast fusion.

The metalloendoprotease inhibitors and substrates appear to block fusion by interacting with a protease, rather than by nonspecific effects, for several reasons. Myoblasts preincubated with 1,10-phenanthroline remained viable and unfused for 3 days. The inhibition by 1,10-phenanthroline was not due to decreased protein synthesis. Phenanthroline inhibition was due to metal chelation, as normal fusion occurred if 1,10-phenanthroline was chelated with an equimolar concentration of $ZnCl_2$. 1,10-Phenanthroline did not prevent fusion by chelating calcium, since 1,10-phenanthroline was removed before calcium was added and increasing the calcium did not alter inhibition by 1,10-phenanthroline. Also, 1,10-phenanthroline has a high critical stability constant for Zn^{2+} ($K = 2 \times 10^{-6}$ M) and a low critical stability constant for Ca^{2+} ($K = 5$ M) (Smith and Martell, 1975). 1,10-Phenanthroline blocks fusion by metal chelation, and this inhibition is not due to calcium chelation, decreased protein synthesis, decreased cell density, or decreased viability.

Inhibition of myoblast fusion by CBZ-L-phenylalanine is unlikely to be nonspecific, since it was stereospecific and many other CBZ amino acids and hippuryl-L-phenylalanine did not inhibit fusion. The ability of dipeptide derivatives to inhibit myoblast fusion matched their ability to bind to characterized neutral metalloendoproteases, with the potency series CBZ-Tyr-Leu-Amide ~ CBZ-Ser-Leu-amide > CBZ-Gly-Phe-amide ~ CBZ-Gly-Leu-amide >> CBZ-Gly-Gly-amide. The dipeptide substrate CBZ-Gly-Phe-amide inhibits fusion and interacts with the metalloendoproteases only if both the carbobenzoxy or amide groups are present. The close correlation between the potency of these dipeptides and their ability to interact with metalloendoproteases makes it likely that they are inhibiting fusion by interacting with the metalloendoprotease. Nonspecific effects are unlikely, since inhibition was reversible and stereospecific in the cases tested.

Several high-affinity inhibitors of thermolysinlike metalloendoproteases, including a hydroxamic acid derivative and phosphoramidon, did not block myoblast fusion. Several possibilities might explain why these compounds had no effect on fusion: (1) a metalloendoprotease is not really required for myoblast fusion, and the other inhibitors were blocking by some other mechanism; (2) these compounds do not inhibit the myoblast metalloendoprotease; and (3) these compounds simply cannot get into the cells and the metalloendoprotease is not exposed to the outside of the cells. The metalloendoprotease would therefore never be exposed to these inhibitors. The next section, characterizing the myoblast metalloendoproteases, further discusses these possibilities.

3.4. Characterization of Myoblast Metalloendoproteases

Metalloendoprotease activity was assayed using the fluorogenic protease substrate 2-aminobenzoyl-Ala-Gly-Leu-Ala-4-nitrobenzylamide

(Couch and Strittmatter, 1983, 1984). This compound inhibits myoblast fusion, indicating that it is recognized by the myoblast metalloendoprotease involved in myoblast fusion.

Metalloendoprotease activity was demonstrated, both in soluble and in particulate fractions in primary cultures of rat myoblasts. Crude fractionation of myoblasts demonstrates two populations of metalloendoprotease activity: soluble and particulate. These two metalloendoproteases seem to be truly separate enzymes based on the following information:

1. Changing membrane preparation by changes in ionic strength, pH, buffer types, and divalent cations has little effect on the proportion of enzyme activity found in the two populations. If the particulate protease was actually a peripheral membrane protein, or the soluble protease was a cleavage product of integral membrane protease, changes in the relative activities should be seen.
2. While many metalloendoprotease inhibitors block both soluble and particulate metalloendoprotease activity, some inhibitors block only the soluble (chymostatin, CBZ-Gly-Leu-amide, and $ClCH_2CO$-(N-OH)-Phe-Ala-Ala-NH_2) or only the particulate (phosphoramidon) metalloendoproteases.

Thus, there seem to be two separate metalloendoproteases in myoblasts, which differ in localization and specificity.

3.4.1. Biochemical Characterization

Two molecular sizing chromatographic methods, HPLC and G-200, were used to characterize the soluble metalloendoproteases in myoblasts (Couch and Strittmatter, 1983, 1984). With either of these methods, one peak of metalloendoprotease activity is observed with a molecular weight of about 70,000–80,000. Two other peaks of proteolytic activity that hydrolyze AAGLAN are not inhibited by the metal chelator 1,10-phenanthroline and thus are not metalloendoproteases.

Column chromatofocusing of the soluble metalloendoprotease activity revealed two peaks of metalloendoprotease activity, with pI values of 5.9 and 4.8. Both soluble metalloendoproteases were inhibited by 1,10-phenanthroline and the active site metalloendoprotease inhibitor 2-(*N*-hydroxycarboxy-amido)-3-phenylpropanoyl-Ala-Gly-NH_2. The soluble metalloendoprotease with a pI of 5.9, however, was not inhibited by CBZ-Gly-Leu-amide, while the metalloendoprotease with pI 4.98 was partially inhibited. The pI 5.9 peak is variable and very unstable. Thus, few experiments were done comparing these peaks of activity. Both seem to be metalloendoproteases (phenanthroline and 2-(*N*-hydroxycarboxamido)-3-phenylpropanoyl-Ala-Gly-NH_2 inhibitable), but they may have different specificities, since CBZ-Gly-Leu-amide inhibited the pI 4.8 activity and not the pI 5.9 activity. The pI 4.8 metalloendoprotease is probably the major activity in soluble protease assays due to the low levels and instability of the pI 5.9 activity.

Myoblasts therefore contain both soluble and particulate metalloendoproteases. The relationship of these proteases to known classes of metalloendoproteases is of interest. The particulate metalloendoprotease resembles the Class EC 3.4.24.11 proteases in that it is particulate (i.e., probably membrane-bound) and is inhibited by both phosphoramidon, a hydroxamic acid derivative, and Homo-Phe-[*N*]-L-Phe-β-Ala. However, whereas these inhibitors block EC 3.4.24.11 enzymes at 10^{-9}–10^{-7} M, they inhibit the myoblast particulate enzyme in the 10^{-5} M range. The myoblast-soluble metalloendoprotease resembles the soluble metalloendoprotease seen in red cell cytosol in its resistance to inhibition by phosphoramidon (Kirschner and Goldberg, 1983).

Various metalloendoprotease inhibitors and substrates block myoblast fusion. The soluble metalloendoprotease was blocked by 1,10-phenanthroline, CBZ-Gly-Leu-amide, and the irreversible inhibitor $ClCH_2CO$(N-OH)-Phe-Ala-Ala-NH_2, all of which also blocked myoblast fusion. However, the particulate metalloendoprotease, while blocked by 1,10-phenanthroline, was not blocked by CBZ-Gly-Leu-amide or the irreversible inhibitor $ClCH_2CO$(N-OH)-Phe-Ala-Ala-NH_2. Furthermore, the particulate protease was blocked by phosphoramidon, which does not block fusion. The soluble metalloendoprotease could therefore be involved in myoblast fusion, whereas the particulate protease is unlikely to be involved in fusion. Several problems exist with this conclusion. The first is that phosphoramidon might not inhibit fusion simply because it does not have access to the metalloendoprotease. The second is that the concentration at which a substrate is cleaved by a protease may not correlate with the concentration at which it inhibits competitively in the protease assay. The third is that inhibition of myoblast fusion by the irreversible inhibitor could be nonspecific. The fourth is that there are still several compounds (chymostatin, the hydroxamic acid derivative, and Homo-Phe-[*N*]-Phe-β-Ala) that clearly inhibit the soluble metalloendoprotease but do not block fusion. However, this observation is consistent with the arguments that these inhibitors do not have access to the protease, as they are unlikely to cross the cell membrane and that a soluble enzyme may not have access to the extracellular space. Specific inhibitors to one metalloendoprotease, but not the other, that have access to intracellular compartments would strengthen the argument for the involvement of a specific metalloprotease in myoblast fusion.

3.5. Conclusion

Several possible roles for metalloendoprotease in myoblast fusion are possible. One, removal of steric hindrance to membrane–membrane interactions seems unlikely, since the protease does not appear to act on the outside of the cell. Posttranslational processing of plasma membrane or secreted proteins could be required at a step between calcium entry and myoblast fusion. This model would be similar to virus fusion, where the HA and F fusion proteins must be processed proteolytically before they can mediate fusion (White *et al.*, 1983). Since fusion begins a short time after calcium addition, the proteolytic substrate would likely be a stored protein, not newly synthesized protein. The metalloendoprotease could cleave proteins that anchor plasma membrane proteins in place, thus allowing the proteins free lateral movement to form protein-free patches seen at the site of myoblast fusion (Kalderon and Gilla, 1979). Some evidence supports such a mechanism in fusogen-induced red cell fusion (Lucy, 1984). A third possibility is that the protease could act as part of a cascade that regulates myoblast fusion. Another possibility is that the metalloendoprotease actually mediates vesicle–plasma membrane fusion rather than plasma membrane–plasma membrane fusion directly. Vesicle–plasma membrane fusion has been suggested as a necessary preliminary step to plasma membrane–plasma membrane fusion (Kalderon, 1980).

4. ADRENAL CHROMAFFIN CELL EXOCYTOSIS

4.1. Receptor-Mediated Membrane Fusion in Exocytosis

The chromaffin cells of the adrenal medulla are part of the sympathetic nervous system. Their main function is the synthesis, storage, and release of catecholamines into the bloodstream in response to acute stress. The medulla receives cholinergic innervation from the sympathetic nervous system directly, via the splanchnic nerve. Stimulation of the nerve results in the release of acetylcholine (ACh), which then interacts with the chromaffin cell to stimulate the release of the catecholamines, norepinephrine and epinephrine. These are contained in granules that were first identified because of the discoloration of slices of adrenal medulla when immersed in chromate salts. Later it was shown that this staining was due to the deposit of chromium in the vesicles, hence the name chromaffin granule.

In addition to catecholamines, the granules contain and release adenine nucleotides [mostly adenosine triphosphate (ATP)], ascorbic acid,

Ca^{2+} (20–30 mM), Mg^{2+}, dopamine β-hydroxylase, and several soluble acidic proteins, the major one being chromogranin A. The granules of epinephrine synthesizing cells also contain Leu- and Met-enkephalin and their precursors (Livett *et al.,* 1982). Secretion from chromaffin cells occurs by exocytosis with granule membrane fusing with the plasma membrane to release the vesicle contents into the extracellular space (Viveros, 1975; Unger and Phillips, 1983). Evidence that the contents are released exocytotically comes from both electron mictroscopic examination, demonstrating granule fusion (Smith *et al.,* 1973; Burgoyne, 1984) and from the demonstration of equal release all granule constituents, both high and low molecular weight (Douglas, 1975).

Secretion in adrenal chromaffin cells is initiated by the binding of ACh to nicotinic cholinergic receptors, opening a receptor-linked ion channel that passes both Na^{+} and Ca^{2+}. The small depolarization triggers voltage-sensitive Na^{+} channels, generating an action potential, opening additional voltage-sensitive Ca^{2+} channels (Fenwick *et al.,* 1982; Baker and Knight, 1981; Kilpatrick *et al.,* 1981*a,b*).

Exocytosis in adrenal chromaffin cells is dependent on calcium (Douglas, 1975). Stimulus-dependent calcium influx has been demonstrated (Holz *et al.,* 1982; Kilpatrick *et al.,* 1982), leading to the conclusion that the intracellular signal for secretion is a rise in the ionized calcium concentration. The central role of calcium in exocytosis has been elegantly shown by the recent work of Knight and Baker (1982), who used a dissociated cell preparation made permeable to low-molecular-weight substances by brief exposure to a high-intensity electric fields. The intracellular calcium concentration in these cells can be precisely controlled by varying the calcium concentration in the media. Increasing the calcium concentration from 0.1 to 2 μM in the absence of cholinergic stimuli releases granule constituents, apparently by exocytosis. Secretion is half-maximal at about 1 μM calcium. Similar results were obtained using the detergent digitonin to permeabilize cells (Dunn and Holtz, 1983; Wilson and Kirshner, 1983). Furthermore, a rise in the free intracellular calcium concentration in response to either ACh or high potassium has been demonstrated using the fluorescent calcium indicator quin-2 (Knight and Kesteven, 1983). The role of calcium influx in initiating exocytosis is still poorly understood.

4.2. Methods Employed in These Studies

4.2.1. Cell Cultures

Bovine adrenal chromaffin cells are purified from whole adrenal glands (Waymire *et al.,* 1983). Cells are maintained either in suspension culture at 200,000 cells/ml or as monolayers in 96 or 24 multiwell Falcon tissue culture dishes.

4.2.2. Catecholamine Release Assay

Catecholamine secretion is measured either in suspension cultures by electrochemical detection or from monolayers preloaded with [^{3}H]-norepinephrine (NE). For the electrochemical assay eight to twelve million cells are placed in a 25-mm 0.45 μm Acrodisc filter assembly (Gelman Sciences) and continuously perfused at a rate 1 ml/min using a P3 model peristaltic pump (Pharmacia Fine Chemicals) with HEPES buffered saline. Carbachol or ACh is injected immediately above the chamber through a side port fitted with a diaphragm. Twenty μl of a 10-mM carbachol solution causes maximal release and is routinely used. The effluent from the chamber flows directly into an electrochemical detector consisting of a Bioanalytical Systems (BAS) LC-4A amperometric detector. To characterize further the catecholamines released from the cells, aliquots of the effluent are collected and injected onto a reverse phase HPLC column. By comparison with standards, the norepinephrine and epinephrine in the effluent are identified and quantitated.

For measuring release of [^{3}H]-NE, monolayer cultures are preloaded with [^{3}H]-NE. Exocytosis is initiated by replacing the preincubation buffer with 100 μl of the same buffer containing 3×10^{-4} M carbachol. After further incubation for 10 min at 25°C, a 50-μl aliquot is removed for liquid scintillation counting. [^{3}H]-NE release was expressed as a percentage of total cellular [^{3}H]-NE:

$$\frac{\text{dpm in supernatant} \times 2}{(\text{dpm in pellet} \times 2) + \text{dpm in supernatant}} = \%\ \text{release}$$

4.2.3. Metalloendoprotease Assays

Purified chromaffin cells are suspended in 10 mM HEPES/150 mM KCl pH 7.4 and freeze-thawed three times in a Dry Ice/acetone bath. The lysate is then homogenized in a Duall homogenizer and centrifuged 2 hr at 100,000 × *g* in an ultracentrifuge in a fixed-angle rotor. The supernatant is collected and the pellet resuspended in the same buffer to wash. The resuspended pellet is again centrifuged at 100,000 × *g* for 2 hr.

4.2.4. Digitonin Permeabilization

Dissociated chromaffin cells plated in 24 multiwell dishes are loaded with [^{3}H]-NE described previously. The permeabilization procedure used was described by Wilson and Kirshner (1983). A second permeabilization method was sometimes used (Dunn and Holz, 1983).

4.3. Effects of Metalloendoprotease Inhibitors on Exocytosis

The nicotinic agonist carbachol causes a time- and concentration-dependent release of catecholamines from dissociated adrenal chromaffin cells. Carbachol-induced catecholamine release in these cells was inhibited when the cells were perfused with buffer containing the metalloendoprotease substrate CBZ-Gly-Phe-NH_2 (Mundy and Strittmatter, 1985). Inhibition by this dipeptide was reversed by washing the suspended cells with buffer containing no dipeptide for 15 min and reinjecting carbachol. The dipeptide CBZ-Gly-Gly-NH_2, which is not a metalloendoprotease inhibitor, had no effect on carbachol-stimulated catecholamine release. In order to evaluate the effects of a larger number of dipeptides on exocytosis, the exocytotic release of catecholamines was also assayed by measuring the release of [^{3}H]-NE into the media by adrenal chromaffin cells preloaded with [^{3}H]-NE. Numerous investigators have shown that this procedure accurately reflects endogenous catecholamine secretion (Kilpatrick *et al.,* 1980). Release is maximal at 10–12 min. Dipeptides that are metalloendoprotease substrates inhibited carbachol-induced [^{3}H]-NE release in a dose-dependent fashion (Mundy and Strittmatter, 1985). The metalloendoprotease inhibitor 1,10-phenanthroline (Barrett, 1977) also prevented carbachol-induced exocytosis with half-maximal inhibition at approximately 20 μg/ml. Addition of excess calcium, equimolar to added 1,10-phenanthroline, did not reverse this inhibition, suggesting that 1,10-phenanthroline inhibited exocytosis by chelating a metal other than calcium. The fluorogenic metalloendoprotease substrate 2-aminobenzoyl-Ala-Gly-Leu-Ala-4-nitrobenzylamide (Kam *et al.,* 1979), used to characterize the adrenal metalloendoprotease, also inhibited receptor-mediated exocytosis in a dose-dependent fashion. These observations demonstrate that exocytosis in adrenal chromaffin cells is blocked by compounds that interact with metalloendoproteases.

4.3.1. Exocytosis Initiated by High Concentrations of Potassium Is Also Blocked by Metalloendoprotease Inhibitors

Catecholamine secretion in adrenal chromaffin cells can be initiated by depolarization of the plasma membrane using high concentrations of K^+, which opens voltage-dependent calcium channels (Kilpatrick *et al.,* 1981*a,* 1982; Cena *et al.,* 1983; Holz *et al.,* 1982). Release is maximal at 55 mM K^+. Both 1,10-phenanthroline and the dipeptide substrates, CBZ-Gly-Leu-NH_2 and CBZ-Gly-Phe-NH_2 blocked catecholamine secretion induced by 55 mM K^+, while CBZ-Gly-Gly-NH_2 did not (Mundy and Strittmatter, 1985). This indicates that these compounds are not acting by interference at the ACh receptor itself but rather act at some point distal to the receptor-mediated events. The dipeptides block K^+-stimulated release with approximately the same dose dependence with which they block ACh-induced release, while higher concentrations of 1,10-phenanthroline are required to inhibit exocytosis in the depolarized cells.

4.4. Characterization of Adrenal Chromaffin Metalloendoprotease Activity

Metalloendoprotease activity in adrenal chromaffin cells was characterized using the fluorogenic metalloendoprotease substrates 2-aminobenzoyl-Ala-Gly-Leu-Ala-4-nitrobenzylamide (AAGLAN) and succinyl-Ala-Ala-Phe-4-aminomethylcoumarin (Mundy and Strittmatter, 1985). Metalloendoprotease activity is defined as that portion of the fluorescent signal that is inhibited by 1,10-phenanthroline (200 μg/ml), which was typically 85% of total activity.

Homogenates of adrenal chromaffin cells contain metalloendoprotease activity that hydrolyzed the fluorogenic metalloendroprotease substrate AAGLAN. Protease activity in the particulate fraction averaged 42.5 nmoles AAGLAN hydrolyzed per hr/mg protein. Protease activity in the soluble fraction was variable, ranging from 0 to 15 nmoles AAGLAN hydrolyzed per hr/mg protein. Metalloendoprotease activity identified in the soluble fraction is inhibited by very low concentrations of phenanthroline (100% inhibition at 10 μg/ml). The pellet enzyme is inhibited by very low concentrations of phosphoramidon; the soluble enzyme is not. These findings indicate the soluble enzyme is distinct from that in the particulate fraction, with similar inhibitor specificities to the soluble and membrane metalloendoprotease in myoblasts.

Hydrolysis of AAGLAN by the particulate-associated protease is linear for at least 20 hr. This protease activity was inhibited by several metalloendoprotease inhibitors (Barret, 1977; Komiyama *et al.,* 1975; Nishino and Powers, 1978): 1,10-phenanthroline (IC_{50} of 20 μg/ml), phosphoramidon (IC_{50} 2×10^{-7} M). The pH optimum for hydrolysis of AAGLAN by the adrenal metalloendoprotease was 6.6.

The fluorogenic metalloendoprotease substrate succinyl-Ala-Ala-Phe-AMC was also hydrolyzed by the particulate fraction of homogenized adrenal chromaffin cells. Hydrolysis of this substrate was completely blocked by the specific metalloendoprotease inhibitor phosphoramidon. Succinyl-Ala-Ala-Ala-AMC, which is not a metalloendoprotease substrate (Mumford *et al.,* 1980), was hydrolyzed minimally. When added to dissociated adrenal chromaffin cells, succinyl-Ala-Ala-Phe-AMC blocked carbachol induced exocytosis while succinyl-Ala-Ala-AMC had little effect on exocytosis.

To determine the localization of the particulate metalloendoprotease, subcellular fractions of adrenal chromaffin cells prepared by homogenization and differential gradient centrifugation. Each fraction is assayed for marker enzymes located in various subcellular fractions: acetylcholinesterase (AChE) for plasma membrane, arylsulfatase for lysosomes, catecholamines for chromaffin granules, succinate dehydrogenase and monoamine oxidase for mitochondria, and NADPH reductase for microsomes. Metalloendoprotease activity was highest in the fractions above the Percoll gradient 20%/40% interface, which was also enriched in the plasma membrane marker enzyme AChE. Membranes containing AChE were further enriched by discontinuous sucrose-gradient centrifugation and assayed for metalloendoprotease activity. Membranes enriched 20-fold in the plasma membrane marker AChE (compared with the starting material) were similarly enriched in metalloendoprotease activity, suggesting that the metalloendoprotease is located primarily in the plasma membrane. Chromaffin granule fractions (enriched with the chromaffin granule catecholamine marker) contained little metalloendoprotease activity.

4.5. Exocytosis Initiated by Direct Introduction of Calcium Is Blocked by Metalloendoprotease Inhibitors

Incubating adrenal chromaffin cells with the detergent digitonin permits direct entry of calcium into the cell from the media, resulting in the release of catecholamines, as shown by Wilson and Kirschner (1983) and Dunn and Holz (1983). This technique permits the initiation of secretion by simply adding calcium to the media, thereby allowing the characterization of those steps required in exocytosis after calcium entry. When digitonin-permeabilized adrenal chromaffin cells are incubated in increasing concentrations of calcium, a biphasic release of catecholamines occurs. A small amount of calcium-dependent secretion occurs at free calcium concentrations less than 100 μM, and a second much larger release occurs at millimolar concentrations of calcium. Between 25 and 40% of the total catecholamines can be released in this manner, and release is complete after 20 min. The metalloendoprotease inhibitor 1,10-phenanthroline blocks the catecholamine release invoked by millimolar concentrations of calcium. Addition of excess calcium does not reverse this inhibition, suggesting that 1,10-phenanthroline does not prevent exocytosis by chelating calcium. Dipeptide substrates of metalloendoproteases similarly blocked exocytosis by the direct introduction of millimolar calcium, while the dipeptide CBZ-Gly-Gly-NH_2, which is not a substrate, did not (Mundy and Strittmatter, 1985). These observations suggest that metalloendoprotease activity is required in exocytosis at a step after calcium entry. However, similar experiments performed with micromolar concentrations of calcium showed that neither 1,10-phenanthroline nor the dipeptide substrates blocked catecholamine secretion. We do not yet have an explanation for these differences.

4.6. Conclusions

Receptor-mediated exocytosis of catecholamines by adrenal chromaffin cells is inhibited by both substrates and inhibitors of metalloendoproteases. 1,10-Phenanthroline and synthetic dipeptides block adrenal chromaffin cell exocytosis at similar concentrations and with the same amino acid specificity as they inhibit metalloendoproteases (Morihara, 1974). In addition, two synthetic substrates for metalloendoproteases, 2-aminobenzoyl-Ala-Gly-Leu-Ala-4-nitrobenzylamide and succinyl-Ala-Ala-Phe-AMC, are hydrolyzed by the adrenal metalloendoprotease, nor does it inhibit receptor-mediated exocytosis. Succinyl-Ala-Ala-Ala-AMC is not hydrolyzed by the chromaffin cell metalloendoprotease nor does it inhibit catecholamine release. These compounds therefore appear to inhibit receptor-mediated exocytosis by interaction with a metalloendoprotease, and not by a nonspecific hydrophobic interaction, supporting the hypothesis that metalloendoprotease activity is required in exocytosis.

Metalloendoprotease activity in adrenal chromaffin cells was assayed and characterized using two sensitive fluorogenic metalloendoprotease substrates. Adrenal cell membranes contain a protease that hydrolyzes these two substrates. This adrenal protease can be further characterized as metalloendoprotease, since it is inhibited by the specific metalloendoprotease inhibitors—1,10-phenanthroline, phosphoramidon, and the active site hydroxamic acid inhibitor 2-(*N*-hydroxycarboamido)-3-phenylpropanyl-Ala-Gly-NH_2. The concentrations of these compounds required to inhibit the adrenal metalloendoprotease are similar to those required to inhibit other metalloendoproteases. The concentration of 1,10-phenanthroline required to inhibit the adrenal metalloendoprotease is also the same as that required to inhibit receptor-mediated exocytosis in the chromaffin cell. Although 1,10-phenanthroline (Morgan, 1983) and CBZ-dipeptide amides (Grinde *et al.*, 1983; Firestone *et al.*, 1982) enter cells and inhibit receptor-mediated exocytosis, phosphoramidon and the hydroxamic inhibitor are polar molecules that do not enter cells and have no effect on adrenal chromaffin cell exocytosis.

Both 1,10-phenanthroline and the CBZ-dipeptide amides were found to block depolarization-mediated catecholamine secretion, suggesting that these inhibitors are not acting at the ACh receptor. The dipeptides block K^+-stimulated release with approximately the same dose dependency as they block carbachol-stimulated release. The inactive membrane metalloendoprotease apoenzyme can be activated by increasing concentrations of zinc. There is a very narrow concentration range for reactivation and at 10^{-9} M free zinc, activity is restored to 80% of control. By 10^{-8} M free zinc, the enzyme is completely inhibited.

We have found that the metalloendoprotease substrates and inhibitors prevent the calcium-dependent secretion of catecholamines from digitonin permeabilized cells at millimolar calcium concentrations (protocol of Wilson and Kirschner, 1983), but not at micromolar calcium concentrations (protocol of Dunn and Holz, 1983). These two procedures appear to cause catecholamine release by different mechanisms because they also differ in (1) dependency on exogenous Mg ATP, (2) inhibition by Mg^{2+}, (3) requirement for EGTA, and (4) rate of loss of cytosolic components as measured by lactate dehydrogenase. Until the differences between these two protocols are understood, it is difficult to determine which procedure more closely mimics the naturally occurring process of calcium-dependent exocytosis.

5. SUMMARY

These observations support the hypothesis that metalloendoprotease activity is required in myoblast fusion and in adrenal chromaffin exocytosis. The protease appears to be required at some step after calcium entry in these cells. Metalloendoproteases can be identified and characterized in these cells using specific and sensitive fluorogenic substrates. Other studies we have conducted have implicated metalloendoprotease in exocytosis from mast cells (Mundy and Strittmatter, 1985), neurons (Baxter *et al.,* 1983; Frederick *et al.,* 1984), and sea urchin fertilization (W. J. Strittmatter, unpublished).

REFERENCES

Adamo, S., Zani, B., Siracusa, G., and Molinaro, M., 1976, Expression of differentiative traits in the absence of cell fusion during myogenesis in culture, *Cell Diff.* **5**:53–67.

Almenoff, J., and Orlowski, M., 1984, Biochemical and immunological properties of a membrane-bound brain metalloendopeptidase: Comparison with thermolysin-like kidney neutral metalloendopeptidase, *J. Neurochem.* **42**:151–157.

Baker, P. F., and Knight, D. E., 1981, Calcium control of exocytosis and endocytosis in bovine adrenal medullary cells, *Philos. Trans. R. Soc. B* **96**:83–103.

Barrett, A. J., 1977, Introduction to proteinases, in: *Proteinases in Mammalian Cells and Tissues* (A. J. Barret, ed.), pp. 1–55, Elsevier, Amsterdam.

Baxter, D. A., Johnston, D., and Strittmatter, W. J., 1983, Protease inhibitors implicate metalloendoprotease in synaptic transmission at the mammalian neuro-muscular junction, *Proc. Natl. Acad. Sci. U.S.A.* **80:**4174–4178.

Bischoff, R., 1978, Myoblast fusion, in: *Membrane Fusion* (G. Poste and G. L. Nicolsen, eds.), pp. 127–178, Elsevier, Amsterdam.

Bischoff, R., and Holtzer, H., 1969, Mitosis and the processes of differentiation of myogenic cells *in vitro, J. Cell Biol.* **41:**188–200.

Bischoff, R., and Lowe, M., 1974, Cell surface components and the interaction of myogenic cells, in: *Exploratory Concepts in Muscular Dystrophy,* Vol. II (A. T. Milhorat, ed.), pp. 17–29, Excerpta Medica, Amsterdam.

Burgoyne, R. D., 1984, Mechanisms of secretion from adrenal chromaffin cells, *Biochem. Biophys. Acta* **779:**201–216.

Burstein, Y., Walsh, K. A., and Neurath, H., 1974, Evidence of an essential histidine residue in thermolysin, *Biochemistry* **13:**205–210.

Cawston, T. E., and Murphy, G., 1981, Mammalian collagenases, *Method. Enzymol.* **80:**711–722.

Cena, V., Nicolas, G. P., Sanchez-Garcia, P., Kirpekar, S. M., and Garcia, A. G., 1983, Pharmacological dissection of receptor associated and voltage-sensitive ionic channels involved in catecholamine release, *Neuroscience* **10:**1455–1452.

Chlebowski, J., and Coleman, J. E., 1976, Zinc and its role in enzymes, in: *Metal Ions in Biological Sciences,* Vol. VI (H. Siegel, ed.), pp. 1–7, Dekker, New York.

Couch, C. B., and Strittmatter, W. J., 1983, Rat myoblast fusion requires metalloendoprotease activity, *Cell* **32:**257–265.

Couch, C. B., and Strittmatter, W. J., 1984, Specific blockers of myoblast fusion inhibit a soluble and not the membrane-associated metalloendoprotease in myoblasts, *J. Biol. Chem.* **259:**5396–5399.

David, J. D., See, W. M., and Higginbotham, C. A., 1981, Fusion of chick embryo skeletal myoblasts: Role of calcium influx preceding membrane union, *Dev. Biol.* **82:**292–307.

Douglas, W. W., 1975, Secretomotor control of adrenal medullary secretion: Synaptic membrane, and ionic events in stimulus–secretion coupling, in: *Handbook of Physiology,* Section 7, *Endocrinology,* Vol. VI (P. O. Greep and E. B. Astwood, eds.), pp. 367–388, American Physiology Society, Washington, D. C.

Dunn, L. A., and Holz, R. W., 1983, Catecholamine secretion from digitonin treated adrenal medullary chromaffin cells, *J. Biol. Chem.* **258:**4989–4993.

Fenwick, E. M., Marty, A., and Neher, E. 1982, A patch clamp study of bovine chromaffin cells and of their sensitivity to acetylcholine, *J. Physiol. (Lond.)* **331:**577–635.

Fibiato, A., and Fabiato, F., 1979, Calculator programs for computing the composition of the solutions containing multiple metals and ligands used for experiments in skinned muscle cells, *J. Physiol. (Lond.)* **75:**463–505.

Firestone, R. A., Pisano, J. M., Bailey, P. J., Sturm, A., Bonney, R. J., Wightman, P., Devlin, R., Lin, C. S., Keller, D. C., and Tway, P., 1982, Lysosomotropic Agents. 4. Carbobenzoxyglycylphenylalanyl, a new protease sensitive masking group for introduction into cells, *J. Med. Chem.* **25:**539–544.

Frederick, J. M., Hollyfield, J. G., and Strittmatter, W. J., 1984, Inhibitors of metalloendoprotease activity prevent K^+-stimulated neurotransmitter release from the retina of Xenopus laevis, *J. Neurosci.* **4:**3112–3119.

Fulcher, I. S., Matsas, R., Turner, A. J., and Kenny, A. J., 1982, Kidney neutral endopeptidase and the hydrolysis of enkephalin by synaptic membranes show similar sensitivity to inhibitors, *Biochem. J.* **203:**519–522.

Grinde, B., Galpin, I. J., Wilby, A. H., and Beynon, R. J., 1983, Inhibition of hepatic protein degradation by synthetic analogues of chymostatin, *J. Biol. Chem.* **258:**10821–10823.

Holz, R. W., Senter, R. A , and Frye, R. A., 1982, Relationship between Ca^{++} uptake and catecholamine secretion in primary dissociated cultures of adrenal medulla, *J. Neurochem.* **39:**635–646.

Kalderon, N., 1980, Muscle cell fusion, in: *Membrane–Membrane Interactions* (N. B. Gilula, ed.), pp. 99–118, Raven Press, New York.

Kalderon, N., and Gilula, N. B., 1979, Membrane events involved in myoblast fusion, *J. Cell Biol.* **81:**411–425.

Kam, C. -M., Nishino, N., and Powers, J. C., 1979, Inhibition of thermolysin and carboxypeptidase A by phosphoramidates, *Biochemistry* **18:**3032–3038.

Kenny, A. J., and Fulcher, I. S., 1983, Microvillar endopeptidase, an enzyme with special topological features and wide distribution, *CIBA Found. Symp.* **95:**12–33.

Kerr, M. A., and Kenny, A. J., 1974, The purification and specificity of a neutral endopeptidase from rabbit kidney brush border, *Biochem. J.* **137:**477–488.

Kilpatrick, D. L., Ledbetter, F. H., Carson, K. A., Kirshner, A. G., Slepetis, R., and Kirshner, N., 1980, Stability of bovine adrenal medulla cells in culture, *J. Neurochem.* **35:**679–692.

Kilpatrick, D. L., Slepetis, R., and Kirshner, N., 1981*a*, Ion channels and membrane potential in stimulus secretion coupling in adrenal medulla cells, *J. Neurochem.* **36:** 1245–1255.

Kilpatrick, D. L., Slepetis, R., and Kirshner, N., 1981*b*, Inhibition of catecholamine secretion from adrenal medulla cells by neurotoxins and cholinergic antagonists, *J. Neurochem.* **37:**125–131.

Kilpatrick, D. L., Slepetis, R. J., Corcoran, J. J., and Kirshner, N., 1982, Calcium uptake and catecholamine secretion by cultured bovine adrenal medulla cells, *J. Neurochem.* **38:**427–435.

Kirschner, R. J., and Goldberg, A. L., 1983, A high molecular weight metalloendoprotease from the cytosol of mammalian cells, *J. Biol. Chem.* **258:**967–976.

Knight, D. E., and Baker, P. F., 1982, Calcium dependance of catecholamine release from bovine adrenal medullary cells after exposure to intense electric fields, *J. Membrane Biol.* **68:**107–140.

Knight, D. E., and Kesteven, N. T., 1983, Evoked transient intracellular free Ca^{2+} changes and secretion in isolated bovine adrenal medullary cells, *Proc. R. Soc. Lond. B* **218:**177–199.

Komiyama, T., Suda, H., Aoyagi, T., Takeuchi, T., and Umezawa, H., 1975, Studies on inhibitory effect of phosphoramidon and its analogs in thermolysin, *Arch. Biochem. Biophys.* **171:**727–731.

Livett, B. G., Day, R., Elde, R. P., and Howe, P. R. C., 1982, Costorage of enkephalins and adrenaline in the bovine adrenal medulla, *Neuroscience* **7:**1323–1332.

Lucy, J. A., 1984, Do hydrophobic sequences cleaved from cellular polypeptides induce membrane fusion reactions *in vivo?, FEBS Lett.* **166:**223–231.

Matsas, R., Fulcher, I. S., Kenny, A. J., and Turner, A. J., 1983, Substance P and Leu-enkephalin are hydrolyzed by an enzyme in pig caudate synaptic membranes that is identical with the endopeptidase of kidney microvilli, *Proc. Natl. Acad. Sci. U.S.A.* **80:**3111–3115.

Morgan, E. H., 1983, Chelator-mediated iron efflux from reticulocytes, *Biochem. Biophys. Acta* **733:**39–50.

Morihara, K., 1974, Comparative specificity of microbial proteinases, in: *Advances in Enzymology* (A. Meister, ed.), pp. 179–243, Wiley, New York.

Mumford, R. A., Strauss, A. W., Powers, J. C., Pierzchala, P. A., Nishino, N., and Zimmerman, M., 1980, A zinc metalloendopeptidase associated with dog pancreatic membranes, *J. Biol. Chem.* **225**:2227–2230.

Mumford, R. A., Zimmerman, M., Broeke, J., Taub, D., Joshua, H., Rothrock, J. W., Hershfield, J. M., Springer, J. P., and Patchett, A. A., 1982, Inhibition of porcine kidney "enkephalinase" by substituted N-carboxymethyl dipeptides, *Biochem. Biophys. Res. Commun.* **109**:1303–1309.

Mundy, D. I., and Strittmatter, W. J., 1985, Requirement for metalloendoprotease in exocytosis: Evidence in mast cells and adrenal chromaffin cells, *Cell* **41**:645–656.

Nishino, N., and Powers, J. C., 1978, Peptide hydroxamic acids as inhibitors of thermolysin, *Biochemistry* **17**:2846–2850.

Nishino, N., and Powers, J. C., 1979, Design of potent reversible inhibitors for thermolysin. Peptides containing zinc coordinating ligands and their use in affinity chromatography, *Biochemistry* **18**:4340–4347.

Ondetti, M. A., Rubin, B., and Cushman, D. W., 1977, Design of specific inhibitors of angiotensin-converting enzyme: New class of orally active antihypertensive agents, *Science* **196**:441–444.

Orlowski, M., Michaud, C., and Chu, T. G., 1983, A soluble metalloendopeptidase from rat brain, *Eur. J. Biochem.* **135**:81–88.

Orlowski, M., and Wilk, S., 1981, Purification and specificity of a membrane-bound metalloendopeptidase from bovine pituitaries, *Biochemistry* **20**:4942–4950.

Przybylski, R. J., Kirby, A. C., and MacBride, R. G., 1982, Calcium content and uptake during chick skeletal muscle development in culture, *J. Cell. Biol.* **95**:373a.

Rash, J. E., and Fambrough, D., 1973, Ultrastructural and electrophysiological correlates of cell coupling and cytoplasmic fusion during myogenesis *in vitro, Dev. Biol.* **30**:166–186.

Rasnick, D., and Powers, J. C., 1978, Active site directed irreversible inhibition of thermolysin, *Biochemistry* **17**:4363–4369.

Shainberg, A., Yagil, G., and Yaffe, D., 1969, Control of myogenesis *in vitro* by Ca^{2+} concentration in nutritional medium, *Exp. Cell Res.* **58**:163–167.

Smith, V., Smith, D. S., Winkler, H., and Ryan, J. W., 1973, Exocytosis in the adrenal medulla demonstrated by freeze etching, *Science* **179**:79–82.

Strittmatter, W. J., Couch, C. B., and Mundy, D. I., 1985, Role of proteins in the fusion of biological membranes, in: *Membrane Fluidity in Biology,* Vol. IV (R. C. Aloia and J. Boggs, eds.), pp. 259–291, Academic Press, New York.

Turner, D. C., Gmur, R., Sieguist, M., Burchkhardt, E., and Eppenberger, H. M., 1976, Differentiation in cultures derived from embryonic chicken muscle, *Dev. Biol.* **48**:258–283.

Unger, A., and Phillips, J. H., 1983, Regulation of the adrenal medulla, *Physiol. Rev.* **63**:787–843.

Vallee, B. L., 1983, Zinc in biology and biochemistry, in: *Zinc Enzymes* (T. G. Spiro, ed.), pp. 3–24, Wiley, New York.

Viveros, O. H., 1975, Mechanism of secretion of catecholamines from adrenal medulla, in: *Handbook of Physiology,* Section 7: *Endocrinology,* Vol. VI (R. P. Greep and E. B. Astwood, eds.), pp. 389–426, American Physiology Society, Washington, D. C.

Waymire, J. C., Bennett, W. F., Boehme, R., Hankins, L., Gilmer-Waymire, K., and Haycock, J. W., 1983, Bovine adrenal chromaffin cells: High yield production and viability in suspension culture, *J. Neurosci. Meth.* **7**:329–351.

White, J., Kielian, M., and Helenius, A., 1983, Membrane fusion proteins of enveloped animal viruses, *Quart. Rev. Biophys.* **16**:151–195.

Wilson, S. P., and Kirshner, N., 1983, Calcium evoked secretion from digitonin permeabilized adrenal medullary chromaffin cells, *J. Biol. Chem.* **258**:4994–5000.

Yaffe, D., 1971, Developmental changes preceding cell fusion during muscle differentiation *in vitro, Exp. Cell. Res.* **66:**33–48.

Yaffe, D., 1973, Rat skeletal muscle cells, in: *Tissue Culture Methods and Applications* (P. F. Kruse, Jr., and M. K. Patterson, Jr., eds.), pp. 106–114, Academic Press, New York.

Chapter 6

Control of Membrane Fusion in Polyethylene Glycol-Resistant Cell Mutants

Applications to Fusion Technology

David S. Roos, Richard L. Davidson, and Purnell W. Choppin

1. INTRODUCTION AND OVERVIEW

Cell fusion has been studied for more than a century (Langerhans, 1868) as a curious morphological phenomenon, occurring in certain developing tissues (Kalderon, 1980) and during the progression of a variety of tumors and viral infections (Roizman, 1962; Guccion and Enzinger, 1972). Only with the use of cell hybrids in somatic cell genetics, however, did the impetus arise to develop fusion technology as an applied art, placing a premium on the control of cell fusion (Davidson, 1977). The development of procedures for hybridoma production has made cell fusion a fundamental tool of the growing biotechnology industry (Köhler and Milstein, 1975), and it is not surprising that improvements in the efficiency of cell hy-

DAVID S. ROOS • Department of Biology, Stanford University, Stanford, California 94305. RICHARD L. DAVIDSON • Center for Genetics, University of Illinois, Chicago, Illinois 60612. PURNELL W. CHOPPIN • Virology Laboratory, Rockefeller University, New York, New York 10021. *Present address:* Howard Hughes Medical Institute, Bethesda, Maryland 20814.

bridization protocols have been a major goal in recent years (de St. Groth and Scheidegger, 1980).

Despite widespread interest in the applications of cell fusion, the precise biochemical mechanisms involved in the fusion of biological membranes remain largely unknown. As an approach to understanding the genetic and biochemical bases of membrane fusion, we have isolated and characterized several series of mutant cell lines which are resistant to the fusion-inducing effects of polyethylene glycol (PEG), a widely used chemical fusogen (Roos and Davidson, 1980; Roos *et al.,* 1983). Biochemical analysis, coupled with functional reconstitution experiments, has demonstrated that cellular lipids play a crucial role in the regulation of fusion in this system (Roos and Choppin, 1985*a,b*). Surprisingly, PEG-resistant cells are also altered with respect to a variety of other fusogenic treatments, including spontaneous cell hybridization and fusion induced by enveloped viruses (Roos, 1984). The manipulation of cellular lipid composition also allows these diverse phenomena to be controlled in a predictable manner, suggesting that common factors may be involved in the generation of hybrid cells by any approach.

This chapter discusses several techniques and observations arising from our studies on fusion-altered cells that have potential application in the further development of fusion technology. The ability to fine-tune the hybridization of cells with specific properties, achieving, in theory, any degree of fusion desired, promises to permit the design of experiments in which the cells that fuse and the extent of fusion are precisely determined.

2. EXPERIMENTAL OBSERVATIONS

Polyethylene glycol is a potent fusogenic compound that, because of its efficiency, reproducibility and ease of preparation has found general acceptance in recent years (Davidson *et al.,* 1976). A wide range of cell types have been fused with polyethylene glycol, but different cells are differentially susceptible to the fusogenic effects of PEG (Pontecorvo *et al.,* 1977; Blow *et al.,* 1977; Wang *et al.,* 1982). Characteristically, for any given cell line efficient cell hybridization is achieved only across a relatively narrow range of PEG concentrations, with little fusion occurring below a threshhold level, and severe toxicity in more concentrated solutions (Davidson *et al.,* 1976). The concentration dependence of PEG-induced fusion in homogeneous cultures of a mouse L-cell line (designated Clone 1D) is illustrated in Figure 1. Virtually all cells remain unfused after treatment for 1 min with 30% PEG, some fusion is observed in 40% PEG-treated cultures, and fusion is extensive in the cells treated with 50% PEG.

Much higher concentrations of PEG cause little cell fusion due to pronounced toxicity of the PEG (Roos and Davidson, 1980).

2.1. Polyethylene Glycol-Induced Fusion and the Selection of PEG-Resistant Mutants

While extensive fusion results from treatment of Clone 1D cells with 50% PEG (cf. Fig. 1), a few cells remain unfused and viable. This observation provided the original suggestion that it might be possible to select for fusion-resistant mutants by isolation and replication of these cells. We have previously reported the derivation of a series of increasingly PEG-resistant cell lines from the highly fusible murine fibroblast cell line Clone 1D (Roos and Davidson, 1980). Briefly, the selection procedure involved treating cells with fusogenic concentrations of PEG, producing many large multinucleates that are incapable of further cell division. Following harvest and replating of these cultures, the remaining giant cells were quickly overgrown by the progeny of those cells that had

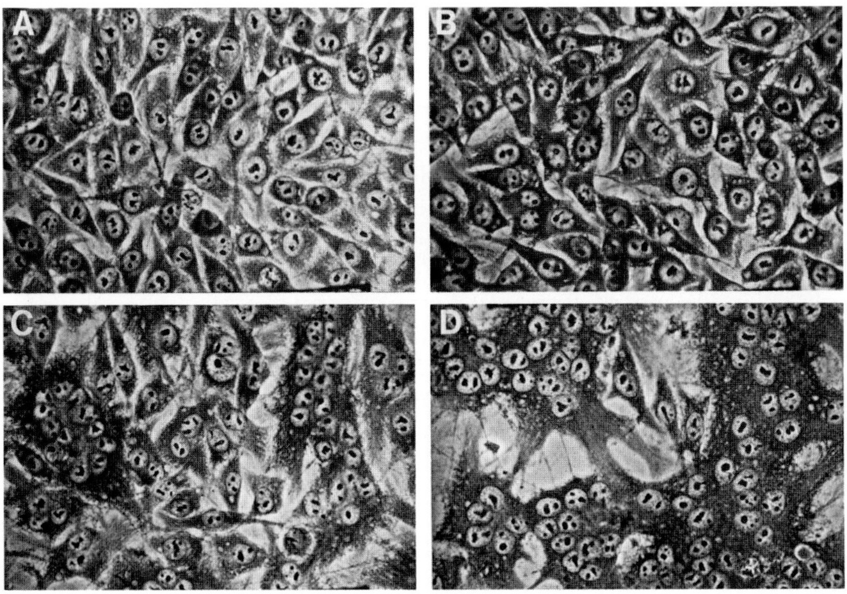

Figure 1. Concentration dependence on PEG-induced cell fusion. Confluent cultures of Clone 1D cells (an LMTK$^-$ murine fibroblast cell line) were treated for 60 sec with various concentrations of PEG-1000 (J. T. Baker Chemical Co., Phillipsburg, NJ) and fixed and stained 90 min after treatment. For further details of fusion protocols, see Robinson *et al.* (1979). (A) Control. (B) 30% PEG. (C) 40% PEG. (D) 50% PEG. Note the sharp threshold for PEG-induced fusion: little fusion is observed below 40% PEG, while treatment at higher concentrations produces massive syncytia.

Table I. PEG-Fusion Resistant Mutants Derived from Several Cell Lines

Cell line	Origin	PEG-fusion yield (%)[a]
Clone 1D	Mouse connective tissue fibroblast	92 ± 3
F_{40}	40-step PEG-resistant mutant of above	19 ± 5
BHK	Hamster kidney fibroblast	78 ± 8[b]
B_{15}	15-step PEG-resistant mutant of above	67 ± 7[b]
CHO	Chinese hamster ovary fibroblast	53 ± 10
C_{11}	11-step PEG-resistant mutant of above	30 ± 8
YAC-1	Leukemic mouse lymphocyte	15 ± 4[c]
Y_8	8-step PEG-resistant mutant of above	7 ± 4[c]

[a]Mean ± standard deviation. Cells were treated with 50% PEG for 60 sec and incubated 2 hr before fixation. Fusion yield was scored as the percentage of all nuclei present which were found in fused cells of any size. For further details of fusion and quantification procedures, see Roos and Choppin (1985*b*).
[b]Cells treated with 55% PEG.
[c]Suspension cells, fused according to the "pancake fusion" technique (O'Malley and Davidson, 1977).

remained unfused in the original culture. These cells were again subjected to PEG treatment and the entire process repeated several times. After as many as 40 cycles of PEG treatment and outgrowth, it has been possible to isolate highly PEG-resistant mutants (Roos *et al.*, 1983).

We have recently applied similar procedures to a number of other cells lines of diverse origin and initial susceptibility to PEG-induced fusion. The fusion yield of some of these isolates is shown in Table 1. With the exception of BHK cells, which are already highly resistant to PEG-induced fusion (Wang *et al.*, 1982), PEG-resistant mutants have been isolated from all cell lines tested. Of particular interest are preliminary observations that fusion-resistant mutants can also be generated from human and murine lymphoid cell lines (e.g., YAC-1). As previously demonstrated for the Clone 1D derivatives (Roos and Davidson, 1980), all these mutants appear to be stable and homogeneous with respect to PEG-fusion, based on the analysis of individual subclones.

2.2. Fusion Response of PEG-Resistant Cell Lines

Although it has been possible to generate cells that are highly resistant to the fusogenic effects of PEG, these cells are not completely refractile to fusion. In fact, PEG-resistant cells can be fused, but only at higher concentrations of PEG, as shown in Figure 2. F_{24} cells, a PEG-resistant line selected from Clone 1D cells by 24 cycles of fusion, is highly resistant to fusion with 50% PEG but still sensitive to fusion when treated with 55%

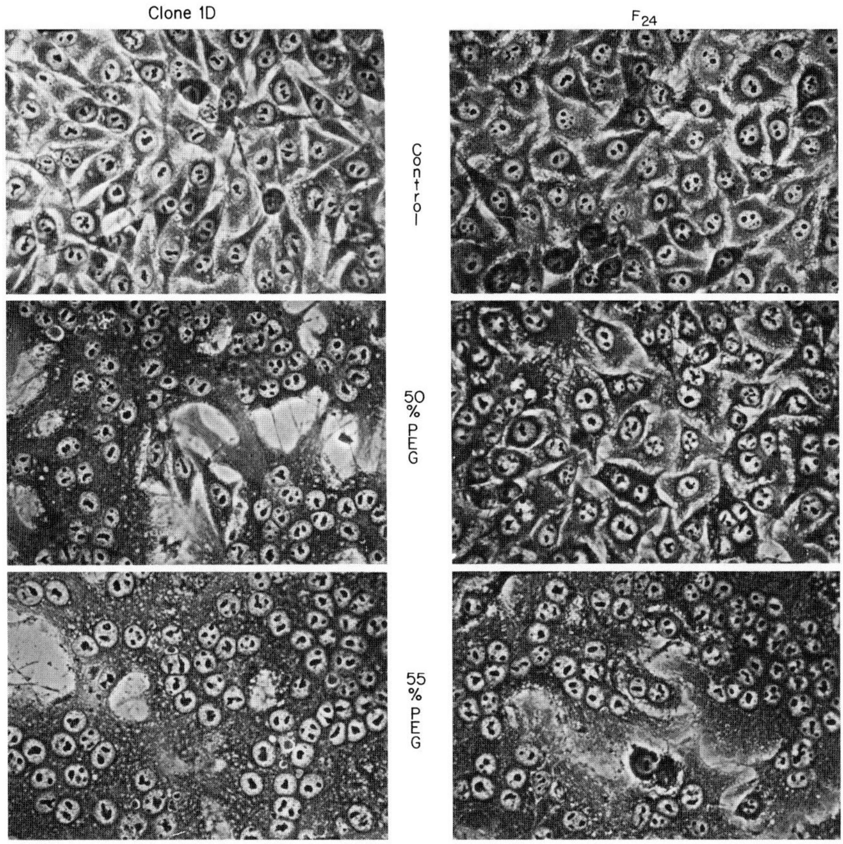

Figure 2. Fusion of resistant cells by higher concentrations of PEG. PEG-resistant cells such as the F_{24} line are highly resistant to fusion after treatment with 50% PEG, but they can be induced to fuse using harsher treatment. In these cells, the threshold for fusion has been shifted to ~55% PEG. In general, the membranes of all cell lines appear to be susceptible to the disruptive effects of PEG above a certain concentration, but in some cells this concentration is so toxic as to preclude the formation of fused syncytia.

PEG. In these cells, the fusion–response curve has simply been shifted to higher PEG concentrations.

In mixed populations containing both fusible and fusion-resistant cells labeled with Latex microbeads (Levine and Cox, 1978), it has been possible to demonstrate directly that fusion depends on the membranes of both partners. These data are summarized in Table II. Homogeneous cultures of Clone 1D cells alone exhibit ~ 90% fusion. When surrounded by F_{40} cells, however, <60% of the Clone 1D cells (identified by prior label-

Table II. Fusion between Clone 1D and F_{40} cells in Mixed Cultures[a]

Culture conditions	Fusion yield (%)	
	Clone 1D	F_{40}
Homogeneous culture of Clone 1D	93	—
Homogeneous culture of F_{40}	—	21
Clone 1D cell line predominant[b]	90	46
F_{40} cell line predominant[b]	59	18

[a]Cells were prelabeled with different sizes of Latex microbeads to permit identification in mixed cultures (Levine and Cox, 1978).
[b]Cultures were inoculated with mixtures of Clone 1D and F_{40} cells at a 10:1 ratio with the indicated cell line in excess. The extent of fusion for each cell line was scored independently.

ing with Latex beads) fused with F_{40}, demonstrating that the extensive fusion normally observed for Clone 1D depends in part on interactions with a fusible partner. Similarly, F_{40} cells alone exhibit only ~20% fusion but, when surrounded fused with Clone 1D cells, nearly 50% of the F_{40} cells fused with Clone 1D, demonstrating that the PEG fusion-resistance characteristic of F_{40} depends on the cells with which they are interacting. Taken together, these experiments indicate that the membranes of both cells contribute to the fusion reaction, a finding that may be particularly useful in the production of hybridomas from difficult-to-fuse preparations.

Having generated a series of cell lines resistant to the fusogenic effects of PEG, it was of interest to examine the response of these cells to other types of fusion as well. Surprisingly, several types of cell fusion are *inversely* correlated with susceptibility to PEG-induced fusion in these cells, as shown in Table III. PEG-resistant cells are unusually susceptible to fusion induced by a wide variety of enveloped viruses, including Sen-

Table III. Fusion Response of PEG-Resistant Cell Lines

Treatment	Fusion yield (%)	
	Clone 1D	F_{40}
Chemically induced fusion[a]	93	20
Virus-induced fusion[b]	35	65
Spontaneous fusion[c]	2	5

[a]Following 1-min treatment with 50% PEG-1000.
[b]Following infection with murine hepatitis virus (Holmes *et al.*, 1981).
[c]Fusion yield in untreated cultures (data compiled from 53 experiments over several years). Standard deviations ± 1% for each.

dai, influenza, and murine hepatitis viruses; (D. S. Roos, K. V. Holmes, and P. W. Choppin, manuscript in preparation). Spontaneous fusion is always infrequent in these cells, but it has been reproducibly observed in experiments over several years that the background level of fusion in untreated cultures of F_{40} is somewhat higher than in the parental cells (Roos and Choppin, 1985*b*). Confirming this observation, when Clone 1D or F_{40} (both TK cell lines) were cocultivated with HGPRT cells, the spontaneous formation of hybrid cells capable of survival in selective medium was sixfold higher for F_{40} than for Clone 1D (unpublished observations).

It is notable that cell lines intermediate in their response to PEG (e.g., F_{16}) are also intermediate in both spontaneous and virus-induced fusion. Whatever the specific alterations that produce this graded series of cell lines may be, these factors appear to affect a number of membrane-associated phenomena in parallel: increasingly PEG-resistant cells are increasingly virus-sensitive and increasingly susceptible to spontaneous fusion.

2.3. Biochemical Analysis of Fusion-Altered Cells

The availability of a series of cell lines that differ so dramatically in their response to PEG has proved very useful in the search for biochemical factors involved in controlling cell fusion. Because the selection procedures employed produced stable cell lines intermediate between the parental cells and the most highly fusion-altered mutants, it has been possible to search for biochemical differences which are progressive across the entire graded series of PEG-resistant cell lines. To date, no significant differences have been detected in the protein composition of any of these mutants. Substantial differences have been found in a variety of cellular and membrane lipids, however, as summarized in Table IV.

The fatty acyl chain composition of Clone 1D and F_{40} cells was examined by gas–liquid chromatography (GLC), as illustrated in Figure 3.

Table IV. Lipid Composition of Fusion-Altered Mutants

	Cell line	
Lipid species	Clone 1D	F_{40}
Ratio of S/P[a] fatty acids	1.0	3.8
Triglycerides[b]	1	5
Neutral ether-linked lipids[b]	1	36
Gangliosides[b]	2.5	1

[a]Ratio of saturated to polyunsaturated fatty acyl chains determined by gas–liquid chromatography.
[b]Relative content determined by quantitative thin-layer chromatography (Roos and Choppin, 1985*a*).

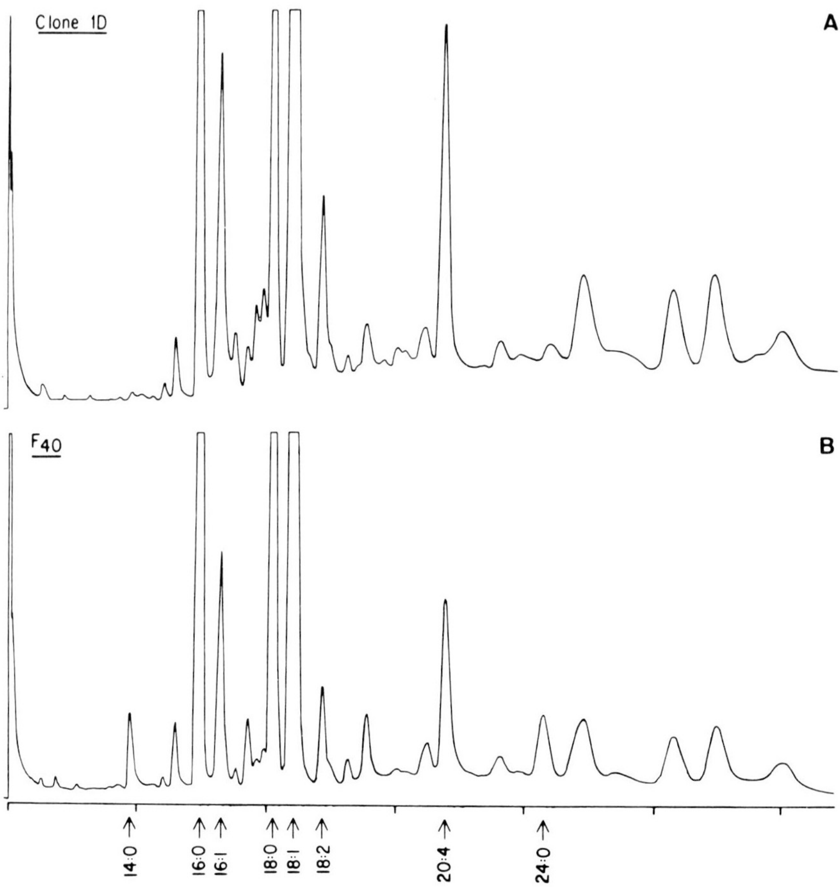

Figure 3. Fatty acid composition of Clone 1D and F_{40} cells. Gas–liquid chromatography of methyl-esterified fatty acyl chains from Clone 1D (A) or F_{40} cells (B). Tick-marks on the horizontal axis indicate 5 min of elution from a 6' × 2-mm column of 10% SP-2330 on 100/120 mesh Chromosorb W-AW (Supelco, Inc., Bellefonte, PA) programmed for 2 min at 150°C, rising at 4°C/min to hold at a final temperature of 200°C (Roos and Choppin, 1985*a*). All saturated fatty acids were more prominent in F_{40} cells (seen most clearly for 14 : 0, 16 : 0, and 24 : 0), while higher concentrations of polyunsaturated fatty acids were found in Clone 1D (cf. 18 : 2, 20 : 4). The relative contributions of specific fatty acid species varied between experiments (e.g., the 14 : 0 peak was not usually as prominent as here), but the overall picture of increased ratios of saturated to polyunsaturated fatty acids in F_{40} cell remained remarkably constant.

Several differences are visible in this representative GLC tracing of methyl-esterified fatty acids from whole cell extracts. Relative to Clone 1D cells, the fatty acids of F_{40} are highly saturated; palmitic and stearic acids (16 : 0 and 18 : 0, respectively*) are increased by 35–60%, and certain less common fatty acids such as myristate (14 : 0) and lignocerate (24 : 0) are >threefold elevated in the F_{40} mutants. These increases in saturated acyl chain content arise largely at the expense of polyunsaturated fatty acids; while there is little change in the content of monounsaturates (e.g., oleic acid, 18 : 1), polyunsaturates (e.g., linoleic acid (18 : 2), and arachidonate (20 : 4)) are markedly reduced in the F_{40} cell line. For convenience, we have expressed the degree of acyl chain saturation as a ratio of saturated to polyunsaturated fatty acids (S/P ratio), which increases from ~1 : 1 in Clone 1D to ~4 : 1 in F_{40} cells. S/P values are intermediate between these two extremes in cell lines intermediate between Clone 1D and F_{40} in fusibility, strengthening the correlation between cell fusion and fatty acid composition (Roos and Choppin, 1985*a*).

Compared with Clone 1D, the neutral lipid composition of F_{40} cells is also unusual, as shown in the thin-layer chromatogram presented in Figure 4. Triglyceride content is slightly increased over the normal levels found in the parental cell line, and a new band appears prominently (double arrows). This latter compound corresponds to a class of neutral ether-linked lipids, the alkyldiacylglycerols (Snyder, 1973; Roos and Choppin, 1985*a*). Consistent with a possible role in mediating membrane fusion, ether–lipid content is intermediate in cell lines that are intermediate in fusibility (Roos and Choppin, 1984). Additional lanes in Figure 4 show the lipid composition of cells harvested at different cell densities ranging from 100 to 5000 cell/mm^2. Some density-dependent variation in neutral glycolipid composition is seen (bracket), but differences in ether–lipid content between Clone 1D and F_{40} are independent of cell density. Neither qualitative nor quantitative differences have been detected in the levels of specific phospholipid species (incompletely separated in the lower third of this chromatogram), cholesterol (the dense band in the middle of the plate), or sterol esters (Roos and Choppin, 1985*a*).

The first series of mutants isolated (culminating in the F_{40} cell line) has been most thoroughly studied with respect to the various aspects of fusion response and lipid composition, but we have also observed similar

*16:0 is a shorthand notation for fatty acid structure, with the number preceding the colon indicating chain length, and the digit following the colon referring to the degree of unsaturation. Thus, 16 : 0 is palmitic acid (hexadecanoate), and arachidonic acid (dodecatetraenoate) is abbreviated as 20 : 4. Unless otherwise indicated, all double bonds are in the cis configuration.

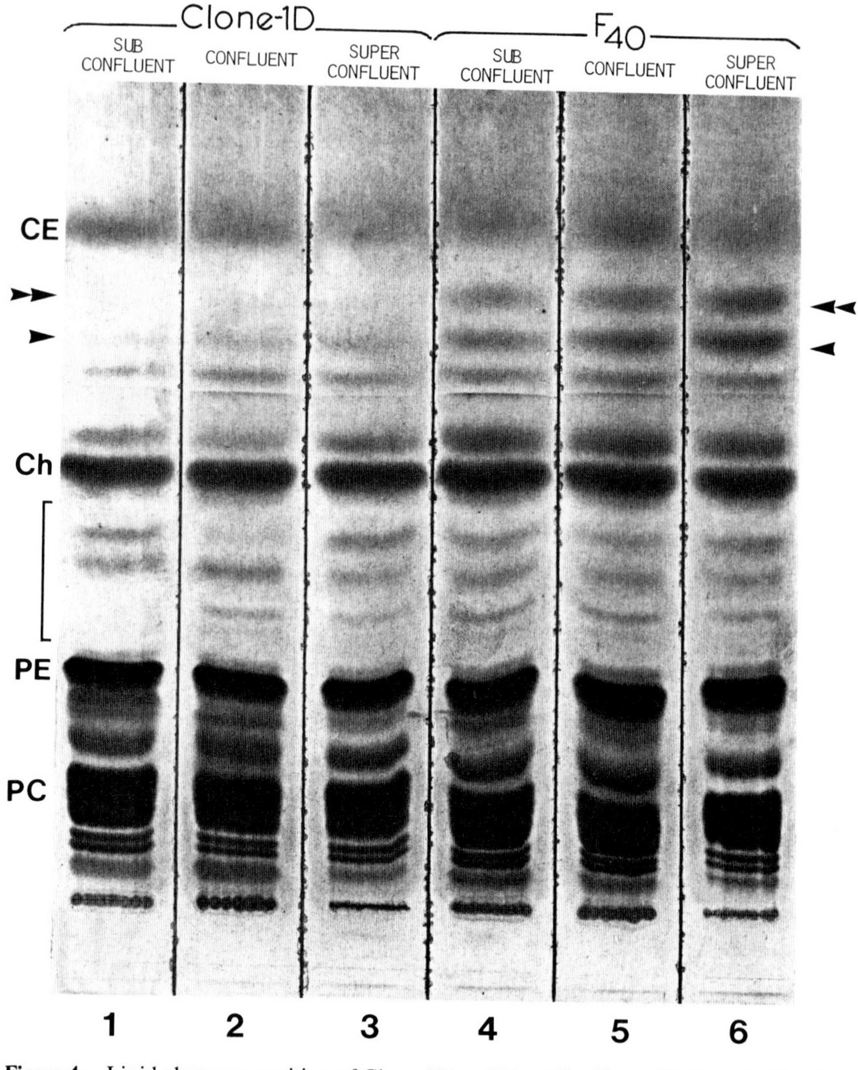

Figure 4. Lipid class composition of Clone 1D and F_{40} cells. Clone 1D (lanes 1–3) or F_{40} cells were harvested from subconfluent, confluent, or overgrown cultures and extracted for thin-layer chromatography (according to solvent schedule C in Roos and Choppin, 1985*a*). Lipids from 10^7 cells were run in each line. Regardless of culture density, F_{40} cells contained elevated triglyceride content (single arrowheads) and high levels of an unusual neutral lipid, alkyldiacylglycerol (double arrowheads). Concentrations of most other lipid compounds, including phosphatidylcholine (PC), phosphatidylethanolamine (PE), cholesterol (Ch), and cholesterol ester (CE), were invariant. Neutral glycolipid composition varied with cell density, but no consistent differences between the two cell lines were noted.

changes in mutants generated from other cell types (see Table I) wherever tested.

Based on several observations, we have previously argued that the two lipid alterations observed in PEG-resistant cells, increased chain saturation and elevated ether–lipid content, are linked (Roos and Choppin, 1985*a*). Neutral lipids, including ether-linked compounds, are characterized by highly saturated acyl chains (Spector *et al.*, 1979). Although acyl chain saturation is elevated in all F_{40} lipids (e.g., phosphatidylcholine, diglyceride), these mutants derive the bulk of their increased saturation from an overall increase in neutral lipid production. Ether-lipids are unusual among neutral lipids in their presence at relatively high concentration in cellular membranes (Pfleger *et al.*, 1968; Schroeder and Vagelos, 1976), suggesting that the high levels of ether–lipid found in F_{40} may function to deliver saturated acyl chains to the surface membranes of these PEG-resistant cells.

2.4. Control of Cell Fusion by Lipid Composition

Both ether–lipid content and fatty acid saturation exhibit a strong correlation with susceptibility to PEG-induced fusion. In order to assess the functional importance of these lipids in relationship to fusion, we have altered the lipid content of cells in culture through the use of appropriate supplements to the growth medium. Clone 1D and F_{40} cells normally maintain a relatively stable degree of acyl chain saturation (albeit different from each other), but their fatty acid composition can be influenced by growth in medium containing high concentrations of either saturated (S) or polyunsaturated (P) fatty acid. For example, when grown in 10^{-4} M saturated fatty acid (supplied either directly as a sodium salt or using serum albumin as a carrier), the S/P ratio of Clone 1D cells rises from the normal level of ~1 : 1 to >3 : 1, similar to the highly saturated level found in control (unsupplemented) cultures of F_{40} cells. Conversely, F_{40} cells grown in high concentrations of polyunsaturated fatty acids acquire a lipid composition resembling that of untreated Clone 1D cells (Roos and Choppin, 1985*b*). The effect of these manipulations on cell fusion is shown in Figure 5, and quantitative analysis is provided in Table V.

Clone 1D and F_{40} normally differ markedly in their response to an identical PEG treatment (Fig. 5A,B). However, decreasing the fatty acid saturation of F_{40} cells to the level normally found in Clone 1D corrects for the fusion defect in these mutants, permitting extensive fusion (D). By contrast, Clone 1D cells, which contain relatively unsaturated fatty acids and are usually susceptible to PEG fusion, become highly fusion-resistant following growth in saturated fatty acids (E). The capacity of lipid sup-

Table V. Control of Cell Fusion by Modification of Fatty Acid Composition

	Lipid supplement[b]		
Cell line	Polyunsaturated FA	None (control)	Saturated FA
	PEG-induced fusion (%)[a]		
Clone 1D	91	93	17
F_{40}	89	20	[13][c]
	Spontaneous fusion (%)		
Clone 1D	<2	2	15
F_{40}	<2	5	[20][c]

[a]Spontaneous fusion subtracted to yield PEG-specific fusion.
[b]10^{-4} M fatty acid supplement for 18 hr prior to treatment. Polyunsaturated FA: linolenic acid (18:3); saturated FA: nonadecanoic acid (19:0).
[c]Substantial cell death observed due to toxicity of saturated fatty acid supplements to F_{40} cells. (Quantitation may be an underestimate of the true extent of fusion.)

plements to control PEG-induced fusion exhibits a convincing dose–response relationship, and the effects of lipid supplementation on cell fusion are remarkably rapid (Roos and Choppin, 1985*b*). Maximal control of fusion can be observed within 24 hr after the addition of lipid supplements to the medium.

The fusibility of cell lines that are intermediate between Clone 1D and F_{40} can be altered predictably in either direction, i.e., made either more or less fusible solely by the manipulation of fatty acids in the growth medium. This observation is demonstrated in Figure 6, which displays the average extent of fusion alteration for several cell lines observed in more than 20 experiments to date. The ability to control cell fusion over a wide range has a number of potential applications to fusion technology.

Certain limits have been observed in the overall ability to control cell fusion, however. In some experiments it has proved possible to control fusion completely between the extremes represented by Clone 1D and F_{40} cells, but in practice our ability to stimulate fusion has proved somewhat variable. Although the fusion yield of all cell lines can readily be suppressed to ~20%, it has sometimes been difficult to stimulate fusion in highly PEG-resistant cells. On average, only 63% of F_{40} cells fuse when treated with PEG after growth in 10^{-4} M polyunsaturated fatty acid (range 46–88%). The reasons for the variation in the ability to enhance fusion are not completely understood.

The observations presented above suggest that further saturation of F_{40} cell lipids might produce an even more highly fusion resistant cell line, but it has not proved possible to suppress fusion below ~20% in either these cells or any other derivative of the parental Clone 1D line. We have

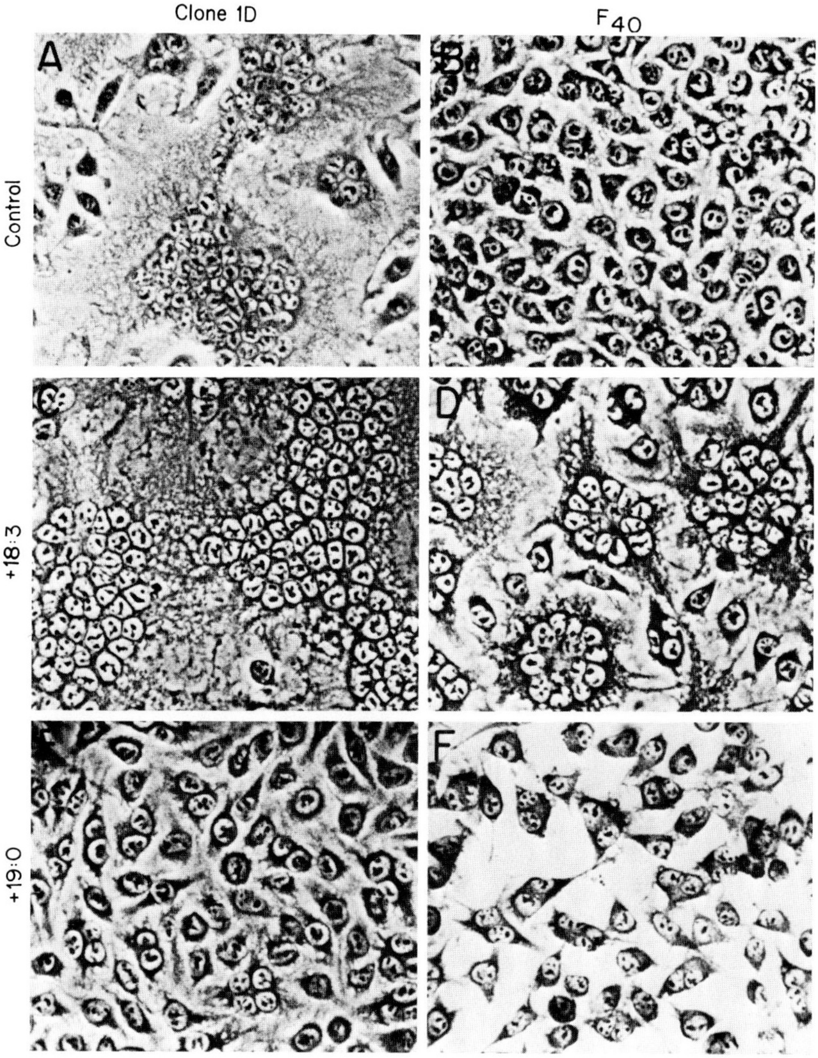

Figure 5. Modulation of PEG-induced fusion by fatty acid supplements. Clone 1D(A, C, E) or F_{40}(B, D, F) cells were grown to confluence in media supplemented for the final 18 hr of culture with 10^{-4} M Na salts of linolenic acid at 18 : 3 (C, D); nonadecanoic acid at 19 : 0 (E, F); or NaOH alone, as control (A, B). Cultures were subjected to identical treatment with 50% PEG and random fields chosen for photography. In unmodified cultures, these two cell lines differ considerably in their response to PEG, but the addition of polyunsaturated fatty acid to F_{40} cells corrected for the fusion-resistant phenotype of these mutants, promoting nearly as extensive fusion as seen in control cultures of Clone 1D. Conversely, the addition of saturated fatty acids to Clone 1D cells inhibited fusion, mimicking the phenotype of untreated F_{40} cells. (From Roos and Choppin, 1985*b*.)

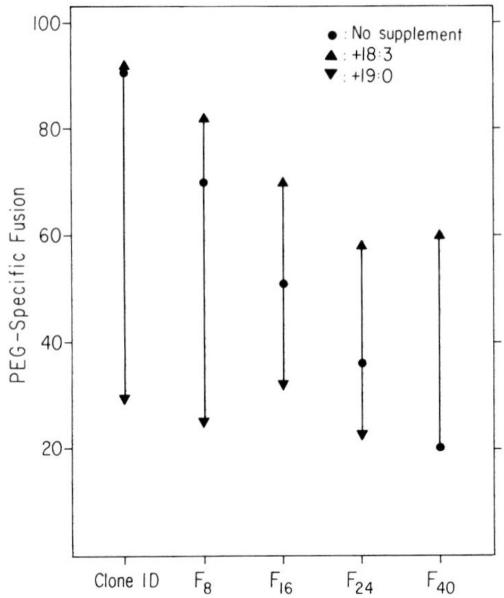

Figure 6. Effects of lipid supplements on cell lines with intermediate resistance to PEG. Cells were grown for 18 hr in the presence of 10^{-4} M 18 : 3(▲)or 19 : 0(▼) fatty acid or without exogenous lipid (●), and treated with PEG. Cell fusion was scored as the percentage of all nuclei present in fused cells of any size, minus the level of spontaneous fusion observed prior to PEG treatment. Data presented are the average of more than 20 seperate experiments. For all cell lines studied, PEG fusion could be controlled over a wide range solely by the addition of fatty acid supplements to the growth medium.

previously speculated that the inability to progress beyond the residual fusibility present in F_{40} cells may reflect limitations in the plasticity of viable cell architecture or metabolism (Roos *et al.*, 1983). Indeed, this now appears to be the case: when F_{40} cells are grown in high concentrations of saturated fatty acid, they simply die rather than becoming more resistant to PEG-induced fusion. Similarly, the S/P ratio of Clone 1D cells can be increased to that normally found in F_{40} (and the fusion yield declines proportionately), but attempts to push these cells beyond the limit of S/P $\approx 4:1$ results in extensive cell death.

As was indicated in Table III, the degree of PEG-resistance in these cell lines is inversely correlated with spontaneous fusion. Nevertheless, spontaneous fusion is also regulated by lipid composition, as shown in Table V. In the presence of high concentrations of saturated fatty acid, substantial fusion was observed, even without PEG treatment. Note that the effects of fatty acid saturation on spontaneous fusion and PEG fusion are in opposite directions: increasing saturation inhibits PEG-induced fusion but stimulates spontaneous fusion. These observations suggest an alternative approach to cell fusion, which may be useful in some applications. The phenomenon of virus-induced fusion in these cell lines, also controllable by lipid composition, will be dealt with elsewhere (D. S. Roos, K. V. Holmes, and P. W. Choppin, manuscript in preparation).

3. APPLICATIONS TO FUSION TECHNOLOGY

The successful production of somatic cell hybrids requires that several conditions be met. Compatible cells must be used. A suitable fusogen must be found that permits fusion of both cell types in question. Conditions for cell fusion must be considered to ensure that possible toxic effects of the fusogen do not interfere with hybrid formation. Culture conditions must be defined, mixing appropriate ratios of the two cell types at an appropriate density to maximize the production of binucleate heterokaryons. Suitable selection and screening protocols must be available to identify and isolate the desired fusion products efficiently. Finally, the hybrid cells must be sufficiently stable to permit their characterization and maintainance in culture.

This discussion focuses on the fusion process, leaving other problems for consideration elsewhere. In practice, the production of hybrid cells has become an increasingly routine process, major difficulties arising only later, in screening the hybrids for the desired combination of properties. Certain cell types, however, most notably human tissues, have proved relatively refractory to hybridization, and increased control of cell fusion would be most welcome for these situations. Even in cases in which fusion is readily achieved, it is likely that particular classes of hybrids are more frequently formed than others, so that the fusion procedure itself may select only a subpopulation of the cells of interest. Thus improved procedures for fusion may increase the spectrum of hybrids attainable, as well as the total yield. The experimental findings discussed above suggest a number of approaches to more efficient fusion protocols, especially in situations in which the starting material is limited, the cells in question are resistant to common fusogens, or the cell types to be hybridized are not readily compatible.

3.1. Fusion of Homogeneous Cultures to Determine Effective Hybridization Conditions

We have achieved good results in predicting the degree of hybridization in mixed cultures by modeling the expected frequencies of fusion from observations of homogeneous cultures. Typically, before attempting any hybridization, we treat homogeneous cultures of each cell type with PEG to get an indication of the relative sensitivity to PEG. In our experience, suitable fusion is most often achieved by treatment for 1 min with a warm solution of 45–50% PEG-1000. PEG-induced cell fusion is characterized by a relatively abrupt concentration threshold for fusion (cf. Fig. 1). This threshold differs in different cell lines, but the shape of the curve is quite consistent (Roos and Davidson, 1980). By examining fusion in just one sample at 45% PEG, it is generally possible to predict the extent of fusion attainable at other concentrations as well. Because extended PEG treatment is toxic to cells (probably due to the destabilization of membranes), we have generally found the manipulation of PEG concentration more useful than altering the length of treatment in attempting to control fusion. In cases in which PEG toxicity is severe even below fusogenic concentrations, higher molecular weight PEG (up to PEG-6000) has proved more suitable for fusion.

3.2. Panels for Fusion Partners with Differing Susceptibility to Fusion

For most purposes, it is desirable to maximize the number of binucleate heterokaryons. To minimize self-hybridization, we generally mix cells in unequal ratios (Davidson and Ephrussi, 1970), plating the less fusible cell line in 10-fold excess. Because the frequency of fusion depends on both partners, as shown in Table II, the fusibility of cells in homogeneous cultures is of great predictive value in maximizing fusion. For example, in designing a fusion protocol, one might mix an excess of cells that exhibit <20% fusion with a minority of cells which fuse >80% following a relatively mild PEG treatment. This situation is unworkable in protocols in which both fusion partners are fixed by experimental design. In many fusions, however, such as when it is desired to immortalize a particular cell culture, the investigator has wide latitude in choosing an immortalizing cell line. In the event that the cell population of interest fuses poorly when treated with PEG, one would choose a relatively PEG-fusible partner for hybridization. When the cells to be immortalized fuse well, one could minimize self-fusion and the production of overly large heterokaryons by mixing a minority of these cells with an excess of fusion-resistant partners. Alternatively, fusible partners could be used and the overall degree of fusion diminished by plating at lower density.

This approach has been used successfully to immortalize several cell types. While we have not systematically pursued the development of appropriate fusion partners for such experiments, one can anticipate the development of a panel of cell lines which differ over a wide range of fusibility. In some cases, cell lines used for immortalization already exist that differ in their sensitivity to PEG-induced fusion (Hämmerling *et al.*, 1981; Kozbor *et al.*, 1983). Where not available, hybridization panels could be generated from appropriate parental stocks by selection according to the procedures used to generate the PEG-resistant cell lines shown in Table 1.

3.3. Use of Lipid Supplements to Promote or Inhibit Fusibility

The surprisingly powerful effects of acyl chain composition on cell fusion (cf. Fig. 5) offer an additional dimension for the control of cell fusion in experimental applications. Using defined lipid supplements, the fusibility of one cell type can be altered temporarily to complement the natural fusion response of the desired fusion partner. For example, if a lymphocyte population of low fusibility is to be immortalized by hybridization but no suitably fusible partner exists, the production of hybrids by PEG may be increased by adding polyunsaturated lipids to the membranes of either cell population (or both). The ability of fatty acid supplements to control PEG-induced fusion provides the flexibility to carefully tailor interactions between fusion partners so as to maximize the production of hybrids, even when panels of genetically stable hybridization partners are not available.

An additional advantage of the use of lipid supplements to control fusion is the possibility of performing fusion at lower (and therefore less toxic) concentrations of PEG. The Clone 1D cell line fuses readily after 50% PEG treatment (cf. Fig. 1), making it a useful partner for hybridization with poorly fusible cells. Yet the toxicity of 50% PEG to some cells diminishes the value of a pairing with Clone 1D; at PEG concentrations in which Clone 1D fuses well, toxicity to the partner may prevent hybrid formation, but lower concentrations of PEG do not allow effective fusion to occur. We have shown that cells that are normally highly resistant to PEG fusion, such as the F_{40} cell line, can be induced to fuse more readily by lipid modification (cf. Fig. 6). Through the use of lipid supplements, Clone 1D cells, which are already fusible at 50% PEG, can also be induced to fuse at lower PEG concentrations, as shown in Figure 7. Clone 1D exhibits little fusion following treatment with 40% PEG (Fig. 7A), but identical treatment after incubation with linolenic acid (18 : 3) results in nearly complete fusion (Fig. 7B).

Two factors contribute to the fusion yield of a given cell type at different concentrations of PEG, a steep increase in fusion above a critical concentration and an increase in toxicity above another (often different) concentration. In our experience, cells that cannot be induced to fuse well often fail to do so because they are particularly sensitive to the toxic effects of PEG. In these cases, it appears that PEG toxicity overwhelms the cells before its fusogenic effects can come into play. In such situations toxicity places a severe restriction on potential fusion protocols, but one which can be circumvented by the choice of a suitable fusion partner, combined

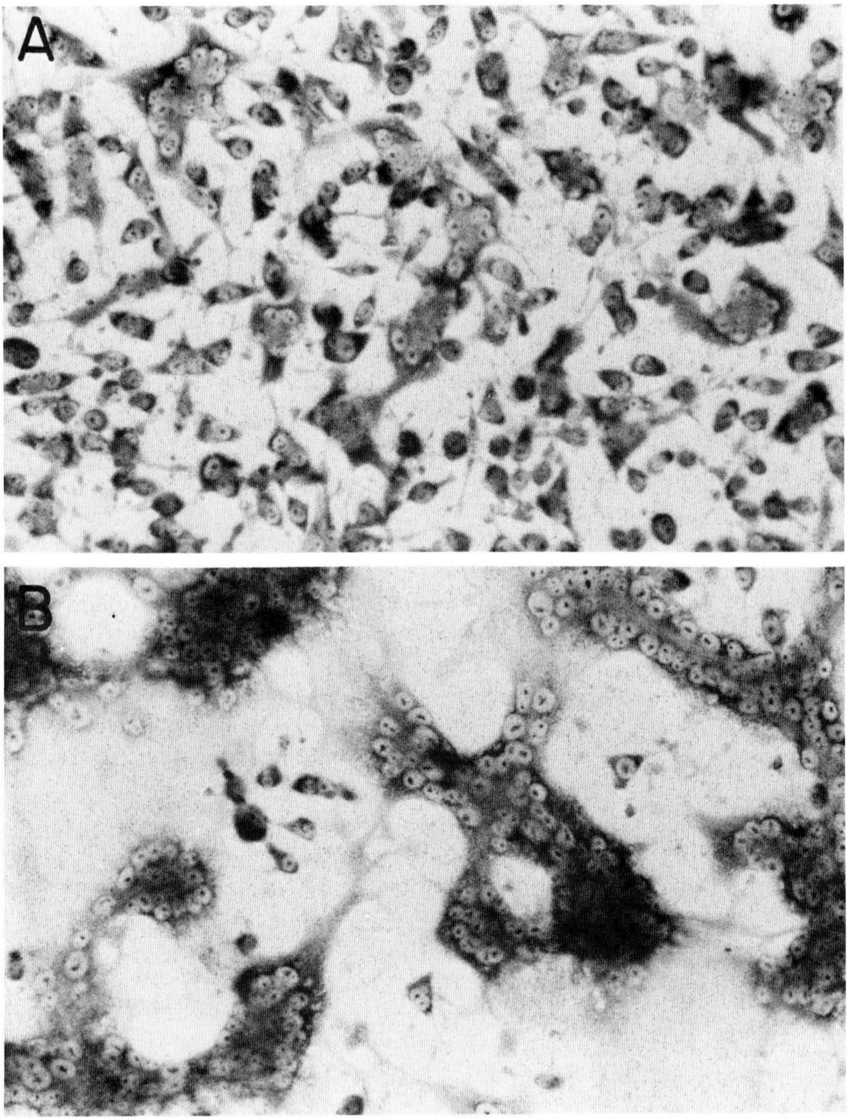

Figure 7. Lowered threshold for PEG-induced fusion in cells treated with polyunsaturated fatty acids. (A) Little fusion was observed in control cultures of Clone 1D cells treated with 40% PEG. (B) Growth of these cells in polyunsaturated fatty acid (18 : 3) prior to PEG treatment permitted extensive fusion, however, demonstrating that fusion can be further stimulated even in relatively fusible cells.

with the engineering of cellular lipid composition to permit fusion at less toxic PEG concentrations.

3.4. Other Approaches to Cell Hybridization: Virus-Induced Fusion and Spontaneous Hybridization

Owing to their ready availability, ease of use, and simplicity of structure, chemical fusogens have proved most widely applicable to problems in cell fusion technology (Klebe and Mancuso, 1981). In certain situations, however, other fusogenic treatments may prove more suitable. Spontaneous cell fusion has been observed to occur naturally at low frequency, and has been associated with a specific chromosomal locus in human cells (Wright and Shows, 1979). Although infrequent, spontaneous fusion may be useful in the hybridization of cells that are particularly fragile in the face of treatment with chemical fusogens (Poste, 1970). It may also be possible to promote this naturally occurring process by treatment with saturated fatty acids, as indicated in Table V.

A wide variety of enveloped viruses can also stimulate cell fusion under appropriate conditions, even after inactivation of the viral genetic material (Holmes and Choppin, 1966; White *et al.,* 1983). Because viruses are capable of binding tightly to cell surfaces, virus-induced fusion may be particularly well suited to the fusion of cells which cannot readily be grown as a monolayer or otherwise held in close contact to each other. In this regard, it is interesting that the lipid modifications which control PEG-induced fusion, are also effective in controlling virus-induced fusion, but in the opposite direction. Thus strategies to maximize cell hybridization similar to those discussed above can be applied to viral as well as chemical fusion systems.

Even the significant effects of lipid alteration on PEG-induced cell fusion may not be sufficient to provide the desired degree of fusion in all cases. In such situations, the observation that susceptibility to virus-induced fusion is inversely correlated with PEG fusion (Table III) suggests that viral fusogens may be more appropriate for cells which cannot readily be fused with PEG. It is interesting to note that the BHK-21F cell line, which has been used extensively in studies of virus-induced fusion because of its extreme sensitivity to viral fusion (Holmes and Choppin, 1966), is among the most highly PEG fusion-resistant cell lines we have encountered to date, requiring treatment with 55% PEG before substantial fusion is observed (Wang *et al.,* 1982). The natural resistance of these cells to PEG-induced fusion is comparable to cell lines specifically selected for PEG resistance (e.g., F_{40}). As might be expected for a naturally occurring equivalent of F_{40} cells (i.e., a PEG-resistant, virus-fusible cell line), BHK cells do not readily submit to selection pressures designed to make them even more highly PEG-resistant (see Table I).

4. SUMMARY AND CONCLUSIONS

The isolation of a series of cell lines that are altered in their susceptibility to chemically induced membrane fusion has provided a useful genetic approach to the study of cell-fusion mechanisms generally. By careful examination of interspecific fusion in cultures of known sensitivity to PEG, we have directly demonstrated that properties of both cell membranes in a fusion pair play a role in determining the overall likelihood of hybridization. This observation has opened the way to a defined approach to cell fusion, where the frequency of hybridization can be maximized based on the susceptibility of the individual cell types to appropriate fusogenic treatments.

Biochemical analyses of cells resistant to PEG-induced fusion revealed differences in their lipid composition, most notably increased saturation of fatty acyl chains and increased content of neutral ether-linked lipids. Although the precise role these lipids play in mediating cell fusion is not yet clear, by manipulation of cellular lipid content *in vitro* it has proved possible to control fusion over a wide range. The ability to exert such control over fusibility in a rapid, transient manner allows a great deal of flexibility in designing fusion protocols to maximize hybridization with minimal toxicity.

While the approaches suggested in this discussion may allow hybridization of cell types previously resistant to fusion, it must be noted that in many applications the ability to fuse cells is not a limiting step. Even when hybrids can be produced readily, however, it is likely that certain subpopulations of cells are more prone to fusion than others. Particularly when dealing with a heterogeneous population, such as spleen cells or circulating lymphocytes, this possibility may skew the spectrum of hybrids formed. In such situations, optimization of fusion protocols may improve not only the number, but also the nature of hybrid cells obtained.

The experimental system and procedures described above should be helpful both in improving our understanding of membrane fusion at a basic level, and at a practical level, in optimizing the design of cell hybridization protocols.

REFERENCES

Blow, A. M. J., Botham, A. J., Fisher, D., Goodall, A. H., Tilcock, C. P. S., and Lucy, J. A., 1978, Water and calcium ions in cell fusion induced by poly(ethylene glycol), *FEBS Lett.* **94:**305–310.

Davidson, R. L., 1977, Genetics of cultured mammalian cells, as studied by somatic cell hybridization, *Natl Cancer Inst. Monog.* **48**:21–30.

Davidson, R. L., and Ephrussi, B., 1970, Factors affecting the defective mating rate of mammalian cells, *Exp. Cell Res.* **61**:222–226.

Davidson, R. L., O'Malley, K. A., and Wheeler, T. B., 1976, Polyethylene glycol-induced mammalian cell hybridization: Effect of polyethylene glycol molecular weight and concentration. *Somatic Cell Genet.* **2**:271–280.

de St. Groth, S. F., and Scheidegger, D., 1980, Production of monoclonal antibodies: strategy and tactics, *J. Immunol. Methods* **35**:1–21.

Guccion, J. G., and Enzinger, F. M., 1972, Malignant giant cell tumors of soft parts, *Cancer* **29**:1518–1529.

Hämmerling, G. J., Hämmerling, U., and Kearney, J. F., eds., 1981, *Monoclonal Antibodies and T-cell Hybridomas: Perspectives and Technical Advances,* Elsevier/North-Holland, Amsterdam.

Holmes, K. V., and Choppin, P. W., 1966, On the role of the response of the cell membrane in determing virus virulence. Contrasting effects of the parainfluenza virus SV5 in two cell types, *J. Exp. Med.* **124**:501–520.

Holmes, K. V., Doller, E. V., and Behnke, J. N., 1981, Analysis of the functions of coronavirus glycoproteins by differential inhibition of synthesis with tunicamycin, in: *Biochemistry and Biology of Coronaviruses* (V. ter Meulen, S. Siddell, and H. Wege, eds.), pp. 133-142, Plenum Press, New York.

Kalderon, N., 1980, Muscle cell fusion, in: *Membrane–Membrane Interactions* (N. B. Gilula, ed.), pp. 99–118, Raven Press, New York.

Klebe, M. J., and Mancuso, M. G., 1981, Chemicals which promote cell hybridization, *Somatic Cell Genet.* **7**:473-488.

Köhler, G., and Milstein, C., 1975, Continuous culture of fused cells secreting antibody of predefined specificity, *Nature (Lond.)* **256**:495–497.

Kozbor, D., Dexter, D., and Roder, J. C., 1983, A comparative analysis of the phenotypic characteristics of available fusion partners for the construction of human hybridomas, *Hybridoma* **2**:7–16.

Langerhans, T., 1868, Über Riesenzellen mit wandständigen Kernen in Tuberkeln und die fibröse Form des Tuberkels, *Virchows Arch. Pathol. Anat. Physiol.* **42**:382–404.

Levine, M. R., and Cox, R. P., 1978, Use of latex particles for analysis of heterokaryon formation and cell fusion, *Somat. Cell Genet.* **4**:507–512.

O'Malley, K. A., and Davidson, R. L., 1977, A new dimension in suspension fusion techniques with polyethylene glycol, *Somat. Cell Genet.* **3**:441–448.

Pfleger, R. C., Anderson, N. G., and Snyder, F., 1968, Lipid class composition and fatty acid composition of rat liver plasma membranes isolated by zonal centrifugation, *Biochemistry* **7**:2826–2833.

Pontecorvo, G., Riddle, P. N., and Hales, A., 1977, Time and mode of fusion of human fibroblasts treated with polyethylene glycol (PEG), *Nature (Lond.)* **265**:257–258.

Poste, G., 1970, Mixed culture polykaryocytosis—A new form of cellular recognition behavior, *Life Sci.* **9**(II):459–463.

Robinson, J. M., Roos, D. S., Davidson, R. L., and Karnovsky, M. L., 1979, Membrane alterations and other morphological features associated with polyethylene glycol-induced cell fusion, *J. Cell Sci.* **40**:63–75.

Roizman, B., 1962, Polykaryocytosis, *Cold Spring Harbor Symp. Quant. Biol.* **27**:327–342.

Roos, D. S., 1984, *Membrane Fusion, Lipid Composition, and Tumorigenicity of Cultured Cells,* Ph.D. thesis, The Rockefeller University, New York.

Roos, D. S., and Choppin, P. W., 1984, Tumorigenicity of cell lines with altered lipid composition, *Proc. Nat. Acad. Sci. U.S.A.* **81**:7622–7626.

Roos, D. S., and Choppin, P. W., 1985*a*, Biochemical studies on cell fusion. I. Lipid composition of fusion-resistant cells, *J. Cell Biol.* **101:**1578–1590.

Roos, D. S., and Choppin, P. W., 1985*b*, Biochemical studies on cell fusion. II. Control of fusion response by lipid alteration, *J. Cell Biol.* **101:**1591–1598.

Roos, D. S., and Davidson, R. L., 1980, Isolation of mouse cell lines resistant to the fusion-inducing effect of polyethylene glycol, *Somat. Cell Genet.* **6:**381–390.

Roos, D. S., Robinson, J. M., and Davidson, R. L., 1983, Cell fusion and intramembrane particle distribution in polyethylene glycol-resistant cells, *J. Cell Biol.* **97:**909–917.

Schroeder, F., and Vagelos, P. R., 1976, Effects of phospholipid base analogs on the subcellular membrane ether composition of suspension cultured LM cells, *Biochim. Biophys. Acta* **441:**239–254.

Snyder, F., 1973, Thin-layer chromatographic behavior of glycerolipid analogs containing ether, ester, hydroxyl, and ketone groupings, *J. Chromatogr.* **82:**7–14.

Spector, A. A., Kiser, R. E., Denning, G. M., Koh, S. -W. M., and DeBault, L. E., 1979, Modification of the fatty acid composition of cultured human fibroblasts, *J. Lipid Res.* **20:**536–547.

Wang, E. W., Roos, D. S., Heggeness, M. H., and Choppin, P. W., 1982, Function of cytoplasmic fibers in syncytia, *Cold Spring Harbor Symp. Quant. Biol.* **46:**997–1012.

White, J., Kielian, M., and Helenius, A., 1983, Membrane fusion proteins of enveloped animal viruses, *Quart. Rev. Biophys.* **16:**151–195.

Wright, C. E., and Shows, T. B., 1979, Genetics of cell fusion: Human chromosome 10 assignment of a gene (FUSE) that promotes polykaryocyte formation, *Somat. Cell Genet.* **5:**503–517.

Chapter 7

Fusion-Permissive Protoplasts

A Plant System for Studying Cell Fusion

Wendy F. Boss

1. INTRODUCTION

While artificial systems are useful for delineating the fusogenic properties of specific lipids or proteins, they cannot encompass the complexity of a living cell membrane. To address basic questions such as what renders the plasma membrane fusion permissive and what regulates the fusion permissive state *in vivo,* a highly fusion permissive system was needed. Since under nonpathological conditions there are only a few times in the life of an organism in which cells are capable of undergoing fusion (e.g., fertilization and muscle maturation), few model systems are available. A new method has been developed for culturing cells of wild carrot (*Daucus carota* L.), so that they yield fusogenic or fusion permissive protoplasts (Boss *et al.,* 1984*a*). The fusion yield (fused protoplasts/fused + nonfused protoplasts) is greater than 50%. Fusion is calcium stimulated, enhanced by the calcium ionophore, A23187, and inhibited by EGTA and calmodulin antagonists (Boss and Grimes, 1985; Grimes and Boss, 1985).

WENDY F. BOSS • Botany Department, North Carolina State University, Raleigh, North Carolina 27695.

Some of the advantages of using the fusogenic carrot protoplasts to study fusion are as follows: (1) the cells can be grown relatively easily in suspension culture; (2) the culture medium is a well-defined mixture of inorganic salts, sugar, and a growth regulator, 2,4-dichlorophenoxyacetic acid (2,4-D), that does not contain calf serum or other nutrient broths or undefined components; (3) the isolated protoplasts are approximately 13 μm in diameter and thus, are visible with the light microscope at low magnification; (4) they do not require a pretreatment period to gain fusion competence, and as soon as the wall is removed, they will fuse; (5) the fact that they are in suspension and not attached to a substrate makes it easier to harvest and wash for short-term experiments; and (6) the protoplasts will regenerate into plants.

To aid in delineating some of the characteristics of the carrot protoplasts that are most critical for the fusion response, we have compared them with other carrot suspension culture cells that yield nonfusogenic protoplasts. These nonfusogenic cells do not have the potential to differentiate and are highly vacuolated. It should be noted that the term nonfusogenic is relative. Plant protoplasts in general are more fusion permissive than animal cells. Although the nonfusogenic protoplasts do not fuse under the mild conditions used for the fusogenic protoplasts (1–10 mM calcium, pH 6.0), they do fuse readily in PEG and with electrofusion (Kanchanapoom *et al.,* 1985; W. F. Boss and K. Kanchanapoom, unpublished results). Plant protoplasts, in general, fuse in about one-half the PEG concentration (Boss, 1983) and at one-tenth the electric field strength required for many animal cells (Zimmermann and Vienken, 1982, Chapter 17). The enhanced fusion potential of plant protoplasts compared with animal cells is at least in part due to the effects of the cell wall digestion process. Both the proteases contaminating the wall digestion enzymes (for review, see Fitzsimmons and Weyers, 1985; see also Chapter 25, this volume) and the mild osmotic stress imposed during this procedure would contribute to altering the structure of the plasma membrane (Taylor and Hall, 1978; Simmonds *et al.,* 1983; Wilkinson and Northcote, 1980; Boss and Grimes, 1985; for review see Galun, 1981).

This chapter is devoted to a description of the *Daucus* fusion system. It will include methods for culturing fusion permissive cells, optimal conditions for fusion, and studies of calcium transport and of lipid metabolism in the protoplasts.

2. CULTURING FUSOGENIC CELLS

The fusion-permissive or fusogenic carrot cells were derived from embryogenic wild carrot cultures. Embryogenic cultures of somatic plant cells can be maintained in a preembryonic state by growing the cells at relatively high auxin concentrations. When transferred at a low inoculum to fresh medium lacking auxin, the cells differentiate to form embryos, hence the term embryogenic.

The embryogenic wild carrot cells were maintained according to Wetherell and Dougall (1976) with biweekly transfers that involved selecting for cells that passed through a 141-μm filter and collected on a 61-μm filter (i.e., the 61-μm fraction), washing the cells for 1 hr in 2,4-D-free medium, and transferring 25 μL settled cell volume into 25ml of wild carrot medium containing 2.3 μM 2,4-D (Wetherell, 1969). To make the cells fusion permissive, the cells in the 61-μm fraction were transferred at a higher inoculum, 0.2 ml per 25 ml, and placed in a medium, denoted fusion-inducing medium (Boss *et al.,* 1984*a*), which contained a higher ammonium to nitrate ratio and lower 2,4-D (0.34 μM). These latter two changes would normally encourage embryo formation; however, the residual 2,4-D and the high inoculum density prevented the cells from differentiating.

Thus, the underlying strategy for initiating fusogenic cultures was to stimulate differentiation but not to allow it to occur. If the cells differentiated into either globular stage embryos or highly vacuolated callus cells, they would not fuse. In addition, treating the cultures with the polyamine biosynthesis inhibitor, difluoromethylarginine, which has been shown to inhibit embryogenesis in carrot cells (Feirer *et al.,* 1984), decreased the fusion potential (Grimes *et al.,* 1985). If the potential to fuse is developmentally regulated, any embryogenic plant cell culture should have a high fusion potential and would therefore be a system of choice for producing new plants through protoplast fusion.

Once placed in the fusion-inducing medium, the cells were subcultured at weekly intervals. For the first 2 weeks, 2 ml of uniformly resuspended medium and cells was transferred; on the third and subsequent weeks, 1.5 ml was transferred. For optimal yield and fusion potential, cells were harvested 4 days after transfer. Cultures that fused well yielded 0.3–0.35 g of fresh weight biomass at 4 days. Too much growth usually suggested the presence of more rapidly growing, vacuolated cells. Highly fusogenic cultures could be maintained for about 7 weeks, after which time there was usually a significant increase in the number of vacuolated, nonfusogenic cells. Protoplasts from the vacuolated cells would not fuse even with the fusogenic protoplasts from the same culture under the mild conditions used for these studies.

The fusogenic cell cultures described in this chapter have been optimized for high fusion potential and culture uniformity and, while the cells no longer readily form embryos, they are densely cytoplasmic and grow in tight clusters with smooth edges that are typical of preembryogenic cultures (Boss *et al.*, 1984*a*). Rapid selection and analysis of the fusion potential of the cultures are based on this cellular morphology.

3. OPTIMIZING CONDITIONS FOR PROTOPLAST FUSION

3.1. Effect of pH and Calcium

The pH of the external solution had little effect on fusion over a broad range of pH 6–8 using 2 mM Mes (2-(*N*-morpholino)ethanesulfonic acid) buffer (A. O. Brightman and W. F. Boss, unpublished results). Since the cells and protoplasts were normally cultured in media at pH 5.5–5.8, the lower range of the pH optimum (pH 6.0) was chosen for the fusion assays. Initially, it was necessary to add 10 mM calcium to see a fusion response. As the culture conditions were refined, the requirement for external calcium decreased; since 10 mM calcium gave a maximum response, however, it has been used routinely as a fusion stimulus.

3.2. Effect of Osmotic Pressure

Protoplast fusion is a dynamic and rapid event best visualized with videomicroscopy. The first morphological response to calcium (10 mM) added under isotonic conditions was crenation or shrinkage of the protoplasts (Boss and Grimes, 1985). Crenation was transient and appeared to precede fusion. This response was not observed in the nonfusogenic protoplasts. The transient shrinkage and swelling of the fusogenic protoplasts when calcium was added suggested that an osmotic driving force was involved in calcium-induced fusion.

Osmotic swelling has been shown to cause fusion of fusion-permissive lipid bilayers (Cohen *et al.*, 1982; Fisher and Parker, 1984). It also was purported to be a driving force for fusion of chromaffin granules with the plasma membrane during exocytosis in adrenal medulla (Creutz and Pollard, 1983). In sea urchin eggs, calcium purportedly caused swelling of the cortical granules and was necessary for granule fusion with the plasma membrane (Zimmerberg and Whitaker, 1985). Fusion of the cortical granules could be manipulated by changing the osmolarity of the surrounding medium. Calcium-dependent exocytosis was prevented by increasing the osmolarity of the surrounding medium; when the osmolarity was lowered, however, granule fusion with the plasma membrane occurred. These data suggested that an osmotic driving force was involved and that calcium might supply such a driving force *in vivo*.

With the fusogenic carrot protoplasts, calcium added under isotonic conditions resulted in fusion only if the protoplasts had adhered while crenated. Once they recovered from crenation (i. e., regained turgor), they would not fuse. Increasing the calcium concentration under isotonic conditions (0.45 osm) did not increase fusion; however, lowering the external osmoticum (from 0.45 to 0.35 osm) in the presence of calcium increased the number of fusion events. Also, by lowering the osmoticum, the rate of cytosolic mixing was enhanced (Kanchanapoom and Boss, 1986). Figure 1 (a–1) shows fusion of two protoplasts and rounding of the fusion product within 140 sec in 0.35 M sorbitol 1 mM Mes (pH 6.0) and 10 mM calcium. Membrane fusion and rounding of the fusion product was more rapid if the external osmolarity was lowered even further. This was demonstrated by bleeding water across the slide. The sequence of fusion events took only 20 sec and is shown in Figure 2 (a–e).

If fusion was osmoregulated, then increasing the external osmolarity should inhibit fusion. When the osmotic strength of the surrounding solution was raised by adding 0.6 M sorbitol and 10 mM calcium, the protoplasts adhered tightly. Apparent fusion, detected by light microscopy as rounding of the fusion product, did not occur until the osmolarity was lowered. EM studies are in progress to monitor membrane fusion under the different osmotic regimes.

4. CHARACTERISTICS OF THE FUSOGENIC PROTOPLASTS

4.1. Regulation of Cytosolic Calcium

Calcium has been shown to play a pivotal role in membrane fusion in many systems (for reviews, see Lucy, 1978; Papahadjopoulos, 1978; Gomperts, 1984; see also Chapters 3, 9, 10, 11, 15, and 25, this book). Much research has focused on the structural role of calcium in bilayer fusion in artificial systems (Fraley *et al.,* 1980; Düzgüneş *et al.,* 1981; see also Chapters 9, 10, and 11, this book). While calcium was found to alter fluidity of the fusogenic protoplasts (Boss and Grimes, 1985), the role of calcium in the fusion process appeared to be more than an effect on surface structure. For instance, negating surface charge with magnesium (100 mM) caused aggregation but did not enhance fusion (Boss *et al.,* 1984*a*). Furthermore, in the absence of external calcium, fusion could be induced by adding the calcium ionophore, A23187, to the fusogenic protoplasts (Grimes and Boss, 1985); however, adding external calcium or A23187 to the nonfusogenic protoplasts caused neither crenation nor fusion. Thus, it appeared that mobilization of intracellular calcium played a role in fusion of the fusogenic protoplasts.

$^{45}Ca^{2+}$ uptake studies were conducted to determine whether there was a difference in the way in which the two cell lines regulated their intracellular calcium stores. There were two major differences in calcium

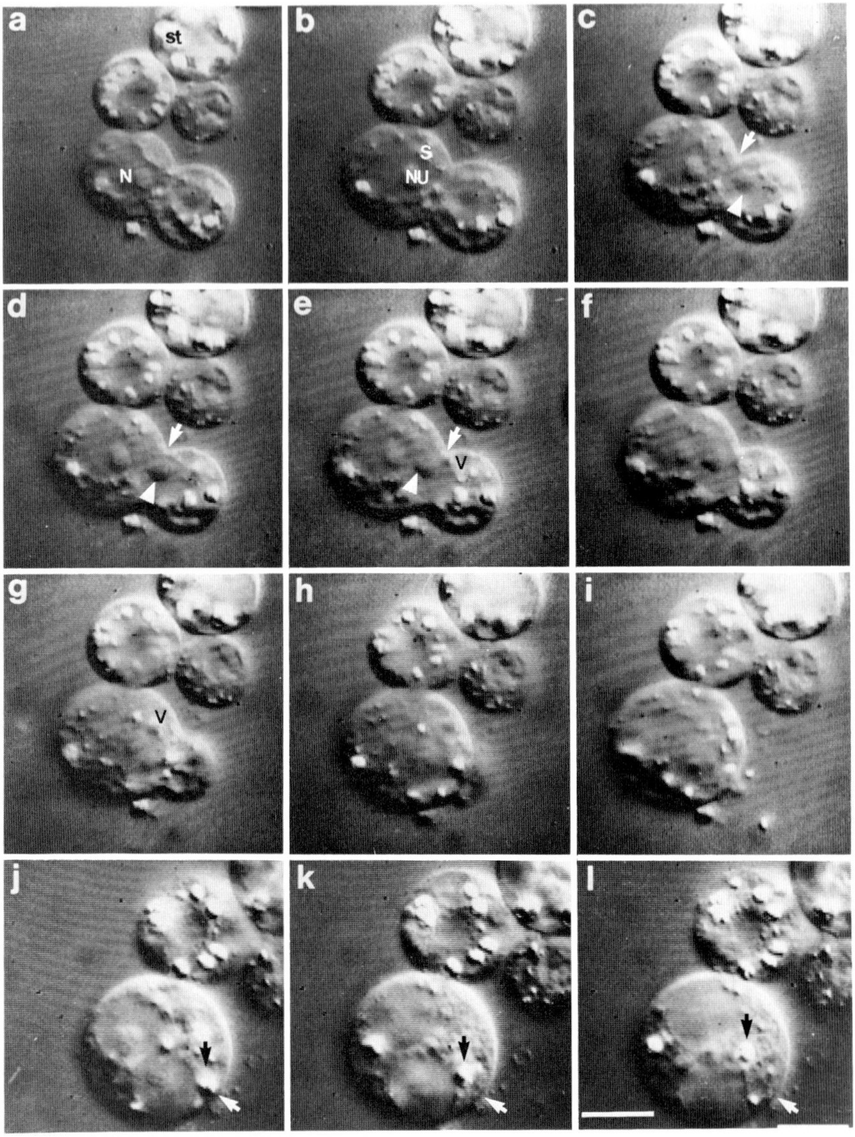

Figure 1. Time-course video images of fusogenic carrot protoplasts suspended in 0.45 M sorbitol and induced to fuse by bleeding across the slide a solution of 10 mM calcium in 0.35 M sorbitol. (a) A few seconds after initial contact of two protoplasts. (b) Lipid bilayers became fused, and the cytoplasmic connection was formed. (c–h) The migration of the nucleolus, vacuoles, spherosomes, and starch granules from one protoplast to another through the cytoplasmic connection. Movement of nucleolus (arrowhead) is shown from (c) to (e) relative to the stationary position of site of membrane fusing (arrow). (i) The spherical fusion product was formed. (j–l) The mixing of cytoplasms between both parental protoplasts were shown. Note the saltation of two nuclei from j to l, also the localization of starch granule (black arrow) relative to plasma membrane margin (white arrow). The actual time sequence from (a) to (i) is 140 sec and from (a) to (l) is 270 sec. n, nucleus; nu, nucleolus; s, spherosome; st, starch granule; v, vacuole. Bar: 20 μm. (From Kanchanapoom and Boss, 1986.)

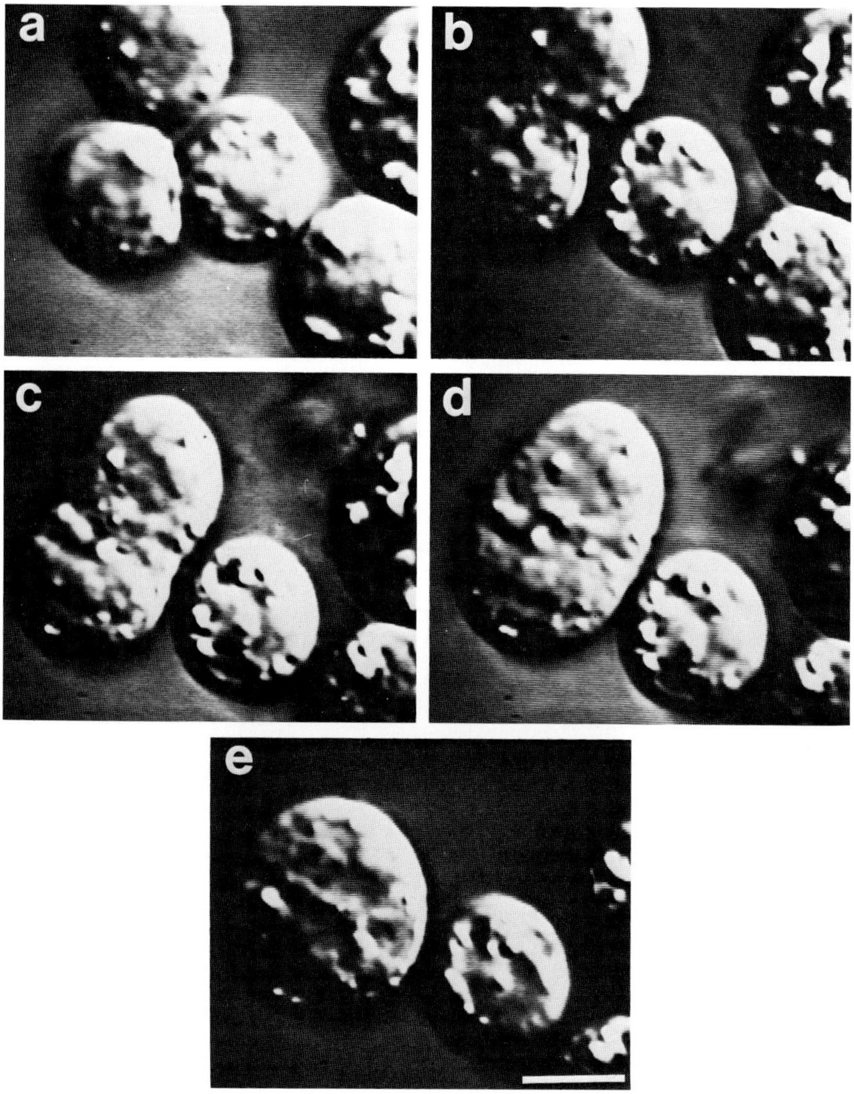

Figure 2. Protoplasts were in 0.45 M sorbitol and water was bled across the slide. The protoplasts adhered (a), the membranes fused (b), the cytoplasmic connections expanded (c), and they formed a round fusion product (d, e) within 20 sec. Bar: 15 μm. (Photos are from a sequence of a video tape by N. S. Allen and W. F. Boss.)

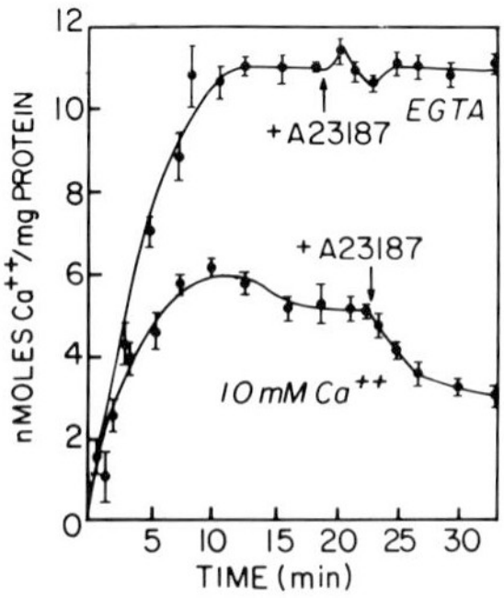

Figure 3. Comparison of $^{45}Ca^{2+}$ uptake profiles from fusogenic protoplasts isolated in 2.5 mM EGTA (upper tracing) versus protoplasts isolated in 10 mM $CaCl_2$ (lower tracing). Note the different response to A23187 between the two treatments. (From Grimes and Boss, 1985.)

uptake in fusogenic and nonfusogenic protoplasts (Grimes and Boss, 1985): (1) the nonfusogenic protoplasts reached $^{45}Ca^{2+}/^{40}Ca^{2+}$ equilibrium in 5 min and the fusogenic in 10 min, and (2) if A23187 was added at equilibrium to the nonfusogenic protoplasts and the external $^{45}Ca^{2+}$ concentration was 3–5 μM, there was an increase in $^{45}Ca^{2+}$ uptake. This was predicted, since the external calcium concentration was greater than the internal concentration and since EGTA (ethyleneglycol-bis-(β-aminoethyl ether)-*N′*, *N′*-tetraacetic acid), which was added during wall digestion (2 hr) to obtain steady-state kinetics, had lowered the calcium stores. By contrast, when A23187 was added to the fusogenic protoplasts released in EGTA, there was a transient increase and then decrease in calcium uptake (Fig. 3, upper trace). The $^{45}Ca^{2+}$ concentration returned to equilibrium within 10 min. These data suggested that the fusogenic protoplasts regulated cytosolic calcium more closely than did the nonfusogenic protoplasts. If the fusogenic protoplasts are released in the presence of 10 mM calcium rather than EGTA, the internal calcium stores were loaded and the addition of A23187 resulted in a net efflux of $^{45}Ca^{2+}$, as would be predicted (Fig. 3, lower trace).

Ca^{2+}/calmodulin-regulated ATPases and Ca^{2+}/H^+antiporters are two mechanisms by which plant cells could regulate cytosolic calcium (Hanson, 1984; Sze, 1985). To investigate a potential role for calmodulin-regulated ATPases in regulating cytosolic calcium during fusion, several calmodulin antagonists were used. When the calmodulin antagonists, trifluoperazine (TFP), chlorpromazine (CPZ), and W-13 (*N*-(4-aminobutyl)-5-chloro-2-naphthalenesulfonamide hydrochloride), were added during the last hour of the wall digestion period, fusion was inhibited (Fig. 4) (Grimes and Boss, 1985). The inactive analogue of W-13, W-12 (*N*-(4-aminobutyl)-naphthalenesulfonamide hydrochloride), had no effect on fusion. The effect of the calmodulin antagonists on fusion was seen if the antagonists were added during wall digestion but not after isolation of the

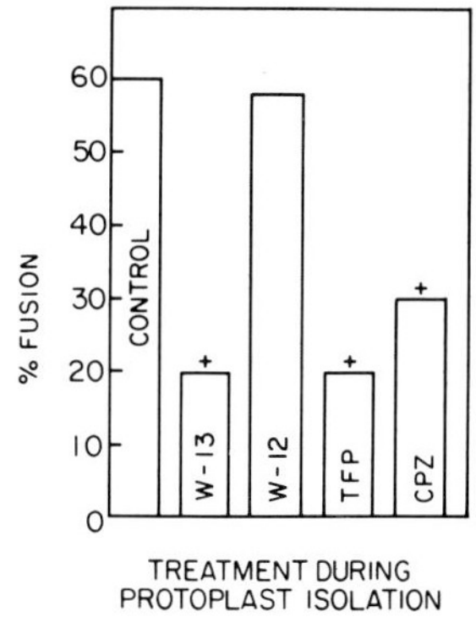

Figure 4. Calmodulin antagonists (indicated by +) strongly inhibited fusion potential. TFP and CPZ were applied at 1 μM for 60 min during protoplast isolation. W-13 and W-12 (the latter an analogue of W-13 with much less affinity for calmodulin) were applied at 86 μM for the last 60 min of protoplast isolation. (From Grimes and Boss, 1985.)

protoplasts. The addition of A23187 to the trifluoperazine-treated protoplasts caused a flux in cytosolic calcium measured as an efflux of $^{45}Ca^{2+}$ (Grimes and Boss, 1985); however, fusion was only marginally affected. Fusion increased from 10 to 25% when the ionophore was added but did not approach control levels of 60%. The calcium efflux was similar to the response observed when protoplasts were released in the presence of calcium, which increased exchangeable calcium in intracellular stores (Fig. 3, lower trace); however, if inhibition was due only to blocking calcium pumps and to increasing intracellular calcium stores, mobilizing calcium with A23187 should have stimulated fusion. The fact that the trifluoperazine-treated protoplasts showed only a marginal response to a change in cytosolic calcium suggested that the plasma membrane was no longer fusion permissive and that calcium–calmodulin-regulated processes might be involved in the biogenesis or maintenance of the fusion-permissive membrane during cell wall digestion.

4.2. Response of Membrane Lipids to Cell Wall Digestion

Before two plant cells can fuse the cell wall must first be removed. Our hypothesis was that the response of the protoplasts to enzymatic cell wall digestion was critical to the maintenance or biosynthesis of the fusion permissive plasma membrane. Since the cell wall digesting enzymes could serve as the initial stimulus in a stimulus–response paradigm, it was important to understand what changes occurred in the membrane lipids during cell wall digestion and if any of these changes correlated with fusion potential.

Other studies have indicated that the composition of membrane lipids could change as a result of cell wall digestion (Webb and Williams, 1984; Hartmann and Hock, 1985). These studies were of photosynthetic tissue and the general trends observed (a decrease in the amount of unsaturated fatty acids and galactolipids) were consistent with the loss of photosynthetic activity and decreased synthesis of chloroplast lipids during protoplast isolation. Webb and Williams (1984) also observed a decrease in phosphatidylcholine (PC) and an increase in triacylglycerol (TAG) as a result of cell wall digestion and suggested that in the leaf tissue there was a shift in metabolism from chloroplast lipids to storage lipids. That is, diacylglycerol (DAG) from PC would be diverted from galactolipid synthesis to triacylglycerol synthesis as a result of the stress of the cell wall digestion.

Since the fusogenic carrots were not photosynthetically competent, changes in metabolism during wall digestion should be different from that of the mesophyll cells. Kleinig *et al.* (1982) showed that cells of carrot suspension cultures contain relatively high levels of TAG and that the amount of TAG tended to vary with the physiological state of the cell. B. Martin, R. Hauptmann, and J. Widholm (B. Martin, personal communication) have found differences in the TAG concentration of cells isolated from mutant lines of carrot, which differ in their fusion potential. They found that fusogenic cells had two- to fivefold more TAG and also about twice the PC of the nonfusogenic cells. The question remained: Were the differences in fusion potential due to differences in the inherent membrane lipid composition of the cells or to changes in the membrane lipid composition caused by cell wall digestion?

4.2.1. Changes in Glycerol Lipid Composition during Wall Digestion

The glycerol lipid composition of cells and protoplasts was compared by prelabeling the cells with 2-[^{3}H]glycerol and placing them in either sorbitol (cells) or sorbitol plus wall digestion enzymes (protoplasts). The only significant difference in the distribution of [^{3}H]glycerol in either neutral or polar lipids was the increase in TAG relative to DAG in the fusogenic protoplasts during cell wall digestion (Boss, 1986). TAG : 1,2-DAG was 1.8 : 1 in the fusogenic protoplasts and 0.8 : 1 in the nonfusogenic protoplasts. The increase in TAG could have resulted from decreased breakdown or from increased synthesis stimulated by the hydrolysis of phospholipids. The fact that 2-hr labeling studies showed an increase in glycerol-labeled TAG relative to 1,2-DAG (Table I) suggested that net synthesis occurred. It is important to note that while lipid hydrolysis occurred in both cell types, as evidenced by increased free fatty acid during wall digestion (Boss and Massel, unpublished results), the TAG : DAG ratio only increased in the fusogenic protoplasts. These data suggested a difference in the regulation of lipid biosynthesis in the two cell types.

Studies quantitating lipids using long-term labeling with radioisotopes were not sensitive to small changes in lipid biosynthesis or turnover, which might occur during cell wall digestion. Therefore, 2-[^{3}H]glycerol was added to unlabeled cells to study changes in glycerol lipid biosynthesis during cell wall digestion. Two things became evident from these short-term labeling studies. First, the amount of glycerol lipids synthesized in the fusogenic cells during cell wall digestion was very low compared with the nonfusogenic cells. The glycerol incorporation in the fusogenic cells was less than one-tenth that of the nonfusogenic cells. Thus, while Webb and Williams (1984) described lipid breakdown as a result of wall digestion, in the nonfusogenic protoplasts, there also was considerable lipid synthesis. Second, the distribution of glycerol among the lipids was different in the two cell types. After only 2 hr there was a distinct difference in labeling of the polar lipids (Table I) (Boss, 1986). PC was the predominant phospholipid labeled in the fusogenic cells and the percentage of PI synthesized *de novo* from glycerol was much smaller in the fusogenic protoplasts during this period (13.3% of the recovered phospholipids versus 23.8% for the nonfusogenic protoplasts). In both types of protoplasts, the recovered phosphatidic acid (PA) was low based on glycerol incorporation. Overall, the distribution of [^{3}H]glycerol in the nonfusogenic protoplasts after 2 hr labeling was very similar to that seen in the protoplasts from prelabeled cells suggesting that by 2 hr they had reached equilibrium with respect to glycerol incorporation.

The low rate of glycerol incorporation characteristic of the fusogenic protoplasts was more similar to postfusion myoblasts than to prefusion myoblasts (Sandra and Ionasescu, 1980). While Sandra and Ionasescu reported increased TAG metabolism associated with prefusion myoblasts, other reports would predict that increased PC or TAG would not favor bilayer fusion (Düzgüneş *et al.,* 1981; Roos and Choppin, 1985*a*; see also Chapter 6 this volume). Thus while the glycerol labeling data suggested a difference in the regulation of lipid biosynthesis in the two carrot cell types, there appeared to be no major qualitative or quantitative differences in the lipids which would logically cause one to be more fusion permissive. In addition, no differences were found in the ratio of saturated to unsaturated fatty acids of similar fusogenic and nonfusogenic carrot cells (B. Martin, personal communication) suggesting that a difference in fatty acid composition was not the key to fusion in this system.

4.2.2. Phosphoinositide Metabolism during Cell Wall Digestion

High rates of lipid turnover, specifically of phosphoinositides, have been correlated with membrane fusion during muscle maturation (Wakelam, 1983; Wakelam and Pette, 1984) and in secretory vesicle fusion with the plasma membrane (Fain and Berridge, 1979; for review, see Laychock and Putney, 1982; Hokin, 1985). The low rate of glycerol lipid synthesis in

Table I. [^{3}H]Glycerol Incorporated into Microsomal Membrane Lipids during 2 hr Cell Wall Digestion as a Percentage of Recovered Neutral or Polar Lipid[a]

	Neutral lipids			Polar lipids			
	TAG	1,3-DAG	1,2-DAG	PA	PI	PC	PE
Fusogenic protoplasts	54.6 ± 3	5.6 ± 2	36.8 ± 7	3.1 ± 0.9	13.3 ± 4	58.9 ± 7	24.3 ± 5
Nonfusogenic protoplasts	40.2 ± 8	3.5 ± 2	56.2 ± 10	3.2 ± 1.7	23.8 ± 3	41.0 ± 7	31.0 ± 8

[a]Values are means ± SD of four or more values from experiments on three different days.

the fusogenic protoplasts did not eliminate the potential for high rates of headgroup turnover; therefore, it was important to investigate whether increased inositol head group metabolism occurred in fusogenic cells.

Inositol incorporation into lipids is typically low in plant tissue culture cells (Moore, 1977; Kleinig and Kopp, 1978), and this was found to be true of the nonfusogenic cells and protoplasts (Boss and Brightman, 1984). The fusogenic cells, however, incorporated inositol into PI at relatively high rates. In the fusogenic protoplasts, inositol incorporation into membrane lipids was consistently five to six times higher than in the nonfusogenic protoplasts. The high rate of inositol incorporation in conjunction with the low rate of glycerol incorporation implied that there was a high rate of inositol turnover relative to glycerol lipid synthesis in the fusogenic protoplasts.

More complete analysis revealed that several inositol-labeled phospholipids were present (Boss and Massel, 1985). In cells labeled for 18 hr with 2-[^{3}H]inositol the following inositol-labeled phospholipids were separated with thin-layer chromatography and identified by comparison with known standards: phosphatidylinositol (PI), lysophosphatidylinositol (LPI), phosphatidylinositol monophosphate (PIP), and phosphatidylinositol bisphosphate (PIP_2). There was an unidentified inositol-labeled spot that migrated between PIP and PIP_2 and occasionally one below PIP_2.

Polyphosphoinositides, specifically phosphatidylinositol-4-phosphate (PIP) and phosphatidylinositol-4,5-bisphosphate (PIP_2), have been shown to play a leading role in signal transmission for several animal hormones, neurotransmitters, and agonist-stimulated responses (for reviews, see Majerus *et al.*, 1984; Nishizuka, 1984; Berridge and Irvine, 1985; Hokin, 1985). In muscle cells, polyphosphoinositide metabolism increased when calcium was added to fusion-competent myoblasts (Wakelam, 1983; Wakelam and Pette, 1984). Lipid breakdown was dependent on external calcium, and PIP and PIP_2 breakdown preceded that of PI. In agonist-stimulated responses, the stimulus causes metabolism of plasma membrane-associated PIP_2 to form inositol trisphosphate (IP_3) and DAG. IP_3 mobilizes calcium from intracellular stores such as the ER (Streb *et al.*, 1983) and generates a cascade of calcium-regulated reactions that affect cellular metabolism, including PI turnover and secretion (Billah and Lapetina, 1982*a*; Fain and Berridge, 1979; Jones and Michell, 1974). In these instances, the breakdown of PIP_2 was independent of exogenous calcium.

Exogenous calcium was the fusion agonist in the myoblast system. Increasing cytosolic calcium with A23187 increased fusion only if added prior to the time the cells were fusion competent (Wakelam and Pette, 1984). While exogenous calcium enhanced fusion of the fusogenic protoplasts, A23187 would enhance fusion in the absence of exogenous calcium. These data suggested that an increase in exogenous calcium was not necessary for the fusion of carrot protoplasts. The agonist in the carrot system might be the stress of cell wall digestion. This stress could cause an increase in cytosolic calcium, which might affect the fusion permissive state of the plasma membrane. A cartoon depicting the potential effects of cell wall digestion on membrane lipid metabolism is shown in Figure 5.

To determine what effects cell wall digestion had on polyphosphoinositides, cells were labeled for 18 hr with 2-[^{3}H]inositol, and the inositol-labeled lipids of cells and protoplasts were compared. There were two major differences in the extracted phosphoinositides of the fusogenic protoplasts. Consistently, the LPI was higher (almost twice the recovered

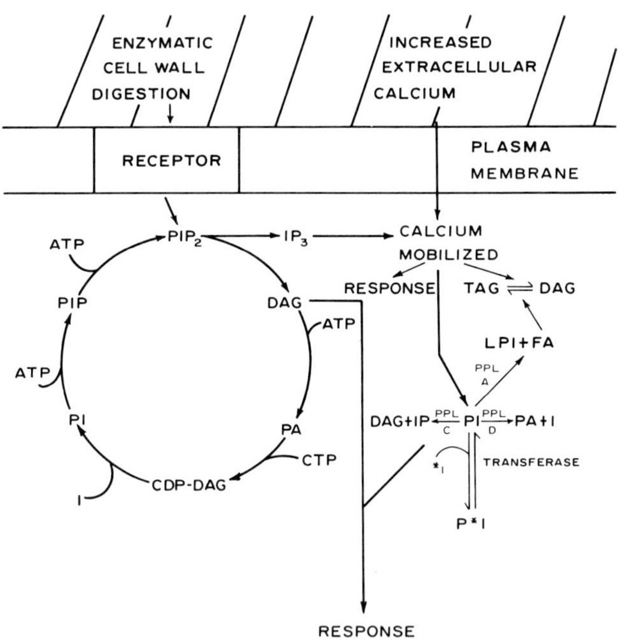

Figure 5. A summary of potential pathways for lipid metabolism in fusogenic cells in response to cell wall digestion and to exogenous calcium. DAG, diacylglycerol; FA, fatty acid; I, inositol; IP_3, inositol trisphosphate; PA, phosphatidic acid; PI, phosphatidylinositol; PIP, phosphatidylinositol monophosphate; PIP_2, phosphatidylinositol bisphosphate; PPL, phospholipase; TAG, triacylglycerol.

Table II. Phosphoinositides as Percentage Total Inositol-Labeled Lipid Recovered from Cells Labeled 18 hr with myo-2-[^{3}H]Inositol and Placed in Either 0.4 M Sorbitol (Cells) or 0.4 M Sorbitol +2% Driselase (Protoplasts)[a]

	Fusogenic			Nonfusogenic		
	Cells (t-o)	Cells (2 hr)	Protoplasts (2 hr)	Cells (t-o)	Cells (2 hr)	Protoplasts (2 hr)
PI	88 ± 4	80 ± 3	68 ± 3	85[b]	85 ± 3	81 ± 3
LPI	6.2 ± 2	14 ± 3	25 ± 4	12.5	11 ± 3	13 ± 4
PIP	2.1 ± 0.8	2.8 ± 1	1.3 ± 0.4	1.2	2 ± 0.6	2.8 ± 1.5
PIP_2	1.5 ± 0.8	0.45 ± 0.2	0.5 ± 0.4	2.0	2 ± 0.7	2.4 ± 1

[a]Mean ± SD of at least four separate experiments.
[b]Data from one experiment done in duplicate.

inositol-labeled lipid) and PIP lower (by about one-half) in protoplasts compared with cells (Table II). In these experiments, comparisons were made of extracts with approximately equal amounts of inositol-labeled PI from cells and protoplasts. This was the first evidence of increased polyphosphoinositide metabolism in plants under a stress condition. The presence of inositol-labeled LPI suggested a more complex pathway than found in the myoblasts and was more typical of the phospholipids present during platelet activation where fusion occurs during secretion (Billah and Lapetina, 1982*b*).

In contrast to the effect found on the fusogenic cells, cell wall digestion did not alter the distribution of inositol-labeled lipids of the nonfusogenic cells (Table II). It might be that due to the low amounts of inositol-labeled lipid recovered from these cells, the small changes in the phosphoinositide metabolism were not detectable with [^{3}H]inositol.

4.3. Effect of Exogenous Calcium on Phosphoinositide Metabolism

EGTA was added routinely during cell wall digestion to decrease spontaneous fusion. EGTA was found not to be toxic to the protoplasts and, when calcium was added, the EGTA-treated protoplasts fused readily (Kanchanapoom and Boss, 1985; Boss *et al.*, 1984*a*). Adding EGTA during wall digestion not only decreased fusion it also decreased the incorporation of inositol. EGTA inhibited inositol incorporation by 63% in the fusogenic protoplasts and only 25% in the nonfusogenic protoplasts (Boss, 1986). Time course studies of the response during the 2-hr period indicated that significant inhibition occurred between the 90 min and 2-hr time point. This was a time during which the control cells and protoplasts showed a marked increase in incorporation of inositol, and it was this increase that appeared to be affected by EGTA (data not shown).

Analysis of the lipids indicated that the decrease in inositol incorporation was due primarily to a decrease in the inositol-labeled PI. The amount of inositol incorporated into LPI and PIP was the same and the PIP_2 was too low in 2 hr to obtain adequate statistics. Thus it appeared that PI metabolism was affected by the levels of exogenous calcium in this system.

The effect of exogenous calcium was further demonstrated by adding 10 mM calcium directly to inositol-labeled protoplasts. Calcium caused a change in the total recovered PI in both cell lines that could be detected within 1 min (Boss and Grimes, 1985). The response was seen primarily in the ER-rich fraction of membranes isolated on density gradients. The inositol-labeled PI in the fusogenic protoplasts increased transiently and then approached control values in about 10 min, while that of the nonfusogenic protoplasts decreased and remained low (Boss and Brightman, 1984). Preliminary thin-layer analysis of the inositol-labeled lipids indicated no change in the polyphosphoinositides in response to exogenous calcium (W. F. Boss, unpublished results). This would be consistent with the hypothesis that PI but not polyphosphoinositide metabolism was calcium regulated. It appeared that, unlike the myoblast system, while calcium was required for the biogenesis of the fusion permissive membrane and while adding exogenous calcium enhanced fusion, exogenous calcium was not the agonist in fusogenic protoplasts.

While the studies of polyphosphoinositide metabolism in plants are in their early stages, calcium has been shown to play an important role in mediating plant development (for review, see Hepler and Wayne, 1985). Whether calcium is a second or third messenger in plant cells has yet to be determined. It should be pointed out, however, that many of the enzymes involved in the PI cycle have been found in plants. Inositol phosphate phosphatase has been isolated from pollen (Gumber *et al.*, 1984), and PI-specific phospholipase C has also been characterized (Irvine *et al.*, 1980; Helsper *et al.*, 1985). IP_3 mobilization of calcium has been demonstrated using microsomal membranes isolated from zucchini hypocotyls (Drobak and Ferguson, 1985) and from fusogenic protoplasts (M. Rincon and W. F. Boss, unpublished results). There is one report of DAG-stimulated phosphorylation in isolated soybean membranes (Morré *et al.*, 1984*a*), and calcium and calcium–calmodulin-regulated protein kinases have been identified (Hetherington and Trewavas, 1984; Ranjeva *et al.*, 1984; Dieter, 1984). There is also evidence that plant hormones affect PI metabolism. Auxin has been shown to stimulate degradation of PI in a cell free system (Morré *et al.*, 1984*b*). Buckhout *et al.* (1981) demonstrated that auxin treatment decreased the number of calcium-binding sites in isolated soybean membranes. The pathway of PI metabolism in these systems has not been established; however, Sexton and Moore (1981) showed that plant cells contain a phosphatidylinositol–inositol exhange enzyme, and phospholipase D was indicated in auxin-stimulated PI metabolism (Morré *et al.*, 1984*b*). In the fusogenic carrots, there was no evidence for increased DAG or phosphatidic acid as a result of wall digestion or calcium treatment based on the distribution of glycerol in protoplasts of labeled cells; however, small changes may not have been detected in the prelabeled protoplasts. Further studies are needed using ^{32}P to follow changes in the headgroup turnover in order to delineate which of these pathways are involved; however, the presence of LPI suggests that phospholipase A plays a role in phosphoinositide metabolism in the carrot cells.

4.4. Is Secretion Involved in Obtaining a Fusogenic Protoplast?

Many agonist-stimulated responses in animal cells that stimulate phosphoinositide turnover involve secretion (Laychock and Putney, 1982). Secretion of peroxidases in plant tissue culture cells has been shown to be enhanced by calcium and decreased by EGTA and calmodulin antagonists (Kevers *et al.*, 1982). Amylase secretion in germinating seeds also appears to be under calcium regulation (Jones and Carbonell, 1984; Akazawa and Mitsui, 1985). If secretion of Golgi-derived vesicles was necessary for the biogenesis of the fusion permissive membrane, blocking secretory vesicle formation should inhibit fusion. Monensin, a K^+/H^+ ionophore, has been shown to cause swelling of the trans cisternae and thus inhibit secretory vesicle formation of the carrot cells as well as other plant and animal cells (for reviews, see Ledger and Tanzer, 1984; Bergen and Bates, 1984). While adding monensin to myoblasts inhibited fusion and secretion of mannose-labeled proteins (Den, 1985), monensin did not affect fusion of the fusogenic carrot protoplasts (W. F. Boss, unpublished results). This appeared to indicate that secretion of Golgi-derived vesicles was not essential for either the synthesis or maintenance of the fusion permissive carrot protoplast membrane; however, electron microscopic analysis of the protoplasts released in the presence of monensin for 2 hr indicated few if any swollen cisternae (Brightman *et al.*, 1985). There was also no effect of monensin on mannose incorporation into TCA-precipitable material. Carrot cells that were in the same solution but that had resisted digestion and retained their cell walls and protoplasts after isolation (i.e., after removing the wall digestion solution) contained many swollen cisternae as a result of monensin treatment. Since swelling of the cisternae was shown (Boss *et al.*, 1984*b*) to be driven by a proton gradient (acid inside), the aberrant response of the cisternae was unusual and may indicate that wall digestion affects the proton gradient across the Golgi cisternae. These results could have far reaching implications with regard to regulation of membrane biogenesis in cells under stress or during exposure to an agonist. Further studies are needed to determine whether the aberrant response of the Golgi apparatus during wall digestion is important with regard to the biogenesis or maintenance of the fusion permissive plasma membrane.

5. SUMMARY AND PERSPECTIVE

In summary, fusion of the fusogenic carrot protoplasts was stimulated by calcium. Under isotonic conditions, calcium appeared to provide the osmotic driving force for fusion as evidenced by the transient crenation. The fusion process could be enhanced by lowering the osmolality of the external solution causing the protoplasts to swell. Calcium transport studies indicated that not only were the fusogenic protoplasts more sensitive to changes in cytosolic calcium with regard to fusion but also that they regulated their calcium stores more closely than the nonfusogenic protoplasts. The effects of calmodulin antagonists were long term and suggested a role of calcium in the biogenesis or maintenance of the fusion permissive membrane during cell wall digestion.

Akazawa and Mitsui (1985) have proposed two pathways for amylase secretion in germinating seedlings: one calcium dependent and one calcium independent. By analogy, a calcium-dependent pathway for membrane lipid synthesis appeared to predominate in the fusogenic protoplasts. It was characterized by high inositol incorporation, which was sensitive to EGTA and calcium. In the nonfusogenic protoplasts, a calcium-independent pathway appeared to predominate. It was characterized by high rates of glycerol incorporation and low sensitivity to EGTA and calcium. Akazawa and Mitsui suggested that the calcium-dependent pathway involved Golgi-derived vesicles and like the myoblast system they found effects of monensin on protein processing and secretion. Monensin, however, did inhibit fusion or mannose incorporation in the fusogenic protoplasts. This may reflect a change in secretory vesicle metabolism during the stress of wall digestion. Alternatively, the calcium-dependent pathway in the protoplasts might not involve Golgi-derived vesicles but rather lipid recycling via other secretory vesicles or lipid-exchange proteins (Bloj and Zilversmit, 1981; Dubacq *et al.,* 1984). The high inositol incorporation relative to glycerol and the increase in TAG relative to 1,2-DAG in fusogenic protoplasts suggested a predominance of membrane lipid recycling relative to *de novo* synthesis.

The biosynthetic pathway for inositol lipid synthesis in the fusogenic protoplasts needs to be investigated. The increase in polyphosphoinositide metabolism during cell wall digestion implies agreement with a stimulus–response paradigm. Whether the increase in polyphosphoinositide metabolism stimulates pathways necessary to maintain or synthesize the fusion-permissive membrane during wall digestion remains to be seen. Wakelam and Pette (1984) indicated that the ratio of PI : PIP : PIP_2 of fusion-permissive myoblasts was high (20 : 1 : 1) compared with that of hepatocytes (100 : 1 : 1) and implied that the higher the rate of polyphosphoinositide metabolism relative to PI metabolism the more likely the cells would fuse. The ratio of inositol-labeled polyphosphoinositide relative to PI in the fusogenic protoplasts is more like the hepatocytes than myoblasts and yet they fuse. Clearly, there is more than phosphoinositides involved in producing a fusion permissive membrane. It is possible that the increase in LPI and as yet unstudied lipids and proteins favor fusion in the Daucus system.

In conclusion, the fusogenic carrot protoplasts are a new system for investigating the mechanism of membrane fusion, for studying how cells can regulate biosynthesis and biogenesis of fusion-permissive membranes, and for producing new hybrids through protoplast fusion.

REFERENCES

Akazawa, T., and Mitsui, T., 1985, Biosynthesis, intracellular transport, and secretion of α-amylase in rice seedlings, in: *New Approaches to Research on Cereal Carbohydrates* (R. D. Hill and L. Munck, eds.), pp. 129–137, Elsevier, Amsterdam.

Bergen, W. G., and Bates, D. B., 1984, Ionophores: Their effect on production efficiency and mode of action, *J. Anim. Sci.* **58**(6):1465–1483.

Berridge, M. J., and Irvine, R. F., 1984, Inositol trisphosphate, a novel second messenger in cellular signal transduction, *Nature (Lond.)* **312**:315–321.

Billah, M. M., and Lapetina, E. G., 1982*a*, Degradation of phosphatidylinositol-4,5-bisphosphate is insensitive to Ca^{2+} mobilization in stimulated platelets, *Biochem. Biophys. Res. Commun.* **109**:217–222.

Billah, M. M., and Lapetina, E. G., 1982*b*, Formation of lysophosphatidylinositol in platelets stimulated with thrombin or ionophore A23187, *J. Biochem. Chem.* **257**(9):5196–5200.

Bloj, B., and Zilversmit, D. B., 1981, Lipid transfer proteins in the study of artificial and natural membranes, *Mol. Cell. Biochem.* **40**:163–172.

Boss, W. F., 1983, Poly(ethylene-glycol)-induced fusion of plant protoplasts: A spin-label study, *Biochim. Biophys. Acta* **730**:111–118.

Boss, W. F., 1986, Glycerol lipid metabolism: The effects of cell wall digestion on fusogenic protoplasts, submitted.

Boss, W. F., and Brightman, A. O., 1984, Protoplast lipid turnover during wall digestion and in response to the fusogen calcium, *Plant Physiol. (Suppl.)* **75**(1):257.

Boss, W. F., and Grimes, H. D., 1985, Dynamics of calcium-induced fusion of fusogenic carrot protoplates, in: *Beltsville Symposia on Agricultural Research.* Vol. IX. *Frontiers of Membrane Research in Agriculture* (J. St. John, P. Jackson, and E. Berlin, eds.), pp. 63–68, Rowman Allanheld, Totowa, New Jersey.

Boss, W. F., and Massel, M. O., 1985, Polyphosphoinositides are present in plant tissue culture cells, *Biochem. Biophys. Res. Commun.* **132**(3):1018–1023.

Boss, W. F., Grimes, H. D., and Brightman, A. O., 1984*a*, Calcium-induced fusion of fusogenic wild carrot protoplasts, *Protoplasma* **120**:209–215.

Boss, W. F., Morré, D. J., and Mollenhauer, H. H., 1984*b*, Monensin-induced swelling of Golgi apparatus cisternae mediated by a proton gradient, *Eur. J. Cell Biol.* **214:**77–82.

Brightman, A. O., Boss, W. F., and Morré, D. J., 1985, An atypical response of some Golgi apparatus of carrot protoplasts to the sodium-selective ionophore monensin, *Plant Physiol. (Suppl.)* **77**(4)**:**388.

Buckhout, T. J., Young, K. A., Low, P. S., and Morré, D. J., 1981, In vitro promotion by auxins by divalent ion release from soybean membranes, *Plant Physiol.* **68**(2)**:**512–515.

Cohen, F. S., Akabas, M. H., and Finkelstein, A., 1982, Osmotic swelling of phospholipid vesicles causes them to fuse with a planar phospholipid bilayer membrane, *Science* **217:**458–460.

Cruetz, C. E., and Pollard, H. B., 1983, Development of a cell-free model for compound exocytosis using components of the chromaffin cell, *J. Autonomic Nerv. Sys.* **7:**13–18.

Den, H., 1985, Effect of monensin on myoblast fusion, *Biochem. Biophys. Res. Commun.* **126**(1)**:**313–319.

Dieter, P., 1984, Calmodulin and calmodulin-mediated processes in plants, *Plant Cell Environ.* **7:**371–380.

Drobak, B. K., and Ferguson, I. B., 1985, Release of calcium from plant hypocotyl microsomes by inositol-1,4,5-trisphosphate, *Biochem. Biophys. Res. Commun.* **130:**1241–1246.

Dubacq, J.-P., Drapier, D., Tremolieres, A., and Kader, J.-C., 1984, Role of phospholipid transfer protein in the exchange of phospholipids between microsomes and chloroplasts, *Plant Cell Physiol. 25:*(7)**:**1197–1204.

Düzgüneş, N., Wilschut, J., Fraley, R., and Papahadjopoulos, D., 1981, Studies on the mechanism of membrane fusion: Role of head-group composition in calcium and magnesium-induced fusion of mixed phospholipid vesicles, *Biochim. Biophys. Acta* **642:**182–195.

Fain, J. N., and Berridge, M. J., 1979, Relationship between phosphatidylinositol synthesis and recovery of 5-hydroxytryptamine-responsive Ca^{++} flux in blowfly salivary glands *Biochem. J.* **180:**655–661.

Feirer, R., Mignon, G., and Litvay, J. D., 1984, Arginine decarboxylase and polyamines required for embryogenesis in the wild carrot, *Science* **223:**1433–1435.

Fisher, L. R., and Parker, N. S., 1984, Osmotic control of bilayer fusion, *Biophys. J.* **46:**253–258.

Fitzsimmons, P. J., and Weyers, J. D. B., 1985, Properties of some enzymes used for protoplast isolation, in: *The Physiological Properties of Plant Protoplasts* (P. E. Pilet, ed.), pp. 12–23, Springer-Verlag, Berlin.

Fraley, R., Wilschut, J., Düzgüneş, N., Smith, C., and Papahadjopoulas, D., 1980, Studies on the mechanism and membrane fusion: Role of phosphate in promoting calcium ion induced fusion of phospholipid vesicles, *Biochemistry* **19:**6021–6029.

Galun, E., 1981, Plant protoplasts as physiological tools, *Annu. Rev. Plant Physiol.* **32:**237–266.

Gomperts, B. D., 1984, Calcium and cellular activation, in: *Biological Membranes* Vol. 5 (D. Chapman, ed.), pp. 290–320, Academic Press, London.

Grimes, H. D., 1985, The role of calcium fluxes, calmodulin, polyamines in the regulation of calcium-induced plasma membrane fusion in fusogenic protoplasts, Ph.D. thesis at North Carolina State University, Raleigh.

Grimes, H. D., and Boss, W. F., 1985, Intracellular calcium and calmodulin involvement in protoplast fusion, *Plant Physiol.* **79:**253–258.

Grimes, H. D., Slocum, R. D., and Boss, W. F., 1985, α-Difluoromethylarginine treatment inhibits protoplast fusion in fusogenic wild carrot protoplasts, *Biochim. Biophys. Acta* **886:**130–134.

Gumber, S. C., Loewus, M. W., and Loewus, F. A., 1984, Further studies on myoinositol-1-phosphatase from the pollen of *Lilium longiflorum* Thunb, *Plant Physiol.* **76**:40–44.

Hanson, J. B., 1984, The functions of calcium in plant nutrition, in: *Advances in Plant Nutrition,* Vol. 1 (P. B. Tinker and A. Lauchli, eds.), pp. 149–208, Praeger, New York.

Hartmann, E., and Hock, K., 1985, Fatty acids in protoplasts, in: *The Physiological Properties of Plant Protoplasts* (P. E. Pilet, ed.), pp. 190–199, Springer-Verlag, Berlin.

Helsper, H. P., de Groot, P. F., Jackson, J. F., and Linskens, H. F., 1985, Phosphatidylinositol phosphodiesterase in lily pollen, *Plant Physiol. (Suppl.)* **77**(4):533.

Hepler, P. K., and Wayne, R. O., 1985, Calcium and plant development, *Annu. Rev. Plant Physiol.* **36**:397–439.

Hetherington, A. M., and Trewavas, A., 1984, Activation of a pea membrane protein kinase by calcium ions, *Planta* **161**:409–417.

Hokin, L. E., 1985, Receptors and phosphoinositide-generated second messengers, *Annu. Rev. Biochem.* **54**:205–235.

Irvine, R. F., Letcher, A. J., and Dawson, R. M. C., 1980, Phosphatidylinositol phosphodiesterase in higher plants, *Biochem. J.* **192**:279–283.

Jones, L. M., and Michell, R. H., 1974, Breakdown of phosphatidylinositol provoked by muscarinic cholinergic stimulation of rat parotid-gland fragment, *Biochem. J.* **142**:583–590.

Jones, R. L., and Carbonell, J., 1984, Regulation of the synthesis of barley aleurone α-amylase by gibberellic acid and calcium ions, *Plant Physiol.* **76**:213–218.

Kanchanapoom, K., and Boss, W. F., 1986, Osmoregulation of fusogenic protoplast fusion, *Biochim. Biophys. Acta* **861**:429–439.

Kanchanapoom, K., Grimes, H. D., Brightman, A. O., and Boss, W. F., 1985, A novel method for monitoring protoplast fusion, *Protoplasma* **124**:65–70.

Kevers, C., Sticher, L., Penel, C., Greppin, H., and Gaspar, Th., 1982, Calcium-controlled peroxidase secretion by sugarbeet cell suspensions in relation to habituation, *Plant Growth Regul.* **1**:61–66.

Kleinig, H., Hara, S., and Schuchmann, R., 1982, Lipid metabolism in plant tissue culture cells. Acetate incorporation; triacylglycerol accumulation, in: *Proceedings of the Fifth International Congress of Plant Tissue and Cell Culture* (Plant Tissue Culture, 1982), pp. 257–258, Japanese Association for Plant Tissue Culture, Tokyo.

Kleinig, H., and Kopp, C., 1978, Lipids, lipid turnover, and phospholipase D in plant suspension culture cells *(Daucus), Biophys. Acta* **773**:99–105.

Laychock, S. G., and Putney, J. W., Jr., 1982, Roles of phospholipid metabolism in secretory cells, in: *Cellular Regulation of Secretion and Release* (P. Michael Conn, ed.), pp. 53–105, Academic Press, New York.

Ledger, P. W., and Tanzer, M. L., 1984, Monensin—A perturbant of cellular physiology, *Trends Biochem. Sci.* **9**(7):313–314.

Lucy, J. A., 1978, Mechanisms of chemically induced cell fusion, in: *Membrane Fusion* (G. Poste and G. L. Nicolson, eds.), pp. 267–304, Elsevier/North-Holland, New York.

Majerus, P. W., Neufeld, E. J., and Wilson, D. B., 1984, Production of phosphoinositide-derived messengers, *Cell* **37**:701–703.

Moore, T. S., Jr., 1977, Phospholipid turnover in soybean tissue cultures, *Plant Physiol.* **60**:754–758.

Morré, D. J., Morré, J. T., and Varnold, R. L., 1984*a,* Phosphorylation of membrane-located proteins of soybean in vitro and response to auxin, *Plant Physiol.* **75**:265–268.

Morré, D. J., Gripshover, B., Monroe, A., and Morré, J. T., 1984*b,* Phosphatidylinositol turnover in isolated soybean membranes stimulated by the synthetic growth hormone 2,4-dichlorophenoxyacetic acid, *J. Biol. Chem.* **259**:15346–15368.

Nishizuka, Y., 1984, Turnover of inositol phospholipids and signal transduction, *Science* **225**:1365–1370.

Papahadjopoulos, D., 1978, Calcium-induced phase changes and fusion in natural and model membranes, in: *Membrane Fusion* (G. Poste and G. L. Nicolson, eds.), pp. 765–790, Elsevier, New York.

Ranjeva, R., Graziana, A., Ranty, B., Cavalie, G., and Boudet, A. M., 1984, Phosphorylation of proteins in plants: A step in the integration of extra and intracellular stimuli?, *Physiol. Veg.* **22**(3):365–376.

Roos, D. S., and Choppin, P. W., 1985*a*, Biochemical studies on cell fusion. I. Lipid composition of fusion-resistant cells, *J. Cell Biol.* **101**(4):1578–1590.

Roos, D. S., and Choppin, P. W., 1985*b*, Biochemical studies on cell fusion. II. Control of fusion response by lipid alterations, *J. Cell Biol.* **101**(4):1591–1598.

Sandra, A., and Ionasescu, V. V., 1980, Alterations in lipid turnover in developing muscle, *Biochem. Biophys. Res. Commun.* **93**(3):898–905.

Sexton, J. C., and Moore, T. S., Jr., 1981, Phosphatidylinositol synthesis by a Mn^{2+}-dependent exchange enzyme in castor bean endosperm, *Plant Physiol.* **68**:18–22.

Simmonds, D. H., Setterfield, G., and Brown, D. L., 1983, Reorganization of microtubules in protoplasts of *Vicia Hajastana, Grossh.* During the first 48 hours of culturing, in: *Sixth International Protoplast Symposium,* Basel, Switzerland, pp. 212–213.

Streb, H., Irvine, R. F., Berridge, M. J., and Schultz, T., 1983, Release of Ca^{2+} from a non-mitochondrial intracellular store in pancreatic acinar cells by inositol-1,4,5-triphosphate, *Nature (Lond.)* **306**:67–68.

Sze, H., 1985, H^+-translocating ATPases: Advances using membrane vesicles, *Annu. Rev. Plant Physiol.* **36**:175–208.

Taylor, A. R. D., and Hall, J. L., 1978, Fine structure and cytochemical properties of tobacco leaf protoplasts and comparison with the source tissue, *Protoplasma* **96**:113–126.

Wakelam, M. J. O., 1983, Inositol phospholipid metabolism and myoblast fusion, *Biochem. J.* **214**:77–82.

Wakelam, M. J. O., and Pette, D., 1984, Myoblast fusion and inositol phospholipid breakdown: Causal relationship or coincidence?, in: *Cell Fusion* (Ciba Foundation Symposium 103), pp. 100–108, Pitman, London.

Webb, M. S., and Williams, J. P., 1984, Changes in the lipid and fatty acid composition of *Vicia faba* mesophyll protoplasts induced by isolation, *Plant Cell Physiol.* **25**(8):1541–1550.

Wetherell, D. F., 1969, Phytochrome in cultured wild carrot tissue, I. Synthesis, *Plant Physiol.* **44**:1734–1737.

Wetherell, D. F., and Dougall, D. K., 1976, Sources of nitrogen supporting growth and embryogenesis in cultured wild carrot tissue, *Physiol. Plant.* **37**:97–103.

Wilkinson, M. J., and Northcote, D. H., 1980, Plasma membrane ultrastructure during protoplast plasmolysis, isolation and wall regeneration: A freeze fracture study, *J. Cell Sci.* **42**:401–410.

Zimmerberg, J., and Whitaker, M., 1985, Irreversible swelling of secretory granules during exocytosis caused by calcium, *Nature (Lond.)* **315**:581–584.

Zimmermann, U., and Vienken, J., 1982, Electric field-induced cell-to-cell fusion, *J. Membrane Biol.* **67**:165–182.

Chapter 8

Experimental Cell Fusion with Selected Siphonocladalean Algal Cells

Seong Yun Lee and R. Malcolm Brown, Jr.

1. INTRODUCTION

Plant cells frequently have been used for cell-fusion experiments as a result of the development of a technique for enzymatic digestion of the cell wall (Cocking, 1960). With this technique, a variety of new genetic manipulations have become possible, such as protoplast fusion, uptake of organelles, and uptake of DNA. With the enzymatic digestion procedure, large populations of genetically and physiologically homogeneous protoplasts are now available (Chaleff, 1981).

The most unique experimental area introduced by the ability to fuse protoplasts is the hybridization of widely divergent taxa that are normally isolated genetically by morphological, developmental, or sexual incompatibility barriers (Salhani *et al.,* 1985). By a method of cell fusion, which involves PEG (polyethylene glycol) and high concentrations of Ca^{2+}, it is possible to construct new and otherwise impossible genetic combinations that can be used for studying the genetic characterization of the plant as well as improving crop quantity and quality.

SEONG YUN LEE and R. MALCOLM BROWN, JR. • Department of Botany, University of Texas at Austin, Austin, Texas 78712.

Exceptional candidates for plant cell-fusion studies are the siphonocladalean algae. The Siphonocladales are marine, tropical green algae that have a distinctive method of cell division, termed segregative cell division (Bold and Wynne, 1978). This process involves cleavage of the protoplasm into many portions of varying size, each of which spherates and forms a protoplast (Enomoto and Hirose, 1972). Each protoplast synthesizes a cell wall and differentiates into the mature form that is a giant coenocytic cell. Among the Siphonocladales, *Boergesenia* and *Valonia* are well known, together with other coenocytic marine algae such as *Bryopsis* and others (Preston, 1968; Burr and West, 1970, 1971*a,b*). These algae have the capacity to regenerate whole cells from protoplasts produced inside the cell when the algae are subjected to adverse mechanical stimulation. By osmotic shock or by wounding a cell with a needle, the cytoplasm of the giant cell begins to aggregate and form many protoplasts (La Claire, 1982). Therefore, protoplast formation can be induced naturally without enzyme treatment. This is a major advantage of Siphonocladalean algae, because it is impossible in vascular plants and many other eukaryotic cells to produce protoplasts without enzymatic treatment. Another major advantage is that Siphonocladalean protoplasts differentiate directly into mature cells and, unlike protoplasts of vascular plants, they do not produce an intermediate callus stage that is undifferentiated. Siphonocladalean algae have a thick cell wall composed largely of highly crystalline and well-ordered cellulose microfibrils. The cells of these algae have been used in many investigations in cell wall and cellulose structure (Preston and Kuyper, 1951; Gardner and Blackwell, 1971).

We know that the cell walls of *Boergesenia* and *Valonia* are different in certain aspects, i.e., cell wall thickness, microfibril orientation, microfibril size, and terminal complex size (Itoh and Brown, 1984, 1985, 1986). Because of the ease of protoplast production, and the difference in morphology, these two genera are ideal candidates for producing hybrid cells. From such studies, it may be possible to learn more about the coexpression of diverse and alien genes that lead to altered cell structure and function. In this study, hybrid cells have been obtained from protoplast fusions between *Boergesenia* and *Valonia* using polyethylene glycol (PEG) as the fusogen.

2. MATERIALS AND METHODS

2.1. Cell Culture

Among the Siphonocladales, *Boergesenia forbesii* (Harvey) Feldmann (West #1651) and *Valonia ventricosa* were used. The cells were cultured from protoplasts to mature cells under aseptic conditions in natural seawater (0.028 ppm salinity). *Boergesenia forbesii* was grown in natural seawater enriched with ES medium (5 ml ES/liter seawater), and *Valonia ventricosa* was cultured in natural seawater enriched with ES medium (10 ml ES/liter seawater). Both genera were grown at 27°C with a 16-hr-photoperiod. The ES medium was prepared according to Provasoli *et al.* (1957). The medium for *Boergesenia* was aerated with filtered air. Cells were transferred to a new medium every 3 weeks. *Boergesenia forbesii* cells approximately 1–1.5 cm in length and *Valonia ventricosa* cells approximately 0.5 cm in diameter were selected and kept in darkness for 4–5 days before use.

2.2. Protoplast Formation

Using a needle, a razorblade, and microscissors, the giant cells were gently wounded and cut in half. Under illumination, protoplasts were allowed to form naturally from the protoplasm, which adhered to the inner surface to the cell wall. Between 2 and 3 hr after wounding, protoplasts had completely formed.

2.3. Protoplast Fusion

After 20 min and 60 min of wounding of *Boergesenia* and *Valonia* semicells, respectively, the contents that had not yet formed protoplasts were released from the cell wall with a gentle stream of medium from a Pasteur pipette within the same small Petri dish (5 cm) and gathered into the center by gentle swirling and shaking. Two or three drops of the mixed protoplasm were transferred into a plastic Petri dish and allowed to settle. The mixed protoplasm was surrounded with several drops of 30% polyethylene glycol (PEG 4000 Fisher) enriched with final concentration of 20 mM $CaCl_2$ as the fusogen, and 100 μM cycloheximide was added to the mix to prevent cell wall formation during the fusion trials. The solution and protoplasm were gently mixed together, and the cells underwent fusion. After 40 min in PEG solution, some hybrid protoplasts were formed, and these were washed with *Boergesenia* medium three times.

2.4. Isolation of Hybrid Cells

Protoplasts of *Boergesenia forbesii* and *Valonia ventricosa* have a different appearance, i.e., a bright green color for *Boergesenia* and a dark brownish-green color for *Valonia.* Under the Zeiss stereomicroscope, the hybrid cells were selected by a distinct hemispheric segregation of the two different colors in a single cell. These hybrids were isolated using a fine needle and Pasteur pipette and cultured in either *Boergesenia* or *Valonia* medium with and without aeration.

2.5. Microscopic Methods

Visualization of the hybrid cell wall was accomplished by the use of the fluorescent brightening agent, Tinopal LPW, at a final concentration of 0.005% (w/v) (Haigler *et al.,* 1980). The excess, nonreactive Tinopal was removed by washing with the culture medium. Fluorescence was observed with a Zeiss photomicroscope equipped for fluorescence microscopy (epifluorescence) with an excitation of 365 nm and observation of emission greater than 397 nm.

3. RESULTS

3.1. Protoplast Formation

Protoplasts of *Boergesenia* were formed by cutting the basal portion of an elongated cell. The sequence of protoplast formation is shown in Figure 1A–F. Note the parent cell of *Boergesenia* just after wounding (Fig. 1A); the cell still has the same appearance after the basal tip has been excised. Figure 1C–F shows the various stages of reticulation and spheration during protoplast formation. By the end of 3 hr, the protoplasts were completely formed. Figure 2A–E shows a similar sequence of protoplast formation in *Valonia ventricosa.* Figure 2A is a hemicylinder of a cell that has been cut. At the time of cutting, the protoplast looks like the parent cell (not shown). Figure 2B–F shows various stages of spheration. By the end of 3 hr, spheration is complete in *Valonia ventricosa.*

3.2. Self-Protoplast Fusions

As a control to test the cell fusion procedure, protoplasts of *Boergesenia forbesii* and *Valonia ventricosa* were fused with themselves by a treatment with PEG as the fusogen. A typical example of cell fusion of *Valonia* is shown in Figure 3A–C. Note that of the three protoplasts in the field of view, two progressively fuse. Fusion is complete at the end of 20 min. This self-fused cell will continue to grow normally into a parent cell. The same is true for *Boergesenia.* It is important to note, however, that the fusion of *Boergesenia* protoplasts occurs much faster than *Valonia,* usually within 10 min after PEG treatment. The self-fusion frequency of *Valonia vetricosa* (up to 10%) is much higher than that of *Boergesenia forbesii* (less than 1%).

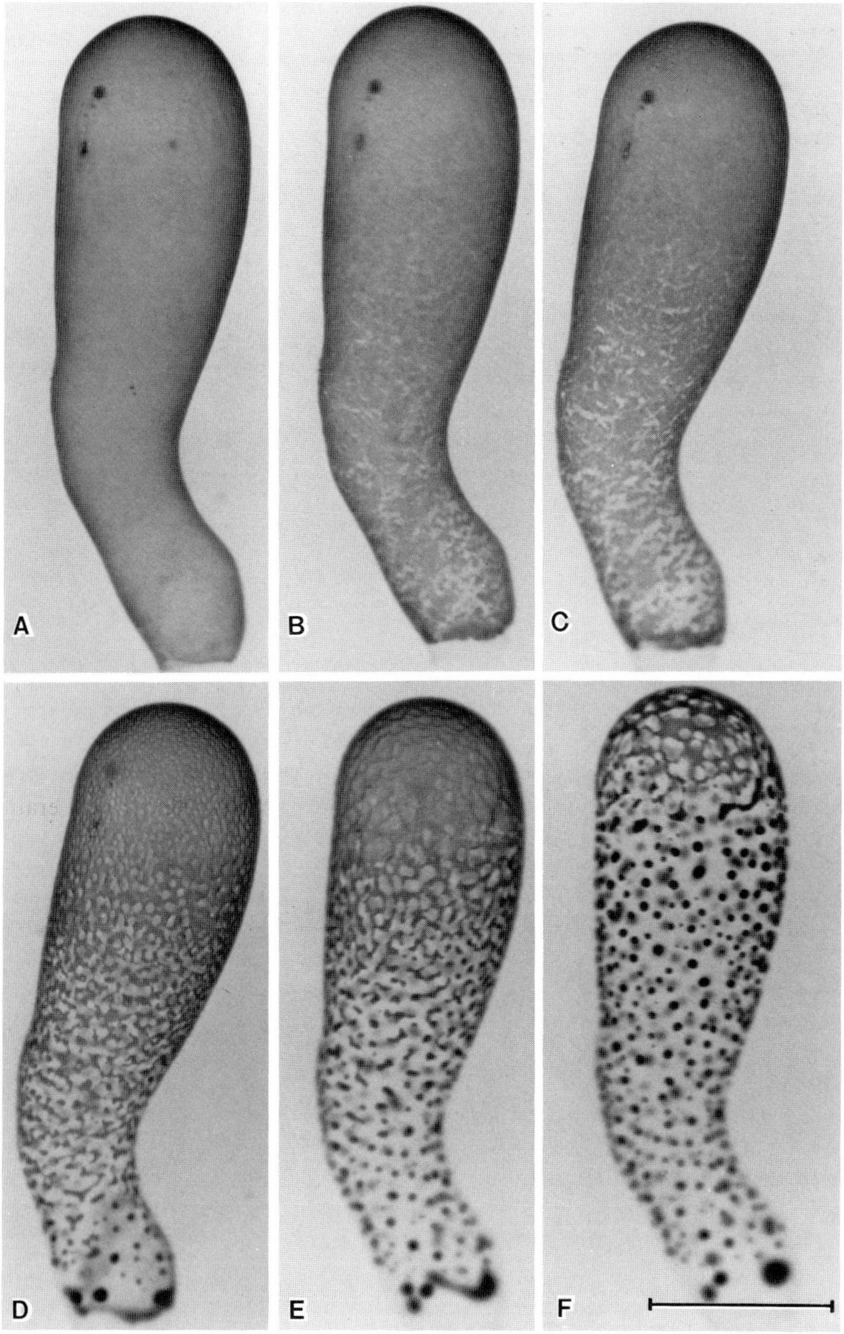

Figure 1. Sequential scenes of protoplast formation of *Boergesenia forbesii.* Protoplasm becomes segregated slightly at 30 min after wounding. Complete protoplast formation occurs within 180 min. Periods: (A) 0 min, (B) 30 min, (C) 60 min, (D) 90 min, (E) 120 min, and (F) 180 min. Bright-field optics, bar: 5 mm.

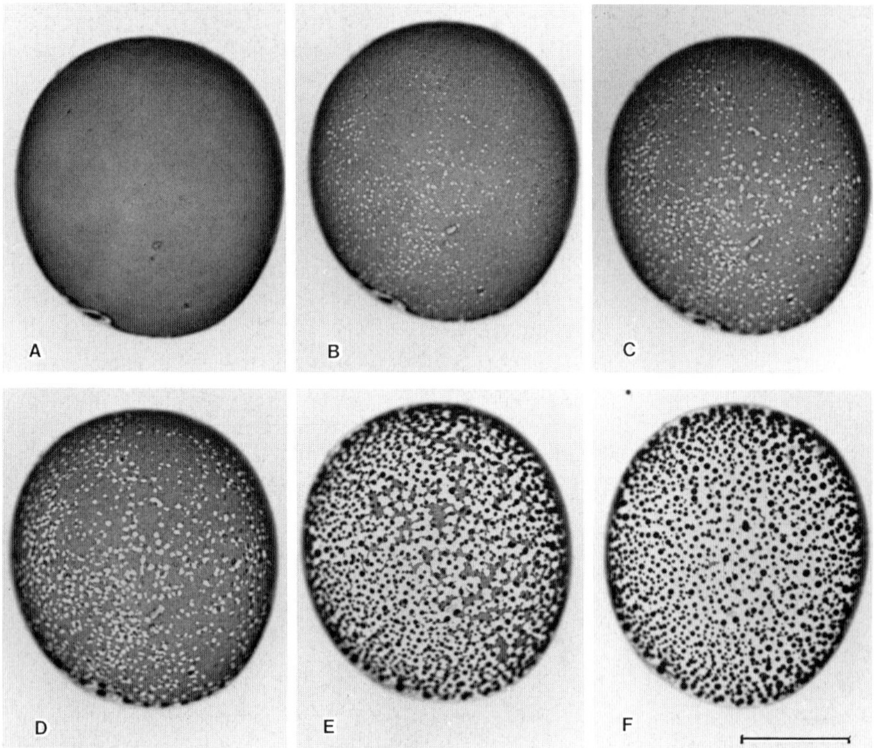

Figure 2. Sequential scenes of protoplast formation of *Valonia ventricosa.* Protoplasts were formed within 180 min after wounding. Periods: (A) 0 min, (B) 60 min, (C) 90 min, (D) 120 min, (E) 150 min, and (F) 180 min. Bright-field optics, bar: 3 mm.

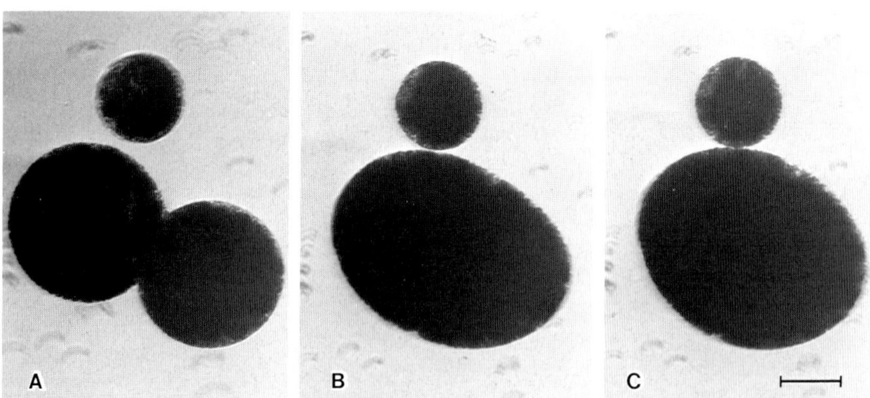

Figure 3. Self-fusion of *Valonia* protoplasts. Protoplasts were treated with 30% PEG-4000. Periods of treatment after adding PEG to the medium: (A) 2 min, (B) 5 min, and (C) 20 min. Bright-field optics, bar: 100 μm.

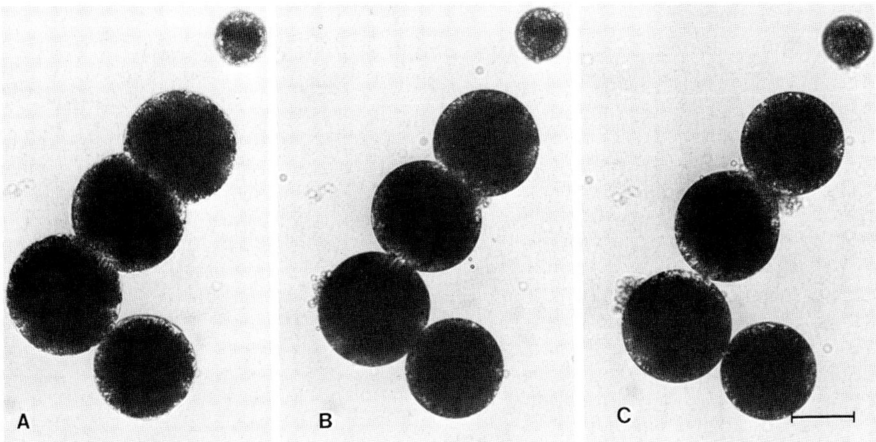

Figure 4. Segregation of incompletely fused protoplasts of *Valonia* in natural seawater (NSW). Periods after removal of PEG and transfer into NSW: (A) 0 min, (B) 20 min, and (C) 120 min. Bright-field optics, bar: 100 μm.

In some cases, protoplasts, which are incompletely fused in the PEG medium, undergo segregation upon removal from PEG medium and transfer to natural seawater (Fig. 4A–C).

3.3. Intergeneric Fusions

The protoplasts of *Boergesenia* and *Valonia* were fused in the presence of PEG. The fused cells were isolated from mixed cell populations of *Boergesenia* and *Valonia* by their differences of color in the cytoplasm. Although these differences are not visible in black and white figures, the colors do not mix (i.e., hybrid cells tend to maintain a segregation of their color). In Figure 5B, the *Boergesenia* cytoplasm appears to be peripheral, while the *Valonia* cytoplasm appears to be located more in the central region of the cell; however, in Figure 5D, there is clearly a polar distribution of *Valonia* cytoplasm (arrow) and *Boergesenia* cytoplasm (arrow). Note the fluorescence photomicrograph of these cells (Fig. 5A,C). The unusual feature of all hybrid cells appears to be the unequal distribution of fluorescence of the cell walls. The area which has the greatest fluorescence appears also to have a thicker cell wall. This seems to come from the area in which the *Boergesenia* cytoplasm appears to be localized.

Note the fluorescence of the cell walls of young *Boergesenia* cells and *Valonia* cells 2.5 hr after wounding (Fig. 6B,D). The fluorescence of the cell walls of the prefused cells is much thinner than that of the hybrid cell, and the fluorescence is much less and more evenly distributed.

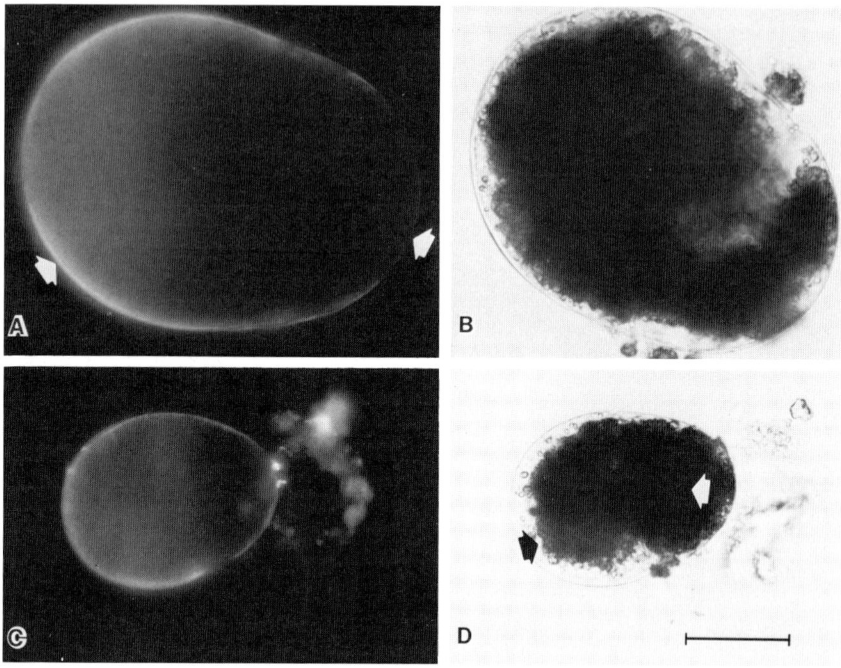

Figure 5. Bright-field and fluorescence microscopy of hybrid cells. The *Boergesenia* and *Valonia* hybrid shows ovoid hybrid cell with brightfield microscopy. (A,C) Fluorescence microscopy of the same cells as in (B,D) Note that the hybrid cells have uneven cell wall (B,D) thickness (arrows). Note uneven distribution of fluorescence in the hybrid cell wall (A,C). Bar: 100 μm.

3.4. Fate of Hybrid Cells

All the hybrid cells (100%) died within 3 weeks. Hybrid cells had thicker cell walls in an early stage than did parental cells (Fig. 5). Some of the hybrid cells had two concentric layers of cell wall (Fig. 7).

4. DISCUSSION

Many fusion studies in higher plants have been accomplished using protoplasts (Evans *et al.,* 1980; Kao and Michayluk, 1974); however, the disadvantage of vascular plant protoplasts is the requirement for enzymatic removal of the cell wall which may affect cell viability and alter normal cell wall biosynthesis. In our experimental system, *Boergesenia forbesii* and *Valonia ventricosa* undergo natural protoplast formation in re-

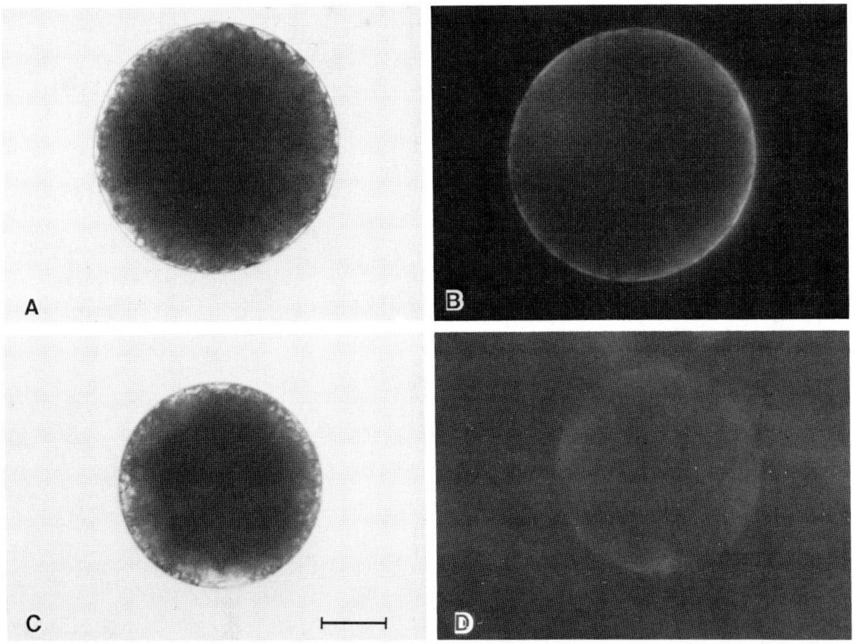

Figure 6. Bright-field and fluorescence microscopy of *Boergesenia* and *Valonia* protoplasts 2.5 hr after wounding. (A,B) *Boergesenia;* (C,D) *Valonia.* Note that the cell wall of *Boergesenia* has already regenerated in contrast to *Valonia.* Bar: 50 μm.

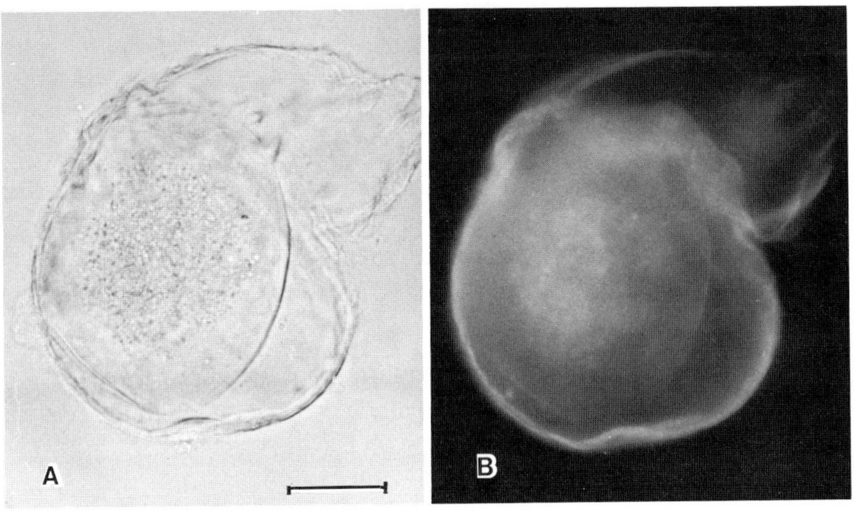

Figure 7. Bright-field and fluorescence microscopy of a hybrid cell 4 weeks after fusion. Note that it has two layers of the cell wall. Bar: 100 μm.

sponse to physical perturbation. With the ease of synchronized protoplast formation, it is possible to obtain large quantities of protoplasts and to study cell wall regeneration and cellulose biosynthesis without alteration due to enzymatic treatments. Cell wall regeneration and cellulose biosynthesis of *Boergesenia* and *Valonia* have been studied, and therefore many characteristics, e.g., cell wall thickness, microfibril assembly, microfibril orientation, microfibril size, and terminal complex size, have been elucidated (Itoh and Brown, 1984). Another advantage of this experimental algal model system is that there is a natural marker to select hybrid cells in the mixed population of parental cells. This marker is the color of the chloroplast and can be distinguished by microscopic examination.

The fusion frequency between *Boergesenia forbesii* and *Valonia ventricosa* is variable, depending on parental cell conditions. Older cells grown in darkness for several days are more apt to fuse than younger and metabolically active cells grown in the light and fresh medium. As a fusogen, the PEG solution was used in the presence of calcium ions and cycloheximide. In this experiment, Ca^{2+} did play an important role to accelerate fusion. Because cycloheximide inhibits cellulose biosynthesis through inhibition of protein synthesis in *Boergesenia forbesii* (Itoh *et al.*, 1986), it was used to prevent early wall deposition so *Boergesenia* cells would be in a condition for fusion with *Valonia* protoplasts.

The ability of cells to undergo a reversal in the fusion process is interesting inasmuch as this reversal may reflect as an internal induction mechanism. Further study of the frequency of reversal may reveal the mechanism of segregative division. Protoplast induction, segregative division, and fusion reversals all may have a common underlying mechanism.

The hybrid cells are not viable for long periods after fusion. All the hybrid cells died within 3 weeks. There might be several possibilities to explain this. Because *Boergesenia* and *Valonia* are two different genera within the Siphonocladales, it is highly probable that there is genetic incompatibility between two species we used. A gradual loss of color of a small portion of *Valonia* cytoplasm incorporated into *Boergesenia* protoplast was observed. The hybrid cells remained viable for 1–2 weeks, but they did not grow well in the same medium in which parental cells grew and enlarged. Aeration also had no effect. It is possible that the hybrid cells might require other ingredients in the medium as well as different culture conditions. In Figure 7, some hybrid cells have two layers of the cell wall. It was not known which cell wall (inner or outer) may come from *Boergesenia* and *Valonia.* One cell wall might be synthesized from the *Boergesenia* genome and the other from *Valonia,* or the two walls might be a result of dual gene expression. It is possible that the outer cell wall that first regenerated from one of the two protoplasms in an early stage of fusion may inhibit the metabolism of the hybrid cell, therefore inhibiting its growth.

These giant algal cells have advantages of natural protoplast formation of great utility to achieve high rates of cell wall regeneration, cellulose biosynthesis, and the ease of obtaining of synchronized protoplasts. The Siphonocladalean algae can be also used as model systems for incorporation of foreign genetic materials and studying the effects of transformation. Fusion of *Agrobacterium* and *Escherichia coli* spheroplasts with *Nicotiana tabacum* protoplasts was reported as a direct gene transfer (Hain *et al.,* 1984). It should be possible to incorporate foreign genetic material, i.e., *Agrobacterium* and Ti-plasmids, into these giant algal cells during protoplast formation after mechanical wounding. Such advancements would pave the way to improve our understanding of the genetic regulation of development and cell wall assembly in these interesting algae.

ACKNOWLEDGMENTS. We thank Dr. John West, of the University of California, Berkeley, for providing the cultures used in the study, Mrs. Pamela Goff for typing the manuscript, and Richard Santos for advice on the cell cultures. We also thank Dr. Arthur Sowers for helpful suggestions with the cell fusion studies.

REFERENCES

Bold, H. C., and Wynne, M. J., 1978, *Introduction to the Algae,* Prentice-Hall, Englewood Cliffs, New Jersey.

Burr, F. A., and West, J. A., 1970, Light and electron microscope observations on the vegetative and reproductive structures of *Bryopsis hypnoides, Phycologia* **9**:17–37.

Burr, F. A., and West, J. A., 1971*a,* Comparative ultrastructure of the primary nucleus in *Bryopsis* and *Acetabularia, J. Phycol.* **7**:108–113.

Burr, F. A., and West, J. A., 1971*b,* Protein bodies in *Bryopsis hypnoides:* Their relationship to wound healing and branch septum development, *J. Ultrastruct. Res.* **35**:476–498.

Chaleff, R. S., 1981, Genetics of higher plants—Application of cell culture, Cambridge University Press, New York, pp. 96–147.

Cocking, E. C., 1960, A method for the isolation of plant protoplast and vacuoles, *Nature (Lond.)* **187**:962–963.

Enomoto, S., and Hirose, H., 1972, Culture studies on artificially induced aplanospores and their development in the marine alga *Boergesenia forbesii* (Harvey) Feldmann (Chlorophyceae, Siphonocladales), *Phycologia* **11**:119–122.

Evans, D. A., Wetter, L. R., and Gamborg, O. L., 1980, Somatic hybrid plants of *Nicotiana glauca* and *Nicotiana tabacum* obtained by protoplast fusion, *Physiol. Plant.* **48**:225–230.

Gardner, K. H., and Blackwell, J., 1971, The substructure of the cellulose microfibrils from the cell walls of the algae *Valonia ventricosa, J. Ultrastruct. Res.* **36**:725–731.

Haigler, C. H., Brown, R. M., Jr., and Benziman, M., 1980, Calcofluor white ST alters the *in vivo* assembly of cellulose microfibrils, *Science* **210**:903–906.

Hain, R., Steinbiss, H. -H., and Schell, J., 1984, Fusion of *Agrobacterium* and *E. coli* spheroplasts with *Nicotiana tabacum* protoplasts: Direct gene transfer from microorganism to higher plant, *Plant Cell Res.* **3**:60–64.
Itoh, T., and Brown, R. M., Jr., 1984, The assembly of cellulose microfibrils in *Valonia macrophysa* Kütz, *Planta* **160**:372–381.
Itoh, T., O'Neil, R., and Brown, R. M., Jr., 1985, Interference of cell wall regeneration of *Boergesenia forbesii* protoplasts by Tinopal LPW, a fluorescent brightening agent, *Protoplasma* **123**:174–183.
Itoh, T., Legge, R. L., and Brown, R. M., Jr., 1986, The effects of selected inhibitors on cellulose microfibril assembly in *Boergesenia forbesii* (Chlorophyta) protoplasts, *J. Phycol.* **22**:224–233.
Kao, K. N., and Michayluk, M. R., 1974, A method for high-frequency intergeneric fusion of plant protoplasts, *Planta* **115**:355–367.
La Claire, J. W., II, 1982, Cytomorphological aspects of wound healing in selected siphonocladales *(Chlorophyceae), J. Phycol.* **18**:379–384.
Preston, R. D., 1968, Plants without cellulose, *Sci. Am.* **218**:102–108.
Preston, R. D., and Kuyper, B., 1951, Electron microscopic investigations of the walls of green algae. I. A preliminary account of wall lamellation and deposition in *Valonia ventricosa, J. Exp. Bot.* **2**:247–256.
Provasoli, L., McLaughlin, J. J. A., and Droop, M. R., 1957, The development of artificial media for marine algae, *Arch. Mikrobiol.* **25**:392–428.
Salhani, N., Vienken, J., Zimmermann, U., Ward, M., Davey, M. R., Clothier, R. H., Balls, M., Cocking, E. C., and Lucy, J. A., 1985, Haemoglobin synthesis and cell wall regeneration by electric field-induced interkingdom heterokaryons, *Protoplasma* **126**:30–36.

II

Fusion in Model Membranes

Chapter 9

Membrane–Membrane Interactions via Intermediates in Lamellar-to-Inverted Hexagonal Phase Transitions

Dave P. Siegel

1. INTRODUCTION

Recent theoretical work (Siegel, 1984, 1986*a–c*) shows that vesicles whose lipids can undergo the L_α/H_{II} (lamellar-to-inverted hexagonal) phase transition should be especially susceptible to three membrane–membrane interactions: fusion, aggregation-induced vesicle leakage, and intervesicular lipid exchange. This chapter describes this model as well as experimental results derived from relevant systems. The model shows that in H_{II}-forming systems the relative rates of the three membrane–membrane interactions may be determined by the lipid-phase behavior and the dynamics of the L_α/H_{II} phase transition, i.e., the rate at which a given lipid system can execute the L_α/H_{II} phase transition under given conditions.

Although the model has been developed for pure one- and two-component lipid systems, and the supporting data have been derived from such systems, the possible relevance of this work to biomembrane

Dave P. Siegel • Miami Valley Laboratories, Proctor and Gamble Company, Cincinnati, Ohio 45247.

behavior will be discussed. At present, the relevance (if any) of nonbilayer structures to biomembrane function can only be inferred from experiments on model systems and from morphological studies of biomembrane preparations.

2. RELEVANCE OF LIPID-PHASE TRANSITIONS TO MEMBRANE–MEMBRANE INTERACTIONS

The mechanisms of lipid bilayer fusion are poorly understood. However, it has long been known that thermotropic (e.g., Lawaczeck *et al.*, 1976; Nicolussi *et al.*, 1982) and cation-induced (e.g., Papahadjopoulos *et al.*, 1977; Portis *et al.*, 1979) liquid crystalline phase transitions are associated with the fusion process in model membrane systems. The most-studied liquid crystalline phase transition relevant to bilayer systems is the melting of the array of lipid acyl chains composing the bilayer, which transforms the bilayer from a gel to a more fluid state. This is called the gel–liquid crystalline or $L_{\beta'}/L_\alpha$ phase transition. Membrane fusion rates are enhanced when vesicle membranes undergo these phase transitions in model membrane systems. It has often been speculated (e.g., Portis *et al.*, 1979) that membrane fusion occurs at structural defects in the two apposed lipid bilayers. The defects in such systems were proposed to be either intermediates in these phase transitions or patches of the second, nascent phase in the membranes—patches of $L_{\beta'}$, or gel phase in membranes in the L_α or liquid crystal phase.

Several years ago, reports appeared (e.g., Verkleij *et al.*, 1979, 1979*a*, 1980; Rand *et al.*, 1981; Mandersloot *et al.*, 1981; Hope *et al.*, 1983) that the rate of fusion and other vesicle interactions is enhanced when the lipids in the system can undergo the lamellar-to-inverted hexagonal (L_α/H_{II}) phase transition. (The structure of the H_{II} phase is depicted in Figure 1.) This implies that ability to form this latter type of liquid crystalline phase is associated with capability to undergo fusion and other intermembrane interactions. This is quite plausible, considering the nature of the L_α/H_{II} transition: It is an inherently interbilayer event, requiring close apposition of L_α-phase bilayers in order to occur (Siegel, 1984, 1986*a*). If vesicles aggregate under circumstances in which their lipids are either in the H_{II} phase at equilibrium or close to the L_α/H_{II} phase boundary, intermediates in this phase transition should form and result in one or more vesicle–vesicle interactions (fusion, intervesicular lipid exchange, aggregation-induced leakage).

This is what is observed in many systems (Section 3). Recent theoretical work (Siegel, 1984, 1986*a–c*) has provided a unifying explanation of this range of behavior in terms of the structure and dynamics of inter-

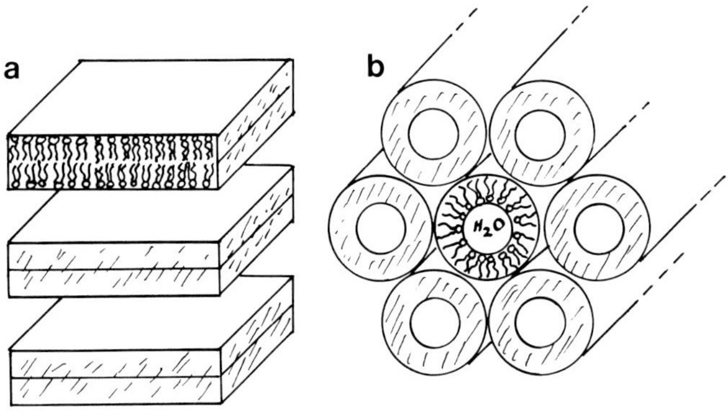

Figure 1. (a) Lamellar (L_α) phase of phospholids, consisting of stacks of liquid crystalline bilayers. (b) Inverted hexagonal (H_{II}) phase, consisting of water rods enclosed in cylindrical lipid monolayers. This phase resembles a hexagonally packed array of cylindrical inverted micelles.

mediates in the L_α/H_{II} phase transition. The predictions of this work have been borne out by studies of the behavior of thermotropic of L_α/H_{II} system. It appears that the L_α/H_{II} phase transition is another liquid crystalline phase transition critical to our understanding of dynamics in biomembrane systems.

I will now briefly describe our current concepts of the structure and dynamics of intermediates in L_α/H_{II} phase transition and how they can be used to predict the dynamics of membrane–membrane interactions.

3. STRUCTURE AND DYNAMICS OF THE INTERMEDIATES IN L_α/H_{II} TRANSITIONS

3.1. Lipidic Particles

Cullis *et al.* (1978) were the first to publish freeze-fracture electron micrographs of beadlike deformations of the bilayers about 10 nm in diameter in systems undergoing L_α/H_{II} transitions. This morphology was observable when the ^{31}P-NMR spectrum obtained from the system contained an isotropic component, indicating that the lipid phosphate groups experienced all orientations on the time scale of the nuclear magnetic resonance (NMR) measurement. These workers noted that such motional averaging is consistent with diffusion of molecules around the interior of inverted micellar structures. Cullis and Hope (1978) had earlier suggested that inverted micellar structures might be involved in the L_α/H_{II} phase transition. These objects were called lipidic intramembranous particles. Initially it was unclear whether this morphology represented structures forming within one bilayer or structures forming between two bilayers (for reviews, see Verkleij, 1984; Siegel, 1984).

It is now fairly clear that lipidic particles are actually interbilayer structures like the one depicted in Figure 2(c). This structure consists of a spherical inverted micelle contained within a semitoroidal (hourglass-shaped) outer monolayer. These have been called inverted micellar intermediates (IMI) (Siegel, 1984, 1986*a*) because of their role in the L_α/H_{II} phase transition (see Section 2.2). Siegel (1984, 1986*a*) showed that IMI should have properties consistent with the available electron microscopic and ^{31}P-NMR data concerning lipidic particles and that the nature and kinetics of the L_α/H_{II} phase transition requires the formation of structures like IMI.

A kinetic model of IMI formation rates (Siegel, 1984, 1986*a*) predicts that IMI should be short-lived structures in dynamic equilibrium with the

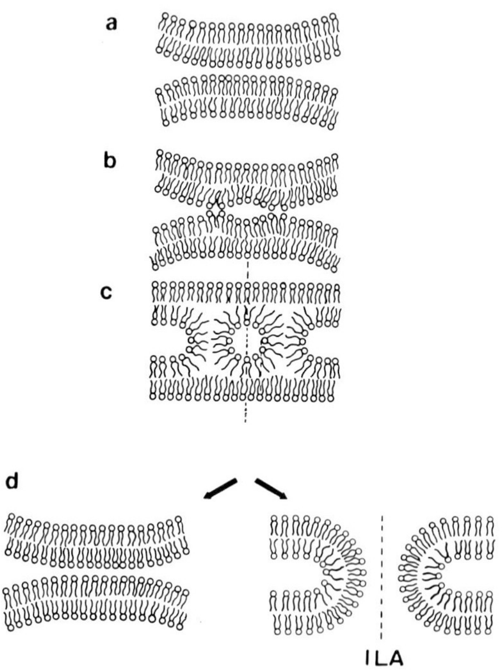

Figure 2. IMI and ILA formation and structure. *IMI Formation*: Two L_α-phase bilayers (a) are closely apposed (b). The interfaces merge at the periphery of a circular patch of lipid and form an IMI (c). The IMI is cylindrically symmetrical about the vertical axis. The inverted micelle in the center is spherical. The external monolayer is hourglass shaped. IMI assemble into the H_{II} phase by coalescing with other IMI that form between the same two apposed bilayers, and are rapidly consumed by this process in systems with rapid L_α–H_{II} phase transitions. (From Siegel, 1968*a*.) *ILA Formation*: (d) IMI are labile structures and can revert to either of two bilayer structures. The IMI can simply re-form the original patch of apposed bilayers (left); this is simply the reverse of the IMI formation process: sequence (c), (b), (a). More rarely, an IMI can form another type of bilayer structure; the ILA (right). The ILA is also hourglass shaped, and is a bilayer connection between the two original apposed membranes (a). ILA formation results in fusion of the original apposed membranes. (From Siegel, 1986*a*.)

L_α phase from which they are formed. The IMI formation process is outlined in Figure 2. Two bilayers, when closely apposed over a small area, can form an IMI when the system is above or in close proximity to the L_α/H_{II} phase transition. The system must be near the L_α/H_{II} phase boundary because the IMI structure is very similar to that of the H_{II} phase (three out of four principal radii of curvature of the IMI are of the same sign and magnitude as in the H_{II} phase and the arrangement of the acyl chains of the lipids is similar). If the H_{II} phase is unstable when the bilayers are apposed, IMI will also be unstable, and the activation energy for IMI formation will be too high. In systems with thermotropic L_α/H_{II} transitions (e.g., phosphatidylethanolamines), it was estimated (Siegel, 1986*a*) that IMI will not form in significant numbers at temperatures more than 10° below the L_α/H_{II} phase-transition temperature T_H.

The average IMI lifetime in unsaturated acyl-chain phosphatidylethanolamine (PE) systems at T_H was estimated (Siegel, 1984, 1986*a*) to be ~1 msec or less. This lifetime is predicted to increase when the system is diluted with lipids that can only adopt the L_α phase under the ambient conditions. Such short IMI lifetimes may explain why IMI morphology (lipidic particles) is not observed in some PE systems (Verkleij, 1984) but is observed in mixtures of PE and phosphatidylcholines (PC) (e.g., Hui *et al.,* 1981). Lipidic particle morphology is very commonly observed in Ca^{2+}-cardiolipin systems, where L_α/H_{II} transitions can occur quite slowly (e.g., De Kruijff *et al.,* 1982). This finding is compatible with the model developed in Siegel (1984, 1986*a*), which shows that IMI lifetimes should be longer in systems with these ionotropic transitions.

3.2. Inverted Micellar Intermediates and the L_α/H_{II} Phase Transition

The appearance of lipidic particles was associated with the L_α/H_{II} phase transition in many of the early reports. It is now fairly clear that the L_α/H_{II} phase transition can only occur when two or more bilayers are apposed, and that IMI are the first structures to form during the transition process (Siegel, 1984, 1986*a*). IMI, diffusing within the planes of apposed bilayers, coalesce with other IMI and form H_{II} phase precursors (reviewed in Verkleij, 1984; Siegel, 1984, 1986*a*). This is essentially a two-dimensional aggregation reaction, and the rate of this process can be estimated. The details of how the H_{II} phase is assembled and the kinetics of this process (Siegel, 1986*a*) are beyond the scope of this review. It is simply important to bear in mind that in rapid L_α/H_{II} phase transitions, IMI that form between apposed bilayers are probably rapidly consumed via H_{II} phase formation. IMI are key intermediates in the L_α/H_{II} phase-transition process because they are the first interbilayer structures to form and because the number of IMI that form per unit area of apposed bilayers determines the rate of the L_α/H_{II} phase transition in a given system (Siegel, 1986*a*). In systems and circumstances in which very few IMI form, the transition should be very slow.

Others have suggested different L_α/H_{II} transition mechanisms (Borovjagin *et al.*, 1982; Hui *et al.*, 1983; Caffrey, 1985). It has been argued (Siegel, 1986*a*) that these routes are very unlikely because they require extremely unstable intermediates. The activation energies for formation of such structures are so high as to produce infinitesimal transition rates. The model of Caffrey also does not account for lipidic particle morphology.

3.3. Interlamellar Attachments

In most H_{II}-forming systems, the transition is rapid and the IMI formed between apposed bilayers are rapidly consumed via coalescence into H_{II} precursors. However, there is a subset of systems predicted (Siegel, 1986*a*,*b*) to have slow, hysteretic transitions. In these systems, another type of metastable intermediate structure is predicted to form from IMI. Morphological evidence exists for such structures. They are particularly interesting because they result in fusion of apposed membranes.

IMI should be in dynamic equilibrium with the patches of apposed bilayers from which they form (Fig. 2c,d). However, it was shown (Siegel, 1986*b*) that under some circumstances IMI could revert to another type of bilayer structure, the interlamellar attachment (ILA) (Fig. 2d, right). An ILA is an hourglass-shaped bilayer attachment between the two original lamellae. Morphology consistent with such a structure is often observed (e.g., Fig. 6 of Verkleij, 1984) in lipid mixtures in which ILA formation is expected to be rapid. Molecules diffusing along the surfaces of an ILA experience all orientations of their head groups with respect to the bilayer normal on a submillisecond time scale, resulting in the isotropic [^{31}P]-NMR resonances often observed in systems exhibiting this morphology. The model developed in Siegel (1986*a*,*b*) predicts that IMI should in general form ILA much less frequently than patches of apposed bilayers (Fig. 2d). In an earlier paper on inverted micellar structures, it was assumed that IMI could revert with equal frequency to patches of planar bilayer or to ILA (Siegel, 1984). The subsequent analysis (Siegel, 1986*a*,*b*) described here shows that this is probably incorrect.

ILA formation rates can be estimated using the equilibrium dimensions of the L_α and H_{II} phase of the lipid system. The key parameter is the ratio of the areas of the lipid head groups at the lipid–water interfaces of the L_α and H_{II} phases at equilibrium. If this ratio is greater than about 1.2, ILA should form very rarely near the L_α/H_{II} phase boundary. This is true in systems with fast L_α/H_{II} transitions, i.e., systems in which the L_α phase can form the H_{II} phase within seconds or less when superheated above the equilibrium transition temperature by a few degrees Kelvin or less (Siegel, 1986*a*). In egg PE, for example, which executes the L_α/H_{II} transition within ~1 sec or less (Cullis and De Kruijff, 1978; Verkleij, 1984) this ratio ≈1.8 (head group area values from Lis *et al.,* 1982; Reiss-Husson, 1967).

In systems with a head group area ratio of about 1.2 or smaller, ILA are predicted to form at a significant rate (≈0.1–1/IMI per second). Small head group area ratios generally correspond to larger H_{II} tube diameters for given L_α phase head group areas. In these systems, the rate of ILA and IMI coalescence will be similar (Siegel, 1986*a,b*). H_{II} structures should form slowly, and many ILA will form. ILA should be metastable (Siegel, 1986*c*) and can pack into long-lived arrays with the attributes of the so-called inverted cubic phase described by others (e.g., Boni and Hui, 1983; Hui *et al.,*1983). They also should exhibit the persistent isotropic [^{31}P]-NMR signals observed (Gagné *et al.,* 1985) in systems with slow and hysteretic L_α/H_{II} transitions. Thus, an aspect of the phase behavior of these lipid systems may be understood in terms of the dynamics of L_α/H_{II} transition intermediates.

It is interesting to note that the systems in which phospholipid cubic phases have been carefully studied are systems that should have the smaller head group area ratios as predicted in Siegel (1986*a,b*). PE–PC mixtures (Boni and Hui, 1983; Hui *et al.,* 1983), will have head group areas in the L_α and H_{II} phases that are more similar than in pure PE due to inclusion of large PC head groups. This tendency toward cubic (isotropic) phase formation or large lipidic particle (i.e., ILA) morphology and persistant isotropic [^{31}P]-NMR signals is generally observed in a wide range of H_{II}-forming lipids containing small mole fractions of lipids that only form lamellar phases (e.g., Brentel *et al.,* 1985; Tilcock *et al.,* 1982; for reviews, see Verkleij *et al.,* 1984; Quinn and Williams, 1983). According to the model in Siegel (1986*a,b*), this corresponds to increasing prevalence of ILA on account of a reduction in head group area ratio and IMI density by the lamellar impurity.

It should be noted that the head group area ratio of a lipid system is a reflection of the intrinsic curvature parameter discussed in the elegant work of Gruner and co-workers (Kirk *et al.,* 1984; Gruner, 1985; Gruner *et al.,* 1985), who have developed a powerful model for the factors determining the relative stability of inverted phases. Their model interprets the stability of amphipathic liquid crystalline systems in terms of an equilibrium radius of curvature of the lipid–water interfaces. The sign and magnitude of this curvature is determined by the geometry and intermolecular forces of the lipid present. In the present model (Siegel, 1986*b*), small values of the head group area ratio correspond to large intrinsic radii of curvature.

3.4. Dynamics in Ionotropic versus Thermotropic Transitions

In some lipid systems, the L_α/H_{II} phase transition is driven by the binding of cations to anionic lipids to form complexes of defined stoichiometry; examples are Ca^{2+}–cardiolipin (Rand and Sengupta, 1972; De Kruijff *et al.*, 1982) Ca^{2+}–phosphatidic acid (Verkleij *et al.*, 1982), and H^+–cardiolipin (Seddon *et al.*, 1983). The kinetics of L_α/H_{II} transitions and relative IMI and ILA formation rates in such systems are only qualitatively described by the model in Siegel (1986*a–c*). This is because the chemical potentials and equilibrium curvatures of different intermediates are sensitive to the local cation concentration (e.g., De Kruijff *et al.*, 1982). Free energies for formation and steady-state populations of the intermediates can be very different under different circumstances. By contrast, in thermotropic systems the chemical potential gradient between L_α and H_{II} phases changes slowly with temperature, which simplifies modeling of transition dynamics (Siegel, 1986*a,b*). Moreover, since anionic lipid–cation complexes are very stable (several kT), and many complexes must dissociate during interconversion of intermediates (e.g., ILA formation from IMI), such transitions are very hysteretic (De Kruijff *et al.*, 1982) and intermediate formation rates are hard to predict. It is likely (Siegel, 1986*b*) that ionotropic systems should form ILA in greater numbers than thermotropic systems with similar head group area ratios.

4. IMPLICATIONS FOR MEMBRANE–MEMBRANE INTERACTIONS

4.1. Summary of the Fate of IMI Forming between Two Apposed Membranes

IMI that form between apposed bilayer are consumed by the following processes:

1. Reversion to apposed planar patches of the original bilayers (Fig. 2d, left-hand side)
2. Coalescence with other IMI into H_{II}-phase precursors (which occurs in seconds or less in most H_{II}-forming systems in which the bilayers can easily be closely apposed)
3. ILA formation (Fig. 2d, right), which is usually an unobservably slow process compared with H_{II} formation, and only has a chance to occur when the head group ratio of the system is in the right range ($\leqslant 1.2$) (see Section 2.3), and the system forms small num-

bers of IMI, e.g., infrequent close apposition of bilayers (see Siegel, 1986*a*)

It should be obvious that if the two apposed bilayers are the walls of two unilamellar vesicles, IMI will mediate three types of vesicle–vesicle interactions (Fig. 3). These are described below and summarized in Table I.

4.3. Intervesicular Lipid Exchange

When an IMI forms, it makes the facing monolayers of the two original bilayers momentarily continuous (Fig. 3); molecules on the interface of one bilayer can diffuse onto the other via the external monolayer of

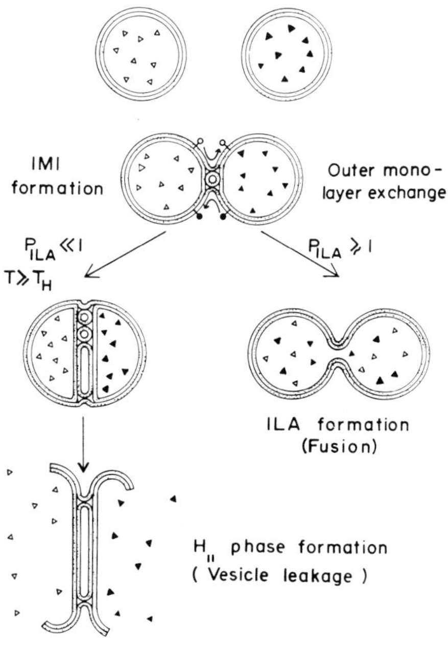

Figure 3. Vesicle–vesicle interactions that can occur via IMI and ILA. (Top) IMI formation between two vesicles make the outer monolayers of the two vesicles momentarily continuous. Lipid molecules on the surface of one vesicle can diffuse then onto the other. Outer monolayer lipid mixing can occur as soon as IMI can form between apposed vesicles. Generally, this should occur at temperatures starting 5°–10°K below T_H and higher. (Middle) When the H_{II} phase is the equilibrium phase of the lipids (temperature $T \geqslant T_H$), the IMI that form between apposed vesicles are consumed by both ILA formation and H_{II} phase formation. If the head group area ratio (Section 2.3) is large, the ILA formation rate (P_{ILA}) is very small ($\ll 1$ sec^{-1} per IMI), and the IMI only assemble into the H_{II} phase (left). The H_{II} phase is more stable than the L_α phase, and the walls of the vesicles will distort to appose as much bilayer area as possible in order to undergo the L_α/H_{II} phase transition. This produces a tension in the vesicle walls that eventually grows big enough to rupture them (lower left), which results in leakage of the vesicle contents. Complete lipid mixing also results via IMI formation and edgewise joining of membrane fragments of ruptured vesicles. If the head group area ratio is small ($P_{ILA} \gtrsim 1$ sec^{-1} per IMI), ILA can form (right) in some proportion of vesicle–vesicle encounters, resulting in fusion (mixing of both lipids and aqueous contents without leakage). The relative extents of these processes is a function of temperature, head group area ratio, and vesicle size (see Table 1). (From Siegel, 1986*b*.)

Table I. Summary of Predicted Vesicle–Vesicle Interactions via L_α/H_{II} Phase-Transition Intermediates: Thermotropic Systems[a]

Temperature (T)	Head group area ratio[b] (Z)	Interactions
$(T_H - 10°K) \leqslant T \leqslant T_H$	Any value	Mixing of lipids in outer monolayers of vesicles Some fusion possible if $Z \leqslant 1.2$[c]
$T \geqslant T_H$	> 1.2	Complete lipid mixing and vesicle leakage after aggregation
$T \geqslant T_H$	≤ 1.2	Fusion in many vesicle–vesicle encounters and leakage in others; the ratio fusion/leakage depends on Z, T, and the average vesicle size
Special cases: (a) Vesicles 0.1 μm in diameter		
$T \geqslant T_H$	> 1.2	Slow fusion in initial vesicle–vesicle encounters until the average vesicle diameter grows >0.1 μm; then leakage
(b) Vesicles about 0.02 μm in diameter		
	Any value	Contact areas between vesicles too small to form IMI or ILA; interactions probably occur via other mechanisms

[a]The behavior described in Sections 3.2–3.4 is summarized (see also Fig. 3).
[b]The head group area ratio, Z, is the ratio of the area per lipid head group in the L_α phase at equilibrium divided by the same area in the H_{II} phase. The critical Z value (1.2) for rapid ILA formation is a theoretical estimate (Siegel, 1986*b*): Z > 1.2 corresponds to small ILA formation rates ($P_{ILA} \ll 1\ sec^{-1}$ per IMI); $Z \leqslant 1.2$ corresponds to rapid ILA formation rates ($P_{ILA} \approx 1\ sec^{-1}$ per IMI).
[c]Note that "fusion" is defined as mixing of lipids and aqueous contents of two aggregated vesicles without leakage of the aqueous contents to the external medium.

the IMI. If the outer monolayer of one vesicle contains labeled lipids, e.g., fluorescence resonant energy transfer probes (Struck *et al.*, 1981), they will be diluted via diffusion onto the other vesicle. Siegel (1986*b*) showed that formation of a single IMI between two vesicles should mix the outer monolayer lipids of vesicles up to ~0.1 μm in diameter.

This sort of lipid exchange should be observed whenever IMI can form, which is whenever the vesicles can be closely apposed (i.e., no strong repulsive forces preventing aggregation) and the system is in the vicinity of the L_α/H_{II} transition. In thermotropic systems like PE with unsaturated acyl chains, it was estimated (Siegel, 1986*b*) that IMI-mediated lipid exchange should be observed at temperatures as much as 10°K below T_H and above and should occur in less than one second after vesicle aggregation at temperatures around T_H.

4.3. Aggregation-Induced Vesicle Leakage via IMI and H_{II} Phase Formation

The H_{II} phase can only form when two or more bilayers are apposed. If two vesicles are apposed when the equilibrium phase of their lipids is the H_{II} phase, IMI will form between them and assemble into the H_{II} phase, just as in multilamellar systems (Fig. 3). The favorable chemical potential differences between the H_{II} and L_α phases will drive close apposition of as much of the surface of the vesicles as possible in order to transform as much bilayer into the H_{II} phase as possible. This develops a tension in the vesicle walls that will eventually become sufficient to rupture them (Siegel, 1986*b*); the result is leakage of the vesicle contents into the suspending medium (Fig. 3). This should occur within a time after vesicle aggregation that is comparable to the bulk phase transition time, seconds or less for unsaturated acyl chain PE systems (e.g., Ranck *et al.*, 1984). It is possible that leakage can occur even before H_{II} tubes form. IMI are high-curvature defects in the original bilayers and may make them much more susceptible to rupture under osmotic or other stresses.

According to the model in (Siegel 1984, 1986*a,b*), the transition occurs via the coalescence of many IMI into H_{II} precursors. Thus, the apposed areas of two vesicles must be large enough to contain, on the average, two or more IMI or the transition (and leakage) cannot occur. It was estimated (Siegel, 1986*b*) that the contact areas between unsaturated acyl chain PE vesicles smaller than about 0.1 μm in radius might be too small to form this many IMI. This would also be true of larger vesicles in systems with slower L_α/H_{II} transitions. In such cases, the aggregated vesicles should not leak rapidly until they grow larger through fusion, (e.g., by ILA; see Section 3.4) and could appose more of their surface areas in vesicle aggregates. There is evidence for this type of behavior (see Section 3.5). Very small (0.02-μm) vesicles probably have contact areas too small for IMI formation.

4.4. Membrane Fusion via ILA Formation

If the head group area ratio of the lipids in a vesicle system is in the right range ($\leqslant 1.2$), IMI that form between apposed vesicles can form ILA. This is most probable for lipid systems in which multilamellar preparations form so-called isotropic or inverted cubic phases discussed in Section 2.3, because it is likely that these structures are formed from ILA. Systems that should have this phase behavior include mixtures of H_{II} lipids with small mole fractions of L_α-phase impurities.

ILA may form at larger rates in unilamellar vesicle dispersions than in large multilamellar liposomes. Most systems should form a steady-state concentration of 10^9–10^{11} IMI/cm^2 of apposed bilayers during the L_α/H_{II} phase transition (Siegel, 1986*b*). The initial area of contact between two 0.1-μm-diameter vesicles is probably on the order of 10^{-11} cm^2 (i.e., patch ~0.04 μm diameter). Thus, in many systems there may be on the average fewer than two IMI existing in the area of contact at any time. Under these circumstances, IMI are not rapidly consumed via H_{II} phase formation, so that ILA formation does not have to compete for the available pool of IMI. Thus, even if the ILA formation rate is very small, fusion may be observable in unilamellar preparations of lipids which don't form many ILA (i.e., cubic or isotropic phases) in multilamellar samples. Fusion via ILA should occur in seconds after vesicle aggregation in systems with small head group area ratios, and less rapidly in systems with larger ratios.

The ILA formation rate should decrease rapidly with decreasing temperature below T_H (Siegel, 1986*b*), and fusion via ILA should be unobservable more than ~10°K below T_H. Note, however, that in some systems there may be confusion about the definition of T_H. In the model in (Siegel, 1986*b,c*), T_H is the temperature at which the chemical potential of the H_{II} phase first becomes equal to that of the L_α phase. IMI can form around this temperature. In systems with small head group area ratios, however, IMI that form at this temperature are predicted to form ILA rapidly and the H_{II} phase more slowly. Hence, the experimental evidence that the system has reached T_H would be persistent isotropic [^{31}P]-NMR resonances and cubic-isotropic phase or large lipidic particle (ILA) morphology. The equilibrium phase of the system at these temperatures may be the H_{II} phase, but the extreme sluggishness of the transition and the relative stability of ILA (Siegel, 1986*b,c*) may prevent the bulk H_{II} phase from forming for days or longer. These systems often form the H_{II} phase rapidly if incubated at higher temperatures (e.g., Gagné *et al.,* 1985), and there is a temptation to call these higher temperatures T_H. H_{II}-phase formation is rapid at high temperatures because more IMI form (Siegel, 1986*a,b*) and because the equilibrium H_{II} tube diameter often decreases with increasing temperature (e.g., Boni and Hui, 1983). This latter effect corresponds to an increase in head group area ratio and a decrease in relative ILA formation rate at higher temperature.

4.5. Experimental Evidence for IMI-Medicated Membrane–Membrane Interactions

4.5.1. Thermotropic Systems

Recent studies on lipid systems with thermotropic L_α/H_{II} transitions have revealed the behavior discussed above the summarized in Table I

Ellens, Bentz, Szoka and co-workers have recently published careful studies of the dynamics of vesicle–vesicle interactions in PE and mixtures of PE with other lipids. The first system they investigated was a mixture of the lipid cholesteryl-hemisuccinate (CHEMS) and egg PE (Ellens *et al.*, 1984, 1985; Lai *et al.*, 1985). CHEMS–PE mixtures adopt the H_{II} phase when the ambient pH is about 5.5 or smaller (Lai *et al.*, 1985; Bentz *et al.*, 1985). At high pH, CHEMS is anionic, and strong electrostatic forces keep liposomes apart. These PE-containing unilamellar or oligolamellar liposomes are stable at pH >6. Liposome–liposome interactions are rapidly triggered by abruptly dropping the pH to 4.5 through addition of an appropriate buffer. This procedure was also used to study pure PE systems (Ellens *et al.*, 1985), since PE is anionic at pH >9. Interactions were studied as a function of temperature around the calorimetrically determined T_H (Lai *et al.*, 1985; Ellens *et al.*, 1985*b*).

In CHEMS–PE systems, Ellens and co-workers showed that at low pH liposomes leaked their contents in an aggregation-dependent manner (Ellens *et al.*, 1985*a*). Lipid mixing rates were determined via a fluorescence resonant energy transfer assay using labeled phospholipids. Lipid mixing occurs both above and below T_H (Bentz *et al.*, 1985*b*). The initial rate of lipid mixing (immediately after the pH jump) reflects the kinetics of intermediate formation between vesicles. In CHEMS–PE and pure PE systems (Bentz *et al.*, 1985; Ellens *et al.*, 1985*b*) the initial rate of lipid mixing at low pH increases rapidly with increasing temperature and reaches a plateau a few degrees below T_H. The initial rate of aggregation-induced leakage also increases rapidly with temperature in the immediate vicinity of T_H. At low pH, no fusion was observed (i.e., via fluorescent aqueous contents-mixing assays) at any temperature in the CHEMS–PE system (Ellens *et al.*, 1985; Bentz *et al.*, 1985). In the pure PE system (Ellens *et al.*, 1985*b*), small fusion rates were observed when the temperature was less than T_H, but these rates were maximal a few degrees below T_H and fell to very small values at temperatures of $\geqslant T_H$. The behavior of these systems is as predicted by the model in Siegel (1986*b*) for systems with fast L_α/H_{II} transitions: lipid mixing via IMI formation a few degrees below T_H, contact-mediated liposome leakage at and above T_H, and no fusion (no simultaneous lipid and contents mixing without leakage) when H_{II} is the equilibrium phase. The small fusion rates observed at temperatures a few degrees below T_H in the pure PE systems may be ILA mediated. Since IMI that form at these temperatures cannot be consumed by H_{II}-phase formation, the rate of ILA formation from IMI may be significant even if Z is not in the optimal range (Section 2.3). However, fusion in this regime may also be occurring via some other mechanism.

The same workers have recently completed a study of the N-methylated dioleoyl-PE (DOPE-Me) system at low pH (Ellens *et al.*, 1986).

Gagné *et al.* (1985) showed that this lipid–water system is in the L_α phase at temperatures below ca. 20°C, forms with H_{II} phase rapidly above ~ 70°C, and yields isotropic [^{31}P]-NMR resonances and mixed large lipidic-particle and H_{II}-like morphology in the intervening region. The large and variable size of these lipidic particles is consistent with ILA structure (Siegel, 1986*c*). Ellens and colleagues found that the rate of aggregation and leakage of DOPE-Me liposomes is trivial at 20°C and below. Between 30° and 60°C, however, the initial rate of liposome fusion (aqueous contents mixing) increases with temperature, goes through a maximum at ~ 60°C, and then decreases. The rate of liposome leakage is small at low temperatures, increases to a plateau-like shoulder in the vicinity of 60°–70°C, and increases sharply at higher temperatures. Almost no fusion is observable above 70°C. This behavior is consistent with more rapid ILA production than H_{II} production from IMI at lower temperatures (i.e., fusion more rapid than leakage), and rapid H_{II} production (leakage much more rapid than fusion) at higher temperatures. These are exactly the temperature-dependent vesicle interactions predicted in Siegel (1986*b,c*) for lipids with this phase behavior (Gagné *et al.,* 1985; see last paragraph of Section 3.4).

Ellens *et al.* (1986) also observed that there was often a lag time between extensive liposome aggregation at low pH and a sudden, massive leakage of aqueous contents to the external medium. Fusion (with retention of aqueous contents) was observed to occur during this interval. The length of this interval decreased with increasing temperature. This behavior is as predicted for systems in which the contact areas between liposomes are too small to contain more than one or two IMI at any given time. The liposomes used in this work were 0.2–0.3 μm in diameter, which is compatible with an area of contact between vesicles (~ 10^{-11} cm^2) that is small compared with the average area/IMI in systems with slow L_α/H_{II} transitions (Siegel, 1986*a*). Leakage is therefore slow until the average vesicle size increases via fusion, as observed. The number of IMI forming per unit area of apposed bilayers increases rapidly with temperature (Siegel, 1986*b*). Thus, the H_{II} phase forms more rapidly (and apposed liposomes leak more rapidly) as the temperature at which vesicles first aggregate increases, as observed by Ellens *et al.* (1986). These workers proposed a model qualitatively similar to that in Siegel (1986*b*) to explain their observations.

Ellens *et al.* (1985) also studied liposome–liposome interactions in the vicinity of L_α/H_{II} transitions in PE that were induced by divalent cation addition at neutral or basic pH. It is not clear if these should be classified as ionotropic: in pure PE systems (Ellens *et al.,* 1985*b*) Ca^{2+} addition induces H_{II} phases with the same T_H as those induced by low pH. High concentrations of Ca^{2+} are required (20 mM). This implies that stable cation–lipid complexes of defined stoichiometry don't form. Ca^{2+} ions may simply reside in the electrical double layer at the lipid–water interface and reduce the average electrostatic repulsion between head groups, which must pack closer together in order to form the high-curvature interface in the H_{II} phase (Fig. 1). In any case, the temperature-dependent behavior of the Ca^{2+}–PE systems around T_H (Ellens *et al.*, 1985*b*) is nearly the same as in PE at low pH and follows the predictions in Siegel (1986*b*) for systems with rapid thermotropic L_α/H_{II} transitions.

Lai *et al.* (1985) and Ellens *et al.* (1985*a*) also studied Ca^{2+}- and Mg^{2+}-induced liposome interactions in CHEMS–PE at neutral pH. These may be the result of an ionotropic L_α/H_{II} transition. Ca^{2+} addition at neutral pH results in formation of an H_{II} phase with a higher T_H than the H_{II} phase formed at low pH in the absence of Ca^{2+}, and only ca. 3 mM Ca^{2+} was required to induce the H_{II} phase (Lai *et al.*, 1985). Ca^{2+} addition didn't result in a marked lateral phase separation of the two components. These observations are compatible with the formation of stable cation–lipid complexes of defined stoichiometry (i.e., an ionotropic transition). Ellens *et al.* (1985*a*) found that Ca^{2+} induces fusion (aqueous contents mixing) and leakage at similar rates, and extremely rapid lipid mixing, in CHEMS–PE at temperatures ~ 30°C below the calorimetric T_H. (Mg^{2+} induced fusion only.) On the basis of the [^{31}P]-NMR spectra, Lai *et al.* (1985) concluded that most of the lipids are in the L_α phase at this temperature. The behavior of this system may be described in terms of an ionotropic L_α/H_{II} transition, where fusion via ILA should be faster than in thermotropic systems (Section 2.4). Fusion may occur over a wide temperature range due to the existence of an isotropic phase like that observed in DOPE-Me (Gagné *et al.*, 1985) and the ionotropic Ca^{2+}–cardiolipin system (DeKruijff *et al.*, 1982) under some circumstances. [^{31}P]-NMR spectra are not always good indices of the proportions of L_α, H_{II}, and isotropic phases present, and the [^{31}P]-NMR data in Lai *et al.* (1985) may be compatible with an L_α-isotropic phase coexistence region. The electron micrograph of a PE–CHEMS dispersion in the presence of Ca^{2+} (Lai *et al.*, 1985; Fig. 9e) resembles micrographs of the isotropic phase in dispersions of Ca^{2+}-cardiolipin (e.g., Fig. 3f of DeKruijff *et al.*, 1982). However, it is possible that the interactions in Ca^{2+}–CHEMS–PE occur via some non-IMI/ILA mechanism. A careful study of the temperature and Ca^{+} concentration-dependent phase behavior could help decide this issue.

Düzgüneş *et al.* (1985) report that liposomes of oleic acid and egg PE interact in much the same way as CHEMS-PE liposomes. Interactions in that system may proceed by the same mechanisms.

4.5.2. Ionotropic Systems

Few studies have been performed relating liposome interactions to phase behavior in lipid systems with ionotropic L_α/H_{II} transitions. It is more difficult to interpret such studies because it is known that fusion can occur in anionic lipid systems exposed to cations by mechanisms that have nothing to do with H_{II}-phase formation. The best-studied example is phosphatidylserine in the presence of Ca^{2+} (Ca^{2+}–PS) (for reviews, see Düzgüneş and Papahadjopoulos, 1983; Nir *et al.*, 1983). This system has not been observed to form the H_{II} phase. In anionic lipid systems that form the H_{II} phase upon exposure to cations, it may be unclear whether fusion occurs via a mechanism related to that in Ca^{2+}–PS system or via ILA.

Wilschut *et al.* (1982, 1985) studied divalent Ca^{2+}-induced fusion in the cardiolipin-PC system, which adopts the H_{II} and isotropic phases upon exposure to divalent cations (e.g., Verkleij *et al.*, 1979; De Kruijff *et al.*, 1979). Wilschut *et al.* (1982) measured liposome leakage and fusion rates as a function of Ca^{2+} concentration and reported that there was a sharp threshold [Ca^{2+}] for the onset of vesicle fusion and leakage in this system. They noted that transient inverted phase intermediates (ILA?) might be involved in this process. They also noted (Wilschut *et al.*, 1985) that the liposome aggregation rate increased rapidly with increasing [Ca^{2+}] above the threshold value and that the rate of lipid mixing was faster than the rate of aqueous contents mixing at higher temperatures (37°C). They postulated a structural rearrangement of the bilayer that facilitated both liposome aggregation and fusion.

This behavior is qualitatively compatible with IMI and ILA-mediated interactions (Siegel, 1984, 1986*a,b*). The threshold [Ca^{2+}] might correspond to the threshold for H_{II} formation at that particular temperature and ionic strength. It was pointed out (Siegel, 1986*b*) that ILA should form more frequently in ionotropic then in thermotropic systems, so that ILA formation (and liposome fusion) might be the first phenomena to occur as a function of increasing [Ca^{2+}]. ILA are bilayer structural rearrangements induced by exposure of Ca^{2+} that make vesicle aggregation irreversible, compatible with the interpretation of Wilschut *et al.* (1985). Finally, IMI formation is thought to be the first step in ILA formation, and IMI result in lipid exchange between liposomes, compatible with the observation made by Wilschut *et al.* that under some circumstances lipid mixing between liposomes occurred at a faster rate than liposome fusion (ILA formation). However, the case for IMI and ILA-mediated liposome–liposome interactions is not as clear in this case as in PE and related systems. It is possible that a Ca^{2+}–PS-like mechanism may be responsible for these observations.

4.5.3. Conclusions: Experimental Support for IMI-Mediated Interactions

The role of IMI and ILA in mediating membrane–membrane interactions has yet to be unambiguously demonstrated. However, the results of careful studies of PE systems (Section 3.5.1) are remarkably compatible with the predictions of this model. This is less clear in the case of lipid systems with ionotropic L_α/H_{II} transitions like the cardiolipin–PC system. Nevertheless, the qualitative predictions of the model (Siegel, 1986*b*) are in fairly good agreement with available data on these systems. More studies relating liposomal system dynamics to lipid-phase behavior are required to test the model more rigorously. Some such experiments are discussed in Section 5.

Some workers (e.g., Bearer *et al.,* 1982) have claimed that intermediates in L_α/H_{II} phase transitions are not involved in fusion because they haven't yet been imaged in the contact regions between vesicles via freeze-fracture electron microscopy. However, as pointed out in Siegel (1984) and as those workers acknowledged (Bearer *et al.,* 1982), it is extremely difficult to image *transient* structures via this technique, and IMI and ILA are expected to be very short-lived under these circumstances. IMI are thought to have half-lives in the submillisecond or millisecond range (Siegel, 1984, 1986*a*). This means that they would revert to bilayer structures during the initial phases of the sample freezing process as the temperature decreased below T_H. Using very rapid freezing techniques (e.g., Heuser, 1977), it is possible to freeze a thin layer of lipid dispersion rapidly without crystalline ice formation in ~ 500 μsec. This time is longer than the lifetime of an IMI experiencing a drop in temperature far below T_H (Siegel, 1984, 1986*a*). This may also be the reason why no transition intermediates of any kind can be imaged via electron microscopy in samples of PEs quenched from the H_{II} phase when T_H is greater than ~ 30°C (Verkleij, 1984); the initial phase of cooling in such samples is probably long enough to permit most intermediates to revert to bilayer structure. Once formed between two vesicles, ILA simply result in a single, fused vesicle like those observed in studies like that of Bearer *et al.* (1982). Finally, very few IMI or ILA must form in order to mediate some of these interactions between vesicles, and the total area of vesicle–vesicle contacts in a preparation is usually a very small fraction of the total membrane area. One would have to examine a very large number of vesicle–vesicle contacts in freeze-fracture electron micrographs to have a significant chance of observing a single intermediate (Siegel, 1984).

5. BIOLOGICAL AND BIOMEDICAL IMPLICATIONS

5.1. Membrane Fusion

Cells may control membrane fusion between their various organelles by adjusting the composition and lipid phase behavior of their membranes. One way this could be achieved is by making specific membranes less stable with respect to H_{II} phase formation. Many biomembranes or biomembrane lipid extracts are in fact already close to the L_α/H_{II} phase boundary under physiological conditions: some adopt the H_{II} phase or form IMI/ILA-like morphology if dehydrated (e.g., Gruner *et al.*, 1982; Crowe and Crowe, 1982; and Gordon-Kamm and Steponkus, 1984), if heated slightly (Ranck *et al.*, 1984; Gounaris *et al.*, 1983*a*), if incubated with divalent cations (Van Venetië and Verkleij, 1982; Albert *et al.*, 1984; and Nicolay *et al.*, 1985), or if incubated at reduced pH (Gounaris *et al.*, 1983*b*).

Small changes in the lipid composition of selected areas of biomembranes could stabilize the H_{II} phase, making fusion via ILA or other IMI-mediated interactions between membranes possible. The stability of bilayers with respect to H_{II} formation is sensitive to the levels of biological lipids like cholesterol (e.g., Cullis *et al.*, 1978*a;* Tilcock *et al.*, 1982; Bally *et al.*, 1983), cholesterol esters (Tilcock *et al.*, 1984), diacylglycerol (Dawson *et al.*, 1984; Das and Rand, 1985), dolichol and dolichol esters (Valtersson *et al.*, 1985), degree of unsaturation or head group structure of galactolipids (Gounaris *et al.*, 1983*c;* Brentel *et al.*, 1985), or unsaturation of PE and PC (Tilcock and Cullis, 1982; Dekker *et al.*, 1983; Cullis and De Kruijff, 1978). Local enrichment in H_{II}-inducing lipids might occur via enzymatic activities controlled by specific cellular stimuli, as in the case of the phosphatidylinositol/diacylglycerol cycle (Majerus *et al.*, 1984).

5.2. Participation of Nonbilayer Lipids in Other Biomembrane Processes

There is some evidence that nonbilayer structures (or at least high concentrations of lipids that can form the H_{II} phase) may be involved in other biomembrane functions. For example, it has been argued (for review, see Quinn and Williams, 1983) that monogalactosyldiglycerides are necessary components of chloroplast thylakoid membranes because they permit optimal packing of photosynthetic system proteins into the thylakoid membrane matrix and proper interaction of the various protein complexes within those membranes. These properties are proposed to arise from the same attributes that make these lipids susceptible to H_{II} phase formation.

Dolichol derivatives can induce nonbilayer structures in model membranes at low concentrations, and some speculate (Valtersson *et al.*, 1985) that nonbilayer structures might be involved in the transmembrane motion of glycosylated lipid intermediates in glycoprotein synthesis. It is not clear how structures like IMI (Siegel, 1984, 1986*a*) could do this, but is conceivable that ILA (Siegel, 1986*b*) could mediate transmembrane mobility of a glycosylated molecule of modest size at low rates per unit area of membrane under some circumstances. Such transport could occur if the membrane was folded upon itself, as are the pleats in an accordion, so that the different regions of the single bilayer are apposed to one another. The membranes in the Golgi apparatus and endoplasmic reticulum, where protein glycosylation occurs (e.g., Alberts *et al.,* 1983), are folded in this manner. Dolichol-induced IMI might form between the folds, and subsequently revert to ILA. During IMI formation, lipids on the facing monolayers (including a small amount of glycosylated lipid intermediate) form the inverted micelle of the IMI (Fig. 2b, c). During ILA formation from the ILA, these lipids from the inverted micelle mix with lipids originally on the "trans" surfaces of the membranes to form the channel lining the ILA (Fig. 2d) (Siegel, 1986*b,c*), resulting in transmembrane lipid transport. It is not known whether enough glycolipid could be transported in this way per unit area of apposed membrane for this to be a possible translocation mechanism *in vivo*.

It is also possible that some fusogens (agents that induce membrane fusion) work by stabilizing ILA-like structures (Cullis and Hope, 1978) in membranes exposed to them. Glycerolmono-oleate seems to form an inverted cubic phase with dioleoyl-PC, for example (Gutman *et al.,* 1984). Such phases may assemble from ILA (Siegel, 1986*c*).

5.3. Implications for Design of Membrane Fusion Assays and Vesicle-Encapsulated Drug-Delivery Systems

IMI formation between apposed liposomal (or other) membranes results in mixing of the lipids of the outer monolayers of the membranes (see Section 3.2). This process is often detected by fluorescence resonance energy transfer assays (Struck *et al.,* 1981) of the extent of mixing between the lipids of separately labeled vesicle populations. This type of assay has also been used by many investigators as an assay for membrane fusion, which is mixing of both lipids and aqueous contents without leakage. From the data discussed in Section 3.5, it shoud be obvious that the process of lipid mixing detected by this assay can be quite different from membrane fusion. In systems close to the L_α/H_{II} phase boundary, lipid mixing may occur via IMI formation, which is a distinct process from fusion; the aqueous contents of interacting vesicles do not mix at all or may leak to the suspending medium (Sections 3.2 and 3.3). The results of such assays should be checked with contents mixing assays for fusion, which measure the degree to which the aqueous contents of vesicles mix (e.g., Wilschut *et al.,* 1980), as has been urged by Bentz *et al.* and Ellens *et al.*

The model described in Siegel (1986*a–c*) also has potential applications in the design of vesicle-encapsulated drug delivery systems. It is believed that when vesicles are endocytosed by cells in culture or *in vivo,* the pH of the endosomal compartment containing them is reduced before the vesicles reach a lysosomal compartment and are degraded. It is desirable for the liposomes to be stable at neutral pH (so as to prevent leakage of their encapsulated drug before they are taken up by the target cells) but to fuse with other membranes rapidly at low pH (so as to release their contents to the cytoplasm before they are degraded by lysosomal enzymes). Some workers have formulated liposomes that, when endocytosed by cultured cells, fuse with the walls of the endocytotic vesicle and release their contents into the cytoplasm (Connor and Huang, 1985; Straubinger *et al.*, 1985; see Chapter 13, this volume).Conner *et al.* (1984) and Düzgüneş *et al.* (1985) characterized the interaction behavior of these liposomes.

The liposomes used by these researchers are composed mostly of unsaturated acyl chain PEs, which adopt the H_{II} phase starting either well below (Connor *et al.*, 1984) or somewhat above (Straubinger *et al.*, 1985; Düzgüneş *et al.*, 1985) physiological temperatures. The liposomes are stabilized at neutral pH by small amounts of an anionic lipid: oleic acid (OA) in the case of Straubinger *et al.* and Düzgüneş *et al.*, palmitoylhomocysteine (PHC) in the case of Connor *et al.* In this respect, they closely resemble the CHEMS–PE system (Ellens *et al.*, 1985*a,b;* Bentz *et al.*, 1985), which also shows vesicle–vesicle interaction behavior as a function of pH and divalent cation concentrations that is very similar. The leakage in the CHEMS–PE system seems to be mediated by IMI, especially at low pH (see Section 3.5.1). Düzgüneş *et al.* (1985) showed that the OA–PE liposomes exhibited faster lipid mixing and contents leakage rates than aqueous contents mixing rates at reduced pH, as had Ellens *et al.* and Bentz *et al.* in CHEMS–PE. Connor *et al.* (1984) reported the PHC–PE vesicles fused at low pH but used lipid mixing and liposome-size measurements to assay for "fusion"; as discussed earlier, these assays may not distinguish between IMI-induced lipid mixing and fusion with retention of contents. (Connor *et al.* reported that the fusion was very leaky.)

The author speculates that the liposomes used by these workers have pH-dependent leakage and fusion properties because the H_{II} phase becomes the equilibrium phase at low pH as in the case of CHEMS–PE: leakage and fusion may be IMI/ILA mediated. The T_H of the PE used by Straubinger *et al.* (1985) and Düzgüneş *et al.* (1985) is 18°C above physiological temperature. However, mixtures of the same PE with CHEMS have a T_H around the experimental temperature used by Düzgüneş *et al.* (see Lai *et al.*, 1985; Bentz *et al.*, 1985).

The phase behavior of the OA–PE and PHC –PE systems should be studied carefully (i.e., after long incubations to overcome transition hysteresis, using X-ray diffraction and rapid-freezing freeze-fracture electron microscopy) to see whether H_{II}, isotropic or cubic phases form at low pH and physiological temperature. If fusion and leakage of such liposomes are IMI/ILA-mediated at low (intraendosomal) pH, their ability to fuse with endosomal membranes may be IMI/ILA mediated as well. This postulate could be tested by studying the interactions of the liposomes at low pH with liposomes of lipids extracted from plasma membranes of the relevant cultured cells. If IMIs form between the PE-containing liposomes and the plasma membrane liposomes (which contain PC), the PC content of the resulting structures should make ILA formation from the IMI (fusion) much more likely than H_{II} formation (leakage), which is consistent with the observed ability of these preparations to fuse with endosomal membranes.

If the lipid phase behavior of PE systems is found to control their ability to deliver drugs to cells via the endosomal route (Connor and Huang, 1985; Straubinger *et al.*, 1985), it will be easier to design such vesicle-encapsulated drug delivery systems for optimal performance. The behavior and efficiency of the system may be predictable from knowledge of the phase behavior of the lipid components.

It also appears that sensitive assays for various antigens can be constructed with liposomes that are destabilized by H_{II} phase formation (Ho and Huang, 1985); the design of such systems would also profit from a better understanding of the membrane–membrane interactions induced by L_α/H_{II} transition intermediates.

6. SUGGESTED TOPICS FOR FURTHER INVESTIGATION

In order to understand the dynamics of membrane–membrane interactions near L_α/H_{II} phase boundaries, we must know more about the structure of the intermediates in L_α/H_{II} phase transitions. The model discussed in this chapter (Siegel, 1986*a–c*) is based on data obtained via indirect techniques such as [^{31}P]-NMR and freeze-fracture electron microscopy. The advent of synchrotron X-ray sources has made time-resolved studies of L_α/H_{II} transitions possible (Ranck *et al.*, 1984; Caffrey, 1985). These studies can yield important information about the transition mechanism and intermediate structures. It may be possible to gain more direct information concerning the structure of intermediates through careful examination of X-ray diffraction patterns obtained from oriented L_α phase samples during different phases of the L_α/H_{II} transition. One might detect intensity scattered from arrays of intermediates like IMI, or conceivably even diffraction patterns from IMI or ILA themselves (Siegel, 1986*a,c*). These studies would also increase our understanding of liquid crystalline phase transitions in general.

The involvement of L_α/H_{II} transition intermediates in membrane–membrane interactions can only be inferred from indirect data. There is good evidence for involvement of IMI and ILA in the thermotropic systems described in Section 3.5, but there is less evidence in ionotropic systems like Ca^{2+}–cardiolipin. A careful study of the leakage and fusion rates as a function of temperature and Ca^{2+} concentration in the vicinity of T_H of the Ca^{2+}–cardiolipin complex might clarify this issue. The complex of Ca^{2+} with cardiolipin extracted from *E. coli* has a T_H below 30°C (Hong *et al.,* 1986). It would be interesting to examine the Ca^{2+}-induced interactions of liposomes of this composition as a function of temperature in the range 0°–40°C. According to the model in Siegel (1986*a,b*), IMI/ILA-mediated interactions are impossible more than ~ 10°C below T_H, and the ratio of initial leakage and fusion rates should increase with decreasing $[Ca^{2+}]$ near the threshold concentration.

Recently, Yager and Chang (1983) and Chang (Chapter 16, this volume) have provided an additional link between the L_α/H_{II} phase transition and membrane–membrane interactions via a method that has great potential. Yager and Chang (1983) established that pressure shifts the L_α/H_{II} phase transition to higher temperatures on account of the smaller density of the H_{II} phase. Chang (Chapter 16) reports a study of the growth of lipid aggregates accompanying the L_α/H_{II} transition as a function of applied pressure. The average size of lipid aggregates is greatly reduced at a given temperature by an applied pressure, consistent with an H_{II}-mediated aggregation process.

The present author proposes that it would be even more informative to measure the rates of interliposome lipid exchange, leakage, and fusion rates as a function of applied pressure and temperature, using assays similar to those employed by Ellens *et al.* (1984, 1985*a,b,* 1986) and by Bentz *et al.* (1985, 1986). By measuring the pressure dependence of these rates, one can measure the activation volumes for lipid mixing, leakage, or fusion. The activation volume is the product of the number of molecules that form the defects responsible for mediating these processes and the average volume change per molecule forming the defect. The activation volume expected if these events are IMI mediated is approximately equal to the product of the number of lipids per IMI estimated via the method in Siegel (1986*a*) and the difference in molar volume between the L and H_{II} phases as measured either via dilatometry or by the technique of Yager and Chang (1983). Comparison of the measured and the predicted activation volume for lipid mixing, for example, could thus help establish whether IMI mediate that process in particular systems. This technique should also be useful in understanding the mechanism of fusion in other (non-H_{II}-forming) systems as well. Measurements of activation volumes should provide more information than is currently available on the structure and size of the defects responsible for fusion.

In studies of membrane fusion, researchers should carefully consider the phase behavior of their systems. The data cited in Section 4.5.1 imply that intervesicular lipid exchange, fusion, and aggregation-induced leakage can be catalyzed by IMI and ILA even when the temperature of a system is a few degrees below T_H; such intermediates can be important even in some circumstances in which the L_α phase is the equilibrium phase. It would be useful to explore the connection between the H_{II} phase and membane–membrane interactions more fully. If the model discussed in this chapter is validated, it should be possible to predict the type and relative rates of membrane–membrane interactions in a broad class of systems simply through knowledge of the lipid–water phase behavior of the components. This would be useful in the design of the liposomal drug delivery systems, for example (Section 5).

REFERENCES

Albert, A. D., Sen, A., and Yeagle, P. L., 1984, The effect of calcium on the bilayer stability of lipids from bovine rod outer segment disk membranes, *Biochim. Biophys. Acta* **771**:28–34.

Alberts, B., Bray, D., Lewis, J., Raff, M., Roberts, K., and Watson, J. D., 1983, *Molecular Biology of the Cell,* Garland, New York.

Bally, M. B., Tilcock, C. P. S., Hope, M. J., and Cullis, P. R., 1983, Polymorphism of phosphatidylethanolamine-phosphatidylserine model systems: Influence of cholesterol and Mg^{2+} on Ca^{2+}-triggered bilayer to hexagonal (H_{II}) transitions, *Can. J. Biochem. Cell Biol.* **61**:346–352.

Bearer, E. L., Düzgüneş, N., Friend, D. S., and Papahadjopoulos, D., 1982, Fusion of phospholipid vesicles arrested by quick-freezing: The question of lipidic particles as intermediates in membrane fusion, *Biochim. Biophys. Acta* **693**:93–98.

Bentz, J., Ellens, H., Lai, M. -Z., and Szoka, F. C., 1985, On the correlation between H_{II} phase and contact-induced destabilization of membranes, *Proc. Natl. Acad. Sci. U.S.A.* **82**:5742–5745.

Bentz, J., Ellens, H., Szoka, F. C., Oliver, J., and Siegel, D., 1986, Membrane fusion, isotropic intermediates and the L_α/H_{II} phase transition, *Biophys. J.* **49**:17a.

Boni, L. T., and Hui, S. W., 1983, Polymorphic phase behaviour of dilinoleoylphosphatidylethanolamine and palmitoyloleoylphosphatidylcholine mixtures: Structural changes between hexagonal, cubic, and bilayer phases, *Biochim. Biophys. Acta* **731**:177–185.

Borovjagin, V. L., Veraga, J. A., and MacIntosh, T. J., 1982, Morphology of intermediate stages of the lamellar to hexagonal lipid transition, *J. Membrane Biol.* **69**:199–212.

Brentel, I., Selstam, E., and Lindblom, G., 1985, Phase equilibria of mixtures of plant galactolipids: The formation of a bicontinuous cubic phase, *Biochim. Biophys. Acta* **812**:816–826.

Caffrey, M., 1985, Kinetics and mechanism of the lamellar gel/lamellar liquid-crystal and lamellar/inverted hexagonal phase transition in phosphatidylethanolamine: A real-time X-ray diffraction study using synchrotron radiation, *Biochemistry* **24**:4826–4844.

Connor, J., and Huang, L., 1985, Efficient cytoplasmic delivery of a fluorescent dye by pH-sensitive immunoliposomes, *J. Cell Biol.* **101**:582–589.

Connor, J., Yatvin, M. B., and Huang, L., 1984, pH-sensitive liposomes: Acid-induced liposome fusion, *Proc. Natl. Acad. Sci. U.S.A.* **81**:1715–1718.

Crowe, L. M., and Crowe, J. H., 1982, Hydration-dependent hexagonal phase lipid in a biological membrane, *Arch. Biochem. Biophys.* **217:**582–587.

Cullis, P. R., and De Kruijff, B., 1978, The polymorphic phase behavior of phosphatidylethanolamines of natural and synthetic origin, *Biochim. Biophys. Acta* **513:**31–42.

Cullis, P. R., and Hope, M. J., 1978, Effects of fusogenic agents on membrane structure of erythrocyte ghosts and the mechanism of membrane fusion, *Nature (Lond.)* **271:**672–674.

Cullis, P. R., Verkleij, A. J., and Ververgaert, P. H. J. Th., 1978*a,* Polymorphic phase behavior of cardiolipin as detected by ^{31}P-NMR and freeze-fracture techniques: Effects of calcium, dibucaine, and chloropromazine, *Biochim. Biophys. Acta* **513:**11–20.

Cullis, P. R., Van Dijck, P. W. M., De Kruijff, B., and De Gier, J., 1978*b,* Effects of cholesterol on the properties of equimolar mixtures of synthetic phosphatidylethanolamine and phosphatidylcholine, *Biochim. Biophys. Acta* **513:**21–30.

Das, S., Rand, R. P., 1985, Diacylglycerol, a product of phosphatidylinositol metabolism, causes major structural perturbations in lipid bilayers, *Biophys. J.* **47:**47a.

Dawson, R. M. C., Irvine, R. F., Bray, J., and Quinn, P. J., 1984, Long-chain unsaturated diacylglycerols cause a perturbation in the structure of phospholipid bilayers rendering them susceptible to phospholipase attack, *Biochem. Biophys. Res. Commun.* **125:**836–842.

Dekker, C. J., Guerts van Kessel, W. S. M., Klomp, J. P. G., Pieters, J., and De Kruijff, B., 1983, Synthesis and polymorphic phase behavior of polyunsaturated phosphatidylcholines and phosphatidylethanolamines, *Chem. Phys. Lipids* **33:**93–106.

DeKruijff, B., Verkleij, A. J., Van Echteld, C. J. A., Gerritsen, W. J., Mombers, C., Noordam, P. C., and De Gier, J., 1979, The occurrence of lipidic particles in lipid bilayers as seen by ^{31}P-NMR and freeze-fracture electron microscopy, *Biochim. Biophys. Acta* **555:**200–209.

DeKruijff, B., Verkleij, A. J., Leunissen-Bijvelt, J., Van Echteld, C. J. A., Hille, J., and Rijnbout, H., 1982, Further aspects of the Ca^{2+}-dependent polymorphism of bovine heart cardiolipin, *Biochim. Biophys. Acta* **693:**1–12.

Düzgüneş, N., and Papahadjopoulos, D., 1983, Ionotropic effects on phospholipid membranes: Calcium/magnesium specificity in binding, fluidity, and fusion, *Membrane Fluid. Biol.* **2:**187–216.

Düzgüneş, N., Straubinger, R. M., Balswin, P. A., Friend, D. S., and Papahadjopoulos, D., 1985, Proton-induced fusion of oleic acid-phosphatidylethanolamine liposomes, *Biochemistry* **24:**3091–3098.

Ellens, H., Bentz, J., and Szoka, F. C., 1984, pH-induced destabilization of phosphatidylethanolamine-containing liposomes: Role of bilayer contact, *Biochemistry* **23:**1532–1538.

Ellens, H., Bentz, J., and Szoka, F. C., 1985*a*, H^{+}- and Ca^{2+}-induced fusion and destabilization of liposomes, *Biochemistry* **24:**3099–3106.

Ellens, H., Bentz, J., and Szoka, F. C., 1985*b,* Destabilization of phosphatidylethanolamine liposomes at the hexagonal phase transition temperature, *Biochemistry* **25:**285–294.

Ellens, H., Bentz, J., and Szoka, F. C., 1986, Fusion of phosphatidylethanolamine liposomes and the mechanism of the L_{α}-H_{II} phase transition, *Biochemistry* **25:**4141–4147.

Gagné, J. L., Stamatatos, L., Diacovo, T., Hui, S. W., Yeagle, P. L., and Silvius, J. R., 1985, Physical properties of membranes containing N-methylated phosphatidylethanolamines, *Biochemistry* **24:**4400–4408.

Gordon-Kamm, W. J., and Steponkus, P. L., Lamellar-to-hexagonal$_{II}$ phase transitions in the plasma membrane of isolated protoplasts after freeze induced dehydration, *Proc. Natl. Acad. Sci. U.S.A.* **81:**6373–6377.

Gounaris, K., Brain, A. P. R., Quinn, P. J., and Williams, W. P., 1983*a,* Structural and functional changes associated with heat-induced phase-separations of non-bilayer lipids in chloroplast thylakoid membranes, *FEBS Lett.* **153:**47–52.

Gounaris, K., Sen, A., Brain, A. P. R., Quinn, P. J., and Williams, W. P., 1983*b*, The formation of structures in total polar lipid extracts of chloroplast membranes, *Biochim. Biophys. Acta* **728:**129–139.

Gounaris, K., Mannock, D. A., Sen, A., Brain, A. P. R., Williams, W. P., and Quinn, P. J., 1983*c*, Polyunsaturated fatty acyl residues of galactolipids are involved in the control of bilayer/non-bilayer lipid transitions higher plant chloroplasts, *Biochim. Biophys. Acta* **732:**229–242.

Gruner, S. M., Rothschild, K. J., and Clark, N. A., 1982, X-ray diffraction and electron microscope study of phase separation in rod outer segment photoreceptor membrane multilayers, *Biophys. J.* **39:**241–251.

Gruner, S. M., 1985, Intrinsic curvature hypothesis for biomembrane lipid composition: A role for nonbilayer lipids, *Proc. Natl. Acad. Sci. U.S.A.* **82:**3665–3669.

Gruner, S. M., Cullis, P. R., Hope, M. J., and Tilcock, C. P. S., 1985, Lipid polymorphism: The molecular basis of nonbilayer phases, *Annu. Rev. Biophys. Biophys. Chem.* **142:**211–238.

Gutman, H., Arvidson, G., Fontell, K., and Lindblom, G., 1984, ^{31}P-NMR and ^{2}H-NMR studies of phase equilibria in the three component system monoolein-dioleoylphosphatidylcholine-water, in: *Surfactants in Solution,* Vol. 1 (K. L. Mittal and B. Lindman, eds.), pp. 143–152, Plenum Press, New York.

Ho, R. J. Y., and Huang, L., 1985, Interactions of antigen-sensitized liposomes with immobilized antibody: A homogeneous solid-phase immunoliposome assay, *J. Immunol.* **134:**4035–4040.

Hong, K., Baldwin, P. A., Allen, T. M., and Papahadjopoulos, D., 1986, Fluorometric detection of bilayer to hexagonal phase transitions in liposomes, *Biochim. Biophys. Acta* (submitted).

Hope, M. J., Walker, D. C., and Cullis, P. R., 1983, Ca^{2+} and pH-induced fusion of small unilamellar vesicles consisting of phosphatidylethanolamine and negatively-charged phospholipids: A freeze-fracture study, *Biochem. Biophys. Res. Commun.* **110:**15–22.

Hui, S. W., and Stewart, T. P., 1981, "Lipidic Particles" are intermembrane attachment sites, *Nature (Lond.)* **290:**427–428.

Hui, S. W., Stewart, T. P., Yeagle, P. L., and Albert, A. D., 1981, Bilayer to non-bilayer transition in mixtures of phosphatidylethanolamine and phosphatidylcholine: Implications for membrane properties, *Arch. Biochem. Biophys.* **207:**227–240.

Hui, S. W., Stewart, T. P., and Boni, L. T., 1983, The nature of lipidic particles and their roles in polymorphic transitions. *Chem. Phys. Lipids* **33:**113–126.

Kirk, G. L., Gruner, S. M., and Stein, D. L., 1984, A thermodynamic model of the lamellar to inverse hexagonal phase transition of lipid membrane–water systems, *Biochemistry* **23:**1093–1102.

Lai, M. -Z., Vail, W. J., and Szoka, F. C., 1985, Acid and calcium-induced structural changes in phosphatidylethanolamine membranes stabilized by cholesterylhemisuccinate, *Biochemistry* **24:**1654–1661.

Lawaczeck, R., Kainosho, M., and Chan, S. I., 1976, The formation and annealing of structural defects in lipid bilayer vesicles, *Biochim. Biophys. Acta* **443:**313–330.

Lis, L., McAlister, M., Fuller, N., Rand, R. P., and Parsegian, V. A., 1982, Interactions between neutral phospholipid membranes, *Biophys. J.* **37:**657–666.

Mandersloot, J. G., Gerritsen, W. J., Leunissen-Bijvelt, J., Van Echteld, C. J. A., Noordam, P. C., and De Gier, J., 1981, Ca^{2+}-induced changes in the barrier properties of cardiolipin/phosphatidylcholine bilayers, *Biochim. Biophys. Acta* **640:**106–113.

Majerus, P. W., Neufeld, E. J., and Wilson, D. B., 1984, Production of phosphoinositide-derived messengers, *Cell* **37:**701–703.

Nicolay, K., Van der Neut, R., Fok, J. J., and De Kruijff, B., 1985, Effects of adriamycin on lipid polymorphism in cardiolipin-containing model and mitochondrial membranes, *Biochim. Biophys. Acta* **819:**55–65.

Nicolussi, A., Massari, S., and Colonna, R., 1982, Effect of lipid mixing on the permeability and fusion of saturated lecithin membranes, *Biochemistry* **21**:2134–2140.

Nir, S., Bentz, J., Wilschut, J., and Düzgüneş, N., 1983, Aggregation and fusion of phospholipid vesicles, *Prog. Surface Sci.* **13**:1–124.

Papahadjopoulos, D., Vail, W. J., Newton, C., Nir, S., Jacobson, N., Poste, G., and Lazo, R., 1977, Studies of membrane fusion. III. The role of calcium-mediated phase changes, *Biochim. Biophys. Acta* **465**:579–598.

Portis, A., Newton, C., Pangborn, W., and Papahadjopoulos, D., 1979, Studies on the mechanism of membrane fusion: Evidence for an intermembrane Ca^{2+}-phospholipid complex, synergism with Mg^{2+}, and inhibition by spectrin, *Biochemistry* **18**:780–790.

Quinn, P. J., and Williams, W. P., 1983, The structural role of lipids in photosynthetic membranes, *Biochim. Biophys. Acta* **737**:223–266.

Ranck, J. L., Letellier, L., Schecter, E., Krop, B., Pernot, P., and Tardieu, A., 1984, X-ray analysis of the kinetics of E. coli lipid and membrane structural transitions, *Biochemistry* **23**:4955–4961.

Rand, R. P., and Sengupta, S., 1972, Cardiolipin forms hexagonal structures with divalent cations, *Biochim. Biophys. Acta* **255**:484–492.

Reiss-Husson, F., 1968, Structure des phases liquide-crystalline de différents phospholipides, monoglycerides, sphingolipides, anhydres ou en presence d'eau, *J. Mol. Biol.* **25**:363–382.

Seddon, J. M., Kaye, R. D., and Marsh, D., 1983, Induction of the lamellar-inverted hexagonal phase transition in cardiolipin by protons and monovalent cations, *Biochim. Biophys. Acta* **734**:347–352.

Siegel, D. P., 1984, Inverted micellar structures in bilayer membranes, *Biophys. J.* **45**:399–420.

Siegel, D. P., 1986*a*, Inverted micellar intermediates and the transitions between lamellar, cubic, and inverted hexagonal lipid phases. I. Mechanism of the L_α-H_{II} phase transitions, *Biophys. J.* **49**:1155–1170.

Siegel, D. P., 1986*b*, Inverted micellar intermediates and the transitions between lamellar, cubic, and inverted hexagonal lipid phases. II. Implications for membrane–membrane interactions and membrane fusion, *Biophys. J.* **49**:1171–1183.

Siegel, D. P., 1986*c*, Inverted micellar intermediates and the transitions between lamellar, cubic, and inverted hexagonal lipid phases. III. Isotropic and inverted cubic state formation via intermediates in transitions between L_α and H_{II} phases, *Chem. Phys. Lipids* (in press).

Straubinger, R. M., Düzgüneş, N., and Papahadjopoulos, D., 1985, pH-sensitive liposomes mediate cytoplasmic delivery of encapsulated contents, *FEBS Lett.* **179**:148–154.

Struck, D. K., Hoekstra, D., and Pagano, R. E., 1981, Use of resonance energy transfer to monitor membrane fusion, *Biochemistry* **20**:4093–4099.

Tilcock, C. P. S., and Cullis, P. R., 1982, The polymorphic phase behavior and miscibility properties of synthetic phosphatidylethanolamines, *Biochim. Biophys. Acta* **684**:212–218.

Tilcock, C. P. S., Bally, M. B., Farren, S. B., and Cullis, P. R., 1982, Influence of cholesterol on the structural preferences of dioleoylphosphatidylethanolamine-dioleoylphosphatidylethanolamine systems: A ^{31}P-NMR and ^{2}H-NMR Study, *Biochemistry* **21**:4596–4601.

Tilcock, C. P. A., Hope, M. J., and Cullis, P. R., 1984, Influence of cholesterol esters of varying unsaturation on the polymorphic phase preferences of egg phosphatidylethanolamine, *Chem. Phys. Lipids* **35**:363–370.

Valtersson, C., Van Duyn, G., Verkleij, A. J., Chojnacki, T., De Kruijff, B., and Dalner, G., 1985, The influence of dolichol, dolichol esters, and dolichol phosphate on phospholipid polymorphism and fluidity in model membranes, *J. Biol. Chem.* **260**:2742–2751.

Van Venetië, R., and Verkleij, A. J., 1982, Possible role of non-bilayer lipids in the structure of mitochondria: A freeze-fracture electron microscopy study, *Biochim. Biophys. Acta* **692**:397–405.

Verkleij, A. J., 1984, Lipidic intramembranous particles, *Biochim. Biophys. Acta* **779**:43–64.

Verkleij, A. J., Mombers, C., Leunissen-Bijvelt, J., and Ververgaert, P. H. J., 1979*a*, Lipidic intramembranous particles, *Nature (Lond.)* **279**:162–163.

Verkleij, A. J., Mombers, C., Gerritsen, W. J., Leunissen-Bijvelt, L., and Cullis, P. R., 1979*b*, Fusion of phospholipid vesicles in association with the appearance of lipidic particles as visualized by freeze fracturing, *Biochim. Biophys. Acta* **555**:358–361.

Verkleij, A. J., Van Echteld, C. J. A., Gerritsen, W. J., Cullis, P. R., and De Kruijff, B. D., 1980, The lipidic particle as an intermediate structure in membrane fusion processes and bilayer to hexagonal (H_{II}) transitions, *Biochim. Biophys. Acta* **600**:620–624.

Verkleij, A. J., De Maagd, R., Leunissen-Bijvelt, J., and De Kruijff, B., 1982, Divalent cations and chloropromazine can induce non-bilayer structures in phosphatidic acid-containing bilayers, *Biochim. Biophys. Acta* **684**:255–262.

Wilschut, J., Düzgüneş, N., Fraley, R., and Papahadjopoulos, D., 1980, Studies on the mechanism of membrane fusion: Kinetics of calcium ion induced fusion of phosphatidylserine vesicles followed by a new assay for mixing of aqueous contents, *Biochemistry* **19**:6011–6021.

Wilschut, J., Holsappel, M., and Jansen, R., 1982, Ca^{2+}-induced fusion of cardiolipin/phosphatidylcholine vesicles monitored by mixing of aqueous contents, *Biochim. Biophys. Acta* **690**:297–301.

Wilschut, J., Nir, S., Scholma, J., and Hoekstra, D., 1985, Kinetics of Ca^{2+}-induced fusion of cardiolipin-phosphatidylcholine vesicles: Correlation between vesicle aggregation, bilayer destabilization, and fusion, *Biochemistry* **24**:4630–4636.

Yager, P., and Chang, E. L., 1983, Destabilization of a lipid non-bilayer phase by high pressure, *Biochim. Biophys. Acta* **731**:491–494.

Chapter 10

Lipid Vesicles

Model Systems to Study Membrane–Membrane Destabilization and Fusion

Francis C. Szoka

1. INTRODUCTION

The biophysical investigator interested in membrane fusion has distilled the complexities of the biological system to the simpler event of fusion between lipid vesicles (liposomes). This approach disregards the diversity and many of the accessory elements suspected to be involved in biological fusions (Poste and Nicolson, 1978) and focuses instead on those aspects thought to be of more universal importance. The development of the lipid vesicle model fusion system has permitted studies concerning the molecular mechanism(s) involved in fusion. In particular, the use of lipid vesicles composed of acidic phospholipids, such as phosphatidylserine, has permitted precise statements to be made concerning the two universal aspects of all fusing systems: the requirement for close apposition of bilayers and the subsequent membrane destabilization preceding fusion (Nir *et al.*, 1983*a*; Wilschut and Hoekstra, 1984). In addition, the role of cations, especially calcium, has been investigated extensively and con-

Francis C. Szoka • Departments of Pharmacy and Pharmaceutical Chemistry, School of Pharmacy, University of California at San Francisco, San Francisco, California 94143.

siderable data have suggested that calcium induces membrane destabilization and fusion only after intermembrane contact (Nir *et al.*, 1983*a*). Such studies have extended the basic physicochemical knowledge of interacting colloidal system and have paved the way for an understanding of the biological factor(s) involved in fusion.

This chapter examines the appropriateness of the lipid vesicle as a model system for the study of fusion and reviews the experimental techniques currently used in this research.

2. DEFINITIONS

Since a number of different terms have been applied to the same phenomena (Verkleij, 1984; Wilschut and Hoekstra, 1984), it is important to define exactly what we mean when referring to membrane aggregation, destabilization, and fusion. Aggregation will be used to describe any reversible interaction between lipid vesicles at either a primary or secondary energy minimum and will imply that membrane fusion has not occurred. Membrane destabilization will be used in those cases in which the organization of the lipids in the membrane is sufficiently perturbed that a measurable alteration in either a bilayer probe or increased permeabilities of an aqueous space probe can be detected. Used in this general sense, increased rates of lipid probe exchange between bilayer membranes would be one example of a membrane destabilization. The term lipid probe exchange is used in those circumstances in which the probe transfers between liposomes through the aqueous phase. In the case of lipid probe exchange, different classes of lipids might exchange at different rates, whereas in the case of membrane fusion, the lipid components of the interacting membranes intermix at similar rates. Lysis or leakage is the term used where aqueous space markers exhibit enhanced permeabilities through the bilayer.

In the case in which only the outer monolayers mix, the term *semifusion* as described by Verkleij (1984) is used. When semifusion occurs between membranes the lipids in the outer monolayers are free to diffuse and intermix, whereas lipids in the inner monolayer are not. In this case, mixing of aqueous contents between the two lipid vesicles does not occur. Membrane fusion is used when the membrane of two lipid vesicles have become one and the aqueous contents in them are free to intermix. We believe that this definition is closest to what is usually implied by those studying biological fusions and is the most rigorous criterion that can be applied to model systems. In selecting these definitions, we have chosen not to define fusion to mean the joining of the outer membranes and the term fission to mean the removal of the lipid barrier formed by the inner monolayer of the two membranes (Palade, 1975; Verkeleij, 1984).

3. APPROPRIATENESS OF LIPID VESICLES AS MODEL SYSTEMS

3.1. Biological Fusions

The various types of biological fusion have been extensively reviewed (Poste and Allison, 1973; Poste and Nicolson, 1978; Düzgüneş, 1985). Such fusions can be classified in a number of ways (Stossel *et al.,* 1978; Palade, 1975); however, for the purpose of model studies a simple classification can be made on the type of interacting membrane, i.e., whether planar, vesicle or both types of membranes are involved. For example, when cells fuse with each other, this is an example of planar membranes interacting. Cell–cell fusions involve the union of plasma membranes, although the fusion proceeds from small regions between the interacting membranes, and at the completion of fusion, large numbers of lipids and proteins can intermix (Knutton and Pasternak, 1979). Agents that can induce cell–cell fusion such as polyethylene glycol have been shown to lead to aggregation and fusion of lipid vesicles (Boni *et al.,* 1981; MacDonald, 1985).

A far greater amount of work has been concerned with the fusion of biological vesicles with plasma membranes or biological vesicles with one another (Papahadjopoulos *et al.,* 1979; Düzgüneş, 1985). It is as a model system for these latter types of fusion that the liposome is often suggested to be well suited (Nir *et al.,* 1983*a;* Wilschut and Hoekstra, 1984; Düzgüneş, 1985).

This fusion of biological organelles with the plasma membrane has been extensively documented from electron microscopic studies. Secretory vesicles release their contents in a process termed exoytosis following fusion of the secretory vesicles with the plasma membrane (DeDuve, 1963; Palade, 1975). Evidence supporting this process has accumulated for a variety of mammalian cell types including the adrenal medulla (see recent review by Ungar and Phillips, 1983; Livett, 1984) acinar cells of the pancreas (Palade, 1959), islets of Langerhans (Orci *et al.,* 1973), luteal cells (Gemmell and Stacy, 1979), mast cells (Lagunoff, 1973) and neutrophils (Chandler *et al.,* 1983). Many other organisms and tissues have also been documented to discharge neurotransmitter or other vesicular contents via an excytosis process (Düzgüneş, 1985). Palade (1975) has classified the membrane events preceding release of contents as following a morphological sequence beginning with attachment of the vesicle to the plasma membrane. This attachment site with its pentalamellar structure is converted into a trilamellar structure. The trilamellar structure is considered to be due to a monolayer fusion where the adjacent monolayers of the plasma membrane and vesicular membrane fuse. Subsequent breakage of the semifused membranes results in release of contents of the vesicle into the extracellular space. In many of the exocytotic systems the release of contents has been correlated with the entry of calcium into the cell (Douglas, 1974; Rubin, 1974).

In one of the better studied systems, the chromaffin cells, acetylcholine (ACh) leads to an increase of intracellular calcium, which results in a release of catecholamines (Douglas, 1974) from chromaffin granules (Winkler and Carmichael, 1982). Although the mechanism of action of calcium is not well understood due to the widespread correlation of calcium influx with release of secretory vesicle contents, it is probable that the ion is directly involved with the fusion step. This has been one of the principal rationales for examining the effects of calcium on liposome fusion. The control of exocytosis in this system certainly involves a number of coordinated elements and cytoplasmic structures connecting the chromaffin granules to the plasma membrane are evident in electron micrographs (Aunis *et al.,* 1979; Ungar and Phillips, 1983; Livett, 1984). Their presence would seem to impede fusion of the vesicle with the plasma membrane. Aunis and collaborators (1979) suggested that they may be contractile proteins that function to redistribute membrane proteins in the region of membrane contact and thus permit exocytosis.

Such electron micrographic studies yield exquisite detail but in most cases the time resolution is poor. Moreover, a number of assumptions must be made to correlate the structures observed with the underlying molecular processes. To unravel some of the factors that control vesicle fusion, one approach has been to use permeabilized cells. Baker *et al.* (1978, 1980) permeabilized plasma membranes to low-molecular-weight substances by exposing the cells to short-duration high-voltage pulses. By using this technique, they were able to correlate catecholamine secretion with calcium levels and obtain a half-maximal release at 1 μm calcium (Baker and Knight, 1978). Similar results have been found when the detergent, digitonin has been used to permeabilize the adrenal cells (Wilson and Kirshner, 1983; Dunn and Holz, 1983); thus it appears that the permeabilized adrenal cell is exquisitely sensitive to ionized calcium. Baker and Knight (1978) examined the permeabilized adrenal cell in greater detail; they found that ATP is required for the calcium-induced exocytosis and that agents implicated in the activation of protein kinase C, such as tetradecanoylphorbol, make the release of catecholamine more sensitive to the effects of calcium. Thus studies on intact cells have identified the phenomena of fusion, whereas work on permeabilized cells has implicated calcium and ATP as important components in the process. To understand the phenomena at a molecular level requires investigation of better defined systems.

3.2. Fusion of Isolated Intracellular Organelles

The need for a better understanding of the factors responsible for biological fusions has prompted researchers to disassemble the cell, isolate secretory vesicles, and examine the effects of various biochemical manipulations on their fusogenic properties. If again, we focus attention on the work on chromaffin granules from adrenal medulla cells, an outline of the factors involved in membrane fusion will become apparent as well as the limitations of this approach.

An impressive amount of research has been completed on the chromaffin granule revealing in great detail its protein, lipid, and low-molecular-weight composition and organization (Winkler and Carmichael, 1982). The diameter of chromaffin granules from most species is in the vicinity of 250 nm, although it ranges from 100 to 800 nm (Winkler and Carmichael, 1982). The overall dry weight composition is 42% protein, 20% catecholamines, 17% nucleotides, 0.6% glycosaminoglycan, and 20% phospholipids/cholesterol. The catecholamine/ATP ratio is 4–8, and the pH of the granule is buffered at a low pH (5.7) (Winkler and Carmichael, 1982; Unger and Phillip, 1983). The intragranule calcium concentration is 20 mM, although it appears to be complexed with other granule constituents, so that the activity is low. The major proteins of the granule are dopamine-β-hydroxylase, chromogranin A, and enkephalins, which are localized in the matrix (Winkler and Carmichael, 1982). The granule membrane has more than 20 associated proteins. The major ones are membrane bound dopamine-β-hydroxylase, a Mg^{2+} ATPase, cytochrome b_{561}, phosphatidylinositol kinase, NADH dehydrogenase, actin, and α-actinin (Winkler and Carmichael, 1982; Ungar and Phillips, 1983). The lipid composition of the granule membrane is quite unusual in that it contains 17% lysophosphatidylcholine, which is localized on the inner membrane. Sphingomyelin and phosphatidylcholine comprise 36% of the membrane, phosphatidylethanolamine is found in a 36% amount, while the acidic phospholipids phosphatidylinositol and phosphatidylserine comprise 10% of the lipids. The phosphatidylethanolamine is located primarily in the outer leaflet as are the acidic phospholipids. In addition, the gangliosides are located on the inner membrane leaflet (Winkler and Carmichael, 1982). The cholesterol/phospholipid molar ratio is 0.6. Secretory vesicles isolated from the adrenal medulla (Dahl *et al.*, 1979; Ekerdt *et al.*, 1981) when exposed to calcium fuse as doublets as evidenced by freeze–fracture electron microscopy. The fracture plane is continuous through both vesicles, with clear evidence that the membranes have merged at the point of contact and the aqueous compartments are continuous (Gratzl *et al.,* 1980). Calcium in the range of 5 μM causes half-maximal fusion for these vesicles while other divalent cations such as Ba^{2+}, Sr^{2+}, and Mg^{2+} do not induce fusion. At calcium concentrations 200 times greater, about 1 mM, multiple fusions between several vesicles are observed (Dahl *et al.,* 1979; Ekerdt *et al.,* 1981). This second type of fusion appears similar to Ca^{2+}-induced fusion of sonicated vesicles. Enzymatic treatment of the adrenal medulla vesicles eliminates the high-sensitivity calcium-induced fusion. Neuraminidase, trypsin, and pronase inhibit the extent of fusion in a concentration dependent fashion (Ekerdt *et al.,* 1981). The sensitivity of the fusion to these enzymes implies that proteins, more specifically, glycoproteins and possibly sialic acid containing glycolipids are involved in the calcium mediated fusion. However, the low-sensitivity calcium-induced fusion is unaffected by enzymatic treatment (Ekerdt *et al.,* 1981). One candidate for an accessory component is the ganglioside G_{M3}, the major glycolipid of the vesicle membrane (Dreyfus *et al.,* 1977). However, incorporation of G_{M3} into liposomes did not alter calcium sensitivity with respect to fusion (Ekerdt *et al.,* 1981). The evidence therefore points to the involvement of proteins or glycoproteins in the high-sensitivity calcium mode of chromaffin granule fusion. This led to an extensive hunt for the actual protein or proteins involved in vesicle fusion.

The location of the membrane glycoproteins in the granule (outside versus inside) is not resolved. Two groups of investigators have claimed that the major glycoprotein are situated with their carbohydrate regions on the inside of the vesicle (Huber *et al.,* 1979; Abbs and Phillips, 1980), whereas Meyer and Burger (1976) claimed that the glycoprotein are exposed on the outside of the vesicle. If the former are correct, the effect of neuraminidase on the high-sensitivity calcium fusion becomes less easily explained. A number of specific proteins have been identified as important components in the process of exocytosis. These include a plasma membrane protein that binds the chromaffin granules to the plasma membrane during exocytosis (Meyer and Burger, 1979*a,*). A releasing factor loosely associated with the plasma membrane has been proposed to play a role in the exocytosis of the granule contents (Izumi *et al.,* 1975, 1977). Also associated with the granule are actin (Burridge and Phillips, 1975; Gabbiani *et al.,* 1976; Meyer and Burger, 1979*b*) and α-actinin (Jockasek *et al.,* 1977). Actin in particular has been suggested to be involved with the movement of the granule to the plasma membrane (Fowler and Pollard, 1982*a*). However, nucleotides, chloride ions, and change in pH are also involved (Fowler and Pollard, 1982*b*).

It has been observed that millimolar concentrations of divalent cations cause granules or their membranes to aggregate (Banks, 1966;

Schober *et al.,* 1977; Morris *et al.,* 1979). The aggregation of the granules leads to exclusion of intramembranous particles from the site of aggregation but can be reversed by the presence of chelating agents (Schober *et al.,* 1977). However, other investigators have presented evidence that granule membranes fuse in the presence of even 0.1 mM calcium (Ekerdt *et al.,* 1981). One possible explanation for this apparent discrepancy is that plasma membrane fragments might have been present where fusion was observed. Konings and DePotter (1981) have demonstrated that incubation of the granules with plasma membrane preparations results in release of the granule contents. In this study, sialic acid-containing molecules were implicated in the fusion. In the presence of divalent cations, the electrophoretic mobility of the granules are reduced (Siegel *et al.,* 1978). This evidence, combined with the fact that calcium is involved in exocytosis, led to the search for calcium binding sites on the granule membrane (Morris and Schober, 1977; Ekerdt *et al.,* 1981). A high-affinity calcium-binding has been localized to membrane proteins, while a second binding site was attributed to the membrane phospholipids by Ekerdt *et al.* (1981). These workers distinguished two modes of fusion in the chromaffin granule using electron micrography; the first is associated with the unperturbed granule and occurs at micromolar calcium levels and resulted in granule doublets. The second was seen in granules that had been treated with enzymes or glutaraldehyde or in lipid vesicles prepared from the granule lipid. This fusion occurs at micromolar calcium levels and was characterized as multiple rounds of fusion.

To complicate the system even further, a number of cytoplasmic proteins have been described that bind to chromaffin granule surface in a calcium-dependent fashion. These include synexin (Cruetz *et al.,* 1978), calelectrins (Sudof *et al.,* 1984), calmodulin (Burgoyne and Geisow, 1981; Creutz *et al.,* 1983), and chromobindins (Cruetz *et al.,* 1981*b;* Cruetz *et al.,* 1983). Synexin was isolated from the adrenal medulla and was found to decrease the calcium concentration required to aggregate chromaffin granules (Cruetz *et al.,* 1978). The action of this protein appears to be specific for calcium. Synexin-induced aggregation does not lead to granule fusion, but in the presence of cis-unsaturated fatty acids rapid fusion of the granule with retention of content ensues (Cruetz, 1981*a*). The chromobindins, a group of proteins that bind to chromaffin granule membranes in a calcium-dependent fashion, include both synexin and the calelectrin. This group of proteins also includes a protein kinase C that is stimulated by diolein and phorbol myristate acetate (Summers and Creutz, 1985). The substrate for this protein kinase is another chromobindin. When the chromobindins are incubated in the presence of chromaffin granules, eight additional proteins are observed to be phosphorylated (Summers and Creutz, 1985). In addition to protein kinase C, a calcium-dependent phosphatidylinositol-specific phospholipase C activity was recently identified in the chromobindin fraction (Creutz *et al.,* 1985). It is interesting that the granule membrane contains relatively high concentrations of phosphatidylinositol and a phosphatidylinositol kinase; the presence of the specific phospholipase thus completes the system for a coordinated cycling of phosphatidylinositol during the exocytosis process. In conjunction with other calcium-sensitive proteins, the presence of the protein kinase C makes possible multiple points of calcium regulation in exocytosis.

In comparison with the work on the chromaffin granule, which is an exocytotic system, there has not been as many studies of fusion with intracellular organelles involved in endocytosis. The earliest study involved the fusion of phagolysosmes from rat liver (Raz and Goldman, 1974). However, a full characterization of this system was not undertaken. The fusion between phagolysosomes from *Acanthamoeba* was studied by Oates and Touster (1976, 1978). The lysosomes fuse at multiple sites in a temperature dependent manner. Millimolar concentrations of KF inhibit this process (Oates and Touster, 1978). A calcium requirement for fusion was demonstrated in vesicles derived from the golgi fraction of the liver (Gratzl and Dahl, 1976; Quinn and Judah, 1978; Judah and Quinn, 1978). A more detailed examination of the fusion of clathrin-coated vesicles isolated from bovine brain, with lysosomes isolated from bovine kidney demonstrated that free calcium in the medium is required for the fusion (Alstiel and Branton, 1983). The fusion occurred in the micromolar calcium range in the presence of magnesium and was detected by the hydrolysis of carboxydiacetylfluorescein by lysosomal enzymes. It was also observed by electron microscopy that removal of the clathrin coat permitted fusion of the endosomal vesicles with each other (Altsteil and Branton, 1983; Unanue *et al.,* 1981). Treatment of the uncoated vesicles with proteolytic enzymes did not inhibit the fusion; thus specific proteins on the surface of the uncoated vesicle do not appear to be involved in fusion. However, the role of proteins on the lysosomal fraction was not explored. Since the external surface of the uncoated vesicle was enriched in phosphatidylethanolamine and phosphatidylserine, Altsteil and Branton (1983) speculate that this fusion is similar to that seen in lipid vesicle systems.

A cell-free endocytotic vesicle fusion assay system, using an enzymatic reaction, was recently described (Davey *et al.,* 1985). It is based on the incorporation of a viral sialidase in one population of endocytotic vesicles and a [^{3}H]labeled sialic acid substrate on the surface of a second virus, which is in a different vesicle population (Davey *et al.,* 1985). The fusion of the two vesicle populations was monitored by the release of [^{3}H]-labeled nontrichloroacetic acid (TCA)-precipitable material. Fusion was observed to occur within 30 min, to require ATP, and to involve about 70% of the substrate enclosed in the vesicles. Lysosomes and plasma membranes were not observed to participate in this fusion. Surprisingly, the only divalent cation present in the reaction medium was magnesium, thus calcium does not appear to play a role in this system (Davey *et al.*, 1985).

This summary of the chromaffin granule system and fusion between intracellular organelles points out the intricacies of biological fusions. It is clear from the composition, number of associated accessory cytoplasmic factors, and the topography of the individual components that exocytosis of the chromaffin granule is quite complex, particularly when viewed in the context of a lipid vesicle, which may only have one or at most a few components. Other granules exhibit a similar complexity (Poisner and Triflano, 1982). This illuminates both the opportunities of the lipid vesicle for unraveling the molecular mechanism of fusion and the interpretational difficulties of relating results obtained with lipid vesicles to biological fusions.

3.3. Lipid Vesicles as Model Systems

The principal advantage of using liposomes as model systems for fusion is the ability to define exactly the composition and morphology of the system and to study the kinetics of the fusion process. The advantages of the lipid vesicle for such studies include the following:

1. It has the ability to form stable well-defined structures with low permeability to water-soluble molecules, so that mixing of contents during the fusion can be followed.
2. The phospholipid composition of the bilayer can be varied, so that influence of the head group, acyl chain composition, lateral phase separations, and phase transitions can be studied in great detail.
3. Other membrane components, such as cholesterol, glycolipids, and proteins, can be incorporated into the liposome.
4. Accessory components, such as cytoplasmic proteins, can be added in defined stoichiometry to the fusion reaction and the molecular mechanisms of their interaction with the system can be examined.
5. A large variety of assay systems and physical techniques have been developed that can yield information both on the kinetics of fusion and on the interactions of the components of the system.
6. Kinetic analyses of the fusion reaction have been developed that permit models of the fusion events to be constructed and tested.

Notwithstanding the advantages of the system it is clear that biological fusions are exceedingly complex. A principal finding to date is that calcium interactions with liposomes formed of acidic phospholipids in the absence of proteins cannot account for the calcium sensitivity found in biological systems (Lucy, 1984; Wilschut and Hoekstra, 1984).

An early objection to the use of liposomes as indicators for calcium-induced fusion is that calcium causes lysis and not fusion of lipid vesicles (Ginsberg, 1978). In this study, Ginsberg pointed out that the morphology of the final product at long periods after addition of calcium was not a vesicle and that the contents had leaked out at long periods after the fusion. This critique has been examined in a number of reviews (Papahadjopoulos *et al.,* 1979; Nir *et al.,* 1983*a;* Wilschut and Hoekstra, 1984; Düzgüneş, 1985), and it is appropriate to reiterate here why it is not a valid criticism of the use of liposome for fusion studies. A large number of systems have now been demonstrated to mix their bilayers and contents prior to this ultimate collapse of the system (Wilschut and Papahadjopoulos, 1979; Struck *et al.,* 1981; Hoekstra, 1982) and that for the phosphatidylserine liposome about 10% of the contents leak per fusion event (Wilschut *et al.,* 1980). Moreover, a number of conditions have been found in which small unilamellar vesicles of phosphatidylserine retain their contents and structure upon fusion induced by calcium, magnesium (Wilschut *et al.,* 1983; Nir *et al.,* 1983*b*), or polypeptides (Wang and Huang, 1984). Other lipid compositions have also been shown to undergo fusion while retaining their contents (Liao and Prestegard, 1979). Finally, freeze-fracture electron microscopy done by rapidly freezing vesicles mixed with calcium has demonstrated the existence of pairs of vesicles joined by a narrow connection indicative of membrane fusion (Bearer *et al.,* 1982). Since these early studies, a large number of other reports have appeared that convincingly show that mixing of contents in a variety of systems precedes the collapse and release of contents of the system. Thus, the evidence documents the lipid vesicle as a suitable model system for examining fusion.

4. CHOICE OF VESICLE PREPARATIONS

The maximum amount of information concerning the fusion of liposomes is obtained with unilamellar vesicles of defined size. With such liposomes, a fusion event can be correlated with both the surface area and the number of liposomes in the assay. Most of the early fusion studies were performed with small unilamellar vesicles (SUV) prepared by sonication (Papahadjopoulos *et al.,* 1979; Nir *et al.,* 1983*a*); more recent work has employed large unilamellar vesicles (LUV) prepared by various techniques (Wilschut and Hoekstra, 1984). The diameter of the various preparations significantly influences the kinetics, extent of fusion, and the susceptibility of the systems to various fusogenic agents (Taupin and McConnell, 1972; Schullery *et al.,* 1980; Wilschut *et al.,* 1981; Nir *et al.,* 1983*a;* Bentz and Düzgüneş, 1985). Thus the selection of the type of liposomes will greatly influence the outcome of the study.

4.1. Preparation of Vesicles

The preparation of liposomes has been reviewed in detail (Szoka and Papahadjopoulos, 1980, 1981); a comprehensive series of volumes describing the myriad methods of liposome preparation and characterization has recently appeared (Gregoriadis, 1984). Given this extensive literature, only the salient features of the most widely used preparations for studies of fusion are discussed. The preparations that have been used to the greatest extent in such studies are the sonicated vesicle and the reverse phase evaporation vesicle (Papahadjopoulos *et al.,* 1979; Nir *et al.,* 1983*a;* Wilschut and Hoekstra, 1984; Düzgüneş, 1985).

The preparation of SUV has been described in numerous publications (Papahadjopoulos and Miller, 1967; Huang, 1969; Bangham *et al.,* 1974; Szoka and Papahadjopoulos, 1980). A dispersion of lipid in the solution to be encapsulated is sonicated under an inert atmosphere until an optical clear preparation is obtained. Both bath-type (Szoka and Papahadjopoulos, 1980) and probe sonicators (Huang, 1969) have been used to prepare SUV. Probe sonicators are more efficient in reducing the lipid dispersion to a minimal size but can be more difficult to regulate in terms of heat production, contaminating oxygen and can contaminate the SUV preparation with metal from the probe. Bath-type sonicators avoid these difficulties but require more attention to the time of sonication, shape of the vessel used to hold the dispersion and the volume of the solution in order to obtain a reproducible SUV preparation (Szoka and Papahadjopoulos, 1980). It is essential in preparing SUV from defined phospholipids that the sonication be performed at a temperature above the gel–liquid crystalline transition temperature of the highest melting component in the mixture. If lipids are used that undergo a lamellar-hexagonal transition to obtain a reproducible preparation, it is necessary to sonicate below this temperature. Sonication results in the formation of vesicles with diameters of 22–70 nm, depending on the lipid composition and conditions of sonication. Owing to the high radius of curvature of SUV there is a greater percentage of lipids in the outer monolayer (60–70%) compared with the inner monolayer. When mixtures of lipids are used, an asymmetric distribution of the components between the inner and outer monolayers can occur. This must be considered when relating the lipid composition to their fusogenic properties. Because of the small diameter, the encapsulation ratio (i.e., liters of aqueous space per mole of lipid) is low with values in the range of 0.2–1.5 liter/M. When fluorescent lipids are the reporter system for the assay the low encapsulation volume is not a difficulty.

In addition to sonication the use of high-pressure extrusion at 20,000 psi through a French press is another convenient method for forming SUV (Barenholz *et al.,* 1979; Hamilton *et al.,* 1980). The size distribution and homogeneity of vesicles prepared in this manner are similar to those prepared by sonication. The most important attribute of the SUV is that homogeneous populations of vesicles can be formed, and any contaminating larger vesicles can be removed by centrifugation at $160{,}000 \times g$ (Barenholz *et al.,* 1977). The small radius of curvature makes such systems metastable and contributes a driving force promoting fusion (Nir *et al.,* 1983*a;* Bentz and Düzgüneş, 1985).

Large unilamellar vesicles can be prepared by a number of techniques, including the reverse-phase evaporation procedure (Szoka and Papahadjopoulos, 1978; Szoka *et al.,* 1980), detergent removal (Marsh *et al.,* 1983; Eidelman *et al.,* 1984), or extrusion techniques (Mayhew *et al.,* 1984). Most work employing LUV for fusion studies has been done with the reverse-phase evaporation vesicles (REV) (see Chapters 11 and 12 this volume) although more recent work exploring the effects of viral envelope proteins on fusion have used various detergent dialysis procedures to prepare the liposomes.

The advantages of using the REV is ease of preparation, the high efficiency of encapsulation of aqueous contents, the ability to prepare small volumes of vesicles, and the characterization of the system as being composed of primarily unilamellar vesicles (Szoka and Papahadjopoulos, 1978; Szoka *et al.,* 1980; Düzgüneş *et al.,* 1983; Parente and Lentz, 1984; Parente *et al.,* 1985). To prepare REV, the lipids are first dissolved in an organic solvent with diethyl ether that has been washed with distilled water or freshly redistilled over bisulfite to remove peroxides. For lipids with a high transition temperature, such as dipalmitoylphosphatidylcholine, diisopropylether is the preferred solvent. When the density of the aqueous phase is high, mixtures (1/1) of the ethers can be made with trifluorotrichloroethane to more closely match the density of the aqueous phase with the density of the organic phase. This will enhance the stability of the subsequent emulsion. The aqueous material is added directly to the phospholipid solvent mixture and the preparation is sonicated for a brief period forming an emulsion. Usually a 3:1 organic–aqueous phase ratio is employed. The emulsification is an important step, and difficulties with the preparation in obtaining high encapsulation in unilamellar vesicles occur often because a good emulsion is not formed. The organic solvents are removed in a two-step procedure in a rotary evaporatory: evaporation at 400 mm Hg vacuum until the mixture becomes a gel followed by evaporation at 150 mm Hg until a homogeneous suspension of liposomes is obtained. Removal of the organic solvent in this two-step procedure keeps the emulsion from flashing into the apparatus as the organic solvent evaporates. When the gel state is reached, the vessel containing the preparation can be removed from the evaporator, additional buffer is added, and the gel can be dispersed by vortexing. Continued evaporation under a high vacuum (100 mm Hg) will then speed the process of forming the liposome suspension. When defined acyl chain phospholipids are used, the formation of the final liposome dispersion occurs much more easily if the remainder of the solvent is removed at a temperature above the transition temperature of the lipid.

After a homogeneous dispersion is obtained, it is best to remove immediately traces of organic solvent remaining in the preparation by dialysis or column chromatography. This will eliminate tendencies of the vesicles to aggregate and will reduce the possibility of forming oxidation products catalyzed by trace solvents remaining in the liposomes. During the removal of the organic solvents, it is necessary to monitor the preparation to ensure that the aqueous phase does not evaporate as well. When this occurs, the unilamellar vesicles collapse and multilamellar dispersions are formed. To facilitate the formation of unilamellar vesicles and reduce the possibility of removing the aqueous phase during the evaporation step, it is best to match the amount of emulsion to the size of the vessel used to prepare them. For small preparations (0.2–1.0 ml aqueous phase), a 12 × 100-mm screwcap glass culture can be used, while larger preparations (3–9 ml aqueous phase) are prepared in vessels with a 50-ml capacity.

Following formation of the vesicles, they are extruded through polycarbonate membranes with a defined pore size (Olson *et al.*, 1979; Szoka *et al.*, 1980). This step removes any larger vesicles formed during the procedure. For fusion studies, we routinely extrude the vesicles through 0.2-μm or, in many cases, through 0.1-μm polycarbonate membranes. The vesicles can be separated from nonencapsulated material by column chromatography, dialysis, or by centrifugation methods (Fraley *et al.*, 1980).

Knowledge of the size and size distribution of lipid vesicles is an important parameter for evaluating the results of a fusion experiment. In many studies, changes in size are the principal criteria for concluding that fusion has occurred. Electron microscopy, nuclear magnetic resonance, gel permeation, and dynamic light scattering have all been employed for this purpose and detailed reviews of these techniques have appeared (Szoka and Papahadjopoulos, 1980; Gregoriadis, 1984). Many laboratories are not equipped to perform the laser light-scattering determinations or to do electron microscopic determinations of vesicle size. In spite of this, a number of simpler techniques yield valuable through indirect estimates of the average size of the vesicle preparations. For instance, by assuming the vesicles are spherical, unilamellar, and the surface area of an individual phospholipid is a particular value, an estimate of the average diameter of the preparation can be achieved by measuring the encapsulation ratio (liters of aqueous space per mole of lipid) (Szoka and Papahadjopoulos, 1978). Other simple formulas for calculating the number of lipid vesicles and their total surface area have been derived by Pidgeon and Hunt (1981).

4.2. Selection of Vesicle Composition

In the early studies with lipid vesicles the choice of lipids for most laboratories was dictated by what was commercially available or what the laboratories could purify or synthesize (Papahadjopoulos *et al.,* 1979). Since that time, a large variety of high-purity phospholipids have become commercially available, and the principal rationale for choosing among them is the specific aspect of fusion in which the investigator is interested. The principal focus of much of the work to date has been the acidic phospholipids, primarily phosphatidylserine, phosphatidic acid, and phosphatidylglycerol and the saturated phosphatidylcholines (see Chapter 11 in this volume and previous reviews). Recently the studies have been extended to vesicles composed of phosphatidylethanolamine (see Chapter 9, this volume). Each of these lipids have some idiosyncrasies that must be taken into account when preparing the liposomes. The necessity to prepare liposomes composed of defined acyl chain phospholipids above the transition temperature has already been mentioned. In the case of the acidic lipids, divalent cations must be rigorously excluded from the buffers. When phosphatidylethanolamine is used as a single component, the liposomes must be prepared at a pH greater than 8.5. If other components are added to the preparation, such as fatty acids, careful attention to the pH of the final preparation is necessary, since the buffering capacity of the fatty acids is quite high and can cause the pH of the mixture to become acidic. In most cases, with an understanding of the chemical form of the starting material; a little forethought can help avoid delays that occur because of the failure of liposome formation. With mixtures of components, particularly those that have not been used before, one runs the risk of phase separation or immiscibilities that may interfere with liposome formation or result in populations of liposomes with different compositions. Given the infinite possibilities, new mixtures must be evaluated on a case-by-case basis. In this regard, interactions with components in the aqueous phase, such as the binding of components of the assay systems, must also be tested, such as done by Wilschut and Papahadjopoulos (1979) or Ellens and colleagues (1985).

5. ASSAYS FOR CHANGE OF SIZE TO FOLLOW FUSION

5.1. Electron Microscopy

Fusion products of liposomes have been examined extensively by electron microscopy (Papahadjopoulos *et al.*, 1979; Nir *et al.*, 1983). Both negative-stain and freeze-fracture electron microscopy have been used to obtain estimates of size distribution of vesicles in fusion studies and were instrumental in documenting that vesicle growth had occurred. Freeze-fracture microscopy is the more informative technique, as it yields evidence for the continuity of the aqueous compartment, gives information on the structure of the bilayer and avoids the aggregation and drying artifacts often observed in negative-stain procedures (Szoka and Papahadjopoulos, 1980; Verkleij and de Gier, 1981). An additional advantage of freeze structure is that rapid freezing techniques can be employed to demonstrate early events in the fusion process (Bearer *et al.*, 1982). This technique is difficult to quantitate and is not very suitable for kinetic studies of fusion. However, it is an excellent technique for confirming that fusion has occurred.

5.2. Light Scattering

Two different methods have been used to follow changes in liposome size during fusion. The first measures turbidity changes at 90°C to measure changes in particle size. Both aggregation and fusion can contribute to the turbidity or light scattering; to document that fusion has occurred, aggregation must be rigorously excluded. In the case of calcium-induced fusion of phosphatidylserine vesicles, the addition of a chelating agent should not reverse the increase in turbidity (Nir *et al.*, 1980). The analysis of turbidity changes in terms of relating it to alterations in particle diameter is complex, depending on both vesicle size and the refractive index of the membranes (Nir *et al.*, 1983*a*). Because of this, it is difficult to apply the theory to changes observed in large-diameter vesicles (diameters >200 nm). However, changes in turbidity can indicate relative alterations in size and, used in conjunction with other information such as fluorescence fusion assays, can provide useful kinetic data (Bentz and Nir, 1981). Dynamic light scattering has also been employed to measure the size of phosphatidylserine vesicle that had been allowed to fuse in the presence of calcium and then treated with EDTA to disaggregate the fused structures (Sun *et al.*, 1978, 1979). In this procedure, the Brownian motion of the vesicles causes temporal fluctuations in the scattered intensity (Goll *et al.*, 1982). The average diffusion coefficient of the vesicles is then obtained from a time-correlation function of the scattered light intensity. Because of the time necessary to obtain sufficient intensity, this procedure cannot be used to obtain the kinetics of fusion directly. By stopping the fusion reaction at various times, the effects of changes in calcium concentration and temperature on the kinetics of fusion have been inferred (Sun *et al.*, 1979). A difficulty in such studies is that one cannot be certain that the observed changes are due to the initial fusion or changes brought about by the subsequent addition of the chelator. For these reasons, dynamic light scattering is best employed on systems that undergo slow rates of change (hours to days) (Wong and Thompson, 1982) rather than on those that undergo rapid fusion such as the phosphatidylserine vesicles.

5.3. Other Methods That Use Changes in Size to Follow Fusion

Gel filtration on large-pore exclusion chromatography medium such as Sepharose, Sephacryl, or BioGel has been used to monitor the increase of vesicle size in a variety of fusing systems (Papahadjopoulos *et al.*, 1974; Prestegard and Fellmeth, 1974; Ginsberg, 1978; Schullery *et al.*, 1980). This technique permits discrimination between vesicles of different size, but it does not yield an unambiguous answer to the question of whether fusion or aggregation has occurred. Given the time necessary to perform the separation, it is not a very useful technique for measuring the kinetics of rapidly fusing systems. The method is useful for demonstrating the presence of vesicles that have fused and mixed their contents (Klausner *et al.*, 1981; Weinstein *et al.*, 1981).

Proton nuclear magnetic resonance (NMR) can also be used to measure the broadening of resonances of the choline methyl, methylene, and chain terminal methyl protons as the vesicles become larger (Prestegard and Fellmeth, 1974; Kantor and Prestegard, 1975). The disadvantage of this approach is that relatively large amounts of lipid are required, and the time required to accumulate a spectrum makes it impractical to do rapid kinetics.

6. ASSAY FOR RELEASE OF ENCAPSULATED COMPOUNDS

As many of the liposome systems that fuse exhibit leakage during the fusion process or ultimately continue on to a collapsed state in which the contents are released, leakage of encapsulated material can be used to follow a process that is kinetically related to fusion. In instances in which leakage is observed, further determinations must be undertaken to ensure the fusion is part of the kinetic pathway to bilayer destabilization, and the process is not a simple case of lysis. Assays to study leakage include enzymatic (Kataoka *et al.*, 1973), electron spin resonance (ESR) (Marsh and Watts, 1981), and fluorescence techniques (Nir *et al.*, 1983*a*). The most widely used assays involve fluorescence. These have the advantage of exquisite sensitivity, rapid response times and the wide availability of spectrofluorometers to follow the rate of change in fluorescence.

The most widely used fluorescence assay to monitor leakage of contents from liposomes involves the dequenching of a self-quenched fluorophore, carboxyfluorescein (Haggins and Yoshikami, 1978; Weinstein *et al.*, 1977). In this assay, carboxyfluorescein is encapsulated at a high concentration (5–10 mM), at which probe–probe interaction results in fluorescence self-quenching (FSQ). When the probe leaks from the liposome, it is diluted into a large volume, the quenching is eliminated and the fluorescence increases up to 30-fold. Since carboxyfluorecein has a high extinction coefficient (77,000 at pH 7.4 measured at 492 nm) and a good quantum yield, it forms the basis for a very sensitive assay for leakage of contents from liposomes (Weinstein *et al.*, 1977). The properties of carboxyfluorescein, methods to purify the compound, and details of its use in FSQ assays for leakage have been excellently covered by Weinstein and colleagues (1984). The fluorescence of carboxyfluorescien decreases as the pH is lowered, while its permeability through the bilayer increases (Weinstein *et al.*, 1984; Szoka *et al.*, 1979). This has led to the use of other fluorophores in the FSQ assay whose properties are less affected by low pH, such as calcein (Allen and Cleland, 1980; Allen, 1984) and lucifer yellow (Eidelman *et al.*, 1984).

Weinstein and co-workers (1981, 1984) and Klausner *et al.* (1981) devised FSQ methods to distinguish between continuous leakage and all-or-none release of vesicle contents. In one version of the method, vesicles are rechromatographed after different periods of incubation. If the quenching in the separated vesicles is the same as the original vesicle suspension, carboxyfluorecein has been released by rupture of a fraction of the vesicle population. If the quenching has been reduced by leakage, the quenching coefficient will be reduced proportionately. This assay is sensitive at lower quenching levels (10–20 mM) (Weinstein *et al.*, 1981, 1984). The FSQ is a particularly sensitive assay, the first to be employed to quantitate leakage of vesicle contents continuously during fusion (Portis *et al.*, 1979); it continues to be a very important tool for examining the destabilization of liposomes during fusion (Düzgüneş *et al.*, 1981).

A variation of the FSQ assay uses a fluorophore (aminonapthalenetrisulfonic acid) and a separate nonfluorescent water soluble quencher (dipyridinium-*p*-xylene) encapsulated in the same liposome (Smolarksy *et al.*, 1977). When both compounds are encapsulated at 50 mM, the fluorescence of the fluorophore is quenched by 85% (Ellens *et al.*, 1984). This combination has an advantage for studying pH-induced fusion and destabilization of lipid vesicles because the fluorescence of the aminonaphthalenetrisulfonic acid is reduced only 12% at pH 4.0 compared with that observed at pH 7.0 at an excitation wavelength of 354 nm (Ellens *et al.*, 1984). If an excitation wavelength of 306 nm is used, no pH quenching is observed at least down to pH 2. The quenching of the fluorophore by the dipyridinium-*p*-xylene is of the collisional type, and the quenching coefficient is a function of pH. Conditions can be arranged so that the assay can be employed down to pH 2 (Ellens *et al.*, 1985). The aminonaphthalenetrisulfonic acid and the dipyridinium-*p*-xylene do not interact with large unilamellar vesicles composed of phosphatidylcholine, phosphatidylethanolamine, or a variety of acidic phospholipids. This fluorophore does not have as high as extinction coefficient as carboxyfluorescein, so that the sensitivity of the assay is not as great as the original FSQ assay (Weinstein *et al.*, 1977). However, release of contents can be measured in 10 μM liposomal lipid containing fluorophore at an initial concentration of 50 mM. This corresponds to a final concentration following release of 0.6 μM, which is sufficiently low to permit a kinetic analysis of the destabilization process without resorting to stopped flow instruments (Ellens *et al.*, 1984; Bentz *et al.*, 1985; Nir *et al.*, 1983*a*).

Alternative methods for following the leakage of contents of liposomes are based on the complexation assay of Wilschut and Papahadjopoulos (1979), which employs the increase in fluorescence observed upon the binding of terbium to dipicolinic acid (Wilschut *et al.*, 1983; Bentz *et al.*, 1983). The release assay has been set up in two different formats. Wilschut and colleagues (1983) measured the increase of fluorescence when terbium was released from vesicle and formed a complex with dipicolinic acid that was in the external medium in the absence of EDTA. The fluorescence signal increased as leakage occurred. These investigators showed that the release kinetics, occurring during the calcium-induced fusion of phosphatidylserine SUV, measured by terbium was identical to that observed for carboxyfluorescein. Bentz and co-workers (1983) started with the terbium–dipicolinic acid complex inside the vesicles and allowed the divalent cations and low levels of EDTA in the medium to dissociate the terbium–dipicolinic acid complex when it was released from the phosphatidylserine vesicle. The fluorescence signal decreased as leakage occurred. That both methods are suitable for following release is due to the ability of EDTA to compete effectively with the dipicolinic acid for the terbium. The results from both studies confirmed that fusion precedes release of aqueous contents from phosphatidylserine SUV.

In summary, a variety of sensitive fluorescent assays are available to follow the kinetics of release of liposome contents.

7. ASSAY FOR MIXING MONOLAYERS

Mixing of vesicle membranes has been followed by calorimetric (Papahadjopoulos *et al.*, 1974, 1976), ESR (Van der Bosch and McConnell, 1975; Maeda and Ohnishi, 1974; Marsh and Watts, 1981), isotopic (Papahadjopoulos *et al.*, 1974), and fluorescence methods (Nir *et al.*, 1983*a*). Of these methods, only the ESR method, which measures the disappearance of spin–spin exchange broadening when a spin-labeled lipid is diluted into a population of vesicles lacking the spin label (Marsh and Watts, 1981) and the fluoresence techniques have the time resolution necessary to follow the kinetics of most fusion reactions. Of these two techniques, fluorescence assays have been the more widely used.

Three different types of membrane fusion assays utilizing fluorescent phospholipids have been devised. These are based upon excimer formation (Owen, 1980), fluorescence lifetime changes (Parente and Lentz, 1985), and resonance energy transfer (Keller *et al.*, 1977; Fung and Stryer, 1978; Gibson and Loew, 1979; Vanderwerf and Ullman, 1980; Uster and Deamer, 1981; Struck *et al.*, 1981; Sunamoto *et al.*, 1983; Pryor *et al.*, 1985). The resonance energy transfer (RET) assay has become the assay of choice for following mixing of membrane lipids and in particular the assay described by Struck and co-workers (1981) has been widely used (Nir *et al.*, 1983*a;* Düzgüneş, 1985). The assay is set up so that the donor phospholipid, *N*-(7-nitro-2,1,3-benzoxadiazol-4-yl)phosphatidylethanolamine (NBD-PE) and the acceptor phospholipid, (lissamine rhodamine B sulfonyl)phosphatidylethanolamine (Rh-PE) are initially in the same bilayer, usually at 1 mol%. When mixed with a liposome lacking the fluorophores and induced to fuse, lateral diffusion of the probes into the newly formed membrane reduces the surface density of the probes; the fluorescence of the donor phospholipid increases, while the fluorescence of the acceptor phospholipid decreases. With this assay, it is possible to calculate the acceptor surface densities from the measured energy-transfer efficiencies. Hence it is possible to estimate the extent of dilution of the probe and thus the level of fusion between vesicle populations (Struck *et al.*, 1981). A convenience of the assay is that the probe resides in the bilayer, so that separation of nonencapsulated material is not necessary. This permits a variety of liposome compositions to be prepared and tested.

In addition, one has a great deal of flexibility in the composition of the aqueous phase, since the liposome does not have to contain enzymes, fluorophores, or other molecules for assaying contents mixing. The assay is also applicable for studying low pH induced fusions an advantage for studying processes such as viral membrane protein induced fusion (Bental *et al.,* 1984).

In almost all cases, the probes used in the RET have been shown to be nonexchangeable and thus a valid measure of membrane intermixing following fusion. However, one case has been reported in which the mixing of the probes between phosphatidylethanolamine/cholesterylhemisuccinate membranes induced to aggregate in the presence of low levels of calcium or magnesium resulted in probe exchange in the absence of release or intermixing of vesicle contents (Ellens *et al.,* 1985). In other cases, the RET assay indicates fusion, whereas the application of a number of other assays do not (Rosenberg *et al.,* 1983; Ababei and Hildenbrand, 1984). This reaffirms the desirability to measure more than a single parameter when studying fusion in model systems.

8. ASSAY FOR MIXING OF AQUEOUS CONTENTS

The field of membrane fusion, using liposomes as model systems was significantly accelerated by the development of assays capable of continuously monitoring aqueous contents mixing. Assays have been developed based on enzymatic, complexation and quenching techniques. Many of these have employed fluorescence probes so they have a high degree of sensitivity.

8.1. Enzymatic Reactions

One of the earliest assays for the mixing of vesicle contents used the enzyme luciferase in one population of vesicles and ATP and magnesium in a second population (Ingolia and Koshland, 1978; Holz and Stratford, 1979). Upon fusion of the two populations of vesicles with each other, a burst of photons was emitted. In both studies, only a small fraction of the expected light emission was observed in the fusion assay. Ingolia and Koshland (1978) interpreted their results as indicating leakage of contents during fusion, while Holz and Stratford (1979) concluded that only a small fraction of the vesicles had fused. Interpretation of the light output during fusion is complicated due to the number of corrections that have to be made to the data. This makes kinetic analysis of the results difficult. To overcome some of the difficulties experienced with this approach, Hoekstra and co-workers (1979) synthesized an internally quenched trypsin substrate where an *o*-aminobenzoyl fluorophore was positioned on the N-terminal side of an arginine-*p*-nitrophenylalanine bond. The *p*-nitrophenylalanine was the quenching residue. Upon cleavage of this substrate by trypsin, the quenching of the aminobenzoyl group was eliminated as the quencher was diluted into the bulk medium. A clever aspect of this assay was the inclusion of a trypsin inhibitor in the external solution to inhibit any reaction between molecules that had leaked from the phosphatidylserine SUV during the calcium induced fusion (Hoekstra *et al.*, 1979). However, the kinetics of the reaction are much slower than reported for other mixing assays. The question of whether the rate-limiting step in the reaction is due to the enzymatic reaction or whether some of the components in the assay system slow the fusion step or stabilize the fused structures after only one or a few rounds of fusion has not been answered. Although many potentially useful enzymatic reactions could be employed in this manner, such as hydrolysis of carboxydiacetylfluorescein (Altseil and Branton, 1983) or the sialidase cleavage method (Davey *et al.*, 1985), the limitations imposed on the assay scheme by the enzymatic nature of the reactions have curtailed their development and use.

8.2. Complexation

A convenient assay to follow the mixing of contents involves the formation of chemical complexes between reagents in separate vesicle populations. Such complexes can be detected by changes in absorbance (Dunham *et al.*, 1977) or fluorescence properties of the reporter molecule (Wilschut and Papahadjopoulos, 1979; Kendall and MacDonald, 1982). Dunham and co-workers (1977) used an Arsenazo III–calcium complex to follow the fusion of phosphatic acid SUV. In this assay, the dye complex is placed in one population of vesicles and EGTA is in a second. Upon fusion, the EGTA interacts with the Arsenazo III–complex resulting in a change of its extinction coefficient and a shift in the absorption maximum. Although the assay was used to show mixing of contents, the facts that the dye binds to calcium and that calcium is used in many systems as a fusogen have limited its widespread use.

The most popular assay for examining the mixing of vesicle contents is the terbium complexation assay introduced by Wilschut and Papahadjopoulos (1979). In this assay, a nonfluorescent terbium–citrate complex is encapsulated in one population of vesicles, and the chelator, dipicolinic acid, is encapsulated at 10 times the concentration of the terbium in a second population of vesicles. Upon intermixing of the contents of the two populations, the terbium forms a fluorescent complex with the dipicolinic acid. The fluorescence of the terbium solution increases more than 10,000-fold upon binding to the dipicolinic acid (Barela and Sherry, 1976), forming the basis of a very sensitive contents mixing assay. The inclusion of EDTA and divalent cations in the assay medium outside of the vesicles effectively competes for the terbium, if any of the terbium–dipicolinic acid complex should leak from the fusing liposomes. This assay has been a major catalyst to our understanding of calcium-induced fusion in acidic phospholipid liposomes (Nir *et al.*, 1983*a;* Wilschut and Hoekstra, 1984; Düzgüneş, 1985; Düzgüneş, Chapters 11 and 12, this volume) and, in conjunction with the kinetic analysis induced by Nir and Bentz (reviewed in Nir *et al.,* 1983*a*), has permitted precise statements to be made about the biophysics of these fusing systems. Unfortunately, below pH 5.0 the dipicolinic acid becomes protonated and the fluorescence from mixtures of terbium and dipicolinic acid decreases rapidly (Barela and Sherry, 1976), presumably due to the dissociation of the terbium from the dipicolinic acid.

An alternative procedure to monitor contents mixing involves the encapsulation of a cobalt–calcein complex in one set of vesicles and EDTA in the second set of vesicles. Upon fusion of the two sets of vesicles, the EDTA effectively competes for the cobalt and, as the cobalt dissociates from the calcein, the fluorescence signal from the calcein increases (Kendall *et al.,* 1982). Using this assay, the leakage of contents from the phosphatidylserine SUV proceeds much more rapidly than do contents mixing between SUV. The reason for the discrepancy between results obtained with the calcein–cobalt assay (Kendall *et al.,* 1982) and the terbium–dipicolinic acid assay (Wilschut *et al.,* 1983) is unknown. However, Wilschut and colleagues (1983) suggested that the dissociation of cobalt from the calcein under the assay conditions may be too slow to monitor the contents-mixing phase of the fusion adequately. The discrepancy appears to be particularly severe when SUV are assayed, since mixing of contents can be observed with the cobalt–calcein assay when phosphatidylserine LUV fuse in the presence of calcium (R. MacDonald, personal communication).

8.3. Quenching of Fluorescence

To explore low pH-induced fusions between vesicles requires an assay that can respond under acidic conditions; carboxyfluorescein, calcein, and the terbium–dipicolinic acid assays are unsuitable for use at pH below 6. Their fluorescence is extinguished by proton titration (Barela and Sherry, 1976; Szoka *et al.,* 1979; Allen, 1984; Weinstein *et al.,* 1984). To avoid this problem, the release assay, employing aminonapthalenetrisulfonic acid as the fluorophore and dipyridinium-*p*-xylene as the quencher (Smolarsky *et al.,* 1977; Ellens *et al.,* 1984), was reconfigured by separating the quencher from the fluorophore in two sets of vesicles (Ellens *et al.,*

1985). Fusion of the two populations of vesicles results in quenching of fluorescence. The rate of leakage of the encapsulated molecules from the intact liposome is negligible, the quenching reaction is fast, and quenching of the fluorophore will only occur upon mixing of the aqueous contents of the vesicles (Ellens *et al.,* 1985). A comparison between the quenching assay and the terbium–dipicolinic acid assay demonstrated that the quenching assay gave comparable, albeit slightly faster rates, for phosphatidylserine LUV fusion in the presence of calcium (Ellens *et al.,* 1985). This assay has been used to study the pH and divalent cation-induced destabilization of phosphatidylethanolamine vesicles (Bentz *et al.,* 1985) and appears to be generally applicable to liposome–liposome fusion.

9. EXPERIMENTAL DESIGN

Since the early studies of fusion, a variety of techniques have been developed to measure almost every aspect of lipid vesicle fusion: bilayer mixing, contents mixing, and subsequent leakage of contents. In our work we have employed all three assay systems to define the events surrounding liposome fusion and collapse (Ellens *et al.,* 1984, 1985; Bentz *et al.,* 1985; Lai *et al.,* 1985). A few simple precautions can assist in the performance of the fluorescence assays. In particular, care must be taken to keep fluorescent materials and detergents from contaminating the buffer systems. In all the fluorescence assays, the indicators are usually highly fluorescent, so that a careless reuse or misuse of a pipet can cause considerable problems. A common difficulty for individuals starting fusion experiments is that detergents are usually added to the cuvette to lyse the vesicles in order to obtain a 100% fluorescence signal. If care is not taken in cleaning the cuvette to remove residual detergent, unexplainable variations can vex the experiments. It is also useful to filter solutions used in the fusion experiments through 0.45-μm membranes. This removes particulates that can nucleate vesicle aggregation at low vesicle concentrations, particularly when a "sticky lipid" such as phosphatidylethanolamine is being studied. Finally, when the results are to be analyzed by the mass–action model of Nir and Bentz (Nir *et al.,* 1982; Bentz *et al.,* 1983), it is important to make precise additions of vesicles or fusogens to a well-stirred cuvette. This is best accomplished from a Hamilton syringe.

The analysis of the data has progressed from the point of merely observing that the vesicle size has increased, to the determination of the rate constants for the aggregation and fusion steps from a mass action model (Nir *et al.,* 1983*a*). The mass-action model of Nir and Bentz separates the aggregation step, which will be a second-order process, from the fusion step modeled as a first-order reaction. At low vesicle concentration, aggregation will be rate limiting to the overall process. At the low vesicle concentrations observable in the fluorescence fusion assays, aggregation can be separated from fusion. At high vesicle concentrations or under conditions that induce aggregation but not fusion, the assay can be set up to focus on the fusion step (Nir *et al.,* 1983*a;* Bentz *et al.,* 1983*b*). The rate constants determined from these two experimental approaches can be used to predict the time course of the fusion products. The details of the mass action model and how to apply it to the fluorescence fusion assay data are described in a recent review by Nir and colleagues (1983*a*) and by Bentz *et al.,* (1983*b*). Criticism of this approach is that the aggregation and fusion occurs so rapidly that a stopped flow spectrofluorometer is required to obtain the data (Morris *et al.,* 1985). However, a comparison of rate constants for phosphatidylserine SUV aggregation and fusion determined by Nir and Bentz mass action method to those obtained in stopped flow experiments (Morris *et al.,* 1985) show a close correspondence. This is particularly reassuring since even small changes in the size distribution of the SUV can have significant effects on the aggregation and fusion rate constants (Bentz and Düzgüneş, 1985).

10. SUGGESTED TOPICS FOR FURTHER DEVELOPMENT

Advances in the study of fusion have been due to the development of experimental methods and kinetic approaches for data analysis. From these methods, more precise statements can be made concerning how the components found in biologic systems mediate and regulate fusion.

The interaction of two membranes that leads to their mutual destabilization, leakage of encapsulated molecules and ultimately fusion, involves a small number of molecules. This process, by its very nature is a nonequilibrium event. The challenge for those interested in this problem is to design an assay with the requisite sensitivity and response time to measure events at the point of contact.

Another useful goal is to design methods or procedures to form asymmetrical vesicles. The external monolayer should be susceptible to fusion while the internal monolayer is not, this should permit studies on more systems that might not collapse but rather under the appropriate conditons continuously fuse. The establishment and validation of the lipid vesicle as a model fusing system have in large part been due to methodological developments, not conceptual breakthroughs, and it seems that for the immediate future this will continue to be the case.

ACKNOWLEDGMENTS. This work was supported in part by grant GM-29514 from the National Institutes of Health. I am grateful to Dr. Joe Bentz, Dr.

Harma Ellens, Dr. Nejat Düzgüneş, and Dr. Demetri Papahadjopoulos for many informative discussions concerning these topics. I thank Elma Belenson for expert typing of this manuscript and the secretarial support of Ms. Georgina Kurtovich.

REFERENCES

Ababei, L., and Hildenbrand, K., 1984, Kinetics of calcium-induced mixing of lipids and aqueous contents of large unilameller phosphatidylserine vesicles, *Chem. Phys. Lipids* **35**:39–48.

Abbs, M. T., and Phillips, T. H., 1980, Organization of the proteins of the chromaffin granule membrane, *Biochim. Biophys. Acta* **595**:200–221.

Allen, T. M., 1984, Calcein as a tool in liposome methodology, in: *Liposome Technology,* Vol. 3 (G. Gregoriadis, ed.), pp. 177–182, CRC Press, Boca Raton, Florida.

Allen, T. M., 1980, Serum-induced leakage of liposome contents, *Biochim. Biophys. Acta* **597**:418–426.

Alstiel, L., and Branton, D., 1983, Fusion of coated vesicles with lysomes: Measurement with a fluorescence assay, *Cell* **32**:921–929.

Aunis, D., Hesketh, J. E., and Devilliers, G., 1979, Freeze-fracture study of the chromaffin cell during exocytosis: Evidence for connections between the plasma membrane and secretory granules and for movements of plasma membrane-associated particles, *Cell Tissue Res.* **197**:433–441.

Baker, P. F., Knight, D. E., and Whitaker, M. J., 1980, Calcium and the control of exocytosis, in: *Calcium Binding Proteins: Structure and Function* (F. L. Siegel, E. Carafoli, R. H. Kretsinger, D. H. MacLennan, and R. H. Wasserman, eds.), pp. 47–55, Elsevier/North-Holland, New York.

Baker, P. F., and Whitaker, M. J., 1978, Influence of ATP and calcium on the cortical reaction in sea urchin eggs, *Nature (Lond.)* **276**:513–515.

Bangham, A. D., Hill, M. W., and Miller, N. G., 1974, Preparation and use of liposomes as models of biological membranes, *Methods Membrane Biol.* **1**:1–68.

Banks, P., 1966, An interaction between chromaffin granules and calcium ions, *Biochem. J.* **101**:18c–20c.

Barela, T. D., and Sherry, A. D., 1976, A simple, one step fluormetric method for determination of nanomolar concentrations of terbium, *Anal. Biochem.* **71**:351–357.

Barenholz, Y., Gibbes, D., Litman, B. J., Goll, J., Thompson, T. E., and Carlson, F. D., 1977, A simple method for the preparation of homogeneous phospholipid vesicle, *Biochemistry* **16**:2806–2810.

Barenholz, Y., Amselem, S., and Lichtenberg, D., 1979, A new method for preparation of phospholipid vesicles (liposomes)—French press, *FEBS Lett.* **99**:210–214.

Bearer, E. L., Düzgüneş, N., Friend, D. S., and Papahadjopoulos, D., 1982, Fusion of phospholipid vesicles arrested by quick-freezing. The question of lipidic particles as intermediates in membrane fusion, *Biochim. Biophys. Acta* **693**:93–98.

Bental, M., Lelkes, P. I., Scholma, J., Hoekstra, D., and Wilschut, J., 1984, Comparison of two liposome fusion assays monitoring the intermixing of aqueous contents and of membrane components, *Biochim. Biophys. Acta* **774**:296–300.

Bentz, J., and Düzgüneş, N. 1985, Fusogenic capacities of divalent cations and the effect of liposome size, *Biochemistry* **24**:5436–5444.

Bentz, J., and Nir, S., 1981, Aggregation of phosphatidylserine vesicles modelled as a dynamic process, *Proc. Natl. Acad. Sci. U.S.A.* **78**:1634–1637.

Bentz, J., Düzgüneş, N., and Nir, S., 1983*a,* Kinetics of divalent cation induced fusion of phosphatidylserine vesicles: Correlation between fusogenic capacities and binding affinities, *Biochemistry* **22:**3320–3330.

Bentz, J., Nir, S., and Wilschut, J., 1983*b,* Mass action kinetics of vesicle aggregation and fusion, *Colloids Surfaces* **6:**333–363.

Bentz, J., Ellens, H., Lai, M. -Z., and Szoka, F. C., 1985, On the correlation between H_{II} phase and the contact-induced destabilization of phosphatidylethanolamine-containing membranes, *Proc. Natl. Acad. Sci. U.S.A.* **82:**5742–5745.

Boni, L. T., Stewart, T. P., Alderfer, J. L., and Hui, S. W., 1981, Lipid–polyethylene glycol interactions: I. Induction of fusion between liposomes, *J. Membrane Biol.* **62:**65–70.

Burgoyne, R. D., and Geisow, M. J., 1981, Specific binding of ^{125}I-calmodulin to and protein phosphorylation in adrenal chromaffin granule membranes, *FEBS Lett.* **131:**127–131.

Burridge, K., and Phillips, J. H., 1975, Association of actin and myosin with secretory granule membranes, *Nature (Lond.)* **254:**526–529.

Chandler, D. E., Bennett, J. P., and Gomperts, B., 1983, Freeze-fracture studies of chemotactic peptide-induced exocytosis in neutrophils: Evidence for two patterns of secretory granule fusion, *J. Ultrastruct. Res.* **82:**221–232.

Creutz, C. E., 1981*a,* Cis-unsaturated fatty acids induce the fusion of chromaffin granules aggregated by synexin, *J. Cell. Biol.* **91:**247–256.

Creutz, C. E., 1981*b,* Secretory vesicle–cytosol interactions in exocytosis: Isolation by Ca^{2+}-dependent affinity chromatography of proteins that bind to the chromaffin granule membrane, *Biochem. Biophys. Res. Commun.* **103:**1395–1400.

Creutz, C. E., Pazoles, C. J., and Pollard, H. B., 1978, Identification and purification of an adrenal medullary protein (synexin) that causes calcium-dependent aggregation of isolated chromaffin granules, *J. Biol. Chem.* **253:**2858–2866.

Creutz, C. E., Dowling, L. G., Sando, J. J., Villar-Palasi, C., Whipple, J. H., and Zaks, W. J., 1983, Characterization of the chromobindins: Soluble proteins that bind to the chromaffin granule membrane in the presence of Ca^{2+}, *J. Biol. Chem.* **258:**14664–14674.

Creutz, C. E., Dowling, L. G., Kygers, E. M., and Ranson, R. C., 1985, Phosphatidylinositol-specific phospholipase C activity of chromaffin granule-binding proteins, *J. Biol. Chem.* **260:**7171–7173.

Dahl, G., Ekerdt, R., and Gratzl, M., 1979, Models for exocytotic membrane fusion, *Symp. Soc. Exp. Biol.* **33:**349–368.

Davey, J., Hurtley, S. M., and Warren, G., 1985, Reconstitution of an endocytic fusion event in a cell-free system, *Cell* **43:**643–652.

de Duve, D., 1963, Endocytosis (footnote), in: *Lysosomes, Ciba Foundation Symposium Lysosomes* (A. V. S. deReuck and M. P. Cameron, eds.), p. 126, J. & A. Churchill, London.

Douglas, W. W., 1974, Involvement of calcium in exocytosis and the exocytosis-vesiculation sequence, *Biochem. Soc. Symp.* **39:**1–28.

Dreyfus, H., Aunis, D., Harth, S., and Mandel, P., 1977, Gangliosides phospholipids of the membranes from bovine adrenal medullary chromaffin granules, *Biochim. Biophys. Acta* **489:**89–97.

Dunham, P., Babiarz, P., Israel, A., Zerial, A., and Weissman, G., 1977, Membrane fusion: studies with the calcium sensitive dye, arsenazo III, in liposomes, *Proc. Natl. Acad. Sci. U.S.A.* **74:**1580–1584.

Dunn, L. A., and Holz, R. W., 1983, Catecholamine secretion from digitonin-treated adrenal medullary chromaffin cells, *J. Biol. Chem.* **258:**4989–4993.

Düzgüneş, N., 1985, Membrane fusion, in: *Subcellular Biochemistry* (D. B. Roodyn, ed.), pp. 195–286, Plenum Press, New York.

Düzgüneş, N., Wilschut, J., Hong, K., Fraley, R., Perry, C., Friend, D., James, T. L., and Papahadjopoulos, D., 1983, Physicochemical characterization of large unilamellar phospholipid vesicles prepared by reverse-phase evaporation, *Biochim. Biophys. Acta.* **732**:289–299.

Eidelman, O., Schlegal, R., Traika, T. S., and Blumenthal, R., 1984, pH-dependent fusion induced by vesicular stomatitis virus glycoprotein reconstituted into phospholipid vesicles, *J. Biol. Chem.* **259**:4622–4628.

Ekerdt, R., Dahl, G., and Gratzl, M., 1981, Membrane fusion of secretory vesicle and liposomes. Two different types of fusion, *Biochim. Biophys. Acta* **646**:10–22.

Ellens, H., Bentz, J., and Szoka, F. C., 1984, pH-induced destabilization of phosphatidylethanolamine-containing liposomes: Role of bilayer contact, *Biochemistry* **23**:1532–1538.

Ellens, H., Bentz, J., and Szoka, F. C., 1985, H^+- and Ca^{2+}-induced fusion and destabilization of liposomes, *Biochemistry* **24**:3099–3106.

Fraley, R., Wilschut, J., Düzgüneş, N., Smith, C., Papahadjopoulos, D., 1980, Studies on the mechanism of membrane fusion: Role of phosphate in promoting calcium ion induced fusion of phospholipid vesicles, *Biochemistry* **19**:6021–6029.

Fowler, V. M., and Pollard, H. B., 1982*a*, Chromaffin granule membrane–F-actin interactions are calcium sensitive, *Nature (Lond.)* **295**:336–339.

Fowler, V. M., and Pollard, H. B., 1982*b*, In vitro reconstitution of chromaffin granule-cytoskeleton interactions. Ionic factors influencing the association of F-actin with purified chromaffin granule membranes, *J. Cell. Biochem.* **18**:295–311.

Fung, B. K. -K., and Stryer, L., 1978, Surface density determination in membranes by fluorescence energy transfer, *Biochemistry* **17**:5241–5248.

Gabbiani, G., and da Parada, M., Richards, G., and Pletscher, A., 1976, Actin association with membranes of monoamine storage organelles, *Proc. Soc. Exp. Biol. Med.* **152**:135–138.

Gemmell, R. T., and Stacy, B. D., 1979, Granule secretion by the luteal cell of the sheep: The fate of the granule membrane, *Cell Tissue Res.* **197**:413–419.

Gibson, G. A., and Loew, L. M., 1979, Phospholipid vesicle fusion monitored by fluorescence energy transfer, *Biochem. Biophys. Res. Commun.* **88**:135–140.

Ginsberg, L., 1978, Does calcium cause fusion or lysis of unilamellar lipid vesicles? *Nature (Lond.)* **275**:758–760.

Goll, J., Carlson, F. D., Barenholz, K., Litman, B. J., and Thompson, T. E., 1982, Photon correlation spectroscopic study of the size distribution of phospholipid vesicles, *Biophys. J.* **38**:7–13.

Gratzl, M., and Dahl, G., 1976, Calcium-induced fusion of Golgi-derived secretory vesicles isolated from rat liver, *FEBS Lett.* **62**:142–145.

Gratzl, M., Schudt, C., Ekerdt, R., and Dahl, G., 1980, Fusion of isolated biological membranes: A tool to investigate basic processes of exocytosis and cell–cell fusion, in: *Membrane Structure and Function,* Vol. 3 (E. E. Bittar, ed.), pp. 59–92, Wiley, New York.

Gregoriadis, G., ed., 1984, *Liposome Technology,* Vol. 1, 2, and 3, CRC Press, Boca Raton, Florida.

Hagins, W. A., and Yoshikami, S., 1978, Intracellular transmission of visual excitation in photoreceptors: Electrical effects of chelating agents introduced into rods by vesicle fusion, in: *Vertebrate receptors* (P. Fatt and H. B. Barlow, eds.), pp. 97–126, Academic Press, New York.

Hamilton, R. L., Goerke, J., Guo, L. S. S., Williams, M. C., and Havel, R. J., 1980, Unilamellar liposomes made with the French pressure cell: A simple preparative and semiquantitative technique, *J. Lipid Res.* **21**:981–992.

Hoekstra, D., 1982, Kinetics of intermixing of lipids and mixing of aqueous contents during vesicle fusion, *Biochem. Biophys. Acta* **692:**171–175.

Hoekstra, D., Yaron, A., Carmel, A., and Scherphof, G., 1979, Fusion of phospholipid vesicles containing a trypsin-sensitive fluorogenic substrate and trypsin. A new method to study membrane fusion activity in a model system, *FEBS Lett.* **106:**176–180.

Holz, R., and Stratford, C., 1979, Effects of divalent ions on vesicle–vesicle fusion studied by a new luminescence assay for fusion, *J. Membrane Biol.* **46:**331–358.

Huang, C., 1969, Studies on phosphatidylcholine vesicles. Formation and physical characteristics, *Biochemistry* **8:**344–352.

Huber, E., Konig, P., Schuler, G., Aberer, W., Plattner, H., and Winkler, H., 1979, Characterization and topography of the glycoproteins of adrenal chromaffin granules, *J. Neurochem.* **32:**35–47.

Ingolia, T., and Koshland, D., 1978, The role of calcium in fusion of artificial vesicles, *J. Biol. Chem.* **253:**3821–3829.

Izumi, F., Kashimoto, T., Miyashita, T., Wada, A., and Oka, M., 1977, Involvement of membrane associated protein in ADP-induced lysis of chromaffin granules, *FEBS Lett.* **78:**177–180.

Izumi, F., Oka, M., Morita, K., and Azuma, H., 1975, Catecholamine releasing factor in bovine adrenal medulla, *FEBS Lett.* **56:**73–76.

Jockusch, B. M., Burger, M. M., daPrada, M., Richards, J. G., Chaponnier, C., and Gabbiani, G., 1977, Actinin attached to membranes of secretory vesicles, *Nature (Lond.)* **270:**628–629.

Judah, J. D., and Quinn, P. S., 1978, Calcim ion-independent vesicle fusion in the conversion of proalbumin to albumin, *Nature (Lond.)* **271:**384–385.

Kantor, H. L., and Prestegard, J., 1975, Fusion of fatty acid containing lecithin vesicles, *Biochemistry* **14:**1790–1794.

Kataoka, T., Williamson, J. R., and Kinsky, S. C., 1973, Release of macromolecular marker (enzymes) from liposomes treated with antibody and complement, *Biochim. Biophys. Acta* **298:**158–179.

Keller, P. M., Person, S., and Snipes, W., 1977, A fluorescence enhancement assay of cell fusion, *J. Cell Sci.* **28:**167–177.

Kendall, D. A., and MacDonald, R. C., 1982, A fluorescence assay to monitor vesicle fusion and lysis, *J. Biol. Chem.* **257:**13892–13895.

Klausner, R. D., Kumar, N., Weinstein, J. N., Blumenthal, R., and Flavin, M., 1981, Interaction of tubulin with phospholipid vesicles with phospholipid vesicles. I. Association with vesicles at the phase transition, *J. Biol. Chem.* **256:**5879–5886.

Knutton, S., and Pasternak, C. A., 1979, The mechanism of cell-cell fusion, *Trends Biochem. Sci.* **4:**220–223.

Konings, F., and DePotter, W., 1981, Calcium-dependent in vitro interaction between bovine adrenal medullary cell membranes and chromaffin granules as a model for exocytosis, *FEBS Lett.* **126:**103–106.

Lagunoff, D., 1973, Membrane fusion during mast cell secretion, *J. Cell. Biol.* **57:**252–259.

Lai, M. Z., Vail, W. J., and Szoka, F. C., 1985, Acid- and calcium-induced structural changes in phosphatidylethanolamine membranes stabilized by cholesteryl hemisuccinate, *Biochemistry* **24:**1654–1661.

Liao, M. -J., and Prestergard, J. H., 1979, Fusion of phosphatidic acid–phosphatidylcholine mixed lipid vesicles, *Biochem. Biophys. Acta.* **550:**157–173.

Livett, B. G., 1984, The secretory process in adrenal medulla cells, in: *Cell Biology of the Secretory Process* (M. Cantin, ed.), pp. 309–358, Karger, Basel.

Lucy, J. A., 1984, Do hydrophobic sequences cleaved from cellular polypeptides induce membrane fusion reactions in vivo?, *FEBS Lett.* **166:**223–231.

MacDonald, R. I., 1985, Membrane Fusion due to dehydration by polyethylene glycol, dextran, or sucrose, *Biochemistry* **24:**4058–4064.

Maeda, T., and Ohnishi, S. -I., 1974, Membrane fusion transfer of phospholipid molecules between phospholipid bilayer membranes, *Biochem. Biophys Res. Commun.* **60:**1509–1516.

Marsh, D., and Watts, A., 1981, ESR spin label studies of liposomes, in: *Liposomes: from physical structure to therapeutic applications* (C. G. Knight, ed.), pp. 139–188, Elsevier/North-Holland, New York.

Marsh, M., Bolzau, E., White, J., and Helenius, A., 1983, Interactions of Semliki Forest virus spike glycoprotein rosettes and vesicles with cultured cells, *J. Cell Biol.* **96:**455–461.

Mayhew, E., Lazo, R., Vail, W. J., King, J. J., and Siciliano, A. A., 1984, Characterization of liposomes prepared using a microemulsifier, *Biochim Biophys. Acta* **775:**169–176.

Meyer, D. I., and Burger, M. M., 1976, The chromaffin granule surface: Localization of carbohydrate on the cytoplasmic surface of an intracellular organelle, *Biochim, Biophys. Acta* **443:**428–436.

Meyer, D. I., and Burger, M. M., 1979*a*, Isolation of a protein from the plasma membrane of adrenal medulla which binds to secretory vesicles, *J. Biol. Chem.* **254:**9854–9859.

Meyer, D. I., and Burger, M. M., 1979*b*, The chromaffin granule surface: The presence of actin and the nature of its interaction with the membrane, *FEBS Lett.* **101:**129–133.

Morris, S. J., and Schober, R., 1977, Demonstration of binding sites for divalent and trivalent ions on the outer surface of chromaffin granule membranes, *Eur. J. Biochem.* **75:**1–12.

Morris, S. J., Chiu, V. C. K., and Haynes, D. J., 1979, Divalent cation-induced aggregation of chromaffin granule membranes, *Membrane Biochem.* **2:**163–201.

Morris, S. J., Gibson, C. C., Smith, P. D., Greif, P. C., Stirk, C. W., Bradley, D., Haynes, D. H., and Blumenthal, R., 1985, Rapid kinetics of Ca^{2+}-induced fusion of phosphatidylserine/phosphatidylethanolamine vesicles, *J. Biol. Chem.* **260:**4122–4127.

Nir, S., Bentz, J., and Wilschut, J., 1980, Mass action kinetics of phosphatidyleserine vesicle fusion as monitored by coalescence of internal vesicle volumes, *Biochemistry* **19:**6030–6036.

Nir, S., Wilschut, J., and Bentz, J., 1982, The rate of fusion of phospholipid vesicles and the role of bilayer curvature, *Biochim. Biophys. Acta* **688:**275–278.

Nir, S., Bentz, J., Wilschut, J., and Düzgüneş, N., 1983*a*, Aggregation and fusion of phospholipid vesicles, *Prog. Surface Sci.* **13:**1–124.

Nir, S., Düzgüneş, N., and Bentz, J., 1983*b*, Binding of monovalent cations to phosphatidylserine and modulation of Ca^{2+}- and Mg^{2+}-induced vesicle fusion, *Biochim. Biophys. Acta* **735:**160–172.

Oates, P. J., and Touster, O., 1976, *In vitro* fusion of *Acanthamoeba* phagolysosomes. II. Quantitative characterization of *in vitro* vacuole fusion by improved electron microscope and new light microscope techniques, *J. Cell Biol.* **79:217**–234.

Olson, F., Hunt, C. A., Szoka, F. C., Vail, W. J., and Papahadjopoulos, D., 1979, Preparation of liposomes of defined size distrubution by extrusion through polycarbonate membranes, *Biochim. Biophys. Acta* **557:**9–23.

Orci, L., Amherdt, M., Malaisse-Lagae, F., Rouiller, C., Renold, A. E., 1973, Insulin release by exocytosis: Demonstration with freeze-etching technique, *Science* **179:**82–84.

Owens, C. S., 1980, A membrane-bound fluorescent probe to detect phospholipid vesicle–cell fusion, *J. Membrane Biol.* **54:**13–20.

Palade, G. E., 1959, Functional changes in the structure of cell components, in: *Subcellular Particles* (T. Hayashi, ed.), pp. 64–80, Ronald Press, New York.

Palade, G., 1975, Intracellular aspects of the process of protein synthesis, *Science* **189:**347–358.

Papahadjopoulos, D., and Miller, N., 1967, Structural characteristics of hydrated liquid crystals, *Biochim. Biophys. Acta* **135:**624–638.

Papahadjopoulos, D., Poste, G., Schaeffer, B. E., and Vail, W. J., 1974, Membrane fusion and molecular segregation in phospholipid vesicles, *Biochim. Biophys. Acta* **352:**10–28.

Papahadjopoulos, D., Vail, W. J., Pangborn, W. A., and Poste, G., 1976, Studies on membrane fusion. II. Induction of fusion in pure phospholipid membranes by calcium and other divalent metals, *Biochim. Biophys. Acta* **448**:265–283.
Papahadjopoulos, D., Poste, G., and Vail, W. J., 1979, Studies on membrane fusion with natural and model membranes, *Methods Membrane Biol.* **10**:1–121.
Parente, R. A., and Lentz, B. R., 1984, Phase behavior of large, unilamellar vesicles composed of synthetic phospholipids, *Biochemistry* **23**:2353–2362.
Parente, R. A., and Lentz, B. R., 1985, Advantages and limitations of 1-palmitoyl-3-((2-(4(-6-phenyl-*trans*-1,3,5,-hexatrienyl)phenyl)ethyl)carbonyl)-2-*sn*-phosphatidylcholine as a fluorescent membrane probe, *Biochemistry* **24**:6178–6186.
Parente, R. A., Hoechli, M., and Lentz, B. R., 1985, Morphology and phase behavior of two types of unilamellar vesicles prepared from synthetic phosphatidylcholines studied by freeze-fracture electron microscopy and calorimetry, *Biochim. Biophys. Acta* **812**:493–502.
Pidgeon, C., and Hunt, C. A., 1981, Calculating number and surface area of liposomes in any suspension, *J. Pharm. Sci.* **70**:173–176.
Poisner, A. M., and Trifaro, J. M., eds., 1982, *The Secretory Granule,* Elsevier, New York.
Portis, A., Newton, C., Pangborn, W., and Papahadjopoulos, D., 1979, Studies on the mechanism fusion: evidence for an intermembrane Ca^{2+}-phospholipid complex, synergism with Mg^{2+}, and inhibition by spectrin, *Biochemistry* **18**:780–790.
Poste, G., and Allison, A. C., 1973, Membrane fusion, *Biochem. Biophys. Acta* **300**:421–465.
Poste, G., and Nicolson, G. L., eds., 1978, *Membrane Fusion,* Elsevier/North-Holland, Amsterdam.
Prestegard, J. H., and Fellmeth, B., 1974, Fusion of dimyristoyl lecithin vesicles as studied by proton magnetic resonance spectroscopy, *Biochemistry* **13**:1122–1126.
Pryor, C., Bridge, M., and Loew, L. M., 1985, Aggregation, lipid exchange, and metastable phases of dimyristoylphosphatidylethanolamine vesicles, *Biochemistry* **24**:2203–2209.
Quinn, P. S., and Judah, J. D., 1978, Calcium-dependent Golgi–vesicle fusion and cathepsin B in the conversion of proalbumin into albumin in rat liver, *Biochem. J.* **172**:301–309.
Raz, A., and Goldman, R., 1974, Spontaneous fusion of rat liver lysosomes *in vitro, Nature (Lond.)* **247**:206–208.
Rosenberg, J., Düzgüneş, N., and Kayalar, C., 1983, Comparison of two liposome fusion assays monitoring the intermixing of aqueous contents and of membrane components, *Biochim. Biophys. Acta* **735**:173–180.
Rubin, R. P., 1974, *Calcium and the Secretory Process,* Plenum Press, New York.
Schober, R., Nitsch, C., Rinne, U., and Morris, S. J., 1977, Calcium-induced displacement of membrane-associated particles upon aggregation of chromaffin granules, *Science* **195**:495–497.
Schullery, S. E., and Schmidt, C. F., Felgner, P., Tillack, T. W., and Thompson, T. E., 1980, Fusion of dipalmitoylphosphatidylcholine vesicles, *Biochemistry* **19**:3919–3923.
Siegel, D. P., Ware, B. R., Green, D. J., and Westhead, E. W., 1978, The effects of calcium and magnesium on the electrophoretic mobility of chromaffin granules measured by electrophoretic light scattering, *Biophys. J.* **22**:341–346.
Smolarsky, M., Teitelbaum, D., Sela, M., and Gitler, C., 1977, A simple method to determine complement-mediated liposome immune lysis, *J. Immunol. Meth.* **15**:255–265.
Stossel, T. P., Bretscher, M. S., Ceccarelli, B., Dales, S., Helenius, A., Heuser, J. E., Hubbard, A. L., Kartenbeck, J., Kinne, R., Papahadjopoulos, D., Pearse, B., Plattner, H., Pollard, T. D., Reutter, W., Satir, B. H., Schliwa, M., Schneider, Y. -J., Silverstein, S. C., and Weber, K., 1978, Membrane dynamics group report, in: *Transport of Macromolecules in Cellular Systems* (S. C. Silverstein, ed.), pp. 503–516, Dahlem Konferenzen, Berlin.

Struck, D. K., Hoekstra, D., and Pagano, R. E., 1981, Use of resonance energy transfer to monitor membrane fusion, *Biochemistry* **20**:4093–4099.

Sudhof, T. C., Ebbecke, M., Walker, J. H., Fritsche, U., and Boustead, C., 1984, Isolation of mammalian calelectrins: A new class of ubiquitous Ca^{2+}-regulated proteins, *Biochemistry* **23**:1103–1109.

Summers, T. A., and Creutz, C. E., 1985, Phosphorylation of a chromaffin granule-binding protein by protein kinase C, *J. Biol. Chem.* **260**:2437–2443.

Sun, S. -T., Day, E. P., and Ho, J. T., 1978, Temperature dependence of calcium-induced fusion of sonicated phosphatidyleserine vesicles, *Proc. Natl. Acad. Sci. U.S.A.* **75**:4325–4328.

Sun, S. T., Hsang, C. C., Day, E. P., and Ho, J. T., 1979, Fusion of phosphatidylserine and mixed phosphatidylserine–phosphatidylcholine vesicles, *Biochim. Biophys. Acta* **557**:45–52.

Sunamoto, J., Iwamoto, K., Tezuka, T., Katsunoi, K., and Kondo, H., 1983, Liposomal membranes XIV. Fusion of liposomal membranes induced by polyisoprenoids as monitored by fluorescence quenching method, *Chem. Phys. Lipids.* **33**:127–133.

Szoka, F., and Papahadjopoulos, D., 1978, Procedure for preparation of liposomes with large internal aqueous space and high capture by reverse-phase evaporation, *Proc. Natl. Acad. Sci. U.S.A.* **75**:4194–4198.

Szoka, F., and Papahadjopoulos, D., 1980, Comparative properties and methods of preparation of lipid vesicles (liposomes), *Annu. Rev. Biophys. Bioeng.* **9**:467–508.

Szoka, F., and Papahadjopoulos, D., 1981, Liposomes: Preparation and characterization, in: *Liposomes: From Physical Structure to Therapeutic Applications* (C. G. Knight, ed.), pp. 51–82, Elsevier/North-Holland, New York.

Szoka, F. C., Jacobson, K., and Papahadjopoulos, D., 1979, The use of aqueous space markers to determine the mechanism of interaction between phospholipid vesicles and cells, *Biochim. Biophys. Acta* **551**:295–303.

Szoka, F., Olson, F., Heath, T., Vail, W., Mayhew, E., and Papahadjopoulos, D., 1980, Preparation of unilamellar liposomes of intermediate size (0.1–0.2 μm) by a combination of reverse phase evaporation and extrusion through polycarbonate membranes, *Biochim. Biophys. Acta* **601**:559–571.

Taupin, C., and McConnell, H. M., 1972, Membrane fusion, in: *Mitochondria/Biomembranes, FEBS Eighth Meeting* (S. G. van den Burgh, P. Borst, L. L. M. van Deenen, J. C. Riemeroma, E. C. Slater, and J. M. Tager, eds.), pp. 219–229, Elsevier/North-Holland, Amsterdam.

Unanue, E. R., Ungewickell, E., and Branton, D., 1981, The binding of clathrin triskelions to membranes from coated vesicles, *Cell* **26**:439–446.

Ungar, A., and Phillips, J. H., 1983, Regulation of the adrenal medulla, *Physiol. Rev.* **63**:787–843.

Uster, P. S., and Deamer, D. W., 1981, Fusion competence of phosphatidylserine-containing liposomes quantitatively measured by a fluorescence resonance energy transfer assay, *Arch. Biochem. Biophys.* **209**:385–395.

Van der Bosch, J., and McConnell, H. M., 1975, Fusion of dipalmitoylphosphatidylcholine vesicle membranes induced by concanavalin A, *Proc. Natl. Acad. Sci. U.S.A.* **72**:4409–4413.

Vanderwerf, P., and Ullman, E. F., 1980, Monitoring of phospholipid vesicle fusion by fluorescence energy transfer between membrane-bound dye labels, *Biochim. Biophys. Acta* **596**:302–314.

Verkleij, A. J., 1984, Lipidic intramembranous particles, *Biochim. Biophys. Acta* **779**:43–63.

Verkleij, A. J., and deGier, J., 1981, Freeze fracture studies on aqueous dispersions of membrane lipids, in: *Liposomes: From Physical Structure to Therapeutic Application* (C. G.

Knight, ed.), pp 83–104, Elsevier/North-Holland, New York.

Wang, C. Y., and Huang, L., 1984, Polyhistidine mediates an acid-dependent fusion of negatively charged liposomes, *Biochemistry* **23:**4409–4416.

Weinstein, J. N., Yoshikami, S., Henkart, P., Blumenthal, R., and Hagins, W. A., 1977, Liposome–cell interactions: Transfer and intracellular release of trapped fluorescent marker, *Science* **195:**489–503.

Weinstein, J. N., Klausner, R. D., Innerarity, T., Ralston, E., and Blumenthal, R., 1981, "Phase transition release," a new approach to the interaction of proteins with lipid vesicles: Applications to liposomes, *Biochim. Biophys. Acta* **647:**270–282.

Weinstein, J. N., Ralston, E., Leserman, L. D., Klausner, R. D., Dragsten, P., Henkart, P., and Blumenthal, R., 1984, Self-quenching of carboxyfluorescein fluorescence: Uses in studying liposome stability and liposome—cell interactions, in: *Liposome Technology,* Vol. 3 (G. Gregoriadis, ed.), pp. 183–204, CRC Press, Boca Raton, Florida.

Wilschut, J., and Hoekstra, D., 1984, Membrane Fusion: From liposomes to biological membranes, *Trends in Biochem. Sci.* **9:**479–483.

Wilschut, J., and Papahadjopoulos, D., 1979, Ca^{2+}-induced fusion of phospholipid vesicles monitored by mixing of aqueous contents, *Nature (Lond.)* **281:**690–692.

Wilschut, J., Düzgüneş, N., Fraley, R., and Papahadjopoulos, D., 1980, Studies on the mechanism of membrane fusion: Kinetics of calcium ion induced fusion of phosphatidylserine vesicles followed by a new assay for mixing of aqueous vesicle contents, *Biochemistry* **19:**6011–6021.

Wilschut, J., Düzgüneş, N., Hong, K., Hoekstra, D., and Papahadjopoulos, D., 1983, Retention of aqueous contents during divalent cation-induced fusion of phospholipid vesicles, *Biochim. Biophys. Acta* **734:**309–318.

Wilson, S. P., and Kirshner, N., 1983, Calcium-evoked secretion from digitonin-permeabilized adrenal medullary chromaffin cells, *J. Biol. Chem.* **258:**4994–5000.

Winkler, G., and Carmichael, S. W., 1982, The chromaffin granule, in: *The Secretory Granule* (A. M. Poisner and J. M. Trifaro, eds.), pp. 3–80, Elsevier, New York.

Wong, M., and Thompson, T. E., 1982, Aggregation of dipalmitoylphosphatidylcholine vesicles, *Biochemistry* **21:**4133–4139.

Chapter 11

Fusion of Phospholipid Vesicles Induced by Divalent Cations and Protons

Modulation by Phase Transitions, Free Fatty Acids, Monovalent Cations, and Polyamines

Nejat Düzgüneş, Keelung Hong, Patricia A. Baldwin, Joe Bentz, Shlomo Nir, and Demetrios Papahadjopoulos

The wrong view of science betrays itself in the craving to be right; for it is not his *possession* of knowledge, of irrefutable truth, that makes the man of science, but his persistent and recklessly critical *quest* for truth.

—Karl R. Popper, *The Logic of Scientific Discovery*

1. INTRODUCTION

Studies on the fusion of phospholipid vesicles have provided considerable insight into the molecular mechanism of membrane fusion and the role of specific molecules in this process. This simple and well-defined system

NEJAT DÜZGÜNEŞ • Cancer Research Institute, School of Medicine, and Department of Pharmaceutical Chemistry, School of Pharmacy, University of California at San Francisco, San Francisco, California 94143. KEELUNG HONG and PATRICIA A. BALDWIN • Cancer Research Institute, School of Medicine, University of California at San Francisco, San Francisco, California 94143. *Present address for* P. A. B.: California Biotechnology, Inc., Mountain View, California 94043. JOE BENTZ • Departments of Pharmacy and Pharmaceutical Chemistry, School of Pharmacy, University of California at San Francisco, San Francisco, California 94143. SHLOMO NIR • Seagram Centre for Soil and Water Sciences, Hebrew University of Jerusalem, Rehovot 76-100, Israel. DEMETRIOS PAPAHADJOPOULOS • Cancer Research Institute and Department of Pharmacology, School of Medicine, University of California at San Francisco, San Francisco, California 94143.

has also helped in the development of fusion assays and the biophysical analysis of the fusion process. Early studies on fusion have been reviewed by Papahadjopoulos *et al.* (1979) and Nir *et al.* (1983*a*). Here we will discuss recent developments concerning the mechanism of membrane fusion induced by divalent cations and protons, and its modulation by the thermotropic properties of the phospholipid bilayer, monovalent cations and polyamines in the medium, and free fatty acids in the membrane.

Phospholipid vesicles, or liposomes (cf. Scherphof, 1983), are used as models of biological membranes in an effort to understand the structure and function of the phospholipid bilayer backbone of these membranes (Bangham, 1968; Papahadjopoulos, 1973). The observation that purified phospholipids could form closed lamellar structures, i.e., vesicles (Bangham *et al.,* 1965; Papahadjopoulos and Watkins, 1967; Huang, 1969), has made it possible to understand the biochemical and biophysical properties of individual phospholipid species in a membrane. The properties of these simple artificial membranes are *a priori* different from the properties of the more complex biological membranes, of which such molecules are a component. It is therefore intuitively obvious that liposomes composed of, say, pure phosphatidylcholine (PC) or phosphatidylserine (PS) will not behave like the red cell membrane with respect to phase transitions, permeability, ion binding or fusion. These vesicles are useful tools to show what individual lipid molecules or a few of them in a cluster might do when embedded in a membrane. Furthermore, vesicles composed of a pure phospholipid have been used to elucidate how membrane proteins reconstituted in them affect their properties and vice versa. Liposomes composed of mixed phospholipid species provide information on the interaction between the individual phospholipids or on how they might interact differentially with a protein.

Similarly, to understand which molecules confer susceptibility or resistance to membrane fusion in biological systems, it is essential to study the properties of particular phospholipids and proteins with respect to fusion, as an *in vitro* study. Single-component model systems are undoubt-

Abbreviations used in this chapter: ANTS, 1-aminonapthalene-3,6,8-trisulfonic acids; CHEMS, cholesterylhemisuccinate; DEPS, dielaidoylphosphatidylserine; DPA, dipicolinic acid; DMPE, dimyristoylphosphatidylethanolamine; DOPC, dioleoylphosphatidylcholine; DOPE, dioleoylphosphatidylethanolamine; DOPS, dioleoylphosphatidylserine; DPPC, dipalmitoylphosphatidylcholine; DPX, *p*-xylylene-bis-pyridinium bromide; LUV, large unilamellar vesicles (approx. 100-nm diameter); NBD-PE, *N*-(7-nitro-2,1,3,-benzoxadiazol-4-yl)phosphatidylethanolamine; PA, phosphatidate (phosphatidic acid); PC, (egg) phosphatidylcholine; PE, (egg) phosphatidylethanolamine; PG, phosphatidylglycerol; PI, phosphatidylinositol; PS, (bovine brain) phosphatidylserine; SUV, small unilamellar vesicles (approx. 30-nm diameter); T_H, temperature range of the lamellar–hexagonal phase transition; TPE, phosphatidylethanolamine synthesized by transphosphatidylation of egg phosphatidylcholine.

edly too simple to account for all the properties of cellular membranes that undergo fusion; they do, nevertheless, provide insight into the molecular factors that govern interactions between membranes. In most studies on membrane fusion using liposomes, the phospholipid composition is chosen for the particular aspect of membrane fusion to be demonstrated, rather than for its analogy to biological membranes. This approach arises from the need to understand which biophysical or biochemical processes are necessary or sufficient for fusion, before trying to decipher the molecular mechanisms of membrane fusion in biological systems. The studies reviewed in this chapter demonstrate the remarkable progress made in this area during the last few years.

2. FUSOGENIC CAPACITIES OF DIVALENT CATIONS

The contribution of individual phospholipid species to the fusion of membranes of which they are a component can be studied by the use of liposomes composed of the pure phospholipid, which provide a well-defined physicochemical system. Clearly, mixed phospholipid membranes have different characteristics than that of the individual components (Düzgüneş *et al.*, 1981*a,b;* Sundler *et al.*, 1981). However, phase separation of lipids induced by temperature (Shimshick and McConnell, 1973; Recktenwald and McConnell, 1981; Stewart *et al.*, 1979; Hui, 1981; Takeuchi and Nikaido, 1981), divalent cations (Ohnishi and Ito, 1974; Papahadjopoulos *et al.*, 1974; Jacobson and Papahadjopoulos, 1975; Galla and Sackmann, 1975; Tokutomi *et al.*, 1981), and proteins (Boggs *et al.*, 1982; Mayer and Nelsestuen, 1981) and the subsequent formation of domains can alter the fusion susceptibility of the membrane.

Fusion between phospholipid vesicles takes place via several kinetically coupled steps: (1) aggregation of the vesicles and close apposition of their bilayers, (2) destabilization and merging of the membranes, and (3) communication of the internal aqueous compartments of the vesicles. Each of these steps appears to have different biophysical and biochemical requirements. These steps can be expressed kinetically as follows (Nir *et al.*, 1980*a*, 1982; Bentz *et al.*, 1983*a*):

$$V_1 + V_1 \underset{D_{11}}{\overset{C_{11}}{\rightleftharpoons}} V_2 \overset{f_{11}}{\rightarrow} F_2 \tag{1}$$

where V_1 denotes the monomeric vesicles, V_2 denotes dimer aggregates, and F_2 denotes fused doublets; C_{11}, D_{11}, and f_{11} are the aggregation, dissociation, and fusion rate constants, respectively.

Early studies on the divalent cation-induced fusion of phospholipid vesicles indicated that the specificity of Ca^{2+} over Mg^{2+} in the fusion of phosphatidylserine (PS) or phosphatidylglycerol (PG) vesicles arises from the differences in their interaction with the phospholipid head groups, such as in condensation of monolayers at the air–water interface, decrease in surface potential and binding (Papahadjopoulos, 1968; Papahadjopoulos *et al.*, 1977) rather than from the simple neutralization and screening of the negative charges as a result of double-layer electrostatic interactions (McLaughlin *et al.*, 1971). These studies have also pointed to the importance for Ca^{2+}-induced fusion, of the negative surface charge of the membrane, the fluidity of the bilayer at the experimental temperature, and the threshold concentration of Ca^{2+}, which is different for different phospholipids. It was shown that Ca^{2+}-induced aggregation is not sufficient to cause fusion, and that fusion is accompanied with drastic changes in the thermotropic behavior of the membranes, i.e., crystallization of the acyl chains (Papahadjopoulos *et al.*, 1976, 1977; Papahadjopoulos, 1978; Newton *et al.*, 1978; Portis *et al.*, 1979).

We have proposed that the amount of divalent cation bound to the anionic phospholipid membrane surface is one of the critical determinants of fusion (Nir *et al.*, 1980*b;* Düzgüneş *et al.*, 1980, 1981*a*). We have monitored the kinetics of fusion defined as the intermixing of internal aqueous contents of liposomes, using the Tb/dipicolinic acid (DPA) fluorescence assay, as shown in Figure 1 (Wilschut *et al.*, 1980; Düzgüneş *et al.*, 1981*a,b*). In this assay, fusion is expressed as the percentage of the maximal fluorescence that would be obtained if all the contents were to intermix. Under ionic conditions in which aggregation is rate limiting, i.e., in 100 mM Na^+, the bulk concentration of divalent cations that induces

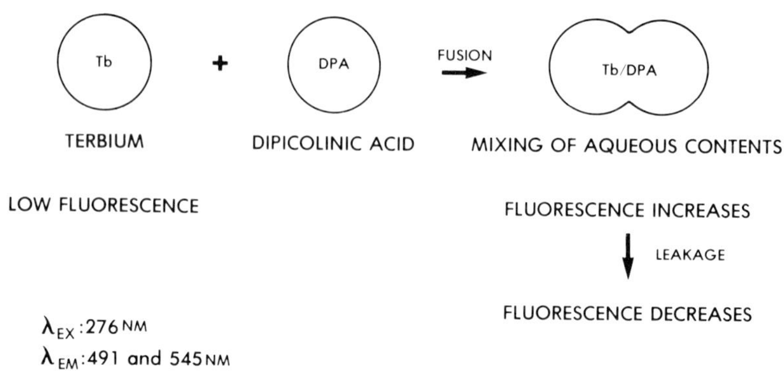

Figure 1. Elements of the Tb/DPA fluorescence assay for fusion. (From Wilschut *et al.*, 1980.)

the fusion of small unilamellar PS vesicles (SUV) at a threshold rate of 10% maximal fluorescence/minute increases in the sequence $Ba^{2+} < Ca^{2+} < Sr^{2+} < Mg^{2+}$ (Bentz *et al.*, 1983*b*). This sequence is the same found for the aggregation of these vesicles in the same electrolyte (Ohki *et al.*, 1982). At higher monovalent salt concentrations, aggregation of the vesicles is rendered very rapid compared to the bilayer destabilization step. This allows for the direct determination of the rate of membrane fusion per se, i.e., f_{11} in Eq. (1). In 500 mM Na^+, in which fusion is rate limiting, Ca^{2+} becomes more effective than Ba^{2+} in inducing the fusion of PS vesicles (Bentz *et al.*, 1983*b*). Furthermore, less Ca^{2+} is bound per PS (~0.15) than Ba^{2+} (~0.2) at the fusion threshold, indicating that Ca^{2+} has a higher fusogenic capacity. Thus, the sequence of effectiveness of the divalent cations for membrane fusion is $Ca^{2+} > Ba^{2+} > Sr^{2+} > Mg^{2+}$.

In Bentz *et al.* (1983*b*), the amount of divalent cation bound has been calculated from binding constants determined by electrophoretic mobility measurements (McLaughlin *et al.*, 1981); this quantity therefore applies to vesicles before they aggregate and fuse. It is known that the binding constant changes upon intermembrane contact (Rehfeld *et al.*, 1981; Ekerdt and Papahadjopoulos, 1982; Nir, 1984). It is important to recognize that at low Na^+ concentrations, the vesicles may have sufficient bound divalent cation to undergo fusion, but, unless the vesicles overcome their mutual electrostatic repulsion and aggregate, no fusion will take place (Bentz *et al.*, 1983*b;* Nir *et al.*, 1983*b*). Thus, at low concentrations of monovalent ion (i.e., <100mM), Ca^{2+}-induced aggregation is always followed by fusion. But at higher concentrations of monovalent ion, e.g., 300 mM, fusion does not necessarily follow aggregation (Bentz *et al.*, 1983*b*).

For large unilamellar PS vesicles (LUV), the aggregation step can be rendered very rapid by the inclusion in the electrolyte of Mg^{2+}, which does not induce fusion by itself (Wilschut *et al.*, 1981); this eliminates the necessity to use high concentrations of Na^+, which would create osmotic problems. As with the SUV, the fusion rate constant increases steeply above a threshold amount of bound divalent cation in the sequence $Ca^{2+} < Ba^{2+} < Sr^{2+}$ (Bentz and Düzgüneş, 1985). The most interesting outcome of this comparison of the threshold amounts of bound Ca^{2+} (or Ba^{2+}) for SUV and LUV is that they are the same, about 0.15 Ca^{2+} (or 0.2 Ba^{2+}) per PS molecule. The dependence of fusion on vesicle size presents itself in the rate of increase of f_{11} [Eq. (1)] with bound divalent cation; i.e., the increase is much steeper for the SUV than the LUV (Fig. 2). For the same binding ratio above threshold, f_{11} for the SUV is considerably larger than for LUV. The difference in the threshold Ca^{2+} concentration between SUV and LUV for the overall fusion reaction (aggregation and fusion per se) arises from the higher electrostatic repulsion of the LUV and their consequent resistance to aggregation (Nir *et al.*, 1983*a*). Thus, when only the

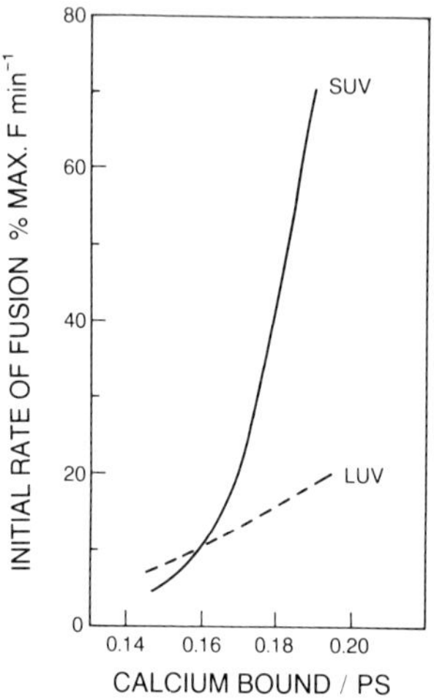

Figure 2. Change in the initial rate of fusion with the amount of calcium bound per PS molecule, under conditions in which the membrane fusion reaction is rate limiting to the overall fusion kinetics. (Data from Bentz *et al.*, 1983*b*; Bentz and Düzgüneş, 1985.)

membrane merging step is considered, SUV do not appear to be intrinsically more unstable than the LUV at the threshold for fusion.

The two remaining divalent cations in the series, Sr^{2+} and Mg^{2+}, are much less effective in inducing the fusion of LUV than SUV composed of PS (Wilschut *et al.*, 1981; Ohki, 1982; Bentz *et al.*, 1983*b;* Düzgüneş *et al.*, 1984*a;* Bentz and Düzgüneş, 1985). The amount of Sr^{2+} bound per PS at the threshold of fusion of SUV is about 0.26, whereas at the threshold of LUV fusion it is around 0.34 (Bentz and Düzgüneş, 1985). Mg^{2+} is completely ineffective in fusing LUV (PS) (Wilschut *et al.*, 1981). This observation has been explained by the inability of Mg^{2+} to dehydrate the membrane surface, since the removal of interfacial water and the resulting hydrophobicity of the membrane surface are thought to be involved in divalent cation-induced fusion of acidic phospholipid membranes (Portis *et al.*, 1979; McIver, 1979; Ohki and Düzgüneş, 1979; Düzgüneş and Papahadjopoulos, 1983).

3. ROLE OF PHASE TRANSITIONS, PHASE SEPARATIONS, AND MEMBRANE FLUIDITY

Lateral phase separation of acidic phospholipids from zwitterionic phospholipids by Ca^{2+} and the formation of domain boundaries between the two phases have been proposed as significant determinants of membrane fusion (Papahadjopoulos *et al.*, 1974, 1977; Papahadjopoulos, 1978). Lateral phase separation manifests itself as the appearance of individual endothermic peaks in differential scanning calorimetry (DSC) thermograms (Papahadjopoulos *et al.,* 1974; Findlay and Barton, 1978), increase in spin–spin interactions in ESR spectra (Ohnishi and Ito, 1974; Galla and Sackmann, 1975; Ohnishi and Tokutomi, 1981), or self-quenching of fluorescence (Hoekstra, 1982). Correlations have been obtained between the fusion susceptibility of PS/PC and PA (phosphatidate)/PC membranes (Papahadjopoulos *et al.,* 1974; Sundler *et al.,* 1981; Düzgüneş *et al.,* 1985*a*) and the lateral phase separation of these mixtures, which is observed in the presence of Ca^{2+}, and not of Mg^{2+} (Papahadjopoulos, 1978; Düzgüneş and Papahadjopoulos, 1983; Ito and Ohnishi, 1974; Ohnishi and Ito, 1974; Jacobson and Papahadjopoulos, 1975). Silvius and Gagné (1984*a*) found that the ability of vesicles composed of either dioleoyl-PS (DOPS)/dioleoyl-PC (DOPC) or dielaidoyl-PS (DEPS)/dielaidoyl-PC to fuse in the presence of Ca^{2+} correlates with the Ca^{2+}-induced phase separation of the individual lipid components, when the PS content is between 40–70 mole%.

We recently observed, however, that fusion occurs in several phospholipid vesicle systems that do not undergo macroscopic lateral phase separation as detected by DSC (Düzgüneş *et al.,* 1984*a*). For example, LUV composed of PS/dimyristoyl-PE (DMPE) (1 : 1) fuse in the presence of Mg^{2+} without a detectable lateral phase separation. Similarly, LUV made of DOPS/dioleoyl-PE (DOPE) and DEPS/dielaidoyl-PE mixtures, in which the PS component is <20 mole%, undergo Ca^{2+}-induced fusion in the absence of any detectable lateral phase separation (Silvius and Gagné, 1984*b*). Although Ca^{2+} does not cause appreciable fusion of LUV composed of PS/dipalmitoyl-PC (DPPC) (1 : 1) at temperatures above the gel–liquid crystalline phase transition, it induces fusion in a limited temperature range within the transition endotherm (see Fig. 3) (Düzgüneş et al., 1984*a*). The latter observation indicates that temperature-induced lateral phase separation of PS from fusion-inhibitory DPPC molecules before the introduction of Ca^{2+} (Stewart *et al.,* 1979) permits membrane fusion at these domains. In this vesicle system, the individual phospholipid components do not undergo phase separation detected by DSC. Thus, macroscopic phase separation does not appear to be necessary for fusion to occur. Formation of microdomains of PS and DPPC of the order of 10-nm diameter in the presence of Ca^{2+}, without macroscopic (detected by DSC) and morphologically distinguishable (by freeze fracture electron microscopy) phase separation, has been observed by [^{31}P]-NMR and microprobe X-ray analysis (Hui *et al.,* 1983).

Tokutomi *et al.* (1981) reported that Ca^{2+} concentrations in the micromolar range induce lateral phase separation of acidic and zwitterionic lipids in an experimental system in which the phospholipid bilayers are in close apposition. Thus, if intermembrane contact is achieved by proteins such as lectins (Düzgüneş *et al.,* 1984*b;* Hoekstra *et al.,* 1985; Sundler and

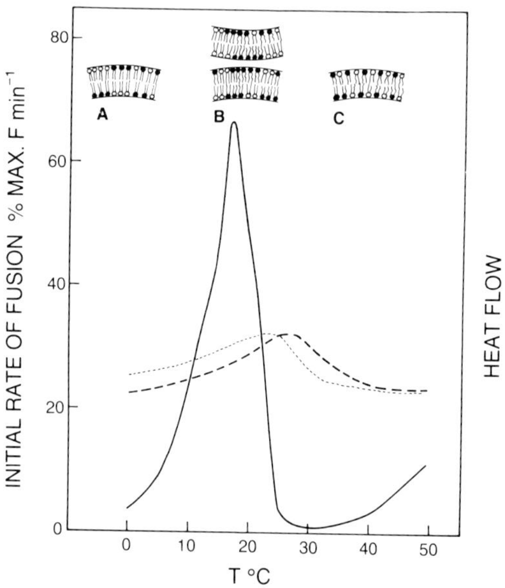

Figure 3. Temperature dependence of the initial rate of fusion of PS/DPPC (1 : 1) LUV induced by 10 mM Ca^{2+} (—) and the DSC thermogram of the vesicles in the presence (---) and absence (...) of 10 mM Ca^{2+}. (Data from Düzgüneş *et al.*, 1984*a*.) A, B, and C: Possible structure of the bilayer at temperatures below, within and above the phase-transition endotherm, respectively. (●) PS; (○) DPPC. (A) The membrane is in the gel phase and is resistant to fusion, although the vesicles aggregate, similar to pure PS vesicles in the gel phase (Wilschut *et al.*, 1985). (B) PS molecules are partially phase separated from the DPPC and can interact with other PS molecules in an apposed membrane. (C) PS and DPPC molecules are mixed uniformly and behave like PS/PC vesicles, which are resistant to fusion (Düzgüneş *et al.*, 1981*b*). Although the initial rate of fusion increases above 40°C, the extent of fusion is very limited.

Wijkander, 1983; Hoekstra and Düzgüneş, 1986; Düzgüneş and Hoekstra, 1986) or other recognition molecules, low concentrations of Ca^{2+} will suffice to phase separate lipids in the zone of contact (Düzgüneş, 1985). The different molecular packing of acidic and zwitterionic phospholipids in the presence of Ca^{2+} or other divalent cations, and the dehydration of the polar group of acidic phospholipids, are thought to produce local defects that expose the hydrophobic interior of the membranes in the region of intermembrane adhesion (see Fig. 4) (Portis *et al.*, 1979; McIver, 1979; Wilschut *et al.*, 1981; Düzgüneş *et al.*, 1981*b*, 1984*a*, 1985*a*; Hoekstra, 1982*b*; Sundler, 1984).

The rate of ionotropic phase separation of phospholipids can be measured by the self-quenching of the fluorescence of NBD-labeled lipids (Hoekstra, 1982*a*). This rate has been found to be slower than the rate of fusion of SUV (PS), suggesting that macroscopic phase separation itself does not induce fusion, but may facilitate it under certain conditions (Hoekstra, 1982*b*). We have found that the initial rate of phase separation of NBD-PE in LUV (PS) in the presence of Ca^{2+} is second order with respect to the vesicle concentration, suggesting that phase separation in this system depends on intermembrane contact (N. Düzgüneş and A. Phan, unpublished data). On the other hand Graham *et al.* (1985) have found

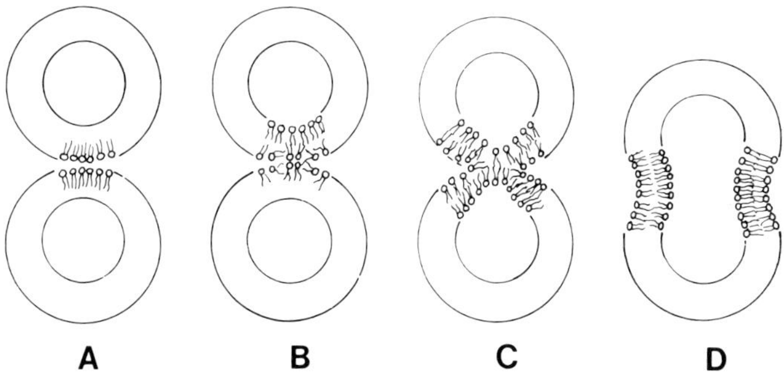

Figure 4. Possible molecular rearrangements during divalent cation-induced fusion of membranes composed of pure acidic phospholipids or their mixtures with zwitterionic phospholipids. Vesicle aggregation (A) is followed by close interbilayer contact and dehydration of phospholipid head groups (B). Differences in molecular packing of the dehydrated phosopholipids and those still in contact with water leads to the formation of point defects, which act as nucleation centers for membrane intermixing (C). This is followed by the establishment of communication between the aqueous contents (D).

that Ca^{2+} induces rapid lateral phase segregation, which precedes contact in the case of PA/PC vesicles.

The induction of isothermal phase transitions in PS vesicles by Ca^{2+}, Mg^{2+}, and H^+ has been considered as a driving force for the destabilization and fusion of the vesicles (Papahadjopoulos *et al.,* 1977, 1978*a,b*). Isothermal phase transitions can be detected as the shift in the temperature of the gel–liquid phase transition of the phospholipids (Träuble and Eibl, 1974; Papahadjopoulos *et al.,* 1977). However, observations on the initial kinetics of fusion have indicated that PS membranes do undergo fusion even in the absence of such transitions. For example, SUV (PS) fuse in the presence of Mg^{2+} even at temperatures above the phase transition temperature of the Mg^{2+}/PS complex, i.e., when no isothermal phase change, as detected by DSC, has occurred, and LUV (PS) do not fuse at all in the presence of Mg^{2+} at temperatures below 40°C (Wilschut *et al.,* 1981, 1985; J. Wilschut, N. Düzgüneş, and D. Papahadjopoulos, unpublished data). Furthermore, Ba^{2+} and Sr^{2+} can fuse LUV (PS) both below and above the phase-transition temperature of the metal ion/PS complex (Düzgüneş *et al.,* 1984*a*). The fusion rate constant, f_{11}, increases approximately linearly over a 20°C temperature range spanning the transition endotherm (Bentz *et al.,* 1985*a*). Thus the ionotropic phase transition, observed macroscopically, does not appear to be the driving force for the initial fusion event. It is not unlikely, however, that phase transitions and phase separations occur locally in the area of intermembrane attachment and that they may not be detected by techniques monitoring bulk properties of the membrane. The phase transition induced by H^+ or Ca^{2+} in acidic phospholipid membranes has a relaxation time of 10–200 msec (Strehlow and Jähnig, 1981); this is certainly within the time scale of the fusion event (Nir *et al.*, 1982, 1983*a,b;* Bentz *et al.*, 1983*a,b;* Miller and Dahl, 1982).

Studies on the temperature dependence of Ca^{2+}-induced fusion of LUV composed of PS have revealed that the vesicles are resistant to fusion below the gel–liquid crystalline transition temperature of the lipid, i.e., in the range 0–5°C, although they aggregate extensively (Wilschut *et al.*, 1985). The initial rate of fusion increases with temperature, especially above 15°C, the temperature to which the phase transition is shifted in the presence of a nonaggregating concentration of Ca^{2+}. This observation suggests that, for fusion to occur, the membrane has to be in a fluid state after Ca^{2+} binding to the outer monolayer. Fluorescence polarization data indicate, however, that at temperatures above the phase transition of the Ca^{2+}/PS complex, the fluidity of the membrane is slightly lower than in the absence of Ca^{2+} (Wilschut *et al.*, 1985). This could mean that some of the outer monolayer lipids are motionally restricted due to the formation of microdomains of partly condensed lipid, in accordance with the condensation of PS in monolayers in the presence of Ca^{2+} (Papahadjopoulos, 1968; Ohki and Düzgüneş, 1979). Thus, laterally fluctuating phase boundaries between condensed and fluid phase lipids could act as nucleation points for membrane fusion (Papahadjopoulos *et al.*, 1977; Wilschut *et al.*, 1985).

The phase transition of SUV (PS) occurs at a considerably lower temperature than the LUV (PS); thus, Ca^{2+} induces a limited extent of fusion of SUV even at 0°C, where the membrane is still in the fluid state. At this temperature, the rate of fusion slows down after an initial rapid phase. This can be explained by the growth of the vesicles to approximately LUV size, thus becoming resistant to further fusion (Wilschut *et al.*, 1985). Mg^{2+}-induced fusion of SUV at temperatures below 20°C is accompanied by an extensive release of contents, consistent with the formation of multilamellar sheets (Papahadjopoulos *et al.*, 1977), whereas at higher temperatures, part of the aqueous contents are retained and fusion proceeds until the vesicles reach a limiting size (Wilschut *et al.*, 1981, 1985). The massive release of contents at lower temperatures correlates with the induction of an isothermal phase transition by Mg^{2+} from the liquid crystalline state of the Na^{+}/PS complex to the gel state of the Mg^{2+}/PS complex. At higher temperatures (>20°C), the Mg^{2+}/PS complex is still in the fluid state, an isothermal transition is not induced, and the vesicles do not release all their contents. They fuse only to a limited extent until they grow to a size equivalent to that of LUV (Wilschut *et al.*, 1981). However, the correlation between leakage and the induction of an isothermal phase transition does not hold for Ba^{2+}- or Sr^{2+}-induced fusion of LUV(PS) mentioned above, since these ions induce extensive leakage above the transition temperature of the cation/PS complex (Düzgüneş *et al.*, 1984*a;* Bentz *et al.*, 1985*a*).

4. ROLE OF THE HEXAGONAL H_{II} PHASE TRANSITION AND LIPIDIC PARTICLES

The last two sections have discussed the effect of lateral phase separation and the gel–liquid crystalline phase transition on membrane fusion. These phase changes do not alter the bilayer structure of the membranes and can promote or inhibit fusion by altering the surface of the membranes at the points of intermembrane contact. However, several phospholipids are capable of undergoing a transition from the lamellar (L_α) to the inverted hexagonal H_{II} phase under certain conditions. For example, various species of PE transform into the H_{II} phase at a characteristic temperature. The thermodynamic properties of this transition have been studied extensively (Cullis and DeKruijff, 1979; Verkleij, 1984; Gruner *et al.*, 1985; Siegel, 1984). The formation of nonbilayer structures under conditions which also induce the fusion of membranes has prompted the speculation that the L_α–H_{II} transition is a driving force for membrane fusion (Cullis and Hope, 1979; Cullis and Verkleij, 1979). Recent studies (Ellens *et al.*, 1986*a,b;* Bentz *et al.*, 1985*b*) have elucidated the effect of the equilibrium L_α–H_{II} phase transition, detected by DSC, on the initial kinetics of contact-mediated destabilization and fusion of phospholipid vesicles.

Large unilamellar vesicles composed of PE prepared by transphosphatidylation of egg PE (TPE) and cholesterylhemisuccinate (CHEMS) are stable in the pH range of 6–8. However, below pH 5.5, where the CHEMS is protonated and acts essentially as cholesterol (Lai *et al.*, 1985), the vesicles leak their contents and the leakage increases as the pH is lowered (Ellens *et al.*, 1984). The kinetic analysis of this process indicates that the close apposition of the vesicles is necessary to trigger bilayer destabilization. Leakage increases dramatically in a temperature range within the broad L_α–H_{II} phase transition endotherm, T_H, measured by DSC (Bentz *et al.*, 1985*b*). The initial rate of lipid mixing, measured by the dilution of fluorescent lipid probes from labelled to unlabelled vesicles (Struck *et al.*, 1981; Rosenberg *et al.*, 1983), increases below the T_H and reaches maximal rates just at the lower end of the T_H. At no temperature is there any intermixing of aqueous contents (Bentz *et al.*, 1985*b*), monitored by the interaction of 1-aminonaphthalene-3,6,8-trisulfonic acid (ANTS) and the collisional quencher *p*-xylylene-bis(pyridinium bromide) (DPX) encapsulated in separate populations of vesicles (see Fig. 5) (Ellens *et al.*,

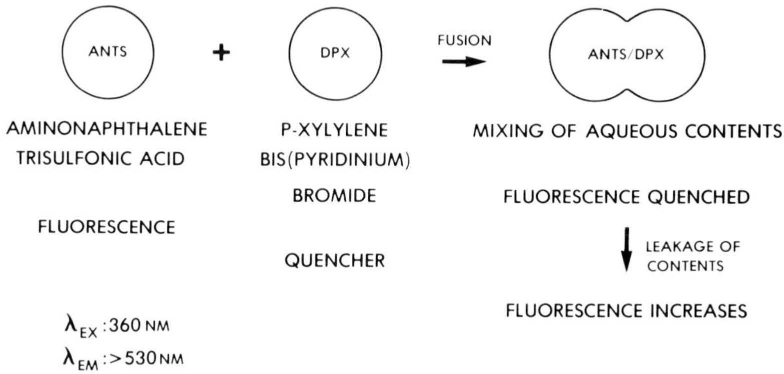

Figure 5. Elements of the ANTS/DPX fluorescence assay for fusion. (From Ellens *et al.*, 1985; Düzgüneş *et al.*, 1985*b*.)

1985; Düzgüneş *et al.*, 1985*b*). Similar studies have been performed with pure PE liposomes (DOPE, egg PE, and TPE) made at pH 9.5. Injection of these liposomes into a pH 4.5 buffer or a 20-mM Ca^{2+} solution leads to a vast enhancement of leakage and lipid mixing at the T_H of the particular PE species, but mixing of aqueous contents (which is observed below the T_H) is abolished (Ellens *et al.*, 1986*a*). Hence the facile transformation of PE to the H_{II} phase does destabilize the bilayers, but it does not lead to fusion in the biologically relevant sense, i.e., mixing of aqueous contents. Furthermore, it is obvious that the transformation of substantial portions of membranes to inverted hexagonal tubes is antithetical to the normal functioning of biologically relevant fusion.

Recent studies show that this phase transition can function as a mechanism of fusion, with the mixing of aqueous contents, but only when the transition is not facile. Ellens *et al.* (1986*b*) found that liposomes composed of DOPE/DOPC (2 : 1) or dioleoylmonomethyl-PE (DOPE with a single methyl group substituted at the head-group amine) undergo fusion with mixing of contents only within the temperature range where an isotropic ^{31}P-NMR signal is found (Gagné *et al.*, 1985). This signal indicates a metastable intermediate state of the lipid dispersion between the lamellar phase (where little or no fusion occurs) and the hexagonal phase (where contact mediated destabilization of liposomes occurs). Thus, these authors claim that the relevance of lipid polymorphism to membrane fusion resides in the existence of this metastable intermediate stage, and not directly in the H_{II} phase per se.

The bilayer-hexagonal H_{II} phase transition does not seem to be necessary for divalent cation-induced fusion of liposomes composed of PE and acidic phospholipids such as PS or phosphatidylinositol (PI). For example, PS/PE vesicles undergo extensive fusion in the presence of Mg^{2+} (Düzgüneş *et al., 1981b*), which does not cause the transformation into the H_{II} phase (Cullis and Verkleij, 1979; Tilcock and Cullis, 1981). Ca^{2+}-induced fusion of vesicles composed of PI and either egg PE or DMPE is dependent on the temperature of the gel–liquid crystalline transition of the mixture and not on the temperature of the L_{α}–H_{II} transition of the individual PE species. The rates of fusion increase with temperature up to 50°C, where they are identical for both types of membrane (Sundler *et al.*, 1981). Fusion is dependent on the presence of PE in this membrane system, since pure PI vesicles do not fuse in the presence of Ca^{2+}, and DMPE does not transform into the hexagonal phase in the physiological temperature range (Harlos and Eibl, 1980). Thus, fusion in this vesicle system does not depend on the formation of the hexagonal phase. Instead, it appears that the divalent cations induce the aggregation of these vesicles by partially neutralizing the negative surface charge, thereby establishing close molecular contact between the PE molecules due to the low level of hydration of the head groups and the partial dehydration of the membrane surface by the divalent cations (Düzgüneş *et al.,* 1985*a*). Molecular packing defects formed as a result of the different packing arrangements of the PE and the acidic phospholipids in the area of intermembrane contact would expose the hydrophobic interior of the bilayer; the lipids of the two vesicles would then be apt to intermix at these points (see Fig. 4). Since PE molecules appear to have a smaller head group than the cross section of the hydrocarbon region, they are likely to participate in the bending of the external monolayer of fusing vesicles, as shown in Figure 4 (Cullis and DeKruijff, 1979; Hauser *et al.,* 1981; Markin *et al.,* 1984; Chernomordik *et al.,* 1985).

Lipidic particles have also been proposed to be intermediate structures in the fusion of various phospholipid vesicles (Verkleij *et al.,* 1979, 1980; Hope *et al.,* 1983; Verkleij, 1984). The rate of formation of inverted micellar intermediates is consistent with the rate of membrane fusion, according to calculations by Siegel (1984). However, lipidic particles are observed in freeze-fracture electron microscopy only after prolonged incubation of certain phospholipid vesicles with Ca^{2+}. No lipidic particles are seen when cardiolipin/PC or PS/PE vesicles are quick frozen immediately after the addition of Ca^{2+}, unless glycerol is present as a cryoprotectant (Bearer *et al.,* 1982). Similar observations have been made by Verkleij *et al.* (1984). It appears, therefore, that lipidic particles as defined by their morphology in freeze fracture electron microscopy are not involved in membrane fusion even in membrane systems in which they are observed after long periods of incubation. Bearer *et al.* (1982), Siegel (1984), and Düzgüneş *et al.* (1985*a*) pointed out, however, that nonbilayer structures may occur at sites of fusion at rates too fast to be visualized by morphological studies or may be confined to a small area in the contact zone between fusing membranes. This intermediate structure could be essentially designated as a local defect in membrane structure that permits the intermixing of membrane components and of internal aqueous contents (Düzgüneş *et al.*, 1984*a;* Düzgüneş, 1985). Some evidence for such structures has been obtained by Hui *et al.* (1981) and Ornberg and Reese (1981). As pointed out by Bearer *et al.* (1982), these nonbilayer intermediates could transform in time to more stable structures, such as lipidic particles, the hexagonal phase or the crystalline bilayer. The experiments in Ellens *et al.* (1986*b*) point out the importance of the relative kinetics of fusion and the accumulation of new morphological structures. Further discussion of these issues can be found in Chapter 9, this volume.

5. EFFECT OF MONOVALENT CATIONS ON DIVALENT CATION-INDUCED MEMBRANE FUSION

The typical effects of monovalent cations on membrane aggregation and fusion are summarized in this section, referring mostly to Li^+, Na^+, and K^+. The effect of H^+ is discussed in Section 6.

Typically, monovalent cations induce reversible, nonleaky aggregation of acidic phospholipid vesicles (Nir *et al.*, 1980*b;* Day *et al.*, 1980; Ohki *et al.*, 1982). When vesicle fusion is induced by divalent cations, an increase in monovalent cation concentration results in a reduced value of the rate constant of fusion per se, due to competition between monovalent and divalent cations for binding to the acidic head groups. Conversely, a reduction in monovalent cation concentration can enhance the rate of fusion induced by divalent cations, up to the stage that aggregation is rate limiting to the overall fusion process. In this case, a further reduction in monovalent cation concentration results in slowing down the overall fusion rate because of the increased electrostatic repulsion between vesicles, which prevents their close approach.

SUV (PS) aggregate in the presence of 500 mM Na^+ (Nir and Bentz, 1978) but do not fuse (Bentz *et al.*, 1983*b*). This aggregation is reversible with respect to Na^+ concentration and temperature changes (Day *et al.*, 1980) and is nonleaky (Nir *et al.*, 1980*b*). Reversible aggregation of the vesicles also occurs in the presence of large concentrations of Li^+ and K^+ (Ohki *et al.*, 1982). The interbilayer repeat distance of PS multilayers decreases to a limiting value of about 20 Å when the Na^+ concentration is increased to 500 mM (Loosley-Millman *et al.*, 1982). Smaller Na^+ concentrations are required for aggregating PA and PA/PE (1 : 1) vesicles than for PS vesicles (Ohki *et al.*, 1984; Yoshimura and Aki, 1985).

Sodium ion concentrations above 100 mM inhibit Ca^{2+}- or Mg^{2+}-induced fusion of PS and PS/PC (4 : 1 and 2 : 1) vesicles (Nir *et al.*, 1980*b*, 1981; Düzgüneş, 1981*a*). Inhibition of divalent cation-induced fusion by increased concentrations of monovalent cations has also been shown by Nir *et al.* (1983*b*) and Bentz *et al.* (1983*b*) for PS vesicles and by Braun *et al.* (1985) for PS/cholesterol vesicles. Since the overall fusion reaction consists of an aggregation step followed by the actual fusion [Eq. (1)], it was proposed, and indeed found, that higher concentrations of monovalent cations reduce the magnitude of the rate constant of fusion per se (Bentz *et al.*, 1983*b*; Nir *et al.*, 1983*b*).

A reduction in the binding of Ca^{2+} to PS in the presence of increased concentrations of Na^+ was found in measurements on pure PS (Newton *et al.*, 1978) and PS/PC vesicles (Düzgüneş *et al.*, 1981). These results were first explained by Nir *et al.* (1978), who proposed that Na^+ actually binds to PS, later experimentally confirmed by Kurland *et al.* (1979). Puskin (1977) found that monovalent cations having smaller nonhydrated radii are more effective in displacing Mn^{2+} from PS vesicles. Monovalent cations cause a reduction in binding of divalent cations due to competition for binding sites as well as to a reduction in the magnitude of the negative surface potential (the screening effect), which in turn results in a smaller enhancement in the concentration of the divalent cations in the vicinity of the membranes. The degree of inhibition of Ca^{2+}- or Mg^{2+}-induced fusion in the presence of 300-mM monovalent cations (Nir *et al.*, 1983*b*) follows the sequence of their binding constants $Li^+ > Na^+ > K^+ >$ tetramethylammonium (Eisenberg *et al.*, 1979). The distance between PS multibilayers in the presence of various monovalent ions is also affected by the different degree of binding to PS; i.e., the distance decreases with increased binding (Loosley-Millman *et al.*, 1982). Ohki *et al.* (1984) pointed out that the order of binding of Li^+ and Na^+ to PS depends on the area per phospholipid molecule, based on monolayer surface potential measurements. At areas per molecule greater than 80 Å^2, the binding sequence is $Na^+ > Li^+ > K^+$, whereas at smaller areas the sequence of Na^+ and Li^+ is inverted.

Nir *et al.* (1980) and Düzgüneş *et al.* (1981*a*) proposed that membrane instability, which is required for the fusion of apposed membranes, depends critically on the extent of divalent (or multivalent) cation bound per acidic headgroup. Bentz *et al.* (1983*b*) and Nir *et al.* (1983*b*) showed that the magnitude of the rate constant of fusion, f_{11}, increases sharply with the amount of divalent cations bound per PS. Braun *et al.* (1985) found a similar result for PS/cholesterol vesicles. In contrast to the fusion rate constant, the overall fusion rate is not necessarily a monotonically decreasing function of monovalent cation concentration. When aggregation is rate limiting to the overall fusion process, an increase in the concentration of monovalent cations can result in faster fusion due to faster aggregation. For instance, when a dilute suspension of PS vesicles is induced to fuse by Ca^{2+}, the overall fusion rate is faster in the presence of 100 mM Na^+ than with 20 mM Na^+. In this respect, the synergistic effect of enhanced binding of Ca^{2+} to PS vesicles in the presence of Mg^{2+} (Portis *et al.*, 1979; Ekerdt and Papahadjopoulos, 1982) has not been found for Ca^{2+} and monovalent cations.

The mode of interaction of monovalent cations with phospholipid head groups is different from that of divalent cations. The fact that the relaxation rates of ^{23}Na bound to PS are only one order of magnitude larger than the solution values (Kurland *et al.*, 1979), may be an indication that the bound Na^+ retains a part of its hydration shell. In addition, each monovalent cation can bind via electrostatic interactions to a single phospholipid head group, whereas the maximization of electrostatic interactions between divalent cations and head groups results in crystallization and may require structural rearrangements of the membranes. In fact, with Ca^{2+} and PS, an additional intermembrane complex has been proposed (Portis *et al.*, 1979). As indicated in Section 6, H^+ plays a special role among the monovalent cations in affecting membrane stability and structure. However, it cannot be ruled out that certain monovalent cations and certain phospholipids interact very strongly. Hauser and Shipley (1983) reported that Li^+ (0.5 M), in contrast to NH_4^+, Na^+, or K^+, induces an isothermal crystallization of bilayers of dimyristoyl-PS. Except in this case, where fusion experiments have not been performed as yet, the trend observed is that monovalent cations stabilize the bilayers and thus reduce their fusion capacity.

6. PROTON-INDUCED MEMBRANE FUSION AND THE ROLE OF DIFFERENT FREE FATTY ACIDS

Protons are the only known monovalent cation to induce membrane destabilization and fusion. Papahadjopoulos *et al.* (1977) showed that SUV (PS) transform to large lamellar sheets at pH 3.5. Ellens *et al.* (1985) found that LUV (PS) fuse, with mixing of aqueous contents, at pH 2. Ca^{2+}-induced fusion of these vesicles is also affected by protons at pH 4. These authors have noted the similarity between the PS headgroup when the carboxyl group is protonated (pH ~ 2) and the zwitterionic PE head group (pH 4–7), facilitating membrane fusion (Düzgüneş *et al.*, 1981*b*, 1985*a,b;* Ellens *et al.*, 1986*a*). The PE headgroup is negatively charged around pH 9, and it is possible to prepare stable liposomes of pure PE at this pH value (Stollery and Vail, 1977; Kolber and Haynes, 1979). Fusion of pure PE vesicles can be induced by increasing the H^+ concentration, i.e., lowering the pH. The protonation of the headgroup presumably permits greater dehydration upon close approach of bilayers.

Addition of oleic acid to PE allows for the preparation of vesicles at neutral pH by imparting a negative surface charge. Vesicles composed of oleic acid/TPE (3 : 7) aggregate, destabilize and fuse below pH 6.5 (Düzgüneş *et al.,* 1983, 1985*b*). Mixing of membrane components and leakage of contents appear to proceed faster than the intermixing of aqueous contents, suggesting that points of destabilization form first, and provide nucleation points for membrane fusion (Düzgüneş and Papahadjopoulos, 1983). Substitution of PS for the oleic acid or of PC for the PE in the liposomes abolishes the sensitivity to pH. The function of oleic acid may be twofold: (1) to provide a negative charge that can be titrated when the pH is lowered, and (2) to act as a fusogen itself. Since pure PE vesicles prepared at pH 9 undergo fusion when the pH is lowered to about 5 (Baldwin *et al.,* 1985; Ellens *et al.,* 1986*a*), the neutralization of the surface charge of oleic acid/PE liposomes would allow the PE component to proceed with membrane fusion.

The role played by the fatty acid in pH-induced fusion of fatty acid/PE liposomes is more complex than simply providing a charge that permits liposome stability at physiological pH and that disappears upon protonation of the fatty acid (Baldwin *et al.,* 1985). This can be seen in Figure 6, which shows the proton-induced fusion of liposomes composed of TPE and different free fatty acids. Clearly, the fusion curves differ depending on the structure of the fatty acid incorporated into the liposomes. This is also the case for Ca^{2+}-induced fusion. These differences are likely to arise from the different effects of the various fatty acids on the gel–liquid crys-

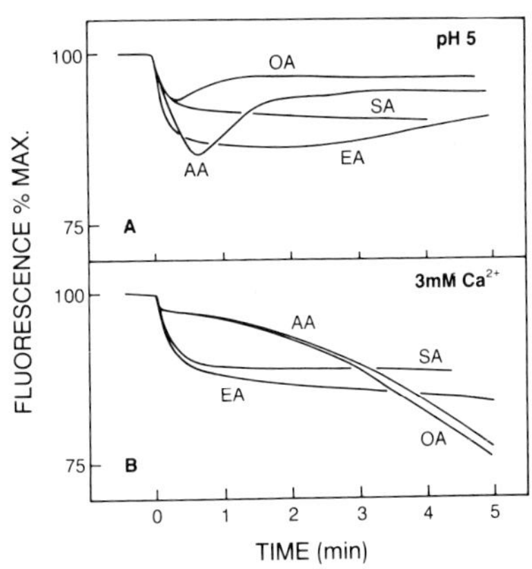

Figure 6. Fusion of liposomes composed of free fatty acids and TPE (3 : 7 mole ratio) induced by shifting the pH from 7.4 to 5 (A), or by 3 mM Ca^{2+} (B), monitored by the ANTS/DPX assay. (From P. A. Baldwin, N. Düzgüneş, and D. Papahadjopoulos, unpublished data.) The fluorescence of the ANTS-containing vesicles is set to 100% maximum. ANTS- and DPX-containing vesicles are mixed at a 1 : 1 ratio in 150 mM NaCl, pH 7.4, at 25°C, and the fusogens are introduced at t = 0. The decrease in fluorescence indicates the quenching of ANTS fluorescence by DPX when the aqueous contents intermix. The subsequent increase in fluorescence results from the leakage of contents into the medium. AA, arachidonic acid; EA, elaidic acid; SA, stearic acid; OA, oleic acid.

talline transition of PE (Baldwin *et al.,* 1986); observed also with other lipids (Eliasz *et al.,* 1976; Usher *et al.,* 1978; Schullery *et al.,* 1981); and on the bilayer to hexagonal transition of PE (Baldwin *et al.,* 1986).

The observation that free fatty acids confer a pH-dependent fusion susceptibility to PE suggests a possible role for free fatty acids formed during the stimulation of secretory cells (Michell, 1975; Michell *et al.,* 1981; Irvine, 1982). Thus, these molecules may be involved in membrane fusion events accompanying intracellular membrane traffic.

Ellens *et al.* (1984) used CHEMS with TPE to make LUV at neutral pH and found that the low pH-induced leakage of aqueous contents from these liposomes requires aggregation. Further studies indicated that H^+ induces lipid mixing but not mixing of aqueous contents (Bentz *et al.,* 1985*b*), whereas Ca^{2+} and Mg^{2+} induce mixing of aqueous contents (Ellens *et al.,* 1985).

The entry of lipid-enveloped viruses, such as vesicular stomatitis virus, Semliki Forest virus, and influenza virus, into the cytoplasm of host cells is thought to take place via endocytosis and subsequent fusion of the viral membrane with the endosomal or lysosomal membrane upon acidification of the lumen of these intracellular compartments (Yoshimura *et al.,* 1982; Marsh, 1984). Intracellular membrane sorting, such as the budding of vesicles from endosomes for receptor recycling to the plasma membrane (Geuze *et al.,* 1983), probably involves fusion of the luminal leaflets of the vesicles in a mildly acidic environment. The discovery that endocytosed liposomes encounter a low-pH compartment (Straubinger *et al.,* 1983, 1985*a*) has prompted an examination of the molecular requirements for membrane fusion at mildly acidic pH, in an attempt to design liposomes that would fuse with or destabilize the endosome membrane, thereby microinjecting their internal aqueous contents into the cytoplasm.

Liposomes composed of oleic acid and TPE have also proved to be an efficient carrier for the introduction of fluorescent probes and macromolecules to the cytoplasm of CV-1 cells (Straubinger *et al.,* 1983*b*, 1985*b*). Liposomes containing the pH-insensitive probe calcein give rise to a diffuse cytoplasmic fluorescence in a high proportion of the cells in culture. Metabolic inhibitors such as azide and deoxyglucose, weakly basic amines such as chloroquine and ammonium chloride, and the ionophore monensin prevent the formation of cytoplasmic fluorescence. Thus, oleic acid/TPE liposomes appear to deliver membrane-impermeant molecules to the cytoplasm via internalization in a compartment that is subsequently acidified. These liposomes can also deliver fluoresceinated dextran (average molecular weight 18,000 or 40,000) into the cytoplasm after a brief treatment of the cells with glycerol (Straubinger *et al.,* 1985*b*). The mechanism of cytoplasmic delivery is unknown; however, the destabilization of the endosome membrane by the pH-sensitive liposomes or fusion of the two membranes are likely routes of delivery.

7. POLYAMINES AS MODULATORS OF MEMBRANE FUSION

The polyamines spermidine and spermine are polycations at physiological pH. Thus, they have pronounced effects on the aggregation and fusion of phospholipid vesicles (Schuber *et al.,* 1983; Hong *et al.,* 1983; Meers *et al.,* 1986). At physiological concentrations, they aggregate vesicles composed of PS or PA, and mixtures of PA with PC, but not pure PC vesicles. The threshold concentration for aggregation is lower for vesicles made of pure acidic phospholipids than for those made of the mixed phospholipids. The aggregation of PS vesicles is reversible, indicating that no fusion has occurred. Indeed the Tb/DPA assay for fusion reveals that polyamines do not induce the fusion of these vesicles. The vesicles also retain their contents. However, addition of polyamines to PA vesicles before the introduction of Ca^{2+} causes a drastic increase in the rate of fusion and reduces the threshold concentration of Ca^{2+} required to initiate fusion (Schuber *et al.,* 1983; Hong *et al.,* 1983). This effect of polyamines is minimal in the case of the PS vesicles. Recent studies have shown that spermine competes with Ca^{2+} binding to PS more effectively than Ca^{2+} binding to PA, providing a possible explanation for the difference in fusion enhancement between the two acidic phospholipids (Meers *et al.,* 1986).

The time of preincubation of vesicles with spermine greatly enhances the initial rate of fusion induced by Ca^{2+}. The phospholipid composition dependence of this effect is in the order PA > PA/PC (1 : 1) > PS. When vesicles are composed of acidic phospholipids, cholesterol, and a high mole fraction of PE, polyamines can induce fusion by themselves (Schuber *et al.,* 1983). Spermine, which is tetravalent, is considerably more effective in inducing fusion than the trivalent spermidine.

Polyamines are expected to reduce the surface charge density and surface potential of negatively charged phospholipid membranes, and thus reduce the electrostatic repulsion between vesicles, allowing the van der Waals attractive forces to dominate the total free energy of interaction. Polyamines are effective in aggregating phospholipid vesicles at lower concentrations than divalent cations. This would be explained by their higher valency, since the surface concentrations of the polyamines are greater at a given bulk concentration. Unlike divalent cations, such as Ca^{2+} (Wilschut *et al.,* 1980) or trivalent cations such as La^{3+} (N. Düzgüneş and J. Bentz, unpublished data), spermidine or spermine does not fuse LUV composed of PS. It is clear that a property other than charge neutralization is necessary for fusion. In the case of vesicles containing a high mole fraction of PE, it appears that aggregation induced by polyamines via their interaction with the acidic component of the membrane is followed by local phase separation of PE and the interaction between PE molecules in the apposed membranes. Molecular packing defects at the boundaries between PE and the polyamine/acidic lipid domains could act as nucleation points for membrane fusion. The propensity of PE molecules to permit close intermembrane contact (Lis *et al.,* 1982) and to transform into nonbilayer phases under certain conditions (Cullis and deKruijff, 1979; Hauser *et al.,* 1981) as well as the low water of hydration of PE (Jendrasiak and Hasty, 1974) could also account for the fusion susceptibility of these membranes in the presence of polyamines.

Our observations on the enhancement of Ca^{2+}-induced fusion of certain phospholipid vesicles by polyamines, and the induction of fusion of certain others, suggest the possible involvement of the membrane action of polyamines in processes such as cell division, cell growth, membrane flow, and exocytosis, which involve membrane fusion (Schuber *et al.,* 1983).

ACKNOWLEDGMENTS. This work was supported by National Institutes of Health grants GM28117 (D.P., N.D., and K.H.) and GM31506 (J.B. and S.N.), a Grant-in-Aid from the American Heart Association, with funds contributed in part by the California Affiliate (N.D.), and a Fellowship from the American Cancer Society (P.A.B.). We thank Dr. Arthur Sowers for useful comments on the manuscript.

REFERENCES

Baldwin, P. A., Düzgüneş, N., and Papahadjopoulos, D., 1985, Free fatty acids affect H^+ and Ca^{2+}-induced fusion of model membranes, *Biophys. J.* **47:**112a.

Baldwin, P. A., Düzgüneş, N., and Papahadjopoulos, D., 1986, The fate of liposome contents during proton-induced fusion is affected by hexagonal phase transitions in liposomes composed of fatty acids and phosphatidylethanolamine, in preparation.

Bangham, A. D., 1968, Membrane models with phospholipids, *Prog. Biophys. Mol. Biol.* **18:**29–95.

Bangham, A. D., Standish, M. M., and Watkins, J. C., 1965, Diffusion of univalent ions across lamellae of swollen phospholipids, *J. Mol. Biol.* **13:**238–252.

Bearer, E. L., Düzgüneş, N., Friend, D. S., and Papahadjopoulos, D., 1982, Fusion of phospholipid vesicles arrested by quick-freezing. The question of lipidic particles as intermediates in membrane fusion, *Biochim. Biophys. Acta* **693:**93–98.

Bentz, J., and Düzgüneş, N., 1985, Fusogenic capacities of divalent cations and effect of liposome size, *Biochemistry* **24:**5436–5443.

Bentz, J., Nir, S., and Wilschut, J., 1983*a,* Mass action kinetics of vesicle aggregation and fusion, *Colloids Surfaces* **6:**333–363.

Bentz, J., Düzgüneş, N., and Nir, S., 1983*b*, Kinetics of divalent cation induced fusion of phosphatidylserine vesicles: Correlation between fusogenic capacities and binding affinities, *Biochemistry* **22:** 3320–3330.

Bentz, J., Düzgüneş, N., and Nir, S., 1985*a*, Temperature dependence of divalent cation induced fusion of phosphatidylserine liposomes: Evaluation of the kinetic rate constants, *Biochemistry* **24:**1064–1072.

Bentz, J., Ellens, H., Lai, M. -Z, and Szoka, F. C., Jr., 1985*b*, On the correlation between H_{II} phase and the contact-induced destabilization of phosphatidylethanolamine-containing membranes, *Proc. Natl. Acad. Sci. U.S.A.* **82:**5742–5745.

Boggs, J. M., Moscarello, M. A., and Papahadjopoulos, D., 1982, Interactions of myelin proteins with phospholipid membranes, in: *The Molecular Biology of Lipid–Protein Interactions,* Vol. 2 (O. H. Griffith and P. Jost, eds.), pp. 1–51, Wiley, New York.

Braun, G., Lelkes, P. I., and Nir, S., 1985, Effect of cholesterol on Ca^{2+}-induced aggregation and fusion of sonicated phosphatidylserine/cholesterol vesicles, *Biochim. Biophys. Acta* **812:**688–694.

Chernomordik, L. V., Kozlov, M. M., Melikyan, G. B., Abidor, I. G., Markin, V. S., and Chizmadzhev, Y. A., 1985, The shape of lipid molecules and monolayer membrane fusion, *Biochim. Biophys. Acta* **812:**643–655.

Cullis, P. R., and DeKruijff, B., 1979, Lipid polymorphism and the functional roles of lipids in biological membranes, *Biochim. Biophys. Acta* **559:**399–420.

Cullis, P. R., and Hope, M. J., 1978, Effects of fusogenic agent on membrane structure of erythrocyte ghosts and the mechanism of membrane fusion, *Nature (Lond.)* **271:**672–674.

Cullis, P. R., and Verkleij, A. J., 1979, Modulation of membrane structure by Ca^{2+} and dibucaine as detected by ^{31}P-NMR, *Biochim. Biophys. Acta* **552:**546–551.

Day, E. P., Kwok, A. Y. W., Hark, S. K., Ho, J. T., Vail, W. J., Bentz, J., and Nir, S., 1980, Reversibility of sodium-induced aggregation of sonicated phosphatidylserine vesicles. *Proc. Natl. Acad. Sci. U.S.A.* **77:**4026–4029.

Düzgüneş, N., 1985, Membrane fusion. *Subcell. Biochemistry* **11:**195–286.

Düzgüneş, N., and Hoekstra, D., 1986, Agglutination and fusion of glycolipid-phospholipid vesicles mediated by lectins and calcium ions, *Stud. Biophys.* **111:**5–10.

Düzgüneş, N., and Papahadjopoulos, D., 1983, Ionotropic effects on phospholipid membranes: Calcium-magnesium specificity in binding, fluidity, and fusion, in: *Membrane Fluidity in Biology* Vol. 2: *General Principles* (R. C. Aloia, ed.), pp. 187–216, Academic Press, New York.

Düzgüneş, N., Hong, K., and Papahadjopoulos, D., 1980, Membrane fusion: The involvement of phospholipids, proteins and calcium binding, in: *Calcium-Binding Proteins: Structure and Function* (F. L. Siegel, E. Carafoli, R. H. Kretsinger, D. H. MacLennan, and R. H. Wasserman, eds.), pp. 17–22, Elsevier/North-Holland, New York.

Düzgüneş, N., Nir, S., Wilschut, J., Bentz, J., Newton, C., Portis, A., and Papahadjopoulos, D., 1981*a*, Calcium- and magnesium-induced fusion of mixed phosphatidylserine/phosphatidylcholine vesicles: Effect of ion binding, *J. Membrane Biol.* **59:**115–125.

Düzgüneş, N., Wilschut, J., Fraley, R., and Papahadjopoulos, D., 1981*b*, Studies on the mechanism of membrane fusion: Role of head-group composition in calcium- and magnesium-induced fusion of mixed phospholipid vesicles, *Biochim. Biophys. Acta* **642:**182–195.

Düzgüneş, N., Straubinger, R. M., and Papahadjopoulos, D., 1983, pH-dependent membrane fusion, *J. Cell Biol.* **97:**178a.

Düzgüneş, N., Paiement, J., Freeman, K. B., Lopez, N. G., Wilschut, J., and Papahadjopoulos, D., 1984*a*, Modulation of membrane fusion by ionotropic and thermotropic phase transitions, *Biochemistry* **23:**3486–3494.

Düzgüneş, N., Hoekstra, D., Hong, K., and Papahadjopoulos, D., 1984*b,* Lectins facilitate calcium-induced fusion of phospholipid vesicles containing glycosphingolipids, *FEBS Lett.* **173:**80–84.

Düzgüneş, N., Wilschut, J., and Papahadjopoulos, D., 1985*a,* Control of membrane fusion by divalent cations, phospholipid head-groups and proteins, in: *Physical Methods on Biological Membranes and Their Model Systems* (F. Conti, W. E. Blumberg, J. DeGier, and F. Pocchiari, eds.), pp. 193–218, Plenum Press, New York.

Düzgüneş, N., Straubinger, R. M., Baldwin, P. A., Friend, D. S., and Papahadjopoulos, D., 1985*b,* Proton-induced fusion of oleic acid-phosphatidylethanolamine liposomes, *Biochemistry* **24:**3091–3098.

Eisenberg, M., Gresalfi, T., Riccio, T., and McLaughlin, S., 1979, Adsorption of monovalent cations to bilayer membranes containing negative phospholipids, *Biochemistry* **18:**5213–5223.

Ekerdt, R., and Papahadjopoulos, D., 1982, Intermembrane contact affects calcium binding to phospholipid vesicles, *Proc. Natl. Acad. Sci. U.S.A.* **79:**2273–2277.

Eliasz, A. W., Chapman, D., and Ewing, D. F., 1976, Phospholipid phase transitions. Effects of n-alcohols, n-monocarboxylic acids, phenylalkyl alcohols and quaternary ammonium compounds, *Biochim. Biophys. Acta* **448:**220–230.

Ellens, H., Bentz, J., and Szoka, F. C., 1984, pH-induced destabilization of phosphatidylethanolamine-containing liposomes. Role of bilayer contact, *Biochemistry* **23:**1532–1538.

Ellens, H., Bentz, J., and Szoka, F. C., 1985, H^+- and Ca^{2+}-induced fusion and destabilization of liposomes, *Biochemistry* **24:**3099–3106.

Ellens, H., Bentz, J., and Szoka, F. C., 1986*a,* Destabilization of phosphatidylethanolamine liposomes at the hexagonal phase transition temperature, *Biochemistry* **25:**285–294.

Ellens, H., Bentz, J., and Szoka, F. C., 1986*b,* Fusion of phosphatidylethanolamine-containing liposomes and the mechanism of the L_α-H_{II} phase transition, *Biochemistry,* **25:**4141–4147.

Findlay, E. J., and Barton, P. G., 1978, Phase behavior of synthetic phosphatidylglycerols and binary mixtures with phosphatidylcholines in the presence and absence of calcium ions, *Biochemistry* **17:**2400–2405.

Gagné, J., Stamatatos, L., Diacovo, T., Hui, S. W., Yeagle, P., and Silvius, J., 1985, Physical properties and surface interactions of bilayer membranes containing N-methylated phosphatidylethanolamines, *Biochemistry* **24:**4400–4408.

Galla, H. J., and Sackmann, E., 1975, Chemically induced lipid phase separation in model membranes containing charged lipids: A spin label study, *Biochim. Biophys. Acta* **401:**509–529.

Geuze, H. J., Slot, J. W., Strous, G. J. A. M., Lodish, H. F., and Schwartz, A. L., 1983, Intracellular site of asialoglycoprotein receptor-ligand uncoupling: Double label immunoelectrin microscopy during receptor-mediated endocytosis, *Cell* **32:**277–287.

Graham, I., Gagné, J., and Silvius, J. R., (1985), Kinetics and thermodynamics of calcium-induced lateral phase separations in phosphatic acid containing bilayers, *Biochemistry* **24:**7123–7131.

Gruner, S. M., Cullis, P. R., Hope, M. J., and Tilcock, C. P. S., 1985, Lipid polymorphism. The molecular basis of nonbilayer phases, *Annu. Rev. Biophys. Biophys. Chem.* **14:**211–238.

Harlos, K., and Eibl H., 1980, Influence of calcium on phosphatidylethanolamine. An investigation of the structure at high pH, *Biochim. Biophys. Acta* **601:**113–122.

Hauser, H., and Shipley, G. G., 1983, Interactions of monovalent cations with phosphatidylserine bilayer membranes, *Biochemistry* **22:**2172–2178.

Hauser, H., Pascher, I., Pearson, R. H., and Sundell, S., 1981, Preferred conformation and molecular packing of phosphatidylethanolamine and phosphatidylcholine, *Biochim. Biophys. Acta* **650:**21–51.

Hoekstra, D., 1982*a,* Fluorescence method for measuring the kinetics of Ca^{2+}-induced phase separations in phosphatidylserine-containing lipid vesicles, *Biochemistry* **21:**1055–1061.

Hoekstra, D., 1982*b,* Role of lipid phase separations and membrane hydration in phospholipid vesicle fusion, *Biochemistry* **21:**2833–2840.

Hoekstra, D., Düzgüneş, N., and Wilschut, J., 1985, Agglutination and fusion of globoside GL-4 containing phospholipid vesicles mediated by lectins and Ca^{2+}, *Biochemistry* **24:**565–572.

Hoekstra, D., and Düzgüneş, N., 1986, *Ricinus communis* agglutinin-mediated agglutination and fusion of glycolipid-containing phospholipid vesicles. Effect of carbohydrate head group size, calcium ions and spermine, *Biochemistry* **25:**1321–1330.

Hong, K., Schuber, F., and Papahadjopoulos, D., 1983, Polyamines: Biological modulators of membrane fusion, *Biochim. Biophys. Acta* **732:**469–472.

Hope, M. J., Walker, D. C., and Cullis, P. R., 1983, Calcium and pH-induced fusion of small unilamellar vesicles consisting of phosphatidylethanolamine and negatively charged phospholipids: A freeze-fracture study, *Biochem. Biophys. Res. Commun.* **110:**15–22.

Huang, C. C., 1969, Studies on phosphatidylcholine vesicles: Formation and physical characteristics, *Biochemistry* **8:**344–352.

Hui, S. W., 1981, Geometry of phase separated domains in phospholipid bilayers by diffraction-contrast electron microscopy, *Biophys. J.* **34:**383–395.

Hui, S. W., Stewart, T. P., Boni, L. T., and Yeagle, P. L., 1981, Membrane fusion through point defects in bilayers, *Science* **212:**921–923.

Hui, S. W., Boni, L. T., Stewart, T. P., and Isaac, T., 1983, Identification of phosphatidylserine and phosphatidylcholine in calcium-induced phase separated domains, *Biochemistry* **22:**3511–3516.

Ito, T., and Ohnishi, S. -I, 1974, Ca^{2+}-induced lateral phase separations in phosphatidic acid-phosphatidylcholine membranes, *Biochim. Biophys. Acta* **352:**29–37.

Jacobson, K., and Papahadjopoulos, D., 1975, Phase transitions and phase separations in phospholipid membranes induced by changes in temperature, pH and concentration of bivalent cations, *Biochemistry* **14:**152–161.

Jendrasiak, G. L., and Hasty, J. H., 1974, The hydration of phospholipids, *Biochim. Biophys. Acta* **337:**79–91.

Kolber, M. A., and Haynes, D. H., 1979, Evidence for a role of phosphatidylethanolamine as a modulator of membrane–membrane contact, *J. Membrane Biol.* **48:**95–114.

Kurland, R., Newton, C., Nir, S., and Papahadjopoulos, D., 1979, Specificity of Na^+ binding to phosphatidylserine vesicles from a $^{23}Na^+$ NMR relaxation rate study, *Biochim. Biophys. Acta* **551:**137–147.

Lai, M-Z., Vail, W. J., and Szoka, F. C., 1985, Acid- and calcium-induced structural changes in phosphatidylethanolamine membranes stabilized by cholesterol hemisuccinate, *Biochemistry* **24:**1654–1661.

Lis, L. J., McAlister, M., Fuller, N., Rand, R. P., and Parsegian, V. A., 1982, Interactions between neutral phospholipid bilayer membranes, *Biophys. J.* **37:**657–666.

Loosley-Millman, M. E., Rand, R. P., and Parsegian, V. A., 1982, Effects of monovalent ion binding and screening on measured electrostatic forces between charged phospholipid bilayers, *Biophys. J.* **40:**221–232.

Markin, V. S., Kozlov, M. M., and Borovjagin, V. L., 1984, On the theory of membrane fusion. The stalk mechanism, *Gen. Physiol. Biophys.* **5:**361–377.

Marsh, M., 1984, The entry of enveloped viruses into cells by endocytosis, *Biochem. J.* **218:**1–10.

Mayer, L. D., and Nelsestuen, G. L., 1981, Calcium- and prothrombin-induced lateral phase separation in membranes, *Biochemistry* **20:**2457–2463.

McLaughlin, S. G. A., Szabo, G., and Eisenman, G., Divalent ions and the surface potential of charged phospholipid membranes, *J. Gen. Physiol.* **58:**667–687.

McIver, D. J. L., 1970, Control of membrane fusion by interfacial water: A model for the action of divalent cations, *Physiol. Chem. Phys.* **11**:289–302.
McLaughlin, S., Mulrine, N., Gresalfi, T., Vaio, G., and McLaughlin, A., 1981, The adsorption of divalent cations to bilayer membranes containing phosphatidylserine, *J. Gen. Physiol.* **77**:445–473.
Meers, P., Hong, K., Bentz, J., and Papahadjopoulos, D., 1986, Spermine as a modulator of membrane fusion: Interaction with acidic phospholipids, *Biochemistry* **25**:3109–3118.
Michell, R. H., 1975, Inositol phospholipids and cell surface receptor function, *Biochim. Biophys. Acta* **415**:81–147.
Michell, R. H., Kirk, C. J., Jones, L. M., Downes, C. P., and Creba, J. A., 1981, The stimulation of inositol lipid metabolism that accompanies calcium mobilization in stimulated cells: Defined characteristics and unanswered questions, *Philos. Trans. R. Soc. Lond. B.* **296**:123–137.
Miller, D. C., and Dahl, G. P., 1982, Early events in calcium-induced liposome fusion, *Biochim. Biophys. Acta* **689**:165–169.
Newton, C., Pangborn, W., Nir, S., and Papahadjopoulos, D., 1978, Specificity of Ca^{2+} and Mg^{2+} binding to phosphatidylserine vesicles and resultant phase changes of bilayer membrane structure, *Biochim. Biophys. Acta* **506**:281–287.
Nir, S., 1984, A model for cation adsorption in closed systems: Application to calcium binding to phospholipid vesicles, *J. Colloid Interface Sci.* **102**:313–321.
Nir, S., and Bentz, J., 1978, On the forces between phospholipid bilayers, *J. Colloid Interface Sci.* **65**:399–414.
Nir, S., Newton, C., and Papahadjopoulos, D., 1978, Binding of cations to phosphatidylserine vesicles, *Bioelectrochem. Bioenerg.* **5**:116–133.
Nir, S., Bentz, J., and Wilschut, J., 1980*a*, Mass action kinetics of phosphatidylserine vesicle fusion as monitored by coalescence of internal vesicle volumes, *Biochemistry* **19**:6030–6036.
Nir, S., Bentz, J., and Portis, A. R., Jr., 1980*b*, Effect of cation concentrations and temperature on the rates of aggregation of acidic phospholipid vesicles. Application to fusion, *Adv. Chem. Ser.* **188**:75–106.
Nir, S., Bentz, J., and Düzgüneş, N., 1981, Two modes of reversible vesicle aggregation: Particle size and the DLVO theory, *J. Colloid Interface Sci.* **84**:266–269.
Nir, S., Wilschut, J., and Bentz, J., 1982, The rate of fusion of phospholipid vesicles and the role of bilayer curvature, *Biochim. Biophys. Acta* **688**:275–278.
Nir, S., Bentz, J., Wilschut, J., and Düzgüneş, N., 1983*a*, Aggregation and fusion of phospholipid vesicles, *Prog. Surface Sci.* **13**:1–124.
Nir, S., Düzgüneş, N., and Bentz, J., 1983*b*, Binding of monovalent cations to phosphatidylserine and modulation of Ca^{2+}- and Mg^{2+}-induced vesicle fusion, *Biochim. Biophys. Acta* **735**:160–172.
Ohki, S., 1982, A mechanism of divalent-ion induced phosphatidylserine membrane fusion, *Biochim. Biophys. Acta* **689**:1–11.
Ohki, S., and Düzgüneş, N., 1979, Divalent cation induced interaction of phospholipid vesicle and monolayer membranes, *Biochim. Biophys. Acta* **552**:438–449.
Ohki, S., Düzgüneş, N., and Leonards, K., 1982, Phospholipid vesicle aggregation: Effect of monovalent and divalent ions, *Biochemistry* **21**:2127–2133.
Ohki, S., Roy, S., Ohshima, H., and Leonards, K., 1984, Monovalent cation-induced phospholipid vesicle aggregation: Effect of ion binding, *Biochemistry* **23**:6126–6132.
Ohnishi, S. -I. and Ito, T., 1974, Calcium-induced phase separations in phosphatidylserine–phosphatidylcholine membranes, *Biochemistry* **13**:881–887.
Ohnishi, S. -I., and Tokutomi, S., 1981, ESR studies of calcium- and proton-induced phase separations in phosphatidylserine-phosphatidylcholine mixed membranes, in: *Biologi-*

cal Magnetic Resonance, Vol. 3 (L. J. Berliner and J. Reuben, eds.), pp. 121–153, Plenum Press, New York.

Ornberg, R. L., and Reese, T. S., 1981, Beginning of exocytosis captured by rapid-freezing of *Limulus* amebocytes, *J. Cell Biol.* **90:**40–54.

Papahadjopoulos, D., 1968, Surface properties of acidic phospholipids: Interaction of monolayers and hydrated liquid crystals with uni- and bivalent metal ions, *Biochim. Biophys. Acta* **163:**240–254.

Papahadjopoulos, D., 1973, Phospholipid membranes as experimental models for biological membranes, in: *Biological Horizons in Surface Science* (L. M. Prince and D. F. Sears, eds.), pp. 159–225, Academic Press, New York.

Papahadjopoulos, D., 1978, Calcium-induced phase changes and fusion in natural and model membranes, in: *Membrane Fusion* (G. Poste and G. L. Nicolson, eds.), pp. 765–790, Elsevier/North-Holland, Amsterdam.

Papahadjopoulos, D., and Watkins, J. C., 1967, phospholipid model membranes. II. Permeability properties of hydrated liquid crystals, *Biochim. Biophys. Acta* **135:**630–652.

Papahadjopoulos, D., Poste, G., Schaeffer, B. E., and Vail, W. J., 1974, Membrane fusion and molecular segregation in phospholipid vesicles, *Biochim. Biophys. Acta* **352:**10–28.

Papahadjopoulos, D., Vail, W. J., Pangborn, W. A., and Poste, G., 1976, Studies on membrane fusion. II. Induction of fusion in pure phospholipid membranes by calcium and other divalent metals, *Biochim. Biophys. Acta* **448:**265–283.

Papahadjopoulos, D., Vail, W. J., Newton, C., Nir, S., Jacobson, K., Poste, G., and Lazo, R., 1977, Studies on membrane fusion. III. The role of calcium-induced phase changes, *Biochim. Biophys. Acta* **465:**579–598.

Papahadjopoulos, D., Portis, A., and Pangborn, W., 1978*a,* Calcium-induced lipid phase transitions and membrane fusion, *Ann. N.Y. Acad. Sci.* **308:**50–66.

Papahadjopoulos, D., Portis, A., Pangborn, W., and Newton, C., 1978*b,* Fusion of artificial membranes with special emphasis on the role of calcium-induced lipid phase transitions, in: *Transport of Macromolecules in Cellular Systems* (S. C. Silverstein, ed.), pp. 413–430, Dahlem Konferenzen, Berlin.

Papahadjopoulos, D., Poste, G., and Vail, W. J., 1979, Studies on membrane fusion with natural and model membranes, *Methods Membrane Biol.* **10:**1–121.

Portis, A., Newton, C., Pangborn, W., and Papahadjopoulos, D., 1979, Studies on the mechanism of membrane fusion: Evidence for an intermembrane Ca^{2+}-phospholipid complex, synergism with Mg^{2+}, and inhibition by spectrin, *Biochemistry* **18:** 780–790.

Puskin, J., 1977, Divalent cation binding to phospholipids: An EPR study, *J. Membrane Biol.* **35:**39–55.

Recktenwald, D. J., and McConnell, H., 1981, Phase equilibria in binary mixtures of phosphatidylcholine and cholesterol, *Biochemistry* **20:**4505–4510.

Rehfeld, S. J., Düzgüneş, N., Newton, C., Papahadjopoulos, D., and Eatough, D. J., The exothermic reaction of calcium with unilamellar phosphatidylserine vesicles: Titration microcalorimetry, *FEBS Lett.* **123:**249–251.

Rosenberg, J., Düzgüneş, N., and Kayalar, C., 1983, Comparison of two liposome fusion assays monitoring the intermixing of aqueous contents and of membrane components, *Biochim. Biophys. Acta* **735:**173–180.

Scherphof, G. L., 1983, A matter of pronunciation, in: *Liposome Letters* (A. D. Bangham, ed.), pp. 319–321, Academic Press, London.

Schuber, F., Hong, K., Düzgüneş, N., and Papahadjopoulos, D., 1983, Polyamines as modulators of membrane fusion: Aggregation and fusion of liposomes, *Biochemistry* **22:** 6134–6140.

Schullery, S. E., Seder, T. A., Weinstein, D. A., and Bryant, D. A., 1981, Differential thermal analysis of dipalmitoylphosphatidylcholine-fatty acid mixtures, *Biochemistry* **20:**6818–6824.

Shimshick, E. J., and McConnell, H. M., 1973, Lateral phase separation in phospholipid membranes, *Biochemistry* **12:**2351–2359.

Siegel, D. P., 1984, Inverted micellar structures in bilayer membranes: Formation rates and half-lives, *Biophys. J.* **45:**399–420.

Silvius, J. R., and Gagné, J., 1984*a,* Calcium-induced fusion and lateral phase separations in phosphatidylcholine-phosphatidylserine vesicles. Correlation by calorimetric and fusion measurements, *Biochemistry* **23:**3241–3247.

Silvius, J. R., and Gagné, J., 1984*b,* Lipid phase behavior and calcium-induced fusion of phosphatidylethanolamine-phosphatidylserine vesicles. Calorimetric and fusion studies, *Biochemistry* **23:**3232–3240.

Stewart, T. P., Hui, S. W., Portis, A. R., Jr., and Papahadjopoulos, D., 1979, Complex phase mixing of phosphatidylcholine and phosphatidylserine in multilamellar membrane vesicles, *Biochim. Biophys. Acta* **556:**1–16.

Straubinger, R. M., Hong, K., Friend, D. S., and Papahadjopoulos, D., 1983*a,* Endocytosis of liposomes and intracellular fate of encapsulated molecules: Encounter with a low pH compartment after internalization in coated vesicles, *Cell* **32:**1069–1079.

Straubinger, R. M., Düzgüneş, N., and Papahadjopoulos, D., 1983*b,* pH-sensitive liposomes: Enhanced cytoplasmic delivery of encapsulated macromolecules, *J. Cell. Biol.* **97:**109a.

Straubinger, R. M., Hong, K., Friend, D. S., Düzgüneş, N., and Papahadjopoulos, D., 1985*a,* Endoctyosis of liposomes and intracellular fate of encapsulated molecules: Strategies for enhanced cytoplasmic delivery, in: *Receptor-Mediated Targeting of Drugs* (G. Gregoriadis, G. Poste, J. Senior, and A. Trouet, eds.), pp. 297–315, Plenum Press, New York.

Straubinger, R. M., Düzgüneş, N., and Papahadjopoulos, D., 1985*b,* pH-sensitive liposomes mediate cytoplasmic delivery of encapsulated macromolecules, *FEBS Lett.* **179:**148–154.

Strehlow, V., and Jähnig, F., 1981, Electrostatic interactions at charged lipid membranes. Kinetics of the electrostatically triggered phase transition, *Biochim. Biophys. Acta* **641:**301–310.

Struck, D. K., Hoekstra, D., and Pagano, R. E., 1981, Use of resonance energy transfer to monitor fusion, *Biochemistry* **20:**4093–4099.

Sundler, R., 1984, Role of phospholipid head group structure and polarity in the control of membrane fusion, *Biomembranes* **12:**563–583.

Sundler, R., and Wijkander, J., 1983, Protein-mediated intermembrane contact specifically enhances Ca^{2+}-induced fusion of phosphatidate-containing membranes, *Biochim. Biophys. Acta* **730:**391–394.

Sundler, R., Düzgüneş, N., and Papahadjopoulos, D., 1981, Control of membrane fusion by phospholipid head groups. II. The role of phosphatidylethanolamine in mixtures with phosphatidate and phosphatidylinositol, *Biochim. Biophys. Acta* **649:**751–758.

Takeuchi, Y., and Nikaido, H., 1981, Persistence of segregated phospholipid domains in phospholipid lipopolysaccharide mixed bilayers: Studies with spin-labeled phospholipids, *Biochemistry* **20:**523–529.

Tilcock, C. P. S., and Cullis, P. R., 1981, The polymorphic phase behavior of mixed phosphatidylserine–phosphatidylethanolamine model systems as detected by ^{31}P-NMR: Effects of divalent cations and pH, *Biochim. Biophys. Acta* **641:**189–210.

Tokutomi, S., Lew, R., and Ohnishi, S. -I., 1981, Ca^{2+}-induced phase separation in phos-

phatidylserine, phosphatidylethanolamine and phosphatidylcholine mixed membranes, *Biochim. Biophys. Acta* **643:**276–282.
Träuble, H., and Eibl, H., 1974, Electrostatic effects on lipid phase transitions: Membrane structure and ionic environment, *Proc. Natl. Acad. Sci. U.S.A.* **71:**214–219.
Usher, J. R., Epand, R. M., and Papahadjopoulos, D., 1978, The effect of free fatty acids on the thermotropic phase transition of dimyristoyl glycerophosphocholine, *Chem. Phys. Lipids* **22:**245–253.
Verkleij, A. J., 1984, Lipidic intramembranous particles, *Biochim. Biophys. Acta* **779:**43–63.
Verkleij, A. J., Mombers, C., Gerritsen, W. J., Leunissen-Bijvelt, L., and Cullis, P. R., 1979, Fusion of phospholipid vesicles in association with the appearance of lipidic particles as visualized by freeze-fracturing, *Biochim. Biophys. Acta* **555:**358–361.
Verkleij, A. J., van Echteld, C. J. A., Gerritsen, W. J., Cullis, P. R., and de Kruijff, B., 1980, The lipidic particle as an intermediate structure in membrane fusion processes and bilayer to hexagonal H_{II} transitions, *Biochim. Biophys. Acta* **600:**620–624.
Verkleij, A. J., Leunissen-Bijvelt, J., de Kruijff, B., Hope, M., and Cullis, P. R., 1984, Nonbilayer structures in membrane fusion, in: *Cell Fusion,* Ciba Foundation Symposium 103, pp. 45–59, Pitman Books, London.
Wilschut, J., Düzgüneş, N., Fraley, R., and Papahadjopoulos, D., 1980, Studies on the mechanism of membrane fusion: Kinetics of Ca^{2+}-induced fusion of phosphatidylserine vesicles followed by a new assay for mixing of aqueous vesicle contents, *Biochemistry* **19:**6011–6021.
Wilschut, J., Düzgüneş, N., and Papahadjopoulos, D., 1981, Calcium/magnesium specificity in membrane fusion: Kinetics of aggregation and fusion of phosphatidylserine vesicles and the role of bilayer curvature, *Biochemistry* **20:**3126–3133.
Wilschut, J., Düzgüneş, N., Hoekstra, D., and Papahadjopoulos, D., 1985, Modulation of membrane fusion by membrane fluidity: Temperature dependence of divalent cation-induced fusion of phosphatidylserine vesicles, *Biochemistry* **24:**8–14.
Yoshimura, T., and Aki, K., 1985, Sodium-induced aggregation of phosphatidic acid and mixed phospholipid vesicles, *Biochim. Biophys. Acta* **815:**167–1732.
Yoshimura, A., Kuroda, K., Kawasaki, K., Yamashina, S., Maeda, T., and Ohnishi, S. -I., 1982, Infectious cell entry mechanism of influenza virus, *J. Virol.* **43:**284–293.

Chapter 12

Protein Modulation of Liposome Fusion

Keelung Hong, Nejat Düzgüneş, Paul R. Meers, and Demetrios Papahadjopoulos

1. INTRODUCTION

Membranes provide the barrier of cellular individuality or intracellular compartmentation. Individuality or compartmentation is unlikely to be traversed except in a well-regulated biological process, membrane fusion. To understand the fundamental mechanism of membrane fusion, simple membrane systems have been studied extensively (reviewed by Papahadjopoulos *et al.,* 1979; Nir *et al.,* 1983; Düzgüneş, 1985). Information obtained from model membranes has been useful not only for defining the role of fusion triggers or modulators, but also for determining the fusion susceptibility of individual components of membranes. However, fusion requirements for these simple systems are often far from those known to be required for biological fusion. For example, fusion between isolated secretory vesicles occurs at much lower Ca^{2+} concentrations than those re-

KEELUNG HONG and PAUL R. MEERS • Cancer Research Institute, School of Medicine, University of California at San Francisco, San Francisco, California 94143. NEJAT DÜZGÜNEŞ • Cancer Research Institute, School of Medicine, and Department of Pharmaceutical Chemistry, School of Pharmacy, University of California at San Francisco, San Francisco, California 94143. DEMETRIOS PAPAHADJOPOULOS • Cancer Research Institute and Department of Pharmacology, School of Medicine, University of California at San Francisco, San Francisco, California 94143.

quired for fusion of liposomes made from the extracted lipids of the secretory vesicles (Gratzl *et al.*, 1980; Ekerdt *et al.*, 1981). A simple explanation for this discrepancy in Ca^{2+} threshold is that other factors are involved in the sensitivity for Ca^{2+} in the fusion event. Among these factors, protein is considered a likely candidate. Therefore, protein-facilitated fusion of liposomes is an important step toward the reconstitution of natural membrane fusion. Besides fusion-inducing viral proteins, no other integral membrane protein has been identified as a possible candidate involved in membrane fusion. Most proteins chosen for the initial studies are water soluble and known to associate with liposomes in the presence of Ca^{2+}. In this chapter we selectively examine the effect of these Ca^{2+}-binding proteins on liposome fusion and hope to establish the specific role of each Ca^{2+}-binding protein in Ca^{2+}-dependent fusion of model membranes. Also discussed are other proteins, which can either associate with liposomes or mediate aggregation of liposomes independently of Ca^{2+}. On the basis of limited information about protein structure and fusion results, our current goal is to identify some important features of protein–lipid interactions in an effort to build the conceptual framework for understanding the role of protein in membrane fusion.

2. FUSION ASSAYS

Assays monitoring membrane fusion can be classified into two categories: those that follow intermixing of aqueous contents, and those that demonstrate mixing of membrane constituents. An assay based on the interaction of liposome-encapsulated Tb^{3+} with dipicolinic acid encapsulated in a separate population of liposomes has been used extensively for studying liposome fusion and its kinetics (Wilschut and Papahadjopoulos, 1979; Wilschut *et al.*, 1980; reviewed by Nir *et al.*, 1983; Düzgüneş *et al.*, 1985). Fusion of liposomes is registered as the increase of Tb fluorescence when the Tb–dipicolinic acid complex is formed inside the aqueous compartment of fused liposomes. This assay is very sensitive, as the fluorescence intensity of the Tb–dipicolinic acid complex is 10,000-fold higher than that of Tb^{3+}. The presence of Ca^{2+} and EDTA in the external medium prevents the formation of the complex outside the vesicles. Therefore, the increase of fluorescence represents a mixing of aqueous contents inside the lumen of fusing liposomes. In some liposome systems, the aqueous contents may be released into the medium faster than they can intermix during fusion (Wilschut *et al.*, 1981; Sundler and Papahadjopoulos, 1981). Thus, the fluorescence signal may be transient or may not occur. In such cases, other assays designed to show mixing of the membrane components are required to demonstrate fusion. An assay that is based on fluorescence energy transfer between a donor and acceptor group of fluorescent phospholipid molecules has been widely used (Struck *et al.*, 1981). In principle, energy-transfer efficiency depends on the surface density of the acceptor for a given donor concentration. When liposomes containing both donor and acceptor fuse with unlabeled liposomes, a reduction in energy transfer efficiency and an increase in the donor fluorescence is observed. Struck *et al.* (1981) showed that head-group-labeled probes do not exchange spontaneously between vesicles. Thus, these types of fluorescent phospholipids are used more frequently. Since mixing of membrane components does not necessarily indicate mixing of aqueous contents (Rosenberg *et al.*, 1983; Ellens *et al.*, 1985), it is a good practice for both assays or other complementary assays to be carried out simultaneously to demonstrate fusion.

3. CALCIUM-BINDING PROTEINS

3.1. Synexin

Synexin, a Ca^{2+}-binding protein originally isolated from bovine adrenal medulla and subsequently found in several other tissues, has been identified as a cytosolic protein. The first indication that synexin may be involved in membrane fusion was its observed capability to promote aggregation of isolated chromaffin granules at Ca^{2+} concentrations above 6 μM (Creutz *et al.*, 1978, 1979). The fusion enhancement of synexin was first demonstrated in phospholipid vesicles (Hong *et al.*, 1981). The activities of synexin are triggered by Ca^{2+}, but not Mg^{2+}. Studies with liposomes composed of various phospholipids suggest that the role of synexin in membrane fusion is also dependent on phospholipid composition (Hong *et al.*, 1981, 1982*a,b*). The phospholipid specificity of the activity of synexin in facilitating vesicle aggregation and membrane fusion is summarized in Table I. Two parameters are used to measure the enhancement of fusion of synexin: the initial rate of fusion and Ca^{2+} threshold for fu-

Table I. Ca^{2+}-Dependent Effects of Synexin on Phospholipid Membranes[a,b]

Lipid composition	Aggregation	Fusion
PS,PS/PE,PA/PE/PS/cholesterol	Enhanced	Enhanced
PS/PC	Enhanced	No effect
PI/PE	Enhanced	Inhibitory

[a]Data from Hong *et al.* (1982*a,b*).
[b]PA, phosphatidate; PC, phosphatidylcholine; PE, phosphatidylethanolamine; PI, phosphatidylinositol; PS, phosphatidylserine.

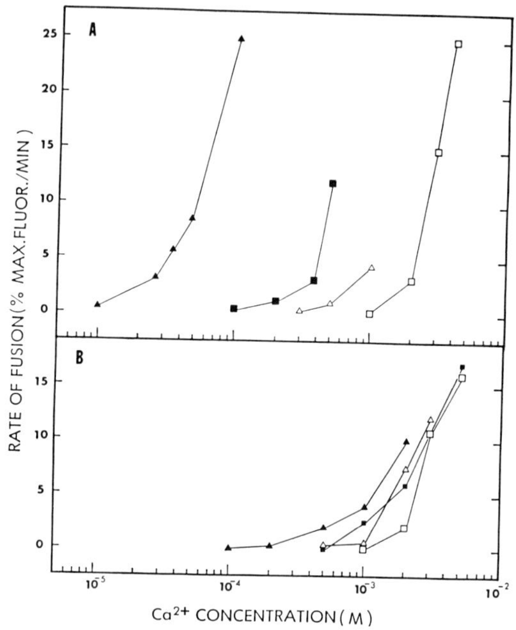

Figure 1. Initial fusion rate of PA/PE (1 : 3) (A) and PA/PE/PC (1 : 2 : 1) (B) liposomes at various divalent cation concentrations. Synexin (final concentration, 6 μg/ml) was added as a concentrated solution to liposome suspension (50 μM phospholipid) before the addition of Ca^{2+}. Ca^{2+} was injected as a concentrated solution at time 0 with constant stirring. Experiments were carried out at 25°C. Mg^{2+} was added to the synexin/liposome suspension before the addition of Ca^{2+}. (□) Ca^{2+} only; (■) synexin/Ca^{2+}; (△) 1.5 mM Mg^{2+}/Ca^{2+}; (▲) synexin/1.5 mM Mg^{2+}/Ca^{2+}. (Data from Hong *et al.*, 1982*b*.)

sion (Hong *et al.*, 1982*a*). The Ca^{2+} threshold is defined as the minimal concentration of Ca^{2+} (mM) required for fusion and is obtained by extrapolating the fusion rate versus Ca^{2+} concentration to zero rate. The dependence of these two parameters on phospholipid composition is demonstrated in Figure 1 and Table II. In general, synexin increases the initial rate of fusion of liposomes containing phosphatidylserine and/or phosphatidate and lowers the Ca^{2+} threshold. By including phosphatidylcholine in liposomes, the facilitation of fusion by synexin is dampened. The most dramatic enhancement by synexin is observed in the case of phosphatidate/phosphatidylethanolamine liposomes. Here, the initial rate of fusion increases by three orders of magnitude, and the Ca^{2+} threshold is reduced from 1.0 mM to 10μM (Fig. 2).

Even though Mg^{2+} does not activate synexin, its presence has a large effect on the Ca^{2+}-dependent action of synexin. For instance, with Ca^{2+} concentration as low as 10 μM, synexin can induce fusion of phosphatidate/phosphatidylethanolamine (1 : 3) vesicles in the presence of 1.5 mM

Table II. Ca^{2+} Threshold for Liposome Fusion[a,b]

Phospholipid composition	Ca^{2+} only		Ca^{2+} (+Mg^{2+})	
	−synexin	+synexin	−synexin	+synexin
PS	2.0	1.0	—	—
PS/PE (1:3)	2.5	1.0	1.0(3.0)	0.5 (3.0)
PA/PE (1:3)	1.0	0.1	0.7(1.0)	0.09(1.0)
PI/PE (1:3)	2.0	3.0	1.0(3.0)	2.0 (3.0)

[a]Tb-dipicolinic acid assay was used. Final concentration of synexin was 5 μg/ml in 50 μM phospholipid. Ca^{2+} or Mg^{2+} concentration (mM).
[b]From Hong *et al.* (1982*a*).

Mg^{2+} (Fig. 1). This high fusion susceptibility of phosphatidate is maintained even when the phosphatidate content is low in liposomes (Table III). In the presence of synexin, phosphatidate/phosphatidylserine/phosphatidylethanolamine/cholesterol (1 : 2 : 3 : 2) vesicles can fuse at Ca^{2+} concentrations similar to the threshold levels required for phosphatide/phosphatidylethanolamine (1 : 3) vesicles. These findings reinforce the idea that large fusogenic domains are not necessary for fusion, provided that there are no inhibitory factors in the potentially fusogenic sites.

The free Ca^{2+} concentration at which exocytosis is induced in a variety of cellular systems is in the micromolar range (Baker and Knight, 1984; Dunn and Holz, 1983; Wilson and Kirshner, 1983), one order of magnitude lower than that for synexin in phospholipid membranes. In addition, synexin in the presence of Ca^{2+} does not mediate the fusion of granule membranes (Creutz *et al.*, 1978). Thus, there are some unknown cofactors that may either assist synexin in inducing fusion of biological membranes or lower the Ca^{2+} threshold to activate synexin. Granules aggregated by Ca^{2+} and synexin can be induced to fuse by the addition of cis-unsaturated fatty acids such as arachidonic acid and oleic acid, but

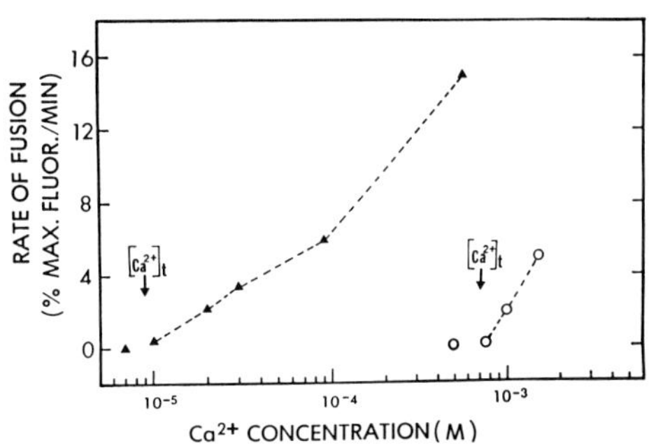

Figure 2. Initial fusion rate of PA/PE (1 : 3) liposomes as a function of Ca^{2+} concentration. Open circle (○) represents the rate of 50 μM liposomes in the presence of 1 mM Mg^{2+} in addition to Ca^{2+}. (▲) Effect of 5 μg/ml synexin on fusion rate of these liposomes. $[Ca^{2+}]_t$, Ca^{2+} threshold required for fusion.

Table III. Fusion Rate of PA/PS/PE/cholesterol (1:2:3:2) Liposomes[a,b]

Ca^{2+} concn.	Fusion rate (% max. Fluor. min^{-1})
0	0.7
0.03	2.7
0.10	7.2
0.40	10.0
1.90	12.0

[a]Tb-dipicolinic acid assay was used. Fusion was measured in the presence of 1.5 mM Mg^{2+} and 6 μg/ml synexin in 50 μM phospholipid; 100% maximal fluorescence in the fusion experiment indicates the fluorescence that would be obtained if all the encapsulated Tb were to react with dipicolinic acid.
[b]From Hong *et al.* (1982*b*).

not by trans- unsaturated or saturated fatty acids (Creutz, 1981). In phospholipid membranes, the activity of synexin measured as the increase of fusion rate is also altered by the presence of cis-fatty acids; e.g., there is more than 50% increase in the fusion rate of phosphatidate/phosphatidylethanolamine (1 : 3) liposomes in the presence of arachidonic acid. The modulation effect of cis-fatty acids suggest that the breakdown products of phospholipids associated with exocytosis (Irvine, 1982) are also involved in membrane fusion.

3.2. Other Ca^{2+}-Binding Proteins

Ca^{2+}-binding proteins other than synexin have been shown to inhibit fusion slightly, e.g., calmodulin from bovine brain, electroplax, and bovine heart or parvalbumin from rabbit muscle (Hong *et al.*, 1982*a,b*; Morris *et al.*, 1982), or to strongly inhibit it, as in the case of prothrombin and its proteolytic fragment 1, which retains the Ca^{2+}-binding activity of the parent protein (Hong *et al.*, 1982*a*).

Several other proteins are also known to bind membranes in a Ca^{2+}-dependent manner. Pollard and Scott (1982) isolated a bovine liver protein, called synhibin, which inhibits the action of synexin. Calelectrin from electric organ aggregates synaptic vesicles and chromaffin granules in the presence of Ca^{2+} (Südhof *et al.*, 1982). Calelectrins of different molecular weights in bovine brain, liver, and adrenal medulla have also been identified (Südhof *et al.*, 1984). A series of proteins (called chromobindins) that bind to immobilized chromaffin granule membranes to varying degrees, depending on Ca^{2+} concentration, has been identified (Creutz *et al.*, 1983), although it is not known how these proteins (except synexin and calmodulin) affect the fusion of phospholipid membrane or secretory vesicles. More recently, a synexinlike protein was identified in human polymorphonuclear leukocyte (Ernst *et al.*, 1985). This synexinlike protein from human tissue acts like bovine synexin in fusion of liposomes (Meers *et al.*, 1986).

3.3. Explanations for the Effects of Proteins on Fusion

To explain the effects of these Ca^{2+}-binding proteins on fusion, the number of active domains of these proteins must be considered. In the absence of liposomes, Ca^{2+} induces protein self-association. Synexin forms extended rods, and prothrombin forms dimers. Even though the nature of this association is unknown, the difference between prothrombin and synexin is clear; i.e., synexin is bipolar and prothrombin is monopolar. Prothrombin and its fragment 1 have only one active site, and no site is available after a dimer is formed. This may explain why binding of prothrombin or fragment 1 to liposomes does not lead to vesicle aggregation (Hong *et al.*, 1982*a*). This indicates that prothrombin or fragment 1 cannot bridge two adjacent vesicles. Therefore, the protein on the membrane surface only serves as a spacer, preventing vesicles from close apposition. By contrast, synexin, with two or more active sites, is able to bridge adjacent vesicles and facilitates membrane close contact. Since synexin promotes fusion of phosphatidate/phosphatidylethanolamine (1 : 3) vesicles at low Ca^{2+} levels where self-association of synexin does not occur (Creutz *et al.*, 1979), it is unlikely that a large synexin polymer is required for fusion. In fact, fusion activity of polymerized synexin is greatly reduced (Table IV).

There are at least two requirements for liposome fusion: (1) aggregation of liposomes and a close intermembrane contact, and (2) a transitional destabilization of closely apposed membranes and mixing of the membrane components. The overall fusion kinetics of liposomes has been modeled as two distinct but kinetically coupled steps: vesicle aggregation and fusion of membranes (Nir *et al.*, 1980, 1983). The role of synexin in these two steps of membrane fusion can be determined by

Table IV. Fusion Activity of Polymerized Synexin[a,b]

Polymerization time[b] (min)	Fusion rate (% max Fluor. min^{-1})
0	4.5
0.4	1.8
1.5	1.0
4.5	0.5

[a]Fusion was measured by the Tb-dipicolinic acid assay. Each sample contained 10 μM PA/PE (1:3) liposomes.
[b]Polymerization time was the time allowed for synexin (2.6 μg) to undergo self-association in the presence of 1 mM Ca^{2+} before the addition of liposomes.

analyzing the kinetics of synexin-facilitated membrane fusion. With a given Ca^{2+} concentration, synexin increases the aggregation rate constant more than 30- to 60-fold, while the fusion rate constant is essentially unchanged (Hong *et al.,* 1983). Under fusion rate-limiting conditions, synexin at low concentration does not change the fusion rate constant appreciably; while at high concentration synexin actually decreases the fusion rate (K. Hong, P. R. Meers, and J. Bentz, unpublished observations).

4. LECTINS

Lectins have also been shown to facilitate the fusion of phospholipid vesicles containing glycolipids. The Ca^{2+} threshold for fusion of phosphatidate/ phosphatidylethanolamine/ phosphatidylethanol-lactobionamide vesicles is reduced by an order of magnitude when preincubated with *Ricinus communis* agglutinin (Sundler and Wijkander, 1983). The preagglutination of the vesicles increases the specificity of Ca^{2+} over Mg^{2+} in this system from less than 2 to greater than 10. Thus, the establishment of interbilayer contact by means of lectin increases both the sensitivity to Ca^{2+} and Ca^{2+}/Mg^{2+} selectivity in this system, similar to the effect of synexin on phosphatidate/phosphatidylethanolamine vesicles. Soybean agglutinin enhances the initial rate of fusion of phosphatidylserine/globoside vesicles induced by Ca^{2+}, and this effect is inhibited by the presence of the haptenic sugar in the medium. Wheat germ agglutinin, likewise, enhances the rate and reduces the threshold Ca^{2+} concentration of fusion of phosphatidate/ phosphatidylethanolamine/ disialoganglioside vesicles (Düzgüneş *et al.,* 1984).

The enhancement of the Ca^{2+}-induced fusion of phosphatidate/ phosphatidylethanolamine/globoside vesicles by soybean agglutinin requires the addition of Ca^{2+} after lectin (Hoekstra *et al.,* 1985). Thus, fusion appears to require the alteration of the bilayer by the lectin before the addition of Ca^{2+}. The addition of Ca^{2+} to preagglutinated glycolipid-containing liposomes substantially reduces the threshold cation concentration for fusion. Quantitatively, Ca^{2+} concentration decreases with decreasing carbohydrate head group size, indicating that the head-group protrusion determines the interbilayer distance within the liposome aggregate (Hoekstra and Düzgüneş, 1986). The headgroup protrusion of the glycolipid in negatively charged liposomes seems to be regulated by Ca^{2+} binding to acidic phospholipids. The dehydration of phospholipid headgroups by Ca^{2+} affects carbohydrate orientation of the glycolipids according to Hoekstra and Düzgüneş (1986).

The significance of preagglutination in the fusion event will depend on the existence of endogenous membrane-localized lectins. The presence of lectins in many animal tissues (Barondes, 1981) raises the possibility that preagglutination is involved in membrane fusion.

5. CLATHRIN

Clathrin, the major protein of coated vesicles, induces fusion of small vesicles composed of synthetic phosphatidylcholines at acidic pH (Blumenthal *et al.,* 1983). This pH-dependent clathrin-induced fusion is not seen in egg phosphatidylcholine vesicles (Hong *et al.,* 1985). Clathrin is not considered an integral membrane protein but, rather, a peripheral structural protein. When associated with PS-containing liposomes it induces the release of aqueous content of liposomes at neutral pH. This clathrin-associated liposome can undergo fusion when pH is lowered below 5.5 in the absence of divalent cations. A strong correlation between the clathrin-induced permeability increase of the bilayer and fusion of liposomes at low pH is observed. The lipid specificity of clathrin-induced fusion is illustrated in Table V (Hong *et al.,* 1985). Like polymerized synexin, self-associated clathrin at low pH can only promote liposome aggregation, but not fusion. It is suggested that clathrin penetration into bilayers destabilizes the closely apposed bilayers and leads to mixing of lipid components.

The pH-dependent fusogenic activity of clathrin is not reduced by phosphatidylcholine in phosphatidylserine/phosphatidylcholine liposomes, as it is with Ca^{2+} or synexin (Düzgüneş *et al.,* 1981; Hong *et al.,* 1982*b*). The lower hydration of phosphatidylethanolamine compared with phosphatidylcholine (Jendrasiak and Hasty, 1974) and its tendency to form nonbilayer structures (Reiss-Husson, 1967; Cullis and de Kruijff, 1978) have been proposed as determinants of the difference in Ca^{2+}-induced fusion of liposomes composed of acidic phospholipids and

Table V. pH-Dependent Clathrin-Induced Fusion of Liposomes[a,b]

Lipid composition	Fusion susceptibility
Egg PC	–
PS/PE (1:2)	+
PS/cholesterol (2:1)	+
PS/PC	++
PS/PE/PC (1:1:1)	++

[a]Fusion was measured by the resonance energy transfer assay. Fusion was induced by lowering pH from 7.4 to 5.0 after a brief incubation of clathrin with liposomes.
[b]From Hong *et al.* (1985).

phosphatidylethanolamine or phosphatidylcholine (Düzgüneş *et al.*, 1981; Hong *et al.*, 1982*a*). The fusion susceptibility of liposomes depends not only on the chemistry of the phospholipid head group, but also on the nature of fusion triggers, i.e., Ca^{2+} dependent versus Ca^{2+} independent. The effect of pH on the interaction of clathrin with liposomes represents an interesting topic for studying protein–lipid interactions. However, the cytoplasmic pH is generally higher than the optimal pH for clathrin-induced fusion of liposomes. Just as in all the *in vitro* studies on pH-dependent assembly of clathrin baskets, the physiological significance of the pH-dependent fusogenic activity of clathrin *in vivo* remains to be determined.

6. SPERM PROTEINS

The fusion of sperm and egg plasma membranes during fertilization is one of the few natural examples of intercellular membrane fusion accessible to experimental study. The gametes of marine invertebrates are especially ideal for *in vitro* studies because they can be obtained in vast quantities, and the fusion event occurs in seawater external to the adult body. For this type of fusion event, species specificity is a very important requirement that nature has imposed on fertilization. During fertilization, the sperm must undergo the acrosome reaction to release granule proteins before binding to and fusing with egg plasma membranes. Morphological evidence suggests that released granule proteins are localized in the area in which membrane fusion is observed. These sperm proteins have recently been studied for their possible role in sperm–egg fusion.

6.1. Bindin

In sea urchins, fusion of the egg plasma membrane with the sperm plasma membrane is restricted to the area of the sperm plasma membrane which arises from the membrane of the acrosome granule after the exocytosis of this granule during the acrosome reaction. Acrosome-reacted sperm is coated with bindin originally contained in the acrosome granule. Bindin is believed to function in the adhesion of sperm to the vitelline layer (Vacquier and Moy, 1977; Glabe and Vacquier, 1977; Glabe and Lennarz, 1979; Glabe *et al.*, 1982). Immunohistochemical evidence suggests that bindin is localized in the restricted region of the sperm that fuses with the egg plasma membrane.

In studies of the interaction of bindin with phospholipid vesicles, Glabe (1985*a*) found that bindin preferentially associates with phospholipid vesicles having gel-phase domains. Association of bindin to sonicated mixed-phase phosphatidylserine-containing vesicles leads to aggregation and fusion of vesicles (Glabe, 1985*b*). The plasma membrane of unfertilized sea urchin eggs contains both gel and fluid lipid domains (Wolf *et al.,* 1981). This mixed-phase membrane may serve as the bindin attachment site for bridging sperm and egg membranes. The ability of bindin to trigger fusion of vesicles containing mixed-phase membrane suggests that bindin may be involved in fusion between sperm and egg plasma membranes.

6.2. Lysin

The eggs of abalone are surrounded by a protective elevated vitelline layer composed of glycoproteins (Lewis *et al.,* 1982). During fertilization, the anterior apex of the sperm binds to the vitelline layer, the acrosome granule opens, and proteins are released, dissolving a hole in the egg vitelline layer (Lewis *et al.,* 1982). The acrosomal proteins are termed lysins because of their ability to dissolve egg investments (Dan, 1967). Present evidence suggests that the mechanisms by which the abalone sperm lysin dissolves the vitelline layer is nonenzymatic; it seems to be based on the hydrophobic competition of lysin for intermolecular hydrophobic bonds among glycoprotein molecules comprising the vitelline layer (Lewis *et al.,* 1982). Recently, a purified lysin of 16,000 M_r was found to induce aggregation and fusion of phosphatidylserine-containing liposomes (Hong and Vacquier, 1985). The general trend of fusion susceptibility for liposomes is as follows: phosphatidylserine/phosphatidylcholine (1 : 2) = phosphatidylserine/phosphatidylcholine/phosphatidylethanolamine (1 : 1 : 1) > phosphatidylserine/phosphatidylethanolamine (1 : 2). Lysin-induced liposome fusion does not requre Ca^{2+}. Although the presence of Ca^{2+} in liposome suspension reduces the initial rate of lysin-initiated fusion, the final extent of fusion is not affected. A computer modeling of the secondary structure of lysin 16,000 M_r from its complete amino acid sequence shows that lysin 16,000 M_r displays clearly separated positively charged and hydrophobic domains (Fridberger *et al.,* 1985). It appears that lysin 16,000 M_r utilizes its existing hydrophobic domain to interact with the hydrophobic region of phospholipid bilayers and facilitates the fusion of membranes, which are in close apposition because of the polycationic feature of lysin.

7. SUMMARY

Fusion of liposomes modulated by proteins reviewed in this chapter can be generalized as either Ca^{2+} dependent or Ca^{2+} independent. A

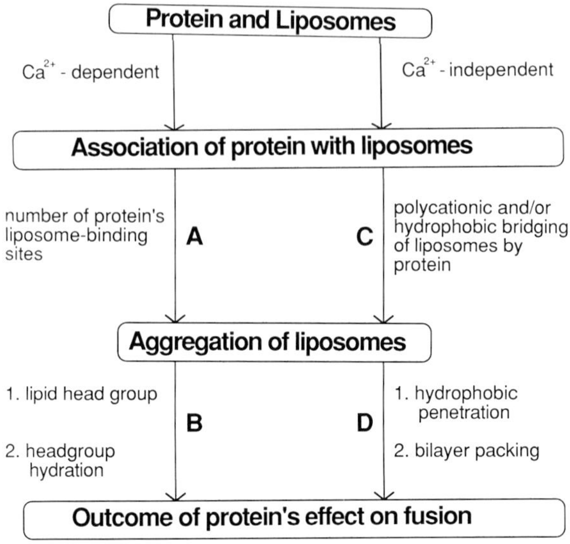

Figure 3. Scheme for establishing role of protein in membrane fusion *in vitro.*

scheme (Fig. 3). is presented for identifying some important factors governing the possible outcome of fusion in each fusion category. This scheme can only be considered as a working hypothesis, as the structures of most proteins are unknown.

After a positive indication that the isolated Ca^{2+}-binding protein of interest can bind to a liposome, one then proceeds to determine whether aggregation of liposomes is enhanced in the presence of protein. As outlined in Figure 3, step A is governed primarily by the number of the liposome-binding sites of the protein. A protein with multiple binding sites will likely bridge adjacent liposomes and thus will facilitate aggregation of liposomes. By contrast, once a protein with a single binding site binds to a liposome, the exposed surface of the protein may not be able to interact with another lipid membrane. Thus, the Ca^{2+}-induced interaction of such a protein with liposomes may not result in liposome aggregation, and consequently fusion may be inhibited. Fusion studies suggest that step B is mainly dependent on the structure and the hydration extent of the lipid headgroups. A bulky and/or highly hydrated head group will prevent liposomes from close intermembrane contact even aggregation is enhanced by protein. The orientation of the protein on the liposome surface will critically affect the intermembrane distance that determines the outcome of the effect of the protein on fusion.

For Ca^{2+}-independent fusion, step C is controlled by the capability of the protein to bind adjacent liposomes simultaneously by either polycationic or hydrophobic bridging. Like step B, step D is essentially dependent on the lipid composition of liposomes. Hydration of the head group in step D is not as critical as in step B. Here, fusion seems to depend on the extent of hydrophobic penetration of the protein, which in turn is largely dependent on the bilayer packing. In aggregated liposomes, a larger hydrophobic perturbation by a protein can facilitate a transitional intermembrane destabilization and consequently lead to fusion.

ACKNOWLEDGMENTS. Our research has been supported by grants GM 28117 and GM 26369 from the National Institutes of Health and a Grant-in-Aid from the American Heart Association. P.R.M. is supported by a Postdoctoral Fellowship (PF-2398) from the American Cancer Society. We thank Dr. Stephen Murray for reading the manuscript. Ms. Rose Antonucci is gratefully acknowledged for preparing the manuscript.

REFERENCES

Baker, P. F., and Knight, D. E., 1984, Calcium control of exocytosis in bovine adrenal medullary cells, *Trends Neurosci.* **7**:120–126.

Barondes, S. H., 1981, Lectins: Their multiple endogenous cellular functions, *Annu. Rev. Biochem.* **50**:207–231.

Blumenthal, R., Henkart, M., and Steer, C. J., 1983, Clathrin-induced pH-dependent fusion of phosphatidylcholine vesicles, *J. Biol. Chem.* **258**:3409–3415.

Creutz, C. E., 1981, *cis*-Unsaturated fatty acids induce the fusion of chromaffin granules aggregated by synexin, *J. Cell Biol.* **91**:247–256.

Creutz, C. E., Pazoles, C. J., and Pollard, H. B., 1978, Identification and purification of an adrenal medullary protein (synexin) that causes calcium-dependent aggregation of isolated chromaffin granules, *J. Biol. Chem.* **253**:2858–2866.

Creutz, C. E., Pazoles, C. J., and Pollard, H. B., 1979, Self-association of synexin in the presence of calcium: Correlation with synexin-induced membrane fusion and examination of the structure of synexin aggregates. *J. Biol. Chem.* **254**:553–558.

Creutz, C. E., Dowling, L. G., Sando, J. J., Villar-Palasi, C., Whipple, J. H., and Zaks, W. J., 1983, Characterization of the chromobindins: Soluble proteins that bind to the chromaffin granule membrane in the presence of Ca^{2+}, *J. Biol. Chem.* **258**:14664–14674.

Cullis, P. R., and deKruijff, B., 1978, The polymorphic phase behavior of phosphatidylethanolamines of natural and synthetic origin. A ^{31}P-NMR study, *Biochim. Biophys. Acta* **513**:31–42.

Dan, J. C., 1967, Acrosome reaction and lysins, in: *Fertilization,* Vol. 1 (C. B. Metz and A. Monroy, eds.), pp. 237–293, Academic Press, New York.

Dunn, L. A., and Holz, R. W., 1983, Catecholamine secretion from digitonin-treated adrenal medullary chromaffin cells, *J. Biol. Chem.* **258**:4989–4993.

Düzgüneş, N., 1985, Membrane fusion, in: *Subcellular Biochemistry,* Vol. 11, (D. B. Roodyn, ed.), pp. 195–286, Plenum Press, New York.

Düzgüneş, N., Wilschut, J., Fraley, R., and Papahadjopoulos, D., 1981, Studies on the mechanism of membrane fusion: Role of head-group composition in calcium- and magnesium-induced fusion of mixed phospholipid vesicles, *Biochim. Biophys. Acta* **642**:182–195.

Düzgüneş, N., Hoekstra, D., Hong, K., and Papahadjopoulos, D., 1984, Lectins facilitate calcium-induced fusion of phospholipid vesicles containing glycosphingolipids, *FEBS Lett.* **173**:80–84.

Düzgüneş, N., Wilschut, J., and Papahadjopoulos, D., 1985, Control of membrane fusion by divalent cations, phospholipid head-groups and proteins, in: *Physical Methods on Biological Membranes and Their Model Systems* (F. Conti, W. E. Blumberg, J. de Gier, and F. Pocchiari, eds.), pp. 193–218, Plenum Press, New York.

Ekerdt, R., Dahl, G., and Gratzl, M., 1981, Membrane fusion of secretory vesicles and liposomes. Two different types of fusion, *Biochim. Biophys. Acta* **646**:10–22.

Ellens, H., Bentz, J. and Szoka, F. C., 1985, H^+- and Ca^{2+}-induced fusion and destabilization of liposomes, *Biochemistry* **24**:3099–3106.

Ernst, J. D., Meers, P., Düzgüneş, N., Hong, K., Scannell, K., Papahadjopoulos, D., and Goldstein, I. M., 1985, A synexin-like protein in the cytosol of polymorphonuclear leukocytes mediates calcium-dependent granule aggregation, *Clin. Res.* **33**:401a.

Fridberger, A., Sundelin, J., Vacquier, V. D., and Peterson, P. A., 1985, Amino acid sequence of an egg-lysin protein from abalone spermatozoa that solubilizes the vitelline layer, *J. Biol. Chem.* **260**:9092–9099.

Glabe, C. G., 1985*a*, Interaction of the sperm adhesive protein, bindin, with phospholipid vesicles, I. Specific association of bindin with gel-phase phospholipid vesicles, *J. Cell Biol.* **100**:794–799.

Glabe, C. G., 1985*b*, Interaction of the sperm adhesive protein, bindin, with phospholipid vesicles, II. Bindin induces the fusion of mixed-phase vesicles that contain phosphatidylcholine and phosphatidylserine *in vitro, J. Cell Biol.* **100**:800–806.

Glabe, C. G., and Lennarz, W. J., 1979, Species-specific sperm adhesion in sea urchins: A quantitative investigation of bindin-mediated egg agglutination, *J. Cell Biol.* **83**:595–604.

Glabe, C. G., and Vacquier, V. D., 1977, Species specific agglutination of eggs by bindin isolated from sea urchin sperm, *Nature (Lond.)* **267**:836–837.

Glabe, C. G., Grabel, L. B., Vacquier, V. D., and Rosen, S. D., 1982, Carbohydrate specificity of sea urchin sperm bindin: A cell surface lectin mediating sperm–egg adhesion, *J. Cell Biol.* **94**:123–128.

Gratzl, M., Schudt, C., Ekerdt, R., and Dahl, G., 1980, Fusion of isolated biological membranes: A tool to investigate basic processes of exocytosis and cell–cell fusion, in: *Membrane Structure and Function,* Vol. 3 (E. E. Bittar, ed.), pp. 59–92, Wiley, New York.

Hoekstra, D., and Düzgüneş, N., 1986, Ricinus communis agglutinin-mediated agglutination and fusion of glycolipid-containing phospholipid vesicles, Effect of carbohydrate headgroup size, calcium ions, and spermine, *Biochemistry* **25**:1321–1330.

Hoekstra, D., Düzgüneş, N., and Wilschut, J., 1985, Agglutination and fusion of globoside GL-4 containing phospholipid vesicles mediated by lectins and calcium ions, *Biochemistry* **24**:565–572.

Hong, K., and Vacquier, V. D., 1986, Fusion of liposomes induced by a cationic protein from the acrosome granule of abalone spermatozoa, *Biochemistry,* **25**:543–549.

Hong, K., Düzgüneş, N., and Papahadjopoulos, D., 1981, Role of synexin in membrane fusion, *J. Biol. Chem.* **256**:3641–3644.

Hong, K., Düzgüneş, N., and Papahadjopoulos, D., 1982*a*, Modulation of membrane fusion by calcium-binding proteins, *Biophys. J.* **37**:297–305.

Hong, K., Düzgüneş, N., Ekerdt, R., and Papahadjopoulos, D., 1982*b*, Synexin facilitates fusion of specific phospholipid vesicles at divalent cation concentrations found intracellularly, *Proc. Natl. Acad. Sci. U.S.A.* **79**:4942–4944.

Hong, K., Ekerdt, R., Bentz, J., Nir, S., and Papahadjopoulos, D., 1983, Kinetics of synexin-facilitated membrane fusion, *Biophys. J.,* **41**:31a.

Hong, K., Yoshimura, T., and Papahadjopoulos, D., 1985, Interaction of clathrin with liposomes: pH-dependent fusion of phospholipid membranes induced by clathrin, *FEBS Lett.* **191**:17–23.

Irvine, R. F., 1982, How is the level of free arachidonic acid controlled in mammalian cells?, *Biochem. J.* **204**:3–16.

Jendrasiak, G. L., and Hasty, J. H., 1974, The hydration of phospholipids, *Biochim. Biophys. Acta* **337**:79–91.

Lewis, C. A., Talbot, C. F., and Vacquier, V. D., 1982, A protein from abalone sperm dissolves the egg vitelline layer by a nonenzymatic mechanism, *Dev. Biol.* **92**:227–239.

Meers, P., Ernst, J., Hong, K., Düzgüneş, N., Goldstein, I. M., and Papahadjopoulos, D., 1986, Synexin-like proteins from the cytosol of human neutrophils mediate aggregation and apparent fusion of specific granules and liposomes, *Biophys. J.* **49**:134a.

Morris, S. J., Hughes, J. M. X., and Whittaker, V. P., 1982, Purification and mode of action of synexin: A protein enhancing calcium-induced membrane aggregation, *J. Neurochem.* **39**:529–536.

Nir, S., Bentz, J., and Wilschut, J., 1980, Mass action kinetics of phosphatidylserine vesicle fusion as monitored by coalescence of internal vesicle volumes, *Biochemistry* **19**:6030–6036.

Nir, S., Bentz, J., Wilschut, J., and Düzgüneş, 1983, Aggregation and fusion of phospholipid vesicles, *Prog. Surface. Sci.* **13**:1–124.

Papahadjopoulos, D., Poste, G., and Vail, W. J., 1979, Studies on membrane fusion with natural and model membranes, *Methods Membrane Biol.* **10**:1–121.

Pollard, H. B., and Scott, J. H., 1982, Synhibin: A new calcium-dependent membrane-binding protein that inhibits synexin-induced chromaffin granule aggregation and fusion, *FEBS Lett.* **150**:201–206.

Reiss-Husson, F., 1967, Structure des phases liquide-cristallines de différents phospholipides, monoglycérides, sphingolipides, anhydres ou en présence d'eau, *J. Molec. Biol.* **25**:363–382.

Rosenberg, J., Düzgüneş, N., and Kayalar, C., 1983, Comparison of two liposome fusion assays monitoring the intermixing of aqueous contents and of membrane components, *Biochim. Biophys. Acta* **735**:173–180.

Struck, D. K., Hoekstra, D., and Pagano, R. E., 1981, Use of resonance energy transfer to monitor membrane fusion, *Biochemistry* **20**:4093–4099.

Südhof, T. C., Walker, J. H., and Obrocki, J., 1982, Calelectrin self-aggregates and promotes membrane aggregation in the presence of calcium, *EMBO J.* **1**:1167–1170.

Südhof, T. C., Ebbecke, M., Walker, J. H., Fritsche, U., and Boustead, C., 1984, Isolation of mammalian calelectrins; A new class of ubiquitous Ca^{2+}-regulated proteins, *Biochemistry* **23**:1103–1109.

Sundler, R., and Papahadjopoulos, D., 1981, Control of membrane fusion by phospholipid head-groups I. Phosphatidate/phosphatidylinositol specificity, *Biochim. Biophys. Acta* **649**:743–750.

Sundler, R., and Wijkander, J., 1983, Protein-mediated intermembrane contact specifically enhances Ca^{2+}-induced fusion of phosphatidate-containing membranes, *Biochim. Biophys. Acta.* **730**:391–394.

Vacquier, V. D., and Moy, G. W., 1977, Isolation of bindin: The protein responsible for adhesion of sperm to sea urchin eggs, *Proc. Natl. Acad. Sci. U.S.A.* **74**:2456–2460.

Wilschut, J., and Papahadjopoulos, D., 1979, Ca^{2+}-induced fusion of phospholipid vesicles monitored by mixing of aqueous contents, *Nature (Lond.)* **281**:690–692.

Wilschut, J., Düzgüneş, N., Fraley, R., and Papahadjopoulos, D., 1980, Studies on the mechanism of membrane fusion: Kinetics of calcium ion induced fusion of phosphatidylserine vesicles followed by a new assay for mixing of aqueous vesicle contents, *Biochemistry* **19**:6011–6021.

Wilschut, J., Düzgüneş, N., and Papahadjopoulos, D., 1981, Calcium/magnesium specificity in membrane fusion: Kinetics of aggregation and fusion of phosphatidylserine vesicles and the role of bilayer curvature, *Biochemistry* **20**:3126–3133.

Wilson, S. P., and Kirshner, N., 1983, Calcium-evoked secretion from digitonin-permeabilized adrenal medullary chromaffin cells, *J. Biol. Chem.* **258**:4994–5000.

Wolf, D. E., Kinsey, W. H., Lennarz, W. J., and Edidin, M., 1981, Changes in the organization of the sea urchin plasma membrane upon fertilization: Indications from the lateral diffusion rates of lipid-soluble fluorescent dyes, *Dev. Biol.* **81**:133–138.

Chapter 13

Acid-Induced Fusion of Liposomes

Jerome Connor and Leaf Huang

1. INTRODUCTION

Liposomes have become an important model system for studying the phenomenon of membrane fusion (Blumenthal, 1985; Gregoriadis, 1984). The simplicity of form of artificial lipid membranes makes them an effective tool for elucidating the actual mechanism of fusion between opposing membranes. The information derived from the study of model liposome fusion is a crucial contribution to the understanding of biologically relevant fusion activities, including endocytosis, exocytosis, and viral infection.

Many studies have been done demonstrating the ability of proteins, such as melittin (Morgan *et al.,* 1983), bovine serum albumin (BSA) (Schenkman *et al.,* 1981), clathrin (Blumenthal *et al.,* 1983), myelin basic protein (Lampe and Nelsestuen, 1982), polymyxin B (Gad and Eytan, 1983), tetanus toxin (Roa and Boquet, 1985), and synexin (Hong *et al.,* 1982) to induce fusion between liposomes. Although these studies have important implications for the mechanisms of the corresponding physiological events, the actual membrane fusion is dependent on the presence

JEROME CONNOR and LEAF HUANG • Department of Biochemistry, University of Tennessee, Knoxville, Tennessee 37996. *Present address for* J. C.: Department of Cell Biology, M. D. Anderson Tumor Institute, Houston, Texas 77030.

of an exogeneous component, and little information can be deduced concerning the membrane to membrane interaction during a fusion event.

Recently, several groups have demonstrated the ability to induce liposome membrane fusion solely by a decrease in the medium pH (Connor *et al.*, 1984; Ellens *et al.*, 1984; Düzgüneş *et al.*, 1985; Hope *et al.*, 1983; Nayar and Schroit, 1985). These systems are all based on the central theme of stabilizing a phosphatidylethanolamine (PE) lipid in a bilayer configuration by the addition of a second lipid component containing a titratable acidic group.

This chapter examines the development and characterization of these acid pH-inducible liposome fusion systems and the application of the pH-sensitive liposomes to cellular delivery of the liposome contents.

2. PROPERTIES OF PHOSPHATIDYLETHANOLAMINE

The propensity of phosphatidylethanolamine (PE) to form nonbilayer structures, i.e., hexagonal II phase, under physiological conditions (Reiss-Husson, 1967; Cullis and DeKruijff, 1979), is an important factor that contributes to its essential requirement as a lipid component of pH-sensitive liposomes. PE molecules prefer the hexagonal phase because the lipid head group is small compared with the acyl chains; this allows for increased interaction between adjacent lipid head groups (Lis *et al.*, 1982), including the possibility of intermolecular hydrogen bonding (Hauser *et al.*, 1981; Papahadjopoulos and Miller, 1967). This PE head-group interaction causes a decrease in the surface hydration that is compensated for by assuming the hexagonal II phase over a bilayer configuration. The transition of PE from the hexagonal II phase to the bilayer phase is easily monitored by ^{31}P-NMR (Cullis and DeKruijff, 1979). It is possible to form a stable bilayer conformation of PE by increasing the pH to greater than 9.0 (Hauser *et al.*, 1981). More importantly, PE can be formed into stable bilayer configurations by the addition of a second lipid component. Phosphatidylcholine (PC), phosphatidylserine (PS), phosphatidylinositol (PI), and phosphatidylglycerol (PG) at a mole% of 20 or greater have the ability to stabilize PE in the bilayer phase, presumably by both decreasing the intermolecular interaction between PE molecules and also by increasing the interfacial hydration level. Interestingly, the addition of Ca^{2+} to bilayers composed of PE : PS or PE : PG causes a phase separation of the negatively charged stabilizing lipid, allowing the PE to revert to the hexagonal II phase (Hope *et al.*, 1983).

An essential factor for the development of pH-sensitive liposomes is the ability of the lipid component, which contains a titratable acidic head group to maintain PE in a bilayer form. For example, the acylated amino acid palmitoylhomocysteine (PHC) (Connor *et al.,* 1984) with a titratable carboxyl group and the fatty acids oleic acid (OA) (Huang and Liu, 1984; Connor and Huang, 1985*c;* Düzgüneş *et al.,* 1985) and palmitic acid (PA) (Hope *et al.,* 1983; Huang and Liu, 1984), both with carboxylic head groups, can all stabilize PE into a bilayer configuration under physiological conditions. In addition, cholesterol derivatized on the hydroxyl group with a succinate group yielding cholesterol hemisuccinate (CHEMS), also with a titratable carboxyl head group, can maintain PE as a bilayer at neutral pH (Ellens *et al.,* 1984).

All these stabilizing lipids contain a negative charge on the head group that provides an electrostatic repulsion blocking the PE intermolecular interaction, which causes it to form the hexagonal II phase. Upon titration of the weakly acidic head group by lowering the pH, the negative charge is neutralized and the PE reverts to the hexagonal II phase. Thus, the bilayer vesicle is pH dependent for its stability. The significance of this transition of PE from a bilayer to hexagonal phase is the observation that membrane fusion in biological systems may involve an inverted micelle (hexagonal II) intermediate structure during the fusion event. Therefore, PE lipid vesicles that undergo a transition to hexagonal phase at acid pH could be used to produce a model membrane system capable of membrane to membrane fusion induced by protons.

3. MEASUREMENT OF FUSION

The fusion of liposomes can be monitored by a variety of methods. Morphologically the fusion of small unilamellar liposomes into larger vesicles can be seen using both negative-stain electron microscopy or freeze-fracture electron microscopy. In order to study the kinetics of pH-induced leakage and fusion of pH-sensitive liposomes, several fluorometric assays have been developed. The intermixing of membrane lipids can be monitored by use of the resonance-energy transfer assay developed by Struck *et al.* (1981). The measurement of aqueous content mixing between fusing liposomes can be determined by using either the 8-aminonaphthalene-1,3,6-trisulfonic acid (ANTS) and *p*-xylyl-enebis (pyridinium) bromide (DPX) (Ellens *et al.,* 1984) or the terbium-dipicolinic acid (Wilschut and Papahadjopoulos, 1979) assays. The ANTS/DPX fluorescence assay has the advantage of being insensitive to pH above 4.0, which makes it an excellent method for measuring acid-induced fusion. Leakage from vesicles due to proton-induced destabilization is monitored by the ANTS/DPX assay, the terbium/dipicolinic acid assay, or the dequenching of entrapped calcein.

In order to study the biophysic phenomenon occurring during proton-induced destabilization and fusion of liposomes both ^{31}P-NMR, indicative of the PE conformational changes, and differential scanning calorimetry (DSC), which reveals broad lipid interactions, are applied to the liposome system. Using both the fluorometric and biophysical methods, the mechanism and results of acid induced fusion can be studied.

4. pH-SENSITIVE LIPOSOME FUSION SYSTEMS

Using freeze-fracture techniques, Hope (1983) demonstrated the fusion of SUV composed of egg transesterified PE/phosphatidic acid (8 : 2 mole/mole) into large unilamellar and oligomellar vesicles. Included in this work are electron micrographs of intermediate vesicles displaying lipidic particles localized to the regions of fusion. Lipidic particles are indicative of PE in a reverse micellular conformation within adjacent membranes; these intermembrane particles have been proposed as being sites of vesicle fusion (Verkeij, 1979). Bondeson (1984) used the resonance-energy transfer method to show fusion of egg transesterified PE/phosphatidate (8 : 2 mole/mole) at slightly acid pHs (6.5–7.0). In this work, though, lectin was incorporated into the liposomes and exogenous agglutinin was added to facilitate liposome to liposome binding. Because fusion is dependent on vesicle collision, the inclusion of the agglutinating factor enhanced the fusion rate.

A well-studied system using resonance energy transfer to study acid-induced liposome fusion, independent of exogenous agglutinin, is the PE/PHC (8 : 2 mole/mole) pH-sensitive liposomes (Connor *et al.*, 1984). As can be seen in Figure 1, these liposomes begin to display fusion activity at a pH value of <7.0 with maximum fusion occurring at pH 4.4. The midpoint of fusion occurs at pH 6.2. Negative-stain electron micrographs of the liposomes before and after the acid addition indicate that approximately six rounds of fusion of the pH-sensitive SUV occur to give the final large vesicles. The PE/PHC liposomes appeared to be very leaky, as determined by calcein dequenching, during the fusion reaction. Inclusion of 40 mole% of cholesterol, although decreasing the total amount of fusion, allows for the retention of entrapped calcein during the fusion event.

In order to characterize fully the relationship between membrane mixing, i.e., resonance energy transfer assay, and aqueous mixing, by means of the ANTS/DPX assay, Düzgüneş *et al.* (1985) established a PE/OA (7 : 3 mole/mole) liposome-fusion system. Aqueous mixing was seen between fusing vesicles as the pH was lowered to 6.3, concurrent with aqueous content mixing, however, was rapid leakage of the entrapped

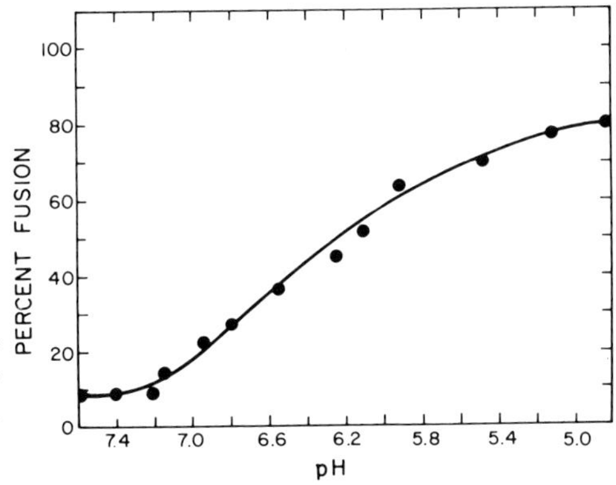

Figure 1. pH dependence of liposome fusion. Phosphatidylethanolamine/palmitoylhomocyteine (8 : 2 mole/mole) liposomes. Fusion was monitored using resonance energy transfer. (From Connor *et al.*, 1984.)

content (Düzgüneş *et al.*, 1985). The acid-induced membrane fusion between PE/OA liposomes was confirmed by resonance-energy transfer and irreversible changes in turbidity (Düzgüneş *et al.*, 1985). Similar to the PE/PHC system, the liposomes are destabilized on protonation and become leaky, possibly from the membrane–membrane contact point, upon fusion. ^{31}P-NMR scans of PE/OA liposomes at slightly acidic pH indicate the presence of PE in the hexagonal II conformation (Connor and Huang, 1985*c*). This destabilizing configuration of PE in the bilayer, although probably responsible for the fusion event, is also a source of rapid aqueous content leakage. Although both PE/PHC and PE/OA liposomes are leaky at fusion pH values, both vesicle types have been successfully demonstrated to be useful vehicles for cellular delivery (see Section 6).

A parallel system has been developed to study acid-induced fusion that employs cholesteryl hemisuccinate (CHEMS) as the stabilizing factor for PE bilayers. Similar to the PE/OA liposomes, the PE/CHEMS vesicles are sensitive to mildly acidic pH values (Ellen *et al.*, 1984). Using the ANTS/DPX fluorescence assay, it was demonstrated that the acid-induced pH leakage of these vesicles occurs subsequent to aggregation and that the destabilization effect of protonation is dependent on bilayer contact (Ellen *et al.*, 1984). The CHEMS molecules are able to stabilize the PE lipid in a bilayer configuration at neutral pH; the negative charge on the succinate head group is important in maintaining bilayer integrity by causing an electrostatic repulsion between liposomes. Upon protonation, the CHEMS acquires a cholesterol-like configuration and is still able to maintain PE in a bilayer form, but the CHEMS-induced repulsion between liposomes is removed, permitting contact between adjacent liposomes, leading to destabilization and leakage of entrapped contents.

In order to characterize the PE/CHEMS interaction Lai *et al.* (1985) studied the biophysical effects of protonation on PE/CHEMS liposomes using differential scanning calorimetry and ^{31}P-NMR. PE alone displays a bilayer to hexagonal phase transition (T_H) at 56°C. The addition of 25 mole% of CHEMS to egg-transesterified PE at neutral pH abolishes the T_H up to a temperature of 72°C, as measured by DSC (Lai *et al.*, 1985). At pH 4.5, PE/CHEMS (75 : 25 mole/mole), the transition again appears but is shifted to a temperature lower than that for egg transesterified PE alone. The increase of the T_H of PE by CHEMS at neutral pH and decrease of T_H of PE at acidic pH was confirmed using ^{31}P-NMR (Lai *et al.*, 1985). Thus, CHEMS is able to stabilize PE by preventing a transition to the hexagonal II phase but, upon protonation, CHEMS actually increases the ability of PE to revert to the inverted micelle structure. CHEMS in a deprotonated state is capable of both stabilizing PE in a bilayer and maintaining separation from adjacent liposomes. In the presence of an acidic environment, the electrostatic repulsion between vesicles is abolished, increasing membrane–membrane contact; in addition, the ability of PE to form the hexagonal II phase is increased for these liposomes. These contact points between membranes are the area of fusion between liposomes, but in the PE/CHEMS vesicles these points also cause contact-induced destabilization and hexagonal II phase formation (Bentz *et al.*, 1985). It appears that in the PE/CHEMS system, a decrease in pH causes leakage and collapse of the liposomes before a fusion event can take place.

In contrast to the finding in the PE/OA fusion system, using the PE/CHEMS liposomes, Ellen *et al.* (1985*a,b*) demonstrated that a mildly acidic condition causes a mixing of bilayer lipids, as measured by resonance energy transfer, but no mixing of aqueous contents, as determined by the ANTS/DPX assay. Thus the liposome–liposome interaction, at acidic pH, permits intermixing of the membrane lipids, but the collapse of the vesicular structure causes the content of the interacting liposome to leak and aqueous mixing cannot occur between vesicles. Whether this liposome composition will be applicable to *in vitro* delivery is yet to be determined.

Recently a system was developed using a succinic derivatized PE molecules as the stabilizing factor to maintain PE in a bilayer (Nayar and Schroit, 1985). DOPE was derivatized with succinic anhydride yielding *N*-succinyl-PE (COPE); this system is unique in that the stabilizing acidic factor contains identical diacyl chain as the PE lipid to be stabilized into a bilayer. Interestingly, liposomes composed of COPE alone are leaky at neutral pH but stable at pH 4.0, while liposomes composed of COPE/DOPE (3 : 7 mole/mole) act in the opposite direction. The pH-induced leakage effect of COPE/DOPE is independent of acyl chain saturation and lipid composition (Nayar and Schroit, 1985). This indicates that the leakage effect is not due to liposome aggregation but probably to packing defects caused by the protonation of the COPE molecules. The apparent difference in the effect of acidic conditions makes the COPE-containing liposomes an excellent system for studying lipid packing in pH-sensitive vesicles. It is also possible that these liposomes have a potential for delivering contents to cells.

5. DIVALENT CATION EFFECT

A second characteristic of the pH-sensitive liposomes of both the PE/ fatty acid or PE/CHEMS types is the divalent cation-induced fusion (Connor and Huang, 1985*b;* Düzgüneş *et al.,* 1985; Lai *et al.,* 1985; Ellens *et al.,* 1985*a*). It has been demonstrated that Ca^{2+} in the range of 2–3 mM and Mg^{2+} in range of 3–4 mM can induce fusion between liposomes composed of PE and a negatively charged stabilizer. Similar to protons, the divalent cations neutralize the negative charge of the stabilizing entity, permitting close contact between PE molecules; in addition, both Ca^{2+} and Mg^{2+} bring about a bridging between adjacent liposomes, increasing the number of bilayer contact points. Divalent cations have also been shown to act synergistically with proton, shifting the pH needed to initiate fusion from 6.5 closer to neutral (Connor and Huang, 1985*b*) (Table I). This synergistic effect of the divalent cation could have an important contribution in cellular delivery experiments, since both Ca^{2+} and Mg^{2+} are present in the growth medium. The divalent cation could permit fusion of liposomes at pH values higher than those required for destabilization and content leakage, increasing the chance of delivery.

The model membrane system presented allows for the precise study of the mechanism of interaction between fusing liposomes. However, the systems deal with liposome–liposome fusion. Two interesting demon-

Table I. Synergistic Effect of Divalent Cations and Protons on pH-Dependent Lipid Mixing of DOPE:OA Liposomes[a]

Divalent cation (mM)	pH_{50}[b]
None	6.9
Ca^{2+} (1 mM)	7.2
Ca^{2+} (2 mM)	7.5
Mg^{2+} (1 mM)	7.0
Mg^{2+} (2 mM)	7.2
Mg^{2+} (3 mM)	7.5

[a]Total lipid concentration is 0.2 mM with an unlabeled/fluorescently labeled liposome ratio of 3. Percentage lipid mixing was determined with equation (1) from Connor *et al.* (1984).

[b]pH_{50} is the pH at which 50% maximum lipid mixing occurs as determined by resonance energy transfer (Connor and Huang, 1985*b*).

strations of acid induced fusion of liposome with biological membrane have been demonstrated recently. Using the resonance energy transfer assay, Huang *et al.* (1985) demonstrated the fusion of liposome composed of PE/OA (8 : 2 mole/mole) with mitoplast or isolated inner membranes of rat liver mitochondria, with a pH optimum at 5.0–5.5. This method may be an effective way to introduce impermeable macromolecules into the matrix of mitoplast.

A second demonstration of fusion is between liposomes composed of PC and high mole% of cardiolipin (CL) and membrane vesicles derived from *Bacillus subtilis* and *Streptococcus cremoris* (Driessen *et al.,* 1985). Using both resonance energy transfer and aqueous delivery of colloidal gold particles, PC/CL (1 : 1 mole/mole), liposomes show a low pH-dependent fusion with the biological membranes. Treatment of the biological membranes with trypsin inhibited the fusion event indicating a protein dependence for the acid induced fusion. This low pH fusion process, though, could be useful in delivering substances to bacterial protoplasts.

6. CYTOPLASMIC DELIVERY BY pH-SENSITIVE LIPOSOMES

Liposomes have been used extensively as a tool to deliver biologically active molecules to cells (Tom and Six, 1980; Gregoriadis, 1983; Ostro, 1983). This has been achieved by use of a fusogenic agent to fuse liposomes to the cell plasma membrane or by allowing cellular endocytosis of the liposomes followed by a passive leakage of the entrapped molecules from the cellular lysosomes. In addition, to accelerate liposome delivery, numerous liposome targeting systems have been developed (Huang *et al.,* 1981, 1983; Martin *et al.,* 1981; Machy *et al.,* 1982; Matthay *et al.,* 1984). In these systems monoclonal antibodies, specific to a cell-surface antigen, are attached to the liposomes by a variety of methods (Gregoriadis, 1983; Connor *et al.,* 1985). These antibody-target liposomes (immunoliposomes), bind to specific cell antigens, are endocytosed via a receptor-mediated endocytosis mechanism, and are delivered, through the endosomes, to the lysosomal compartment. An obvious drawback to immunoliposome delivery is the inability of most biologically active molecules to escape the endosome/lysosome pathway into the cytoplasm; thus, unless the molecule to be delivered can escape the degradative lysosome vacuole, conventional immunoliposomes are a poor method for cytoplasmic delivery. By contrast, pH-sensitive immunoliposome can solve the problem by exploiting the inherent cellular endocytic pathway. Following receptor-mediated endocytosis, the liposomes encounter an acidic pH 5.0–6.5 in the endosome vacuoles (Straubinger *et al.,* 1983);

therefore, upon endocytosis, the pH-sensitive liposomes are confronted with a condition that readily activates the fusion activity of the liposomes. Two examples of pH-sensitive liposomes, one targeted and one nontargeted, have been demonstrated for efficient cytoplasmic delivery of an entrapped molecule by exploiting the endocytic pathway.

6.1. Nontargeted Delivery

Calcein and FITC-dextran were entrapped in PE/OA (7 : 3 mole/mole) extrusion vesicles. Calcein has been demonstrated to be an impermeable, pH-insensitive aqueous space fluorescent marker (Straubinger *et al.,* 1983). Dye-entrapped liposomes were incubated with CV-1 African green monkey kidney cells, and delivery was monitored by fluorescence microscopy. Cells incubated with PE/OA liposomes containing either calcein or FITC-dextran displayed diffused cytoplasmic fluorescence indicating cytoplasmic delivery. Control liposomes composed of either PS/PE (3 : 7 mole/mole) or OA/PC (3 : 7 mole/mole) containing the same fluorescent marker displayed only punctate fluorescence when incubated with the CV-1 cell, indicating delivery only to the endosome/lysosome system. In addition pretreatment of CV-1 cell with chloroquine, a permeable weak base that raises the internal pH of the endosome/lysosome, inhibited the cytoplasmic delivery of calcein by the pH-sensitive liposomes. Therefore, the cytoplasmic delivery depended on the acidification of the endosomes.

6.2. Targeted Delivery

Calcein was entrapped in PE/PHC liposomes prepared by the reverse-phase evaporation method and bearing an attached anti H-2K^k antibody. Anti H2K^k antibody is a monoclonal antibody against the murine histocompatibility H-2K^k antigen. These immunoliposomes were incubated with both L-929, H-2K^k antigen-expressing target cells and A-31, H-2K^d antigen-expressing control cells; delivery was monitored by fluorescence microscopy (Connor and Huang, 1985*a*). pH-sensitive immunoliposomes incubated with L-929 cells displayed diffused cytoplasmic fluorescence, while A-31 cells displayed no fluorescence at all, as the vesicles do not bind with the control cells. Control immunoliposomes composed of PC showed only punctate fluorescence when incubated with L-929 cells indicative of lysosomal delivery only. Similar to the nontargeted liposome experiments, chloroquine was able to block the cytoplasmic delivery of calcein by pH-sensitive immunoliposomes (Connor and Huang, 1985*a*).

To clarify further the ability of pH-sensitive immunoliposomes to deliver their contents to the cytoplasm, a cytotoxic drug was used to monitor delivery. In these experiments, arabinoside C(ara-C) was chosen as the cytotoxic agent, as it has previously been demonstrated that ara-C delivered with pH-insensitive immunoliposomes was ineffective in killing cells due to its delivery to, and subsequent degradation in, the lysosomes (Huang *et al.*, 1983; Rustum *et al.*, 1981). ara-C was encapsulated in PE/OA pH-sensitive and PC (pH-insensitive) immunoliposomes and incubated with L-929 cells. Cell killing was monitored by inhibition of [^{3}H]thymidine incorporation into DNA.

ara-C entrapped in pH-sensitive immunoliposomes was five times more effective in killing L-929 cells than was free ara-C. Both untargeted pH-sensitive liposomes and pH-insensitive immunoliposomes were ineffective in killing the L-929 cells (Fig. 2). As in the previous work, pretreatment of cells with chloroquine or NH_4Cl, another lysosomotropic agent, inhibited the cytoplasmic delivery (Connor and Huang, 1986).

The ability to deliver protein via pH-sensitive immunoliposomes has also been demonstrated in this model system. Diphtheria toxin A fragment (DTA) is unable to cross lipid membranes and is therefore nontoxic to the L-929 cells. When DTA is encapsulated in PE/OA pH-sensitive immunoliposomes, it is effectively delivered to the cytoplasm resulting in

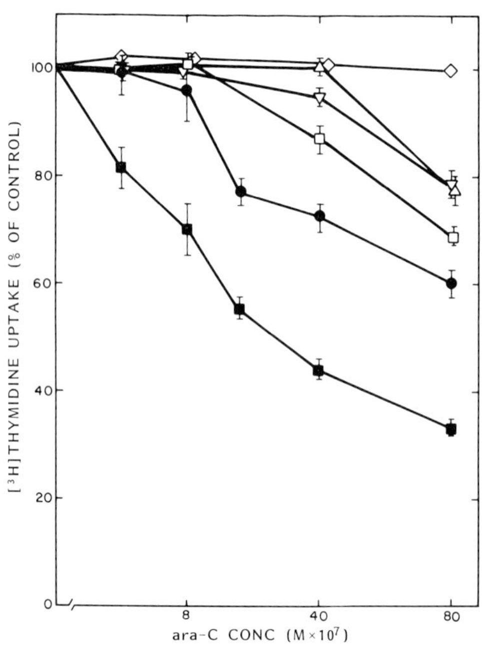

Figure 2. Cytotoxicity of arabinoside C (ara-C) in L929 cells. Cells were incubated with DOPE/OA (8 : 2 mole/mole) immunoliposomes (■); DOPC pH-insensitive immunoliposomes (▽); DOPE/OA (8 : 2 mole/mole) nontargeted liposomes (□); and DOPC/nontargeted liposomes (△), all containing ara-C. ara-C as free drug (●) and empty DOPE/OA (8 : 2 mole/mole) immunoliposomes (◇) were also incubated with L-929 cells. Cytotoxicity was assayed as inhibition of [^{3}H]thymidine incorporation. (From Connor and Huang, 1986.)

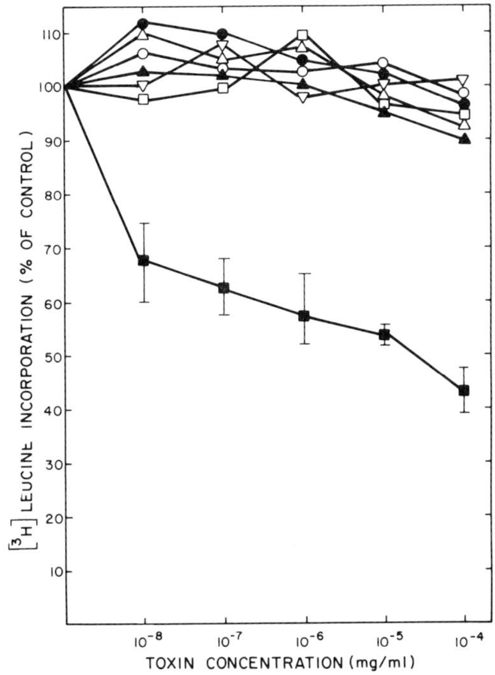

Figure 3. Inhibition of protein synthesis by liposome-encapsulated diphtheria toxin A frament. L-929 cells were incubated with PE/OA (8 : 2 mole/mole) immunoliposomes (■); DOPC immunoliposomes (●); PE/OA (8 : 2 mole/mole) nontargeted liposomes (□); DOPC nontargeted liposomes (○), all containing diphtheria toxin A fragment. Free diphtheria toxin (▲), free diphtheria toxin A fragment (Δ), and empty PE OA (8 : 2 mole/mole immunoliposomes (∇) were also incubated with L-929 cells. Protein synthesis was assayed by inhibition of [^{3}H]leucine incorporation. (From Collins and Huang, 1986.)

killing the L-929 cells (Fig. 3). Delivery and killing were monitored by inhibition of protein synthesis (Collins and Huang, 1986). DTA entrapped in PC immunoliposomes or administered alone is ineffective in killing the target cells. Once again, both chloroquine and NH_4Cl effectively inhibit the delivery of DTA by the pH-sensitive immunoliposomes.

An additional example of pH-sensitive liposome-mediated delivery has been demonstrated with plant protoplasts. Liposomes composed of DOPE/cholesterol/OA (4 : 4 2 mole/mole) with entrapped calcein demonstrated enhanced cytoplasmic delivery to carrot protoplast over several conventional pH-insensitive liposomes (Wang *et al.,* 1985).

To test the potential of employing pH-sensitive immunoliposomes for *in vivo* delivery, our laboratory tested their stability in mouse serum. PE/OA liposomes quickly aggregate and leak their contents upon exposure to serum (J. Connor, unpublished data); thus, in order to develop an effective *in vivo* pH-sensitive targeting system, the lipid composition must be altered to increase stability. It may be possible to add a third lipid component to the pH-sensitive liposomes, such as cholesterol, which may inhibit the serum-induced leakage yet allow the vesicles to retain their pH sensitivity.

6.3. Mechanism of Action

Clearly, pH-sensitive immunoliposomes can effectively deliver a variety of compounds (i.e., dyes, cytotoxic drugs, and proteins) to the cellular cytoplasm that cannot be delivered by conventional (pH-insensitive) immunoliposomes. The precise mechanism of delivery has not been elucidated, but the delivery occurs prior to lysosomes and is dependent on acidication of the cellular endosomes. Several possibilities are illustrated in Figure 4. It is possible that the acid-triggered event is a fusion reaction between the liposome and the adjacent endosome membrane (Fig. 4, step 5). Alternatively, the immunoliposomes could discharge their entrapped contents in the endosome interior where, at high concentrations, the molecules could leak out of the endosomes (step 6). Whereas this is a possibility with calcein and cytotoxic drugs, it is not likely for DTA to move across membranes by itself; therefore, the enhanced delivery of DTA by pH-sensitive immunoliposomes cannot be explained by this mechanism. Another possible mode of cytoplasmic delivery is the rupturing of the endosome vacuole (step 7). The destabilizing effect of the DOPE/OA vesicles undergoing a bilayer to hexagonal phase transition could rupture the endosome membrane, releasing its contents into the cytoplasm. This possibility cannot be ruled out at this point. The propen-

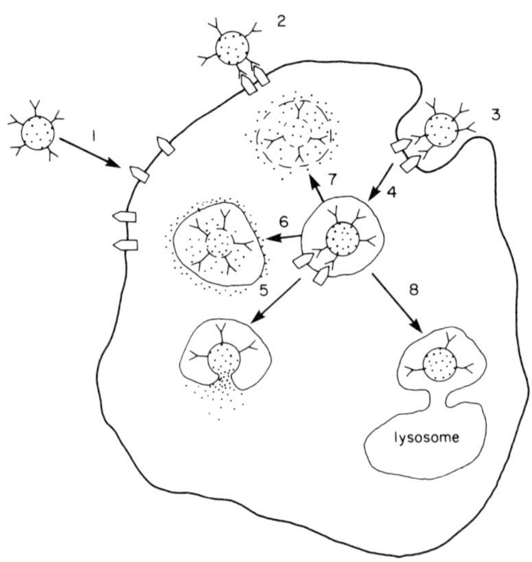

Figure 4. Schematic presentation of the interactions of the immunoliposomes with target L-929 cells: 1, immunoliposomes bearing antibody (Y) are incubated with target cells expressing antigens (◇); 2, binding of the immunoliposomes by the cell-surface antigen; 3, initiation of the receptor-mediated endocytosis at the coated pits; 4, internalization of the immunoliposomes into the acidic endosomes; 5, fusion of the pH-sensitive immunoliposomes with the endosome membrane with the liposomal contents released into the cytoplasm; 6, leakage of the liposomal contents into the endosome with subsequent escape into the cytoplasm; 7, rupturing of the endosomal vacuole causing release of liposome contents into the cytoplasm; and 8, fusion of the endosome with lysosomes and the delivery of the pH-insensitive immunoliposomes into the lysosomes.

sity of pH-sensitive liposomes to undergo an acid-induced fusion and content mixing with adjacent membrane gives strongest support to the hypothesis of a fusion event between the pH-sensitive immunoliposome and the endosome membrane. Although further work must be done to clarify the mode of delivery, the obvious potential for pH-sensitive liposomes as a delivery system for biologically active molecules, such as immunomodulators, proteins, drugs, and nucleic acids, is enormous. These applications are currently being explored in our and other laboratories.

ACKNOWLEDGMENTS. Work done in this laboratory has been supported by grant CA 24553 from the National Institutes of Health. L. H. is a research career development awardee (CA 00718).

REFERENCES

Bentz, J., Ellens, H., Lai, M. Z., and Szoka, F. C., 1985, On the correlation between H_{II} phase and the contact-induced destabilization of phosphatidylethanolamine-containing membranes, *Proc. Natl. Acad. Sci. U.S.A.* **82**:5742–5745.

Blumenthal, R., Henkart, M., and Steer, C. J., 1983, Clathrin induced pH dependent fusion of phosphatidylcholine vesicles, *J. Biol. Chem.* **258**:3409–3415.

Blumenthal, R., 1985, Membrane fusion, *Curr. Topics Membranes Transp.* **22**:172–251.

Bondeson, J., Wijkander, J., and Sundler, R., 1984, Proton-induced membrane fusion role of phospholipid composition and protein-mediated intermembrane contact, *Biochim. Biophys. Acta* **777**:21-27.

Collins, D., and Huang, L., 1986, Delivery of diptheria toxin A fragment to the cytoplasm of toxin-resistant cells by pH-sensitive immunoliposomes, *Cancer Research,* in press.

Connor, J., and Huang, L., 1985*a,* Efficient cytoplasmic delivery of a fluorescent dye by pH-sensitive immunoliposomes, *J. Cell Biol.* **101**:582–589.

Connor, J., and Huang, L., 1985*b,* The effects of pH and divalent cations on the phase behavior of liposomes composed of phosphatidylethanolamine and fatty acid, *Biophys. J.* **47**:428a.

Connor, J., and Huang, L., 1986, pH-sensitive immunoliposomes as an efficient and target-specific carrier for antitumor drugs, *Cancer Res.* **46**:3431–3435.

Connor, J., Sullivan, S., and Huang, L., 1985, Monoclonal antibodies and liposomes, *Pharm. Ther.* **28**:341–365.

Connor, J., Yatvin, M. B., and Huang, L., 1984, pH-Sensitive liposomes: Acid induced liposome fusion, *Proc. Natl. Acad. Sci. U.S.A.* **81**:1715–1718.

Cullis, P. R., and DeKruijff, B., 1979, Lipid polymorphism and the functional roles of lipids in biological membranes, *Biochim. Biophys. Acta* **559**:399–420.

Driessen, A. J. M., Hoekstra, D., Scherphof, G., Kalicharan, R. D., and Wilschut, J., 1985, Low pH induced fusion of liposomes with membrane vesicles derived from Bacillus subtilis, *J. Biol. Chem.* **260**:10880–10887.

Düzgüneş, N., Straubinger, R. M., Baldwin, P. A., Friend, D. S., and Papahadjopoulos, D., 1985, Proton-induced fusion of oleic acid-phosphatidylethanolamine liposomes, *Biochemistry* **24**:3091–3098.

Ellens, H., Bentz, J., and Szoka, F. C., 1984, pH-induced destabilization of phosphatidylethanolamine-containing liposomes: Role of bilayer contact, *Biochemistry* **23:**1532–1538.

Ellens, H., Bentz, J., and Szoka, F. C., 1985*a,* H^+ and Ca^{2+} induced fusion and destabilization of liposomes, *Biochemistry* **24:**3099–3106.

Ellens, H., Bentz, J., and Szoka, F. C., 1985*b,* Destabilization of phosphatidylethanolamine liposomes at the hexagonal transition temperature, *Biophys. J.* **47:**169a.

Gad, A. E., and Eytan, G. D., 1983, Chlorophylls as probes for membrane fusion polymyxin B-induced fusion of liposomes, *Biochim. Biophys. Acta* **77:**170–176.

Gregoriadis, G., 1983, *Liposome Technology,* CRC Press, Boca Raton, Florida.

Hauser, H., Pascher, I., Pearson, R. H., and Sundell, S., 1981, Preferred conformation and molecular packing of phosphatidylethanolamine and phosphatidylcholine, *Biochim. Biophys. Acta* **650:**21–51.

Hong, K., Düzgüneş, W., Ederdt, R., and Papahadjopoulos, D., 1982, Synexin facilitates fusion of specific phospholipid membranes at divalent cation concentrations found intracellularly, *Proc. Natl. Acad. Sci. U.S.A.* **79:**4642–4644.

Hope, M. J., Walker, D. C., and Cullis, P. R., 1983, Ca^{2+} and pH induced fusion of small unilamellar vesicles consisting of phosphatidylethanolamine and negatively charged phospholipids: A freeze-fracture study, *Biochem. Biophys. Res. Commun.* **110:**15–22.

Huang, L., and Lui, S. S., 1984, Acid induced fusion of liposomes with inner membranes of mitochondria, *Biophys. J.* **45:**72a.

Huang, A., Kennel, S. J., and Huang, L., 1983, Interaction of immunoliposomes with target cells, *J. Biol. Chem.* **258:**14034–14040.

Huang, L., Liu, S. S., Ding, Y. Z., and Gao, F. H., 1985, Acid induced fusion of liposomes with inner membranes of mitochondria, *Biochemistry,* submitted.

Lai, M. Z., Düzgüneş, N., and Szoka, F. C., 1985*a,* Effects of replacement of the hydroxyl group of cholesterol and tocopherol on the thermotropic behavior of phospholipid membranes, *Biochemistry* **24:**1646–1653.

Lai, M. Z., Vail, W. J., and Szoka, F. C., 1985*b,* Acid and calcium-induced structural changes in phosphatidylethanolamine membranes stabilized by cholesteryl hemisuccinate, *Biochemistry* **24:**1654–1661.

Lampe, P. D., and Nelsestuen, G. L., 1982. Myelin basic protein-enhanced fusion of membranes, *Biochim. Biophys. Acta* **693:**320–325.

Lis, L. J., McAlister, M., Fuller, N., Rand, R. P., and Parsegian, V. A., 1982, Interaction between neutral phospholipid bilayer membranes, *Biophys. J.* **37:**657–666.

Machy, P., Pierres, M., Barbet, J., and Lesserman, L. D., 1982, Drug transfer into lymphoblasts mediated by liposomes bound to distinct sites on H-2 encoded I-A, I-E and K molecules, *J. Immunol.* **129:**2098–2101.

Martin, F. J., Hubbell, W. L., and Papahadjopoulos, D., 1981, Immunospecific targeting of liposomes to cells: A novel and efficient method for covalent attachment of ab fragments via disulfide bonds, *Biochemistry* **20:**4229–4236.

Matthay, K. K., Heath, T. D., and Papahadjopoulos, D., 1984, Specific enhancement of drug delivery to AKR lymphoma by antibody-targeted small unilamellar vesicles, *Cancer Res.* **44:**1880–1886.

Morgan, C. G., Williamson, H., Fuller, S., and Hudson, B., 1983, Mellitin induces fusion of unilamellar phospholipid vesicles, *Biochim. Biophys. Acta* **732:**668–674.

Nayar, R., and Schroit, A., 1985, Generation of pH-sensitive liposomes: Use of large unilamellar vesicles containing N-succinyldioleoylphosphatidylethanolamine, *Biochemistry* **24:**5967–5971.

Ostro, M. J., 1983, *Liposomes,* Marcel Dekker, New York.

Papahadjopoulos, D., and Miller, N., 1967, Phospholipid model membranes: Structural characteristics of hydrated liquid crystals, *Biochim. Biophys. Acta* **135**:624–638.

Reiss-Husson, F., 1967, Structure de phase liquid crystalline de différents phospholipides, monoglycérides, sphingolipides, anhydrés ou en presence d'eau, *J. Molec. Biol.* **25**:363–382.

Roa, M., and Boquet, P., 1985, Interaction of tetanus toxin with lipid vesicles at low pH, *J. Biol. Chem.* **260**:682–6835.

Rustum, Y. M., Mayhew, E., Szoka, F., and Cambell, J., 1981, Inability of liposome encapsulated l-β-D-ara-binofuranosylcytosine nucleotides to overcome drug resistance in L1210 cells, *Eur. J. Clin. Oncol.* **17**:809–817.

Schenkman, S., Araujo, P. S., Dijkman, R., Quina, F. H., and Chaimovich, H., 1981, Effect of temperature and lipid composition on the serum albumin-induced aggregation and fusion of small unilamellar vesicles, *Biochim. Biophys. Acta* **649**:633–641.

Straubinger, R. M., Hong, K., Friend, D. S., and Papahadjopoulos, D., 1983, Endocytosis of liposomes and intracellular fate of encapsulated molecules: Encounter with a low pH compartment offer internalization in coated vesicles, *Cell* **32**:1069–1079.

Straubinger, R. M., Düzgüneş, N., and Papahadjopoulos, D., 1985, pH-sensitive liposome mediated cytoplasmic delivery of encapsulated macromolecules, *FEBS Lett.* **179**:148–154.

Struck, D. K., Hoekstra, D., and Pagano, R. E., 1981, Use of resonance energy transfer to monitor membrane fusion, *Biochemistry* **20**:4093–4099.

Szoka, F., Jacobson, K., and Papahadjopoulos, D., 1979, The use of aqueous space markers to determine the mechanism of interaction between phospholipid vesicles and cells, *Biochim. Biophys. Acta* **551**:295–303.

Tom, B. H., and Six, H. R., 1980, *Liposomes and Immunology,* Elsevier/North-Holland, New York.

Verkleij, A. J., Mombers, C., Leunissen-Birjuelt, J., and Ververgaert, P. H. J. Th., 1979, Lipid intramembranous particles, *Nature (Lond.)* **279**:162–163.

Wilschut, J., and Papahadjopoulos, D., 1979, Ca^{2+} induced fusion of phospholipid vesicles monitored by mixing of aqueous content, *Nature (Lond.)* **279**:162–163.

Wang, C. Y., Hughs, C. W., and Huang, L., 1986, Improved cytoplasmic delivery to plant protoplasts via pH-sensitive liposomes, *Plant Physiology* **82**:179–184.

Chapter 14

The Mechanism of Polyethylene Glycol-Induced Fusion in Model Membranes

L. T. Boni and S. W. Hui

1. INTRODUCTION

Polyethylene glycol (PEG) has been used frequently as an ideal cell fusogen (Kao and Michayluk, 1974; Davidson and Gerald, 1977; Galfre *et al.,* 1977; Power *et al.,* 1978; Robinson *et al.,* 1979; Knutton, 1979; Oi *et al.,* 1980; Fazekas de St. Groth and Scheidegger, 1980; Krahling, 1981; Wojcieszyn *et al.,* 1983). PEG is capable of fusing a wide variety of cells, including interspecific and interkingdom cell types. PEG exhibits a very low degree of cytotoxicity; it is only toxic at high concentrations (greater than 50%) and prolonged incubation (minutes for most cell lines). PEG is a water-soluble fusogen unlike nonpolar or amphipathic chemical fusogens, such as fatty acids, retinol, and lysolecithin, that act by entering the bilayer. Thus, it is relatively easy to remove PEG from the cell medium by washing. The use of PEG has several additional advantages over other fusogens: reproducibility in fusion capabilities, high fusion efficiency, inexpensiveness, availability, and ease of handling.

L. T. Boni • Department of Physiology and Biophysics, Harvard Medical School, Boston, Massachusetts 02115. S. W. Hui • Department of Biophysics, Roswell Park Memorial Institute, Buffalo, New York 14263.

Several similar protocols are employed in PEG-induced fusion. The protocols typically employ a short (1–3 min) incubation of cells in a concentrated PEG solution (usually 50% w/w of 6000 M_r), followed by washing away the PEG by dilution or dilution and then centrifugation to pellet the cells to complete the fusion. Fusion between cells is observed only after the removal of PEG. These fusion protocols have been applied successfully for over a decade. Only recently has the mechanism of PEG-induced fusion been studied by physical and chemical methods.

Essential to understanding the mechanism of PEG-induced fusion is a comprehension of the factors that govern membrane stability. The existence of an intact biological membrane depends on a favorable interaction between the constituent lipids and proteins and their aqueous environment (Poste and Allison, 1973). Perturbing the organization of these membrane components and/or changing the nature of the aqueous phase can potentially render the membrane structurally unstable. Protein clustering, for example, can lead to a loss of membrane integrity and a change in the lateral organization of the lipids and proteins (Ahkong *et al.,* 1975; Power *et al.,* 1978). The manner by which PEG induces reorganization of membrane components, forces cell or vesicle contact, and disrupts bilayer structure and continuity that results in fusion is the subject of this chapter.

Intramembrane particles (IMP) have been seen by freeze-fracture electron microscopy to segregate during incubation of cells in PEG and thus to create bare membrane patches (Maul *et al.,* 1976; Robinson *et al.,* 1979; Knutton, 1979; Krahling, 1981; Wojcieszyn *et al.,* 1983; Hui *et al.,* 1985). IMP represent integral membrane proteins. The effect of PEG on the IMP distribution in erythrocytes is shown in Figure 1.

PEG also causes extensive cell agglutination; the site of cell contact and subsequent fusion is associated with the IMP-denuded regions. The fact that cell lines resistant to PEG-induced fusion do not form IMP clusters (Roos *et al.,* 1983) argues for the necessity of direct contact of pure lipid regions for fusion to proceed. Therefore, in describing the possible mechanisms of PEG-induced membrane fusion, its interaction with pure lipid vesicles is discussed, followed by a description of additional factors pertinent to the fusion of biological membranes.

2. PHYSICAL PROPERTIES OF POLYETHYLENE GLYCOL

PEG is a polymer of ethylene oxide that contains repeating OCH_2CH_2-monomer units. It is highly water soluble (Bailey and Koleske, 1967). The PEG typically employed in fusion studies is of 1000–6000 M_r (PEG 1000 and PEG 6000, the latter of which refers to PEG of an average

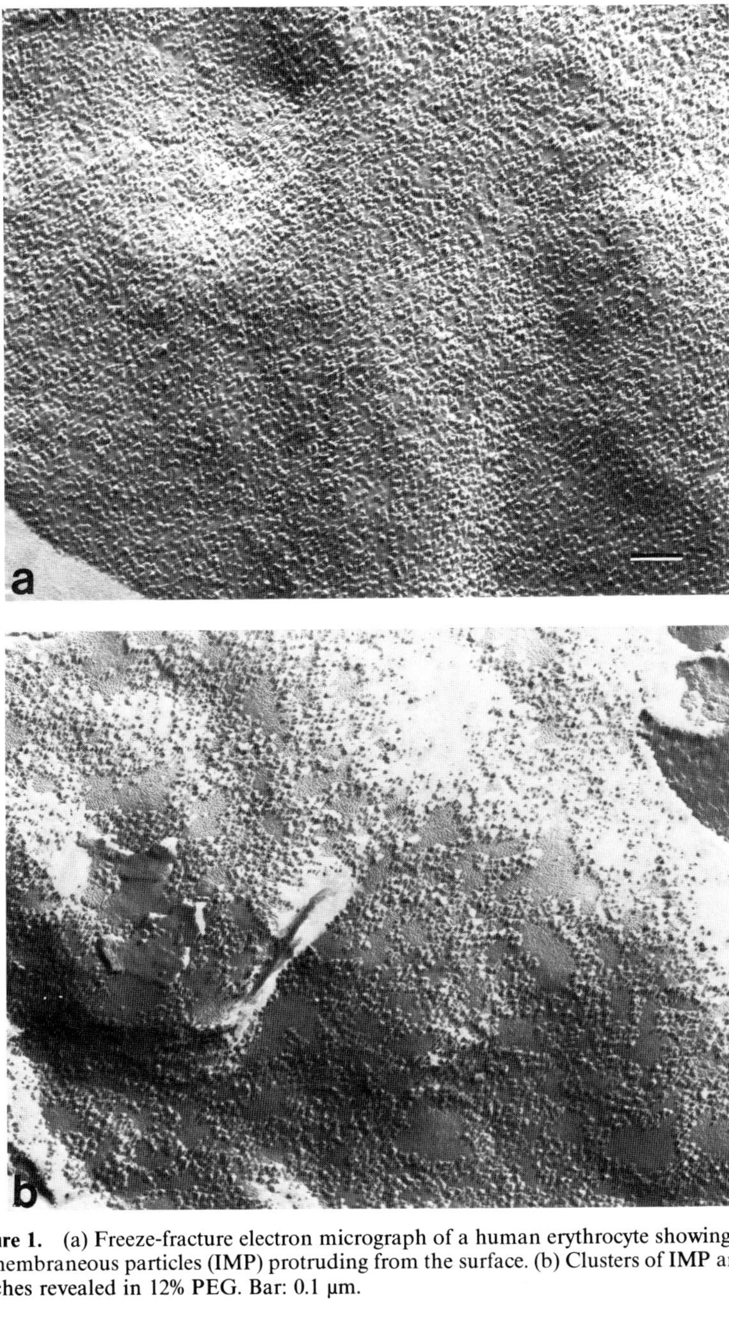

Figure 1. (a) Freeze-fracture electron micrograph of a human erythrocyte showing the intramembraneous particles (IMP) protruding from the surface. (b) Clusters of IMP and bare patches revealed in 12% PEG. Bar: 0.1 μm.

molecular weight of 6000–8000) and at aqueous concentrations of 25–50% w/w. It is of interest to note that w/w and w/v concentrations of PEG are almost equivalent (Tilcock and Fisher, 1982).

An important property of PEG is its ability to bind and structure water. The high viscosities observed for aqueous PEG solutions of molecular weight 6000 or greater have been attributed to the unusual hydration of the polymer along with the possible existence of a tightly held gel-like network (Bailey and Koleske, 1967). Negative values for the heat and entropy of dilution are indicative of a high degree of orientation of water molecules with respect to the polymer chain. Baran *et al.* (1972) used high-resolution proton nuclear magnetic resonance (NMR) to determine the mobility of water molecules at various PEG concentrations. They compared linewidths and intensities of NMR spectra as a function of temperature and concluded that each PEG monomer unit bound three to four water molecules while structuring up to 16. This corresponds to all the water being bound at 38–45% PEG and all the water being structured by 13% PEG. Ultrasonic attenuation measurements have indicated the direct hydrogen bonding by water to the ether oxygen of PEG or even protonation of the ether oxygen (Hammes and Schimmel, 1967). In addition, further ordering of water was attributed to hydrophobic water structuring around the ethyl groups of the polymer.

Using differential scanning calorimetry (DSC), Blow *et al.* (1978) observed two endothermic peaks for solutions less than 45% (w/w) PEG 6000. A 260–270°K peak was attributed to the melting of free water, while the second peak at 255°K was believed to be due to a polymer-hydrate melt. The free water peak disappeared at 45% PEG 6000. Additional DSC studies by Tilcock and Fisher (1982) showed an increase in the number of water molecules bound per polymer unit from 1.8 for PEG 200 to 2.7 for PEG 6000, with no detectable free water for 48.1% w/w solution of PEG 6000. The greater amount of bound water with increasing molecular weight PEG is attributed to the random coiling of the polymer in aqueous solution, which provides additional sites for water entrapment.

The binding of water by PEG also decreases the total volume of solution available for other macromolecules. On the basis of their DSC data, Tilcock and Fisher (1982) proposed a model for the space occupied by the hydrated PEG complex. For PEG of $>1500\ M_r$, the hydrated polymer was calculated to induce a maximum degree of exclusion of a second molecular species when in a globular configuration. This model is in agreement with that of Hammes and Roberts (1968), in which the maximum size of the cooperative polymer–solvent unit occurs with PEG 3400 or greater.

The ability of PEG to dehydrate other macromolecules and cells in aqueous media is best described by the increase in osmolarity with in-

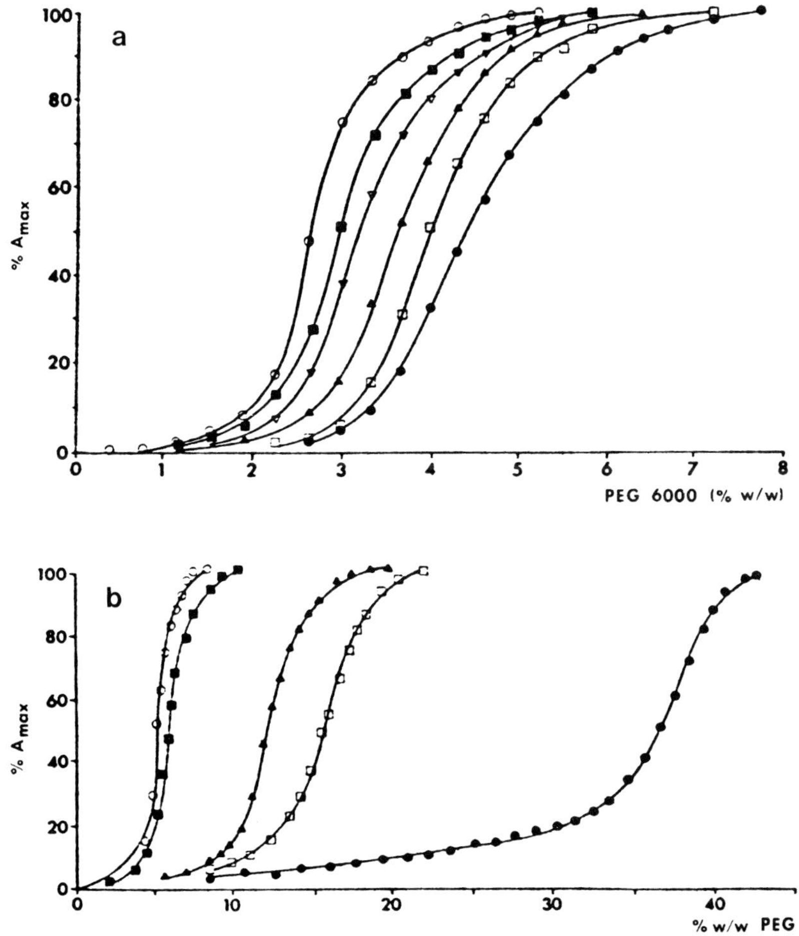

Figure 2. (a) The aggregation of dimyristoylphosphatidylcholine (DMPC) vesicles by PEG 6000. Values for the normalized absorbance (% A_{max}) of sonicated dispersions of DMPC in the presence of PEG 6000 are shown for six concentrations of lipid: 4.75 mg/ml (○), 1.58 mg/ml (■), 0.810 mg/ml (▽), 0.372 mg/ml (▲), 0.352 mg/ml (□), and 0.176 mg/ml (●). (b) The aggregation of DMPC vesicles by PEG 200–4000. Values for the normalized absorbance of sonicated dispersions of DMPC in the presence of PEG are shown for PEG 200 (●), PEG 400 (□), PEG 600 (▲), PEG 1500 (■), and PEG 4000 (○), the lipid concentrations varied between preparations; values were 3.07, 1.94, 2.80, 1.81, and 1.54 mg/ml, respectively. All measurements were performed at 32°C. Each point is the mean +SD of at least four determinations; the values of SD are within the spread of the symbols. (From Tilcock and Fisher, 1982.)

Steric hydration forces are thought to dominate the repulsion between bilayers at distances less than 3.0 nm (LeNeveu *et al.,* 1977; Cowley *et al.,* 1978). The stabilizing hydration layer surrounding the lipid surface of membranes represents the primary barrier to membrane fusion (McIver, 1979). The results on vesicle aggregation are consistent with the ability of PEG to bind water, remove the hydration shell surrounding lipid molecules, and exclude a second molecular species.

3.2. Binding and Influence on Lipid Motion

Direct binding of 0.088 mol PEG/mol PC has been noted through use of radiolabeled PEG 6000 (Boni *et al.,* 1984*a*). A lesser degree of dextran binding to lipid has been reported—18.4 g dextran/mol DPPC for dextran 150,000 (Minetti *et al.,* 1979). Additional information concerning dextran binding was obtained using the Nuclear Overhauser Enhancement Effect (NOEE). By selectively irradiating the *N*-methyl protons of PC SUV, Yeagle *et al.* (1975) observed an enhancement in the intensity of the phosphorous resonance. A decrease in the enhancement effect for PC vesicles was observed in the presence of high molecular weight dextrans (Minetti *et al.,* 1979). This indicates a change in the packing of head groups. Since PEG binds to and dehydrates lipid to a greater extent than does dextran, a greater effect on the packing of the bilayer could be expected. In another NOEE experiment in which the PEG was directly irradiated, a transfer of resonance energy from the PEG polymer to the choline moiety of the phospholipid was noted, which is consistent with a close proximity of the polymer (S. W. Hui and J. L. Alderfer, unpublished results). In addition, infrared spectroscopy has revealed a strong influence of PEG on the carbonyl region of the PC molecules (S. W. Hui and H. H. Mantsch, unpublished results). This finding is consistent with the particular perturbing nature of PEG at the lipid–water interface of the bilayer.

In NMR studies of PEG interactions with sodium dodecyl sulfate (SDS), a change in chemical shift with PEG concentration occurred for only the C1, or carboxyl, carbon resonance of the SDS molecule (Cabane, 1977). This indicates that the hydrated part of the SDS molecule is bound to the polymer and that the polymer does not enter the micelle core. An increase in $^{23}Na^{+}$-NMR relaxation rates with increasing concentrations of PEG was proposed by Cabane (1977) to explain the binding of sodium ions to the PEG–SDS complex. A drastic reduction in the choline group intensity by ^{1}H-NMR was observed for egg PC SUV in 6% PEG 6000 if the buffer contained 10 mM NaCl. No salt effect was seen for dextran (Boni, 1983). The partitioning of multilamellar vesicles (MLV) in aqueous two-phase systems of PEG/dextran has shown that vesicles favor the PEG phase and act as if they possessed a net positive surface charge and PEG a negative charge (Eriksson and Albertsson, 1978). These results support the model proposed by Kao and Michayluk (1974) in which a cation forms a salt bridge between the polymer unit and lipid.

Ohno *et al.* (1981*a*) observed a decrease in motion of the choline methyl moiety over the methylene resonances of the hydrophobic core for DPPC SUV upon incubation with as low as 0.2% PEG. This is consistent with a direct interaction of PEG with the liposome surface, particularly at the head-group region. The lesser broadening of the methylene region can be attributed to a head-group–acyl chain interaction. PEG-induced broadening of egg PC SUV ^{31}P-NMR resonances was shown not to be due to a viscosity increase because dextran and glycerol of the same viscosities did not induce significant broadening (Boni *et al.,* 1984*a*). The broadening, indicative of a restriction in lipid motion, must be due to a decrease in the lateral diffusion of the lipid, thus further substantiating a specific PEG-lipid head-group interaction. Ohno *et al.* (1981*b*), using the fluorescent probe ANS, reported a decrease in membrane fluidity at 0.1% PEG. A decrease in membrane fluidity of DPPC vesicles in 40% PEG, but not with intact human erythrocyte ghosts, was also noted by Herrmann *et al.* (1983), who used the spin-label probe 4-(2*n*-undecyl-3-oxyl-4,4-dimethyloxazolidin-2-yl)butyric acid [I(10,3)]. Boss and Mott (1982), however, observed a decrease in fluidity of the plasma membrane of carrot protoplasts with the electron spin resonance (ESR) probe 5-nitroxy steric acid only in the presence of both 15% PEG 14,000 and 10 mM calcium chloride. An enhancement in the fusion efficiency of PEG due to the presence of calcium (Kao and Michayluk, 1974) or sodium (Krahling, 1981) has been noted.

^{1}H-NMR spin-lattice relaxation times for egg PC SUV in 6% PEG decreased from 0.32 to 0.20 sec for the *N*-methyl, 0.39–0.25 sec for the methylene and 0.58–0.33 sec for the terminal methyl protons. These results provide additional evidence for restricted motion of the lipid at low PEG concentrations (Boni *et al.,* 1984*a*; Boni, 1983). A decrease in the spin-lattice relaxation times noted for ^{1}H-NMR but not for ^{31}P-NMR indicates a decrease in the frequency of trans-gauche isomerization of the acyl chains (Horwitz *et al.,* 1973).

Thus, it can be concluded that a low concentration (less than 10%) of PEG interacts with lipid bilayers both indirectly by changing their surface potential and competing with the available water and directly by binding to the membrane surface. The direct interaction with the polar headgroups changes bilayer packing, weakening the association between the phospholipids in the bilayer (Minetti *et al.,* 1979). A shearing force may then rupture the bilayer and facilitate the exposure of the hydrophobic region of the bilayer (Tilcock and Fisher, 1979). The resultant decrease in mobility of the lipids, possibly due to a reduction in the trans-gauche isomerization of the methylene carbons, indicates that the lipids are more gel-like. Such an effect is usually associated with the dehydration of lipid bilayers. Furthermore, the presence of cations appears to enhance this effect. The coexistence of liquid-crystalline and gel domains may be induced by other fusogens; it has been proposed that fusion occurs at the boundaries between these domains (Papahadjopoulos *et al.,* 1978; Jain and White, 1978; McIver, 1979; Tilcock and Fisher, 1982). Since the volume occupied by lipid molecules in the liquid-crystalline phase is greater than that in the gel phase, a domain of gel phase contained within a liquid-crystalline phase would be under tension (Lee, 1977), and these regions would be highly unstable. Thus, PEG may destabilize the bilayer indirectly by forming phase-separated domains.

3.3. Lipid Release and Exchange

For fusion to proceed, two events must occur—PEG facilitated bilayer contact and subsequent bilayer destabilization. A transient destabilization of the membrane is a necessary step in the fusion process (Papahadjopoulos *et al.,* 1974). Upon treatment of cells with 50% PEG 6000, freeze-fracture electron microscopy showed what appeared to be protein-free lipid droplets blebbing off (Knutton, 1979). Biochemical studies have shown that cells treated with 40% PEG 1000 for five minutes released twice the amount of [^{3}H]glycerol (a lipid precursor) into the culture medium during a 10-hr period following washing as that released from control cells (McCammon and Fan, 1979). A decrease in [^{3}H]glucosamine-labeled glycoproteins and glycolipids and [^{14}C]isoleucine-labeled proteins indicated a preference toward release of membrane components rich in lipids and low in protein and glycolipids.

The release of the apolar fluorescent probe pyrene from erythrocyte ghosts in varying concentrations of PEG was recorded by Arnold *et al.* (1983). The steady decrease in the ratio of excimer-to-monomer intensity of pyrene at 10–30% PEG is indicative of loss of probe to the less polar PEG medium. A smaller loss of probe was observed when the longer chained, more hydrophobic dodecerylpyrene was used.

The partitioning of the ESR spin-label 1-oxyl-2,2,6,6-tetramethyl piperidine (TEMPO) between DPPC vesicles and the external aqueous phase decreased with increasing PEG concentration. The decrease was most dramatic between 5 and 20% PEG (Herrmann *et al.,* 1983). This suggests a reduction in the polarity of the aqueous PEG phase. A complete transfer of lipid components between DMPC and DPPC MLV was shown by DSC to occur within 25 min for 45% PEG 400, as indicated by the coalescing of the two DSC endotherms (Tilcock and Fisher, 1982). This could be a result of exchange through the PEG media via direct contact of

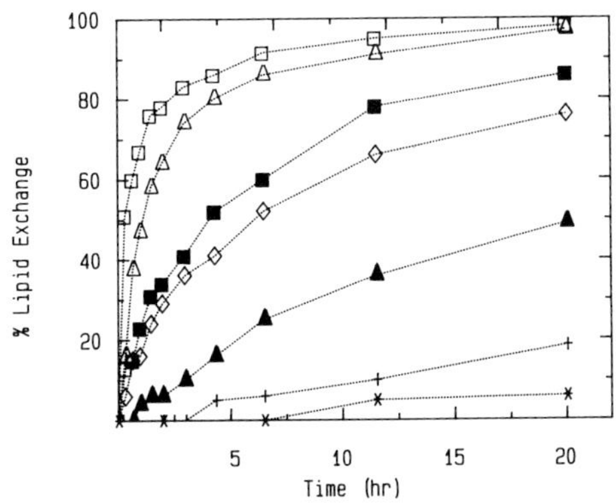

Figure 3. Time dependence of transfer of fluorescent phospholipid probes from 1 mM egg PC/NBD-PE/rhodamine-PE (98.8/0.4/0.8) SUV to 9 mM egg PC SUV at 20°C in the following reagents: 2% PEG 6000 (*), 3% PEG 6000 (+), 4% PEG 6000 (▲), 5% PEG 6000 (■), 6% PEG 6000 (△), 10% PEG 6000 (□), and 10% dextran 70,000 (◇). The buffer used was 7 mM Tris-HCl, pH 7.4.

the aggregated vesicles, by breakage and resealment of vesicles, or by fusion. Morgan *et al.* (1983) observed the transfer of the fluorescent triacylglycerol and possibly phospholipid probes between DPPC vesicles upon incubation in 5% PEG 6000.

A direct observation of transfer of phospholipid between SUV is shown in Figure 3 (L. T. Boni, unpublished results). The lipid mixing assay of Struck *et al.* (1981) was used. Phospholipid vesicles containing the fluorescent phospholipids *N*-4-nitrobenzo-2-oxa-1,3-diazole PE (NBD-PE) and lissamine rhodamine B sulfonyl dioleoyl PE (Rhod-PE) were at concentrations such that the Rhod-PE partially quenched the fluorescence of the NBD-PE. A release of quenching upon incubation with SUV devoid of probe in various concentrations of both PEG and dextran is indicative of lipid exchange. This occurs only at aggregating concentrations of PEG 6000; a large increase occurs between 3 and 6% PEG. When the PEG incubation buffers contained an additional 25 mM sodium chloride, up to a twofold enhancement of transfer efficiency was observed. Transfer also occurred when the vesicles were incubated in 10% and 14% dextran 70,000. Fusion of SUV to larger entities is not observed under these conditions by light scattering (Boni *et al.,* 1981*a,* 1984*a*) and gel filtration (L. T. Boni, unpublished results). These results offer an explanation for the observations by light microscopy for the spread of fluorescent lipid probes from membranes of erythrocytes to culture cells in the presence of fusogenic and purified nonfusogenic 44% PEG 8000 (Wojcienzyn *et al.,* 1983). In this experiment, the transfer occurred in less than 30 seconds for the dihexadecylinodo-carbocyanine (diI) and NBD-PE label and appeared to be due to direct contact since probe was not detected in the PEG solution. The proteins of the erythrocyte membranes did not diffuse into the culture cell membranes.

Ordered water lowers the thermodynamic barrier against the transfer of apolar groups to the aqueous phase (Hafeti and Hanstein, 1974). Destabilization of the bilayer structure must occur in a way that permits the lipid hydrocarbon moieties to merge without necessitating a large amount of work to remove water from the polar head group (Cowley *et al.*, 1978). This contact-mediated transfer of lipid is indicative of the ability of dehydrating agents to allow exposure of the hydrophobic core of the bilayer, a necessary step in the fusion process (Hui *et al.*, 1981).

3.4. Fusion

The ability of PEG to fuse pure lipid bilayers was first reported by Boni *et al.* (1981*a*). Egg PC or bovine PS SUV was incubated in 50% PEG 6000 for 1 min and washed by dilution and centrifugation. Fusion to MLV took place during the incubation in PEG, as shown in Figure 4. The wormlike texture on the fracture plane of the fusion product in PEG (Fig. 4b) is a consequence of dehydration (Kleman *et al.*, 1977; Ranck *et al.*, 1974). The turbidity decreased linearly after dilution of 1.6 mM egg PC SUV from varying initial PEG concentrations for 28% PEG 6000 and greater (Boni *et al.*, 1981*a*). This indicates an inability of vesicles to deaggregate, consistent with fusion to larger entities. A more detailed observation of the fusion capability of PEG was obtained by measuring the increase in the size of egg PC SUV by dynamic light scattering. Vesicles were incubated in varying concentrations of different molecular weight PEG, dextran, glycerol, and sucrose, all of which are dehydrating agents. Light scattering measurements were taken following dilution to 0.1% final reagent concentration to allow for vesicle deaggregation. As shown in Figure 5a, significant fusion occurred at 20–30% PEG 6000 with a decrease in fusion for the lower-molecular-weight PEG. No deviation from this curve was observed when recrystallized or recrystallized and dialyzed PEG 6000 was employed, indicating that impurities do not affect SUV fusion. This finding is contrary to the results of several PEG cell fusion studies that depended on PEG purity (Honda *et al.*, 1981; Wojcieszyn *et al.*, 1983). Figure 5a also reveals minimal fusion for the other dehydrating reagents. These light scattering results are confirmed by both NMR and negative-stain electron microscopy (Boni *et al.*, 1984*a*). Fusion of SUV in 40% PEG 6000 has also been observed by Saez *et al.* (1982), Morgan *et al.* (1983), and MacDonald (1985). It is interesting to note that MacDonald performed parallel experiments in which SUV were in direct contact with

PEG, dextran or dialyzed against these reagents. Lipid mixing was observed in all cases by the fluorescence energy transfer technique, indicating that direct contact of PEG with the vesicles was not required for fusion. The high degree of dehydration in the dialyzed cases could have promoted the observed fusion, and lipid exchange without fusion could make interpretations of the results difficult. There appears to be a strong dependence of lipid exchange on the dialysis conditions that mediate the homogeneity of dehydration, such as the time of dialysis and the excess volume of the external dialysis medium (L. T. Boni, unpublished results).

The kinetics of the fusion event for SUV is extremely rapid. This is seen for the 1-min incubation of egg PC SUV in 50% PEG 6000 (Fig. 4) and by a freeze-fracture experiment where an SUV–PEG mixture was sampled immediately after mixing (Boni *et al.*, 1981*a*). Egg PC SUV were seen by light scattering to fuse to completion within 5 min following dilution from 25 or 45% PEG 6000 (Boni, 1983). This is consistent with the PEG-induced rapid lipid mixing observed by Morgan *et al.* (1983) and MacDonald (1985).

The effect of the lipid phase on the fusion efficiency of dimyristoyl PC (DMPC) SUV is shown in Figure 5b. A dramatic degree of fusion is seen below the phase transition at as low as 6% PEG 6000. No fusion was observed for DMPC SUV incubated at 10°C for 1 hr in the absence of PEG. This is consistent with the instability of bilayers below the phase transition where structural defects are thought to exist (Larrabee, 1979). The result also agrees with the enhanced degree of fusion or lipid exchange for vesicles below the phase transition observed by Tilcock and Fisher by DSC (1982). It could explain the lipid mixing observed by Morgan (1983) for DPPC vesicles in 5% PEG 6000. The enhanced fusion of DMPC SUV at 10°C as compared with 20°C is not just a temperature effect; a decrease in the fusion efficiency for egg PC SUV in PEG was noted at 10°C (Boni, 1983). The necessity of charge neutralization of vesicles for PEG-induced fusion was shown for bovine phosphatidylserine (PS) SUV by dynamic light scattering (Boni *et al.*, 1984*a*). In a low salt medium, fusion did not occur even at 45% PEG 6000, whereas incubation in 100 mM sodium chloride permitted extensive fusion.

The effect of vesicle size on fusion efficiency has been studied by Aldwinckle *et al.* (1982). Fusion of MLV of PC/phosphatidic acid/cholesterol (7 : 2 : 1) to larger vesicles in 30% PEG did occur, but over a much slower (hours) time course than that seen for SUV. A slow lipid exchange in MLV was also noted by MacDonald (1985). This could be due to fusion or as a consequence of structural defects in the membrane. In a freeze-fracture study employing LUV prepared by the ether injection technique (Boni *et al.*, 1984*a*), incubation in 45% PEG 6000 resulted in a shift in the

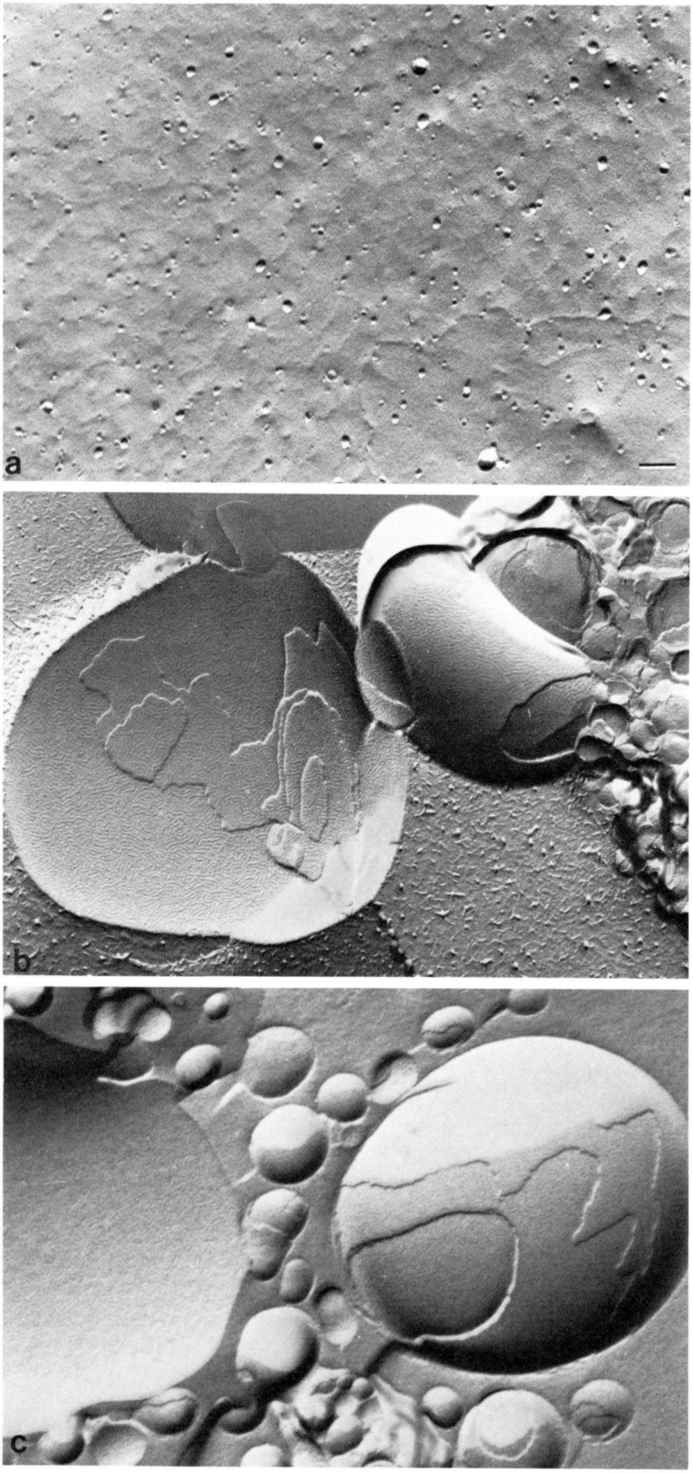
a
b
c

size distribution of the LUV from a mode diameter of 120–200 nm and median diameter of 140–220 nm. Thus, fusion does occur between larger vesicles of low curvature but not as efficiently as for SUV.

3.5. Leakage and Lytic Effect

Fusion of two vesicular systems requires both the fusion of their membranes and the intermixing of their internal contents. This should ideally occur with minimal leakage of their contents into the external medium.

The leakage of $^{45}Ca^{2+}$ into erythrocyte ghosts after a 1-min incubation with PEG 1500 and 6000 rapidly increased above 30% PEG (Blow *et al.*, 1978). This increase in permeability coincided with an increase in fusion efficiency. The increased fusion efficiency can be attributed to calcium-induced changes in the cytoplasmic side of the plasma membrane and/or to a calcium-dependent degradation of membrane proteins. PEG-in-

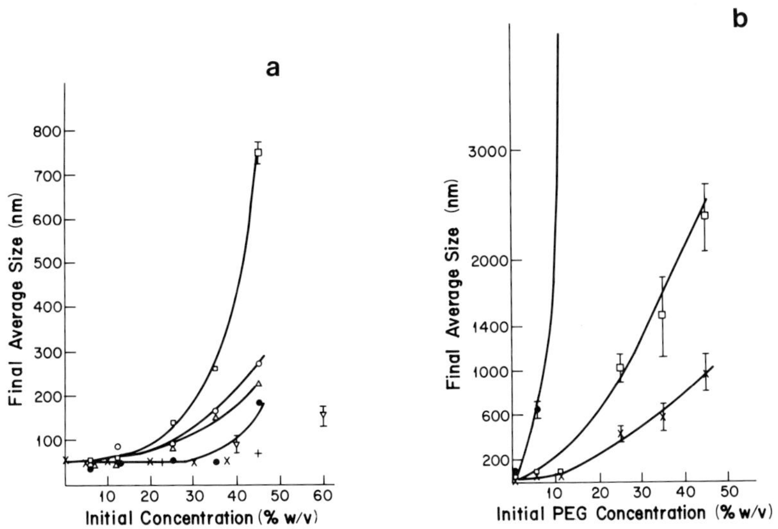

Figure 5. (a) Fusion induced by poly(ethylene glycol) and analogues, as shown by dynamic light scattering after dilution of 10 mM egg PC SUV from the given percentage (w/v) of PEG 6000 (□), PEG 1000 (○), PEG 600 (△), PEG 200 (●), ethylene glycol (+), dextran 200,000–300,000 (×), and glycerol (▽). No fusion was observed for up to 60% sucrose. All data were recorded at 20°C. (b) Dynamic light scattering showing the temperature dependence on fusion for 10 mM DMPC SUV at 10 (●), 20 (□), and 38 (×)°C following dilution from the given initial percentage PEG 6000. (From Boni *et al.*, 1984*a*.)

Figure 4. Freeze-fracture electron micrographs of egg PC at various stages of PEG-induced fusion: (a) SUV before the addition of PEG; (b) after mixing in 50% PEG 6000 for 1 min; and (c) after dilution and removal of PEG. Bar: 0.1 μm. (From Boni *et al.*, 1981*a*.)

duced leakage of vesicular internal contents to the external aqueous medium was observed for egg PC SUV that contained the water-soluble probe 6-carboxyfluorescein at self-quenching concentrations (Saez *et al.*, 1982). An increase of fluorescence, indicating leakage of probe, occurred at concentrations as low as 10% PEG 1000. Leakage from SUV was much more pronounced than from MLV. It was suggested that the fusion process in SUV is by a mechanism of vesicle lysis and reassembly.

Liposomal permeability to calcium, shown in Figure 6a, increased with time and concentration of PEG 6000. A large increase occurred at 20–25% (Aldwinkle *et al.*, 1982). Leakage diminished after increasing the cholesterol content of the liposomes. It was proposed that the leakage was a result of local discontinuities in bilayers rather than partial or total lysis of the membrane because the activation energy for loss was too high to be attributed to a lytic effect. The simultaneous release of isotopically labeled PEG 4000 and Arsenazo III from MLV with varying PEG 6000 concentrations is depicted in Figure 6b. Not only does this indicate the existence of a sufficiently large rupture in the bilayer for passage of a macromolecule, but it also shows that at a concentration of 20% PEG 6000 or higher, PEG can leak into vesicles. Leakage of low-molecular-weight compounds and ions into erythrocytes has been noticed by Knutton and Pasternak (1979).

Thus, a transient bilayer instability can be created. The increase in permeability may enable molecules from the external medium to enter cells and possibly enhance the fusion efficiency. Loss of internal contents does occur, but it can be minimized by a short incubation in PEG.

3.6. Formation of Defects

PEG has been shown to interact directly with pure lipid species and cause them to fuse. PEG also alters properties of MLV. The effect of PEG on pure lipid dispersions may shed light on the nature of any bilayer instability that may render the membrane less resistant to fusion.

DSC studies of Tilcock and Fisher (1979) reveal both a broadening and an upward shift of the main gel to liquid-crystalline transition (T_C) for DPPC dispersions in PEG from molecular weight 400 to 6000. These effects were more pronounced for the higher-molecular-weight PEG. In addition, a loss of the pretransition for DPPC occurred at as low as 18% w/w PEG 6000. An upward shift and broadening of the T_C are characteristic of dehydrated lipid. The concurrent changes in both enthalpy and entropy of the transition above 30% PEG are consequences of a decrease in the cooperativity of the transition—an indication of packing distortions in the bilayer. Similar DSC effects have been observed for egg PC, DMPC, soy PE, and bovine PS dispersions (Boni *et al.*, 1981*b;* Boni *et al.*, 1984*b*). The single peak for the DSC endotherms implies that the dehydration effect applies to all layers of the MLV equally. A penetration of the polymer into the bilayer is also possible. Aldwinckle *et al.* (1982) noted the release of encapsulated [^{14}C]-labeled PEG 4000 from vesicles incubated with PEG 6000 (Fig. 6b).

X-ray diffraction has identified the effect of dehydration on the lamellar repeat spacings. DMPC dispersions exhibited a reduction in spacings from 6.4 to 5.1 nm when incubated in 50% PEG 6000 at 40°C (Boni *et al.,* 1984*b*). Similar PEG-induced decreases in lamellar repeat spacings have been observed for dispersions of egg PC, bovine PS, soy PE, and egg PC/cholesterol (Le Neveu *et al.,* 1977; Boni *et al.,* 1981*b*; Boni *et al.,* 1984*b*) also observed a decrease in the repeat spacing for egg PC MLV suspended in dextran.

Various structural defects have been observed in lipid dispersions in 50% w/w PEG 6000 by freeze-fracture electron microscopy (Boni *et al.,* 1981*b,* 1984*b*). The wormlike texture seen in Figure 7a (and Fig. 4b) has been noted for egg PC dispersions that were 10–20% hydrated (Ranck *et al.,* 1974; Kleman *et al.,* 1977; Costello and Gulik-Krzywicki, 1976).

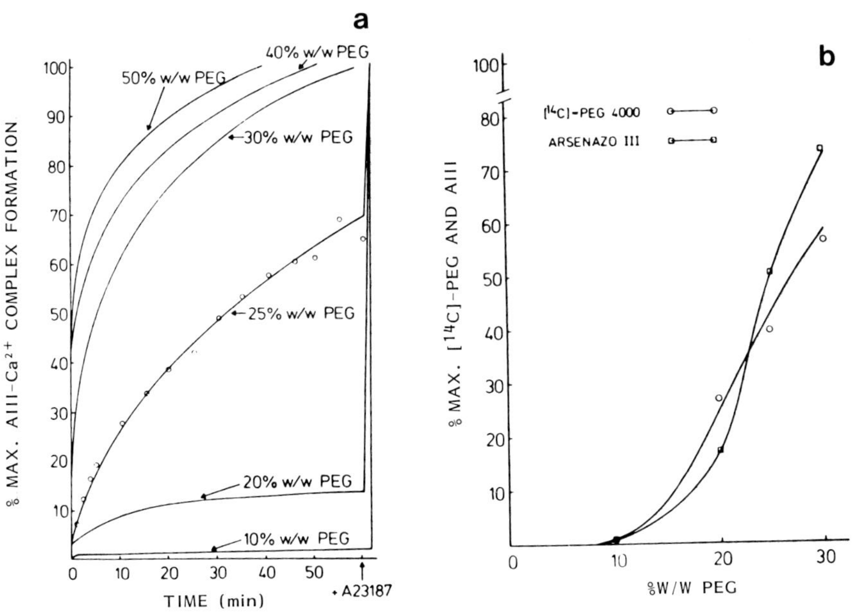

Figure 6. (a) Formation with time of the Arsenazo III–Ca^{2+} complex when multilamellar vesicles, encapsulated with Arsenazo III (phosphatidylcholine/phosphatidic acid/cholesterol, 7 : 2 : 1), were incubated at 25°C in solutions of poly(ethylene glycol) 6000 containing Ca^{2+}. (b) A comparison of the simultaneous extraliposomal release of isotopically labeled poly(ethylene glycol) 4000 and of Arsenazo III from multilamellar vesicles that were incubated with poly(ethylene glycol) 6000. (From Aldwinckle *et al.,* 1982.)

Scalloped deformations with protrusions along the edges, possible sites of interlamellar contact and exposure of hydrophilic portions of the bilayer, can also be seen in Figure 7a. Evenly spaced rows and pits visible in Figure 7b for DMPC MLV have also been observed for egg PC and bovine PS dispersions above their T_C (Boni *et al.*, 1981*b*, 1984*b*). Similar defects have been reported by Kleman *et al.* (1977) for egg PC MLV partially hydrated (16% water). These phenomena can be attributed to a relaxation strain between large confocal domains. As defined by de Gennes (1974), confocal domains are deformations in liquid crystals (smectic mesophases) that are adjustments to uneven surface conditions, such as varying degrees of local hydration. The stacked layers bend and slip over each other to keep the interbilayer thickness constant, resulting in regions of high curvature. The core or dimple of these domains connects distant lamellae. These pits and protrusions are not seen in DMPC and bovine PS MLV below their phase transitions (Boni *et al.*, 1984*b*). These systems may be too rigid to bend, or other microscopic defects may exist to accommodate the external strain.

Dispersions of bovine PS typically form loosely packed MLV in aqueous buffer and give a diffuse X-ray pattern (of a 7.3-nm lamellar spacing). When dispersions of bovine PS are incubated in 50% PEG 6000, two superimposed sharp diffraction patterns are seen. These correspond to a 6.4-nm spacing, whose pattern became less intense with increasing temperature, and a 5.35-nm spacing, which is consistent with PS MLV being maximally dehydrated (Papahadjopoulos *et al.*, 1978). The two different spacings indicate the coexistence of two discrete domains. Thus, PEG reduces the negative interbilayer charge repulsion in PS MLV, a necessary action for the promotion of cell fusion.

Screw dislocations, or links between adjacent bilayers, are shown in Figure 7c. These connections between hydrophobic regions of neighboring lamellae allow for the rapid diffusion of lipids and could explain the rapid PEG-induced exchange between MLV observed by Tilcock and Fisher (1982).

When egg PC MLV were incubated in 50% PEG 600 and 1000, the reduction in the lamellar repeat distance was less than that induced by 50% PEG 6000 (Boni, 1983). Confocal domains were present but less frequent than seen for egg PC MLV in 50% PEG 6000.

Egg PC dispersions were examined in other dehydrating agents by X-

Figure 7. Freeze-fracture electron micrograph of (a) egg PC MLV in 50% PEG 6000 at 20°C exhibiting a wormlike texture (*) and scalloped deformations (arrows); (b) DMPC MLV in 50% PEG 6000 quenched from 40°C revealing rows of peaks (arrows); (c) Soy PE dispersion in 50% PEG 6000 quenched from 24°C; arrow indicates screw dislocations. Bar: 0.1 μm. (From Boni *et al.*, 1981*b*, 1984*b*.)

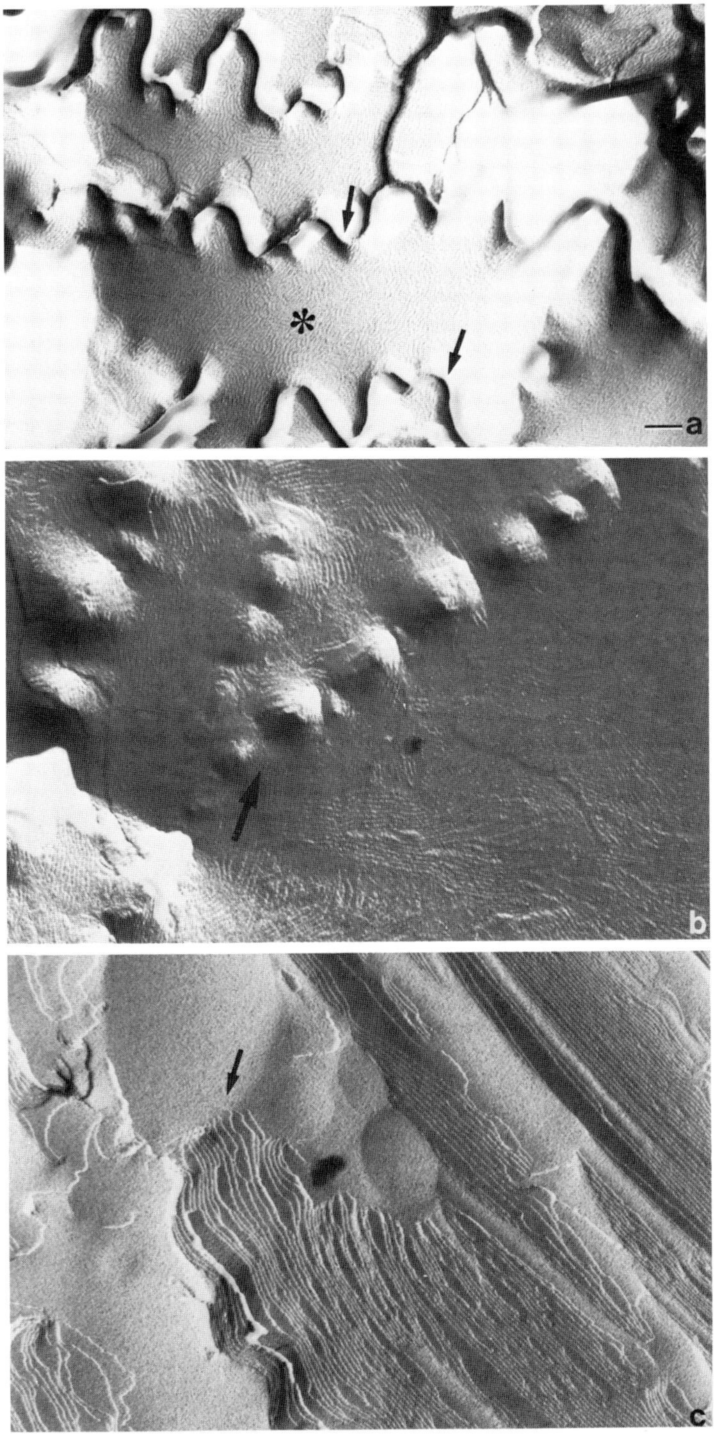
a
b
c

ray diffraction and freeze-fracture electron microscopy (Boni, 1983). The reduction of lamellar repeat spacings in 40% dextran 200,000, 25% polyvinyl alcohol 16,000, and 40% polyacrylic acid 5000 was comparable to that found for 50% PEG 6000. Freeze-fracture electron microscopy revealed bumpy surface textures for these agents, possibly entrapped pockets of water, but membrane defects found for 50% PEG 6000 were absent. The background for all the analogues was either rough without a distinct texture or filled with isolated spheres on the order of 10 nm (Boni, 1983). The background texture of 50% PEG 6000 has an interwoven threadlike appearance (Fig. 4b) and may represent the tightly held PEG–water network proposed by Bailey and Koleske (1967) and Atha and Ingham (1981).

Two points are clear. First, DSC endotherms, freeze-fracture electron microscopy, and X-ray diffraction all show that PEG has a dehydrating effect on lipid dispersions. The decrease in lamellar spacings indicates a reduction in the repulsion between bilayers. Second, the presence of particular bilayer defects facilitates the exposure of hydrophobic portions of the bilayer. These defects may be formed between neighboring membranes and thus provide the next stage of the fusion process following the aggregation, or bilayer contact stage—the actual breakdown of a discrete bilayer structure. All bilayer defects described in this section can be reversed by PEG removal through dilution, and membrane integrity is restored. The sites of PEG-created defects between apposed bilayers are then sealed together and the fusion event is completed.

4. EFFECTS ON RECONSTITUTED SYSTEMS AND CELL MEMBRANES

There are two major differences between PEG-induced fusion of lipid vesicles and that of cell membranes. First, the concentrations of PEG used for the fusion of biological membranes are usually in the 35–50% range (Davidson and Gerald, 1977), much higher than that required to fuse pure lipid bilayers. Second, lipid vesicles fuse in the presence of PEG, while the fusion of biological membranes or cells requires exposure to PEG followed by an incubation period in the absence of PEG. There must be additional factors in biological membranes that resist the PEG-induced fusion. Several studies have identified these factors. A logical approach is to study the fusion process in reconstituted or simplified membranes. The erythrocyte ghost membrane and its derivatives offer good models for the study of the PEG-induced fusion process in biological membranes.

4.1. Redistribution of Intramembrane Particles

Freeze-fracture electron microscopy studies have shown that PEG treatment results in the creation of large areas free of IMP in the plasma membrane. Such areas persist even after the membrane has been washed free of PEG. The creation of IMP-free areas seems to be common to all PEG-induced membrane fusion processes. Roos *et al.* (1983) found that cells resistant to PEG-induced fusion also lack the ability to form IMP-free patches on their plasma membranes. If the IMP-free areas are indeed protein-free lipid bilayers, one function of PEG must be the clearing away of proteins to facilitate bilayer–bilayer contact. IMP-free regions have been noted where apposed cell membranes are in tight contact (Knutton, 1979). Induction of the formation of defects in these lipid-rich regions allows for fusion to occur.

In order to verify that the creation of PEG-induced IMP-free areas on the membrane is a physicochemical process and not a cellular process dependent on other cellular components, such as cytoskeleton, Hui *et al.* (1985) treated dilute suspensions of proteoliposomes reconstituted with egg PC and glycophorin, an erythrocyte glycoprotein that spans the bilayer. At low PEG 6000 concentrations (less than 12.5%), the IMP representing glycophorin in the proteoliposomes aggregated and gave rise to IMP-free areas without visible evidence of membrane–membrane contact. At higher vesicle or PEG concentrations, the vesicles were attached at the IMP-free regions. These IMP-free areas were also observed in the inner layers of multilamellar vesicles (Fig. 8a) at a PEG concentration below that expected for leakage through lamellae (Fig. 6b). Thus, IMP-free regions apparently form without the need for direct PEG-membrane contact. Similar vesicle behavior was observed when cytoskeleton-free membrane vesicles budded from erythrocytes were used. Figure 8b,c,d shows IMP patches in these vesicles in 6% PEG. No contact between vesicles is observed. Patching disappears if the PEG is removed by dilution. If the PEG concentration reaches 35%, however, IMP-free patches remain after PEG dilution. Since IMP-free areas are created prior to membrane contact, the theory of electrostatic displacement of IMP upon close apposition of cells proposed by Knutton and Pasternack (1979) is not applicable.

Is the ability of forming IMP-free areas unique to PEG or is it a general protein precipitation phenomenon due to progressive dehydration of the environment? An experiment using erythrocytes was performed using dextran (500,000-M_r) and glycerol at concentrations that were of equivalent or greater osmolarity than 35% PEG. Neither dextran

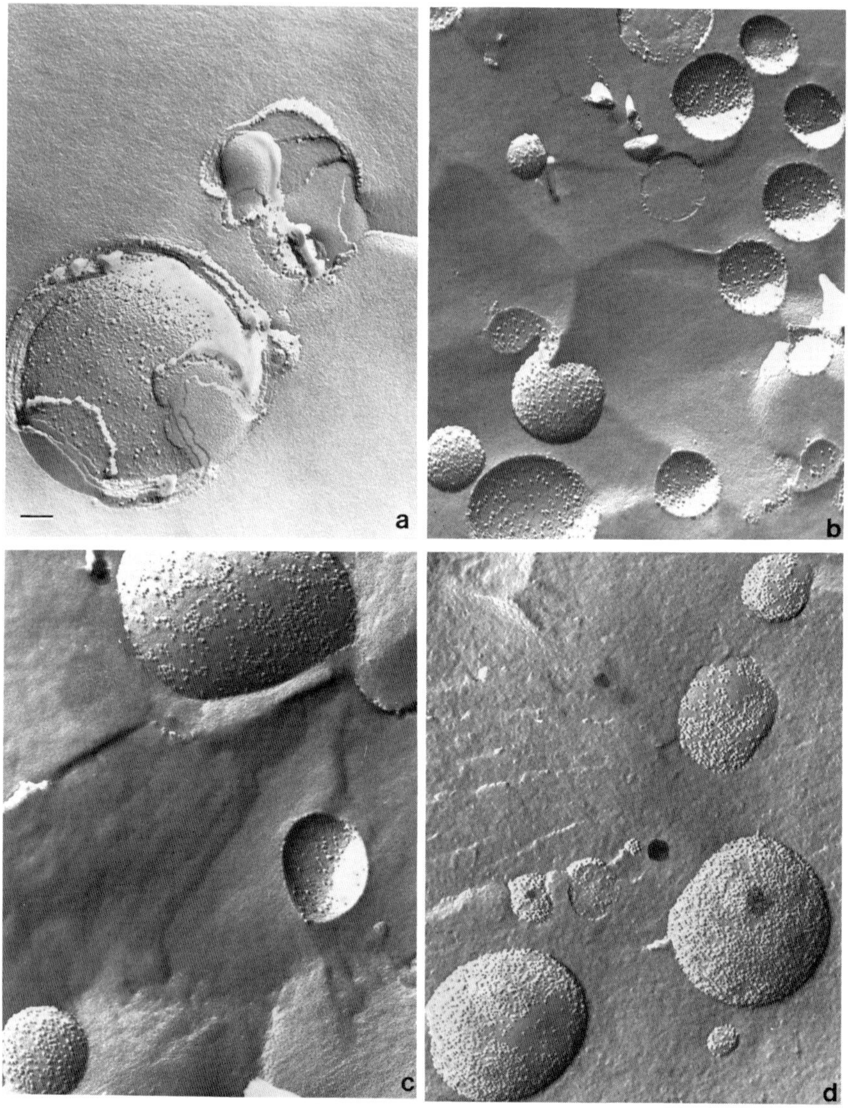

Figure 8. Freeze-fracture electron micrographs of (a) egg PC/glycophorin-reconstituted vesicles in 12% PEG 6000; (b) cytoskeleton-free erythrocyte vesicles; (c) cytoskeleton-free vesicles in 6% PEG 6000; and (d) cytoskeleton-free vesicles diluted from 35 to 2% PEG 6000. All samples were frozen from 20°C. Bar: 0.1 μm. (From Hui *et al.,* 1985.)

nor glycerol caused the extensive IMP-free regions that are created by PEG (Hui *et al.,* 1985). Thus, the ability of PEG to create IMP-free areas on the membrane surface is not based solely on its ability to dehydrate the membrane. Other properties, such as the change in polarity of the medium, are important. A more hydrophobic solvent increases the sensitivity of proteins to electrolytes and can thus induce their coagulation (Glasstone and Lewis, 1960).

The nucleation of IMP-free areas may be created by PEG–lipid binding, PEG-protein interactions, or PEG displacing or structuring the water at the membrane surface. ESR experiments using spin-labeled proteins yielded an irreversible immobilization in erythrocyte ghosts due to an increase in the external osmotic pressure (d'Avila Nunes, 1981). Similar results have been reported as a result of PEG treatment (Herrmann *et al.,* 1983). The deformation and rearrangement of membrane components due to osmotic pressure were related to protein conformational changes and structural dissociation from the membrane lipids (d'Avila Nunes, 1981). Conformational changes may be due to the different solubilities of the various amino acids in PEG (Ingham, 1977). Differential effects on membrane components by PEG may cause proteins to cluster to more favorably accommodate them in the membrane (Israelachvili, 1977). This would alter the protein–lipid interaction. The resultant protein-free areas then expand upon membrane contact.

It appears that PEG-induced fusion of biomembranes is a two-step process. First IMP-free areas of "bare" lipid bilayers are exposed and then the lipid bilayers are attached and fused in the PEG environment. For erythrocytes, the first step requires a higher percentage of PEG (35%), while the latter needs 25% PEG. It has been shown that pure lipid vesicles can be fused at 25% PEG 6000 (Boni *et al.,* 1981*a,* 1984*a*). If treatments can induce the formation of IMP-free patches before or during the PEG treatment, the threshold PEG concentration should then be lowered considerably. This is indeed the case. Table I indicates that when human erythrocytes were subjected to treatments of sublytic solutions of DMSO, lysolecithin, polylysine, spermidine, trichloro-acetic acid, and ethanol, some of these treatments led to the formation of IMP-free regions on the plasma membranes either prior to or during the PEG co-treatment (Huang and Hui, 1986). Those agents that caused the formation of IMP-free patches also led to cell fusion. The fusion efficiency, however, was different for each agent. Those agents that did not affect the distribution of IMP failed to induce cell fusion in 25% PEG. This simple correlation indicates that the availability of IMP-free areas is an indispensable step in PEG-induced fusion.

The need for IMP-free regions for PEG-induced fusion is consistent with the observed enhancement of the PEG fusion efficiency by a pretreatment of cells with a protease preparation (Hartmann *et al.*, 1976). It is possible, however, that fusogenic peptides are formed by the protease treatment and they are responsible for the increased fusion efficiency. Indeed, protein-denuded areas have also been noted in cell fusion induced by calcium phosphate (Zakai *et al.*, 1977), Sendai virus (Bachi *et al.*, 1977), the divalent cation ionophore A23187 (Vos *et al.*, 1976), uranyl acetate (Majumdar *et al.*, 1980), and electrical breakdown (Stenger and Hui, 1986). Although the aggregation of IMP does not necessarily result in fusion, as observed by Schneider *et al.* (1982) in cholesterol-enriched inner mitochondrial membranes, it often precedes fusion.

Lectin treatment of certain cell membranes can lead to IMP patching and formation of IMP-free bubbles (Huang and Hui, 1986). It is expected that this treatment may facilitate PEG-induced cell fusion (Szoka *et al.*, 1981). When turkey erythrocytes were treated with concanavalin A or wheat germ agglutinin, they fused even at PEG concentrations as low as 5% (Huang and Hui, 1986). Without the lectin treatments, turkey erythrocytes cannot be induced to fuse by PEG, even up to lytic concentrations (Arnold and Lucy, 1983).

4.2. Osmotic Swelling

Another difference between fusion of biological membranes and fusion between pure lipid bilayers is that bilayer vesicles fuse in the presence of PEG while biological membranes fuse only after the treated membranes are washed free of PEG and incubated in a PEG-free medium (Robinson *et al.*, 1979). Although lipid exchange may take place during PEG treatment, the mixing of the bulk membrane components occurs only after the removal of PEG (Wojcieszyn *et al.*, 1983). This observation raises the question: What is the function of the washing and incubation step in the fusion of biological membranes?

Unsealed (leaky) human erythrocyte ghosts cannot be fused by the typical PEG treatment protocols. When resealed human erythrocyte ghosts were stored in an isotonic medium (such as balanced salt solution), treated in PEG, and then washed and incubated in the same isotonic solution, no fusion was observed (Hui *et al.*, 1985). The fusion efficiency was much improved if the samples were incubated in a hypotonic solution following the PEG treatment. Increasing the osmolarity of the incubation medium reduced the PEG fusion efficiency. Similar observations were reported for fusion induced by Sendai virus (Knutton and Bachi, 1980). This corroborates the observation by Zimmerberg *et al.* (1980) that fusion between lipid vesicles and a planar bilayer is facilitated by an osmotic gradient. It seems that an additional "push" by the osmotic pressure is necessary to break the fusion barrier between biological membranes.

Table I. Combined Effects of PEG and Other Membrane Active Agents on Intact Human Erythrocytes[a]

Chemicals	Concn. used	IMP aggregation[b] Without PEG	IMP aggregation[b] In PEG	Fusion (%)
DMSO	2% (v/v)	−	−	0
Lysolecithin	0.15 mM	−	−	0
Polylysine	0.05% (w/v)	−	−	0
Spermidine	0.05% (w/v)	−	++	29.4
Trichloroacetic acid	100 mM	−	++	6.6
Ethanol	4% (v/v)	+	++	11.9

[a]Cells were washed in Hanks' balanced salt solution (BSS), incubated at 20°C in 25% PEG 6000 in BSS for 5 min then washed and incubated at 37°C for one hour. Fusion percentages were measured by dark-field light microscopy. The degree of IMP aggregation was assessed by rapid freeze-fracture electron microscopy.
[b]The degree of IMP aggregation is indicated by the number of plus (+) signs. A minus (−) sign indicates no aggregation.

What is the nature of the additional fusion barrier in biological membranes? The results of a fusion experiment between erythrocytes and fibroblasts indicate that an intact cytoskeleton may be the structural force impeding the completion to fusion while still in PEG (Wojcieszyn *et al.*, 1983). When erythrocyte vesicles were incubated in isotonic or hypotonic media following PEG treatment, the response to osmotic-assisted fusion was the same regardless of whether the vesicles contained cytoskeletal elements (Hui *et al.,* 1985). Possibly the cytoskeletal network was not well preserved in the vesicles even though they still contained cytoskeletal proteins. A possible role of the cytoskeletal network is the preservation of the "random" distribution of the intramembrane proteins, thereby impending the close contact between lipid bilayers. Cytoskeletal elements seem to play a rather passive role in the PEG-induced fusion process and remain intact (Robinson *et al.,* 1979), although one report has shown an increase in microtubules and stress fibers following PEG treatment (Fuseler *et al.,* 1978).

5. CONCLUSION

The ability of PEG to bind and structure water has two membrane destabilizing effects. PEG dehydrates membranous systems by competing for free water and it changes the dielectric properties of water. Thus, PEG facilitates the "solvating" of lipid molecules and leads to the creation of bilayer defects. Dehydration alone is not sufficient to induce fusion of membranes as indicated by the inability of dextrans and other dehydrating agents to fuse vesicles or cells. The large PEG polymer does not sterically interfere with the fusion process. PEG differs from other dehydrating agents by its abilities to structure water, bind to phospholipids, form salt complexes, aggregate membrane proteins, and/or even by possibly possessing a different exchange rate between free and bound water.

The mode of action of PEG may be postulated based on studies on model lipid systems and cellular systems and the various parameters that either enhance or inhibit the fusion event. Close approach of cells or vesicles occurs due to exclusion of the cells or vesicles from the PEG–water network. Local dehydration, the binding of PEG to the bilayer, and the change in ionic environment may lead to a clustering of membrane proteins. Membrane proteins (IMP) that inhibit close approach (Maroudas, 1975) are forced into clusters, leaving bare patches of high lipid content. Local dehydration and the changing of the dielectric constant of water allow for defects in bilayer packing, permitting the exposure of the hydrocarbon chains. This can be facilitated by the formation of domains of different phase and composition in the bilayer.

The decrease in the electrostatic repulsion between bilayers due to the decrease in polarity of the medium would facilitate the contact and fusion of the bilayers and/or the rapid transfer of membrane components between bilayers. Cytoplasmic streaming would increase the probability of contact junctions between membranes, but it would also limit the lifetime of these contacts (Lucy, 1970). This, along with the weak van der Waals attractive forces, make these contacts unstable.

The final step of removing the PEG by dilution is necessary to "seal" bilayer defects at contact points and for the expansion of the contact area via cell or vesicle swelling (Knutton and Bachi, 1980; Pinto da Silva, 1980; Hui *et al.,* 1985). In addition, osmotic shock may facilitate further defect formation. Too high a percentage of PEG, on the other hand, would deplete cells of water content and lead to leakage of components and be detrimental to cell viability. Cells with extensive contact points would remain fused and continue on to complete cell fusion, defined by cytoplasmic mixing.

REFERENCES

Ahkong, Q. F., Fisher, D., Tampion, W., and Lucy, J. A., 1975, Mechanisms of cell fusion, *Nature (Lond.)* **253:**194–195.

Aldwinckle, T. J., Ahkong, Q. F., Bangham, A. D., Fisher, D., and Lucy, J. A., 1982, Effects of poly(ethylene glycol) on liposomes and erythrocytes. Permeability changes and membrane fusion, *Biochim. Biophys. Acta* **689:**548–560.

Applegate, H. G., 1960, Freezing-point depressions of Hoagland's 'carbowax' systems, *Nature (Lond.)* **186**:232–233.

Arnold, K., and Lucy, J. A., 1983, *Cell Fusion,* Ciba Foundation Symposium 103, Pitman, London.

Arnold, K., Herrmann, A., Pratsch, L., and Gawrisch, K., 1985, The dielectric properties of aqueous solutions of poly(ethylene glycol) and their influence on membrane structure, *Biochim. Biophys. Acta* **815**:515–518.

Arnold, K., Pratsch, L., and Grawrisch, K., 1983, Effect of poly(ethylene glycol) on phospholipid hydration and polarity of the external phase, *Biochim. Biophys. Acta* **728**:121–128.

Atha, D. H., and Ingham, K. C., 1981, Mechanism of precipitation of proteins by polyethylene glycols, *Biol. Chem.* **256**:12108–12117.

Bachi, T., Deas, J. E., and Howe, C., 1977, Virus-erythrocyte membrane interaction in: *Cell Surface Reviews,* Vol. 2 (G. Poste and G. L. Nicolson, eds.), pp. 83–127, North-Holland, Amsterdam.

Bailey, F. E., and Koleske, J. V., 1967, Configuration and hydrodynamic properties of the polyethylene chain in solution, in: *Nonionic Surfactants* (M. J. Schick, ed.), pp. 794–822, Marcel Dekker, New York.

Baran, A. A., Solomentseva, I. M., Mank, V. V., and Kurilenko, O. D., 1972, Role of the solvation factor in stabilizing disperse systems containing water-soluble polymers, *Dokl. Akad. Nauk SSSR* **207**:363–366.

Blow, A. M. J., Botham, G. M., Fisher, D., Goodall, A. H., Tilcock, C. P. S., and Lucy, J. A., 1978, Water and calcium ions in cell fusion induced by poly(ethylene glycol), *FEBS Lett.* **94**:305–310.

Boni, L. T., 1983, Ph.D. thesis, State University of New York at Buffalo.

Boni, L. T., Hah, J. S., Hui, S. W., Mukherjee, P., Ho, J. T., and Jung, C. Y., 1984*a*, Aggregation of fusion of unilamellar vesicles by poly(ethylene glycol), *Biochim. Biophys. Acta* **775**:409–418.

Boni, L. T., Stewart, T. P., Alderfer, J. L., and Hui, S. W., 1981*b*, Lipid-polyethylene glycol interactions II. Formation of defects in bilayers, *J. Membrane Biol.* **62**:71–77.

Boni, L. T., Stewart, T. P., Alderfer, J. L., and Hui, S. W., 1981*a*, Lipid–polyethylene glycol interactions. I. Induction of fusion between liposomes, *J. Membrane Biol.* **62**:65–70.

Boni, L. T., Stewart, T. P., and Hui, S. W., 1984*b*, Alterations in phospholipid polymorphism by polyethylene glycol, *Membrane Biol.* **80**:91–104.

Boss, W. F., and Mott, R. L., 1980, Effects of divalent cations and polyethylene glycol on the membrane fluidity of protoplast, *Plant Physiol.* **66**:835–837.

Cabane, B., 1977, Structure of some polymer-detergent aggregates in water, *J. Phys. Chem.* **81**:1639–1645.

Costello, M. J., and Gulik-Krzywicki, T., 1976, Correlated x-ray diffraction and freeze-fracture studies on membrane model systems. Perturbations induced by freeze-fracture preparative procedures, *Biochim. Biophys. Acta* **455**:412–432.

Cowley, A. C., Fuller, N. L., Rand, R. P., and Parsegian, V. A., 1978, Measurement of repulsive forces between charged phospholipid bilayers, *Biochemistry* **17**:3163–3168.

d'Avila Nunes, M., 1981, A spin label study of erythrocyte membranes during simulation of freezing, *J. Membrane Biol.* **60**:155–162.

Davidson, R. L., and Gerald, P. S., 1977, Mammalian somatic cell hybridization by polyethylene glycol, in: *Methods in Cell Biology* (D. Prescott, ed.), p. 325, Academic Press, London.

De Gennes, P. G., 1974, *The Physics of Liquid Crystals,* Clarendon Press, Oxford.

Eriksson, E., and Albertsson, P. A., 1978, The effect of the lipid composition on the partition of liposomes in aqueous two-phase systems, *Biochim. Biophys. Acta* **507**:425–432.

Fazekas De St. Groth, S., and Scheidegger, D., 1980, Production of monoclonal antibodies: Strategy and tactics, *J. Immunol. Methods* **35**:1–21.

Fuseler, J. W., Miller, C. L., and Fuller, G. M., 1978, Alteration of cytoskeletal morphologies and growth patterns in human fibroblasts treated with PEG, *Cytobiologie* **18**:345–359.

Galfre, G., Howe, S. C., Milstein, C., Butcher, G. W., and Howard, J. C., 1977, Antibodies to major histocompatibility antigens produced by hybrid cell lines, *Nature (Lond.)* **266**:550–552.

Glasstone, S., and Lewis, D., 1960, Surface chemistry and colloids, in: *Elements of Physical Chemistry,* pp. 559–599, D. Van Nostrand, New York.

Hafeti, Y., and Hanstein, W. G., 1974, Destabilization of membranes with chaotropic ions, *Methods Enzymol.* **31**:770–79.

Hammes, G. G., and Roberts, P. B., 1968, Cooperativity of solvent–macromolecule interactions in aqueous solutions of polyethylene glycol and polyethylene glycol–urea, *J. Am. Chem. Soc.* **90**:7119–7122.

Hammes, G. G., and Schimmel, P. R., 1967, An investigation of water–urea and water–urea–polyethylene glycol interactions, *J. Am. Chem. Soc.* **89**:442–446.

Hartmann, J. X., Galla, J. D., Emma, D. A., Kao, K. N., and Ganborg, O. L., 1976, The fusion of erythrocytes by treatment with proteolytic enzymes and polyethylene glycol, *Can. J. Genet. Cytol.* **18**:503–512.

Herrmann, A., Pratsch, L., Arnold, K., and Lassmann, G., Effect of poly(ethylene glycol) on the polarity of aqueous solutions and on the structure of vesicle membranes, 1983, *Biochim. Biophys. Acta* **733**:87–94.

Honda, K., Maeda, Y., Sasakawa, S., Ohno, H., and Tsuchida, E., 1981, The components contained in polyethylene glycol of commercial grade (PEG-6,000) as cell fusogen, *Biochem. Biophys. Res. Commun.* **101**:165–171.

Horwitz, A. F., Michaelson, D., and Klein, M. P., 1973, Magnetic resonance studies on membrane and model membrane systems. III. Fatty acid motions in aqueous lecithin dispersions, *Biochim. Biophys. Acta* **298**:1–7.

Huang, C., and Hui, S. W., 1986, Chemical co-treatments and IMP patching in PEG-induced fusion of turkey and human erythrocytes, *Biochim. Biophys. Acta.,* in press.

Hui, S. W., and Isac, T., Boni, L. T., and Sen, A., 1985, Action of polyethylene glycol on the fusion of human erythrocyte membranes, *Membrane Biol.* **84**:137–146.

Hui, S. W., Stewart, T. P., Boni, L. T., and Yeagle, P. L., 1981, Membrane fusion through point defects in bilayers, *Science* **212**:921–923.

Ingham, K. C., 1977, Polyethylene glycol in aqueous solution: solvent perturbation and gel filtration studies, *Arch. Biochem. Biophys.* **184**:59–68.

Israelachvili, J. N., 1977, Refinement of the fluid–mosaic model of membrane structure, *Biochim. Biophys. Acta* **469**:211–225.

Jain, M. K., and White, H. B., 1978, Long-range order in biomembranes, *Adv. Lipid Res.* **15**:1–60.

Johansson, G., 1970, Partition of salts and their effects on partition of proteins in a dextran–poly(ethylene glycol)–water two-phase system, *Biochim. Biophys. Acta* **221**:387–390.

Kao, K. N., and Michayluk, M. R., 1974, A method for high-frequency intergeneric fusion of plant protoplasts, *Planta* **115**:355–367.

Kleman, M., Williams, C. E., Costello, M. J., and Gulik-Krzywicki, T., 1977, Defect structures in lyotropic smectic phases revealed by freeze-fracture electron microscopy, *Philo. Mag.* **35**:33–56.

Knutton, S., 1979, Studies of membrane fusion. III. Fusion of erythrocytes with polyethylene glycol, *J. Cell Sci.* **36**:61–72.

Knutton, S., and Bachi, T., 1980, The role of cell swelling and haemolysis in Sendai virus-induced cell fusion and the diffusion of incorporated viral antigens, *J. Cell Sci.* **42**:153–167.

Knutton, S., and Pasternak, C., A., 1979, The mechanism of cell-cell fusion, *Trends Biochem. Sci.* **4**:220–223.

Krahling, H., 1981, Discrimination between two fusogenic properties of aqueous polyethylene glycol solutions, *Z. Naturforsch.* **36**:593–596.

Lagerwerff, J. V., Ogata, G., and Eagle, H. E., 1961, Control of osmotic pressure of culture solutions with polyethylene glycol, *Science* **133**:1486–1487.

Larrabee, A. L., 1979, Time-dependent changes in the size distribution of distearoylphosphatidylcholine vesicles, *Am. Chem. Soc.* **18**:3321–3326.

Lee, A. G., 1977, Lipid phase transitions and phase diagrams. I. Lipid phase transitions, *Biochim. Biophys. Acta* **472**:237–281.

LeNeveu, D. M., Rand, R. P., Parsegian, V. A., and Gingell, D., 1977, Measurement and modification of forces between lecithin bilayers, *Biophys. J.* **18**:209–230.

Lucy, J. A., 1970, The fusion of biological membranes, *Nature (Lond.)* **227**:815–817.

MacDonald, R. I., 1985, Membrane fusion due to dehydration by polyethylene glycol, dextran, or sucrose, *Biochem. J.* **24**:4058–4066.

Maggio, B., Ahkong, Q. F., and Lucy, J. A., 1976, Poly(ethylene glycol), surface potential and cell fusion, *Biochem. J.* **158**:647–650.

Maggio, B., and Lucy, J. A., 1978, Interactions of water-soluble fusogens with phospholipids in monolayers, *FEBS Lett.* **94**:301–304.

Majumdar, S., Baker, R. F., and Kalfra, V. K., 1980, Fusion of human erythrocytes induced by uranyl acetate and rare earth metals, *Biochim. Biophys. Acta* **598**:411–416.

Maroudas, N. G., 1975, Polymer exclusion, cell adhesion and membrane fusion, *Nature (Lond.)* **254**:695.

Maul, G. G., Steplewski, Z., Weibel, J., and Koprowski, H., 1976, Time sequence and morphological evaluations of cells fused by polyethylene glycol 6000, *In Vitro* **12**:787–796.

McCammon, J. R., and Fan, V. S. C., Release of membrane constituents following polyethylene glycol treatment of HEp-2 cells, 1979, *Biochim. Biophys. Acta* **551**:67–73.

McIver, D. J. L., 1979, Control of membrane fusion by interfacial water: A model for actions of divalent cation, *Physiol. Chem. Phys.* **11**:289–302.

Minetti, M., Aducci, P., and Viti, V., 1979, Interaction of neutral polysaccharides with phosphatidylcholine multilamellar liposomes. Phase transitions studied by the binding of fluorescein-conjugated dextrans, *Biochemistry* **18**:2541–2548.

Morgan, C. G., Thomas, E. W., and Yianni, Y. P., 1983, The use of fluorescence energy transfer to distinguish between poly(ethylene glycol)-induced aggregation and fusion of phospholipid vesicles, *Biochim. Biophys. Acta* **728**:356–362.

Ohno, H., Maeda, Y., and Tsuchida, E., 1981a, ^{1}H-NMR study of the effect of synthetic polymers on the fluidity, transition temperature and fusion of dipalmitoyl phosphatidylcholine small vesicles, *Biochim. Biophys. Acta* **642**:27–36.

Ohno, H., Sakai, T., Tsuchida, E., Honda, K., and Sasakawa, S., 1981*b*, The interaction of human erythrocyte ghost or liposomes with polyethylene glycol detected by fluorescence polarization, *Biochem. Biophys. Res. Commun.* **102**:426–431.

Oi, V. T., and Herzenberg, L. A., 1980, Immunoglobulin-producing hybrid cell lines, in: *Selected Methods in Cellular Immunology* (B. B. Mishell and S. M. Shigii, eds.), pp. 351–372, W. H. Freeman, San Francisco.

Papahadjopoulos, D., Portis, A., and Pangborn, W., 1978, Calcium induced lipid phase transitions and membrane fusion, *Ann. N. Y. Acad. Sci.* **308**:50–66.

Papahadjopoulos, D., Poste, G., Schaeffler, B. E., and Vail, W. J., 1974 Membrane fusion and molecular segregation in phospholipid vesicles, *Biochim. Biophys. Acta* **352**:10–28.

Pinto da Silva, P., Kazufumi, S., and Parkinson, C., 1980, Fusion of human erythrocytes by Sendi virus: Freeze-fracture aspects, *J. Cell Sci.* **43**:419–432.

Poste, G., and Allison, A. C., 1973, Membrane fusion, *Biochim. Biophys. Acta* **300**:421–465.

Power, J. B., Evans, P. K., and Cocking, E. C., 1978, Fusion of plant protoplasts, in: *Cell Surface Reviews,* Vol. 5 (G. Poste and G. L. Nicolson, eds.), pp. 369–385, North-Holland, Amsterdam.

Ranck, J. L., Mateu, L., Sadler, D. M., Tardieu, A., Gulik-Krzywicki, T., and Luzzati, V., 1974, Order-disorder conformational transitions of the hydrocarbon chains of lipids, *J. Mol. Biol.* **85:**249–277.

Robinson, J. M., Roos, D. S., Davidson, R. L., and Karnovsky, M. J., 1979, Membrane alterations and other morphological features associated with polyethylene glycol-induced cell fusion, *J. Cell. Sci.* **40:**63–75.

Roos, D. S., Robinson, J. M., and Davidson, R. L., 1983, Cell fusion and intramembrane particle distribution in polyethlyene glycol-resistant cells, *J. Cell Biol.* **97:**909–917.

Saez, R., Alonso, A., Villena, A., and Goni, F. M., 1982, Detergent-like properties of polyethyleneglycols in relation to model membranes, *FEBS Lett.* **137:**323–326.

Schneider, H., Hochli, M., and Hackenbrock, C. R., 1982, Relationship between the density distribution of intramembrane particles and electron transfer in the mitochondrial inner membrane as revealed by cholesterol incorporation, *J. Cell Biol.* **94:**387–393.

Singer, S. J., and Nicolson, G. L., 1972, The fluid mosaic model of the structure of cell membranes, *Science* **175:**720–730.

Smith, C. L., Ahkong, Q. F., Fisher, D., and Lucy, J. A., 1982, Is purified poly(ethylene glycol) able to induce cell fusion?, *Biochim. Biophys. Acta* **692:**109–114.

Stenger, D. A., and Hui, S. W., 1986, Kinetics of ultrastructural changes during electrically-induced cell fusion of erythrocytes, *J. Membrane Biol.,* in press.

Struck, D. K,. Hoekstra, D., and Pagano, R. E., 1981, Use of resonance energy transfer to monitor membrane fusion, *Am. Chem. Soc.* **20:**4093–4099.

Szoka, F., Magnusson, K. E., Wojcieszyn, J., Hou, Y., Derzko, Z., and Jacobson, K., 1981, Use of lectins and polyethylene glycol for fusion of glycolipid containing liposomes with eukaryotic cells, *Proc. Natl. Acad. Sci. U.S.A.* **78:**1685–1689.

Tilcock, C. P. S., and Fisher, D., 1979, Interaction of phospholipid membranes with poly(ethylene glycol)s, *Biochim. Biophys. Acta* **577:**53–61.

Tilcock, C. P. S., and Fisher, D., 1982, The interaction of phospholipid membranes with poly(ethylene glycol) vesicle aggregation and lipid exchange, *Biochim. Biophys. Acta* **688:**645–652.

Vos, J., Ahkong, Q. F., Botham, G. M., Quirk, S. J., and Lucy, J. A., 1976, Changes in the distribution of intramembranous particles in hen erythrocytes during cell fusion induced by the bivalent-cation ionophore A23187, *Biochem. J.* **158:**651–653.

Wojcieszyn, J. W., Schlegel, R. A., Lumley-Sapanski, K., and Jacobson, K. A., 1983, Studies on the mechanism of polytheylene glycol-mediated cell fusion using fluorescent membrane and cytoplasmic probes, *J. Cell Biol.* **96:**151–159.

Yeagle, P. L., Hutton, W. C., Hunag, C. H., and Martin, R. B., 1975, Headgroup formation and lipid-cholesterol association in phosphatidylcholine vesicles: a $^{31}P[^{1}H]$ nuclear Overhauser effect, *Proc. Natl. Acad. Sci. U.S.A.* **72:**3477–3481.

Zakai, N., Kulka, R. G., and Loyter, A., 1977, Membrane ultrastructural changes during calcium phosphate-induced fusion of human erythrocyte ghosts, *Proc. Natl. Acad. Sci. U.S.A.* **74:**2417–2421.

Zimmerberg, J., Cohen, F. S., and Finkelstein, A., 1980, Micromolar Ca^{2+} stimulates fusion of lipid vesicles with planar bilayers containing a calcium-binding protein, *Science* **210:**906–908.

Chapter 15

Physicochemical Factors Underlying Lipid Membrane Fusion

Shinpei Ohki

1. INTRODUCTION

Membrane fusion is one of the essential events involved in many biological cellular processes (Poste and Allison, 1973; Rasmussen, 1970). In spite of extensive research on membrane fusion, its molecular mechanism is not well understood. This is probably due to the many possible molecular pathways for various membrane fusion events and also because of the complex nature of biological membranes (Poste and Nicolson, 1978).

In order to gain insight into the mechanism involved in such biological membrane fusion events, membrane fusion studies have utilized model lipid membranes with defined molecular components and simpler molecular structures. Proposed mechanisms for membrane fusion observed in such membrane systems so far include the following:

1. Membrane micellization by incorporating short chain fatty acids into membranes. Short-chain fatty acids may produce micelle type structures within the membrane that are thought to be the initia-

SHINPEI OHKI • Department of Biophysical Sciences, School of Medicine, State University of New York at Buffalo, Buffalo, New York 14214.

tion sites for membrane fusion of the two closely apposed membranes (Lucy, 1970).

2. Membrane instability occurring possibly at the phase boundary of gel–liquid crystalline phases induced by divalent cations (Papahadjopoulos *et al.,* 1977). These phase boundaries serve as the initiation sites of membrane fusion.
3. Membrane expansion due to increased temperature or osmotic pressure gradients across membranes (Breisblatt and Ohki, 1975; Miller *et al.,* 1976; Cohen *et al.,* 1980; Ohki, 1984).
4. Molecular phase defects in membranes produced by lowering temperature (Schullery *et al.,* 1980) or caused by mixed molecular species in membrane (Hui *et al.,* 1981).
5. The inverted micelle structure within membranes as an intermediate structure for membrane fusion (Verkleij *et al.,* 1980).
6. Surface energy (surface tension) increase in the two apposed membrane surfaces as a result of divalent or polyvalent cation binding (Ohki, 1982, 1984).
7. Dehydration of water from membrane surfaces by large organic molecules (e.g., polyethyleneglycol) (Herrmann *et al.,* 1983; Knutton, 1979).
8. Electric field induced membrane fusion, where membrane fusion is induced by destabilizing membrane molecules in close contact by applying an electric breakdown potential across the two membranes (Zimmermann, 1982).

Although the physical basis of molecular mechanisms for these fusion events has not been established, this author recently proposed a unified theory of membrane fusion. In this theory the sufficiently increased surface free energy or increased hydrophobicity of the membrane surface is a main driving force of membrane fusion (Ohki, 1982).

This chapter describes experiments on cation-induced phosphatidylserine vesicle fusion and certain associated membrane properties (interfacial tension, vesicle electrophoretic mobility, and vesicle aggregation). An attempt is made to give a physical basis for a general theory of membrane fusion in terms of molecular interaction energy. It was found experimentally that the degree of membrane fusion correlated with the degree of increase in the interfacial tension of the membrane. For a theoretical study, the interaction energy for two closely apposed membranes was calculated in terms of molecular interaction energy as well as the surface hydration energy.

It was found that the dehydration energy of the membrane, the magnitude of which depends on the nature of surface hydrophilic layers, plays a decisive role in membrane fusion. On the other hand, the van der

Waals interaction energy between the two membranes at a given separation distance, which depends on the nature of the surface hydrophilic layer as well as its thickness, contributes to the membrane fusion forces as a secondary factor. When the surface hydrophilic layer becomes sufficiently similar in nature to the bulk hydrocarbon phase of the membrane (i.e., the interfacial tension of the membrane surface increases), the dehydration energy decreases and the attractive interaction energy increases. Consequently the membrane in close contact could fuse and become one. Physical properties of the membrane, such as interfacial tension, which is related to membrane fusion, is discussed in relation to the proposed theory, and a molecular mechanism of phospholipid membrane fusion is proposed.

2. EXPERIMENTS: MATERIALS AND METHODS

2.1. Chemicals

Phosphatidylserine was purchased from Avanti Biochemical Co. (Birmingham, Al). Some of the purchased lipids were further purified through a DEAE (diethylaminoethyl) cellulose column by a slightly modified method of Rouser *et al.* (1961). Both samples showed a single spot on silica gel thin-layer chromatographic plates. Each lipid sample gave identical experimental results within experimental error.

NaCl was of ultrapure grade, purchased from Alfa Chemical Co. Before use, the NaCl was heated to 400°–500° for 2 hr to eliminate possible organic contaminants. $TbCl_3$•6 H_20 (99.9% pure) and dipicolinic acid (pyridine-2,6-dicarboxylic acid) were obtained from Alfa and Sigma Chemical Companies, respectively. Buffers used were a mixture of L-histidine (Calbiochem, A grade) and TES [N-Tris(hydroxymethyl)methyl-2-aminoethanesulfonic acid] (Calbiochem, Ultrol grade). Hexane was used as the lipid solvent and was obtained from Fluka (Purum Grade) and further purified through activated alumina and silica gel (Fisher Scientific Co.) columns to remove possible contaminating surfactants. Spermine, spermidine, and putrescine (as the hydrochloride) were obtained from the Sigma Chemical Co. $LaCl_3$ (gold grade) was obtained from Aldrich Chemical Co. Other chemicals used in the experiments were of reagent grade and were obtained from Fischer Chemical Co. Some solutions contained a small amount of EDTA (0.01 mM) to remove possible divalent and polyvalent cation contaminants in the experimental solutions. The water used was distilled three times, including an alkaline permanganate process.

2.2. Vesicle Preparation

Small unilamellar phosphatidylserine vesicles were prepared in either (1) 100 mM NaCl, (2) 10 mM $TbCl_3$ and 100 mM sodium citrate, or (3) 100 mM dipicolinic acid (DPA) (sodium salt) solution. All solutions contained 2 mM L-histidine and 2 mM TES as buffer. The pH of these salt solutions was adjusted to 7.4 with NaOH. The phosphatidylserine was added to each one of the above solutions at a concentration of 10μmole lipid/ml, then vortexed for 10 min, sonicated for 1 hr, and centrifuged at 100,000 × g as described in an earlier paper (Düzgüneş and Ohki, 1977). The yield of small (~300 Å diameter) unilamellar vesicles was about 95% of the total lipid used for preparation. Vesicles prepared in solutions (2) and (3) were separated from nonencapsulated materials using a Sephadex G-75 column according to the method of Wilschut *et al.*, (1980). The elution buffer (0.1 M NaCl/2 mM histidine/2 mM TES) contained 0.05 mM EDTA except in the La^{3+} case. Multilamellar vesicles were prepared by dispersing phosphatidylserine in the NaCl buffer solution at the same lipid concentration as that of the above small unilamellar vesicles and by vortexing the suspension for 15 min.

2.3. Assay of Vesicle Fusion

The fusion of small unilamellar lipid vesicles (SUV) induced by divalent and polyvalent cations was followed by a Tb/DPA assay, monitoring the fluorescent intensity changes (SLM-8000 Fluorometer, SLM Instruments, Ann Arbor, MI) due to mixing of the internal aqueous contents of unilamellar vesicles (Fig. 1). The details are described in an earlier paper (Ohki, 1982). The temperature of the sample solution was main-

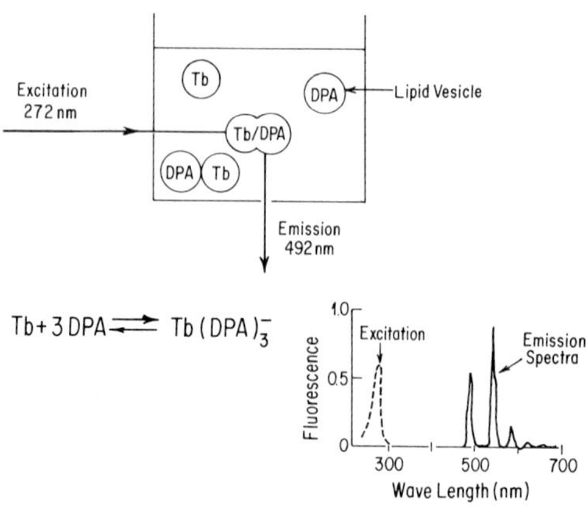

Figure 1. Schematic diagram of the Tb/DPA fusion assay. Two kinds of phospholipid vesicles (0.1 μmoles lipid each) which are encapsulated with 10 mM Tb/100 mM sodium citrate and 100 mM dipicolinic acid (DPA), respectively, were suspended in 2 ml of 0.1 M NaCl/2 mM histidine/2 mM Tes/ 0.01 mM EDTA (0 mM EDTA for La^{3+})/pH 7.4. Fusion was followed by measuring the enhanced intensity of Tb at 492 nm, upon mixing of the internal aqueous contents and thus forming the Tb/DPA complex. Fluorescence intensity was measured with an SLM-8000 spectrofluorimeter.

tained at 23°C. After $TbCl_3$ and DPA-encapsulated vesicles of an equimolar amount (0.1 μmoles phospholipid each for small vesicles and 0.05 μmoles each for larger vesicles) were suspended in 2 ml of 0.1 M NaCl buffer solution in quartz cuvettes, concentrated salt solutions of various cations were injected into the vesicle suspension in small increments, and the solution was shaken well to give a homogeneous mixture. It took about 10 sec for the divalent ion concentration in the experimental solution to change. The fluorescence intensity was measured about 15 sec after each change in ion concentration. The excitation wavelength was 275 nm and the emission fluoresence was measured at 492 nm. The contribution of scattered light to the signal was subtracted by performing the same experiment without Tb. The value for 100% fusion was determined in the presence of 50 μm dipicolinic acid by release of the contents from Tb-containing vesicles (the same amounts as in the experiment, except free from EDTA during the Sephadex G-75 column chromatography) in the EDTA-free NaCl buffer solution with 0.5% (wt/wt) sodium cholate.

2.4. Surface Tension Measurements

Lipid monolayers were prepared by placing an aliquot of the lipid spreading solution (approx. 1 mM lipid in Hexane) by means of a microsyringe on an aqueous surface of a constant area (64 cm^2 in a glass dish) (Davies and Rideal, 1962). Aqueous subphase solutions were 0.1 M NaCl/1 mM histidine/1 mM TES/0.01 mM EDTA pH 7.4. The surface tension of a monolayer-coated aqueous interface (i.e., the interfacial tension of a monolayer) was measured with an electronic balance (Beckman, LM 500), using either a microscopic cover glass (18 × 18 × 0.2 mm) or a thin Teflon plate (11 × 11 × 1 mm) as a Wilhelmy plate (accuracy of ±0.2 dynes/cm). The detailed procedures for measuring the surface tension were described previously (Ohki, 1982). The experiments were designed to measure the change in the surface tension of lipid monolayers as a function of various metal cation or polyamine concentrations in the subphase solution. The metal cation or polyamine concentrations in the subphase solution were altered successively by injecting a small amount of concentrated metal salt solution or concentrated polyamine solutions.

2.5. Measurements of Lipid Vesicle Aggregation

For vesicle aggregation, the turbidity of phospholipid vesicle suspensions was measured (at 400 nm with a Hitachi 100-60 spectrophotometer) as a function of the concentration of various types of polyamines and metal cations. The vesicles were suspended at a concentration of 0.2 μmole lipid/ml of 0.1 M NaCl buffer solution. In the case of $LaCl_3$, the suspended lipid vesicle concentration was 0.02 μmole lipid/ml. The metal cation or polyamine concentration was raised in increments. The absorbance was recorded 2 min after changing the cation or polyamine concentration. When the observed absorbance was plotted against cation or polyamine concentration, a maximum increase in the rate of absorbance change was obtained at a certain concentration of cation or polyamine. This was defined as the threshold concentration of cation or polyamine to induce lipid vesicle aggregation.

The electrophoretic mobility of the lipid vesicles was measured with a microelectrophoresis apparatus (Mark II, Rank Bros., Bottisham, Cambridge). Great care was taken to focus on the stationary layer. The observed mobility, u, and ζ-potential are related by the following Helmholtz-Smoluchowski equation:

$$u = \zeta\, \varepsilon_r \varepsilon_o / \eta$$

where η is the viscosity of the aqueous phase and ε_r, ε_o are the permittivities of the aqueous medium and a vacuum, respectively.

All experiments mentioned above were done at 23 ± 1°C.

Threshold cation concentration for vesicle fusion: For polyvalent cation-induced vesicle fusion, the fluorescence intensity measured 15 sec after each change of cation concentration was plotted as a function of polyvalent cation concentration. The concentration at which the fluorescence intensity–ion concentration curve gave the sharpest increase, was determined. This was defined as the *threshold concentration* of cation-induced vesicle fusion. A reduction in fluorescence intensity with time was observed at cation concentrations above the threshold value, but this was not observed below the threshold concentration of cations.

For H^+-induced vesicle fusion, the turbidity was measured with respect to various pH values of 0.1 M NaCl solution. Vesicles aggregated at a certain pH were redispersed by raising pH of solution to pH 7.4. The residual amount of turbidity of the suspension after the redispersion was considered due to vesicle fusion and plotted with respect to pH of the solution. The sharp increase in the residual turbidity at a given pH, defined the threshold of H^+ concentration that induced vesicle fusion. Sodium citrate was used as a buffer, and the pH of the solution was adjusted with either HCl or NaOH to desired value. The pH was checked before and after the experiments.

3. EXPERIMENTAL RESULTS

3.1. Vesicle Fusion

Figure 2 shows the changes in fluorescence intensity of phosphatidylserine vesicle suspensions in 0.1 M NaCl with respect to various poly-

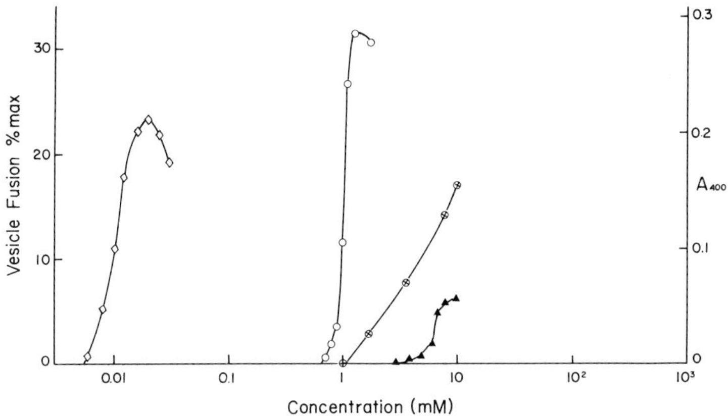

Figure 2. Fusion of phosphatidylserine vesicles in 0.1 M NaCl with various polyvalent cation concentrations. The left ordinate refers to the fluorescence intensity at 492 nm and the abscissa refers to polyvalent cation concentrations. The fluorescence intensity was corrected for the background scattered light by subtracting the scattered light intensity from the total light intensity at 492 nm. The intensity is expressed in percentage of the maximum fusion. The wavelength of excitation light was 272 nm. H^+-induced fusion was determined by measuring the nonreversal residual turbidity of the vesicle suspension. The right ordinate refers to the turbidity. (◇) La^{3+}; (○) Ca^{2+}; (▲) Mg^{2+}; (⊕) H^+.

valent cation concentrations in which two kinds of phosphatidylserine vesicle encapsulated with $TbCl_3$ or dipicolinic acid were suspended. All the experimental NaCl solutions contained 2 mM histidine, 2 mM TES, and 0.01 mM EDTA (0 mM EDTA in the La^{3+} case). The pH of the solution was adjusted with NaOH to 7.4. The change in fluorescence intensity represents the degree of fusion of the two kinds of vesicles (Wilschut and Papahadjopoulos, 1979; Ohki, 1982). The threshold concentrations of cation that induced vesicle membrane fusion were 0.7 mM for Ba^{2+}, 1.0 mM for Ca^{2+}, 1.4 mM for Sr^{2+}, 6 mM for Mg^{2+}, and 20 μM for La^{3+}. Data for Ba^{2+} and Sr^{2+} are not shown in Figure 2. The threshold values obtained for Ca^{2+} and Mg^{2+} in this experiment correspond to those reported earlier (Wilschut *et al.,* 1980, 1981). The threshold concentration for La^{3+} is much smaller than that (0.1 mM) observed by Hammoudah *et al.* (1979); however, they used other techniques to determine membrane fusion. For the La^{3+} case, the concentration of the vesicles in the suspension solution, was one-twentieth of that for other divalent cation cases.

For H^+-induced vesicle fusion, the threshold concentration was about pH 2.5. The concentration corresponding to the threshold was not as narrow a range as those observed for polyvalent cation cases. Other monovalent cations (Na^+, Li^+, and K^+) did not induce fusion of the small unilamellar phosphatidylserine vesicles up to concentration 2 M. It was also found that polyamines such as spermine, spermidine, and putrescine

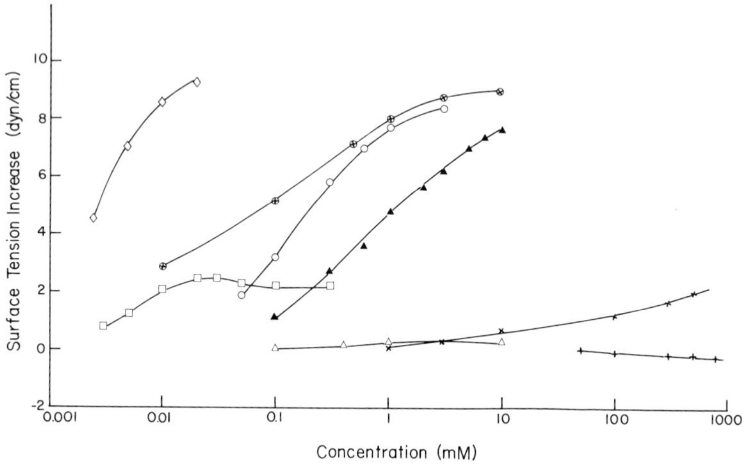

Figure 3. The interfacial tension of a phosphatidylserine monolayer (70 Å^2 per molecule) formed at the air–water (0.1 M NaCl/1 mM histidine/1 mM Tes/0.01 mM EDTA (0 mM EDTA for La^{3+})/pH 7.4) interface with respect to various metal cation and polyamine concentrations. (◇) La^{3+}; (○) Ca^{2+}; (▲) Mg^{2+}; (⊕) H^+; (×) Li^+; (+) Na^+; (□) spermine^{4+}; (△) spermidine^{3+}.

did not cause fusion of the phosphatidylserine vesicles. Similar observations were made by Schuber *et al.* (1983).

3.2. Surface Tension

Figure 3 shows the surface tension increase in a phosphatidylserine monolayer (70 Å^2 per molecule) as a function of various cation and polyamine concentrations. Trivalent cation, La^{3+}, and divalent cations (Ca^{2+}, Mn^{2+}, Ba^{2+}, Sr^{2+}, and Mg^{2+}) caused large changes in the surface tension of the phosphatidylserine monolayer in the 5–20 μM and the 0.5–10 mM ranges, respectively. Data for Mg^{2+}, Ba^{2+}, and Sr^{2+} are not shown in Figure 3. Also, H^+ resulted in gradual but large changes in the surface tension of the monolayer in the range of pH 5–2. The surface tension increases showed saturation at about 10 dynes for all cases. On the other hand, Na^+, Li^+, and K^+ did not cause such increases in surface tension of the phosphatidylserine monolayer up to 1.0 M.

Spermine up to 25 μM caused a monotonic increase in the surface tension of the phosphatidylserine monolayer. The surface tension increase was the greatest (2.5 dynes/cm) at about 25 μM of spermine; at concentrations above 35 μM, the surface tension increase was slightly less than that at 25 μM spermine as the concentration increased. However, other polyamines (spermidine and putrescine) did not cause any significant change in surface tension of the phosphatidylserine monolayer at any concentration (Fig. 3).

For the case of spermidine, the change in surface tension of the monolayer with respect to the change in spermidine concentration was somewhat similar to those of the spermine case, but the magnitude of surface tension increase was much smaller and the concentrations of spermidine needed to cause such surface tension changes were much greater than those of spermine.

Putrescine did not cause any appreciable change in surface tension of the phosphatidylserine monolayer up to its concentration of 0.2 M.

3.3. Vesicle Electrophoretic Mobility

Figure 4 shows the electrophoretic mobility of multilamellar phosphatidylserine vesicles (~5 μm in diameter) with respect to various cation and polyamine concentrations. Vesicle suspension solutions were all 0.1 M NaCl/2 mM His/2 mM Tes/0.01 mM EDTA, pH 7.4. Even at a very low concentration of spermine, the ζ potential of the vesicle was greatly reduced from that in the absence of spermine (Fig. 4). This suggests that the quarternary valent positively charged spermine binds strongly to the phosphatidylserine membrane surface and reduced its net negative charge. The spermine concentration causing the same reduction in the ζ

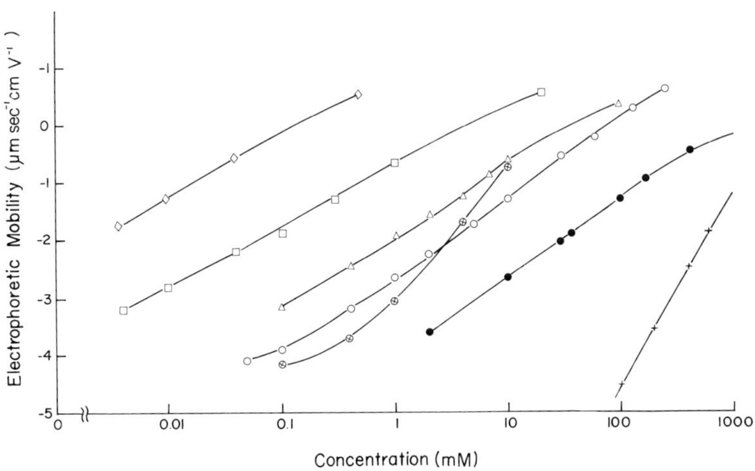

Figure 4. Electrophoretic mobility of multilamellar phosphatidylserine vesicles suspended in 0.1 M NaCl/2 mM histidine/2 mM TES/0.01 mM EDTA (0 mM EDTA for the La^{3+} case)/ pH 7.4 as a function of various metal cation and polyamine concentrations. (◇) La^{3+}; (○) Ca^{2+}; (⊕) H^+; (+) Na^+; (□) spermine^{4+}; (△) spermidine $^{3+}$; (●) putrescine^{2+}.

potential as that caused by the presence of 1 mM Ca^{2+} in the 0.1 M NaCl buffer solution was about 50 μM.

The relationship of the change in vesicle mobility with respect to the logarithm of the spermine concentration was almost linear up to 0.5 mM of the spermine concentration. At the spermine concentrations above 1 mM, the rate of reduction in the vesicle mobility became less and the relationship between the change in vesicle mobility and the logarithm of spermine concentration became nonlinear. At about 6 mM of spermine, the vesicle showed zero mobility; at the concentration above 6 mM, the vesicle mobility became positive in sign. This shows that the positive charges due to adsorption of polyamines overcome the original negative surface charges of the phosphatidylserine membrane. Results similar to those mentioned above with spermine have been obtained by Chung *et al.* (1985). The trend of vesicle mobility change, with respect to spermidine and putrescine concentrations was similar to that for spermine. However, the concentrations of spermidine and putrescine needed to reduce the surface charges of the phosphatidylserine vesicle to the value in the case of 1 mM Ca^{2+} in 0.1 M NaCl were greater than the spermine case: 0.4 mM for spermidine and 10 mM for putrescine, respectively.

For La^{3+}, the slope of the vesicle mobility with respect to the logarithm of its concentration was straight and slightly smaller (1.0 $\mu m \cdot sec^{-1}\ cm \cdot V^{-1}$ per 10-fold change in salt concentration) than that for the Ca^{2+} case (1.5 $\mu m \cdot sec^{-1}$ per $cm \cdot V^{-1}$ per 10-fold change in salt concentration). In both cases, there were the reversal points of the vesicle mobilities: 150 μM for La^{3+} and 90 mM for Ca^{2+}. At concentrations above these points, the vesicle mobility changed in sign. For Na^+, Li^+, and H^+, the relationships between the vesicle mobility and the logarithm of their concentrations were also linear and the slope of the relationship was larger (~3.0–3.5 $\mu m \cdot sec^{-1}\ cm \cdot V^{-1}$ per 10-fold change in salt concentration) than the others, but the mobility did not alter the sign even at high Na^+ or Li^+ concentration (3 M) or H^+ (0.1 M).

3.4. Vesicle Aggregation

Cation- or polyamine-dependent phosphatidylserine vesicle aggregation was followed by measuring the turbidity of lipid vesicle suspensions. Turbidities of small unilamellar phosphatidylserine vesicles with respect to various cation (La^{3+}, Ca^{2+}, Na^+, Li^+, H^+) or polyamine (spermine, spermidine, and putrescine) concentrations are shown in Figure 5. The sharp increase in turbidity of phosphatidylserine vesicles suspended in 0.1 M NaCl occurred at about 8 μM La^{3+}, 90 μM spermine, 1 mM Ca^{2+}, 3 mM spermidine, 3 mM H^+, 0.1 M putrescine, 0.7 M Na^+, and 0.8 M Li^+. Data for Li^+ is not shown in Figure 5. Hong *et al.* (1983) obtained a similar result

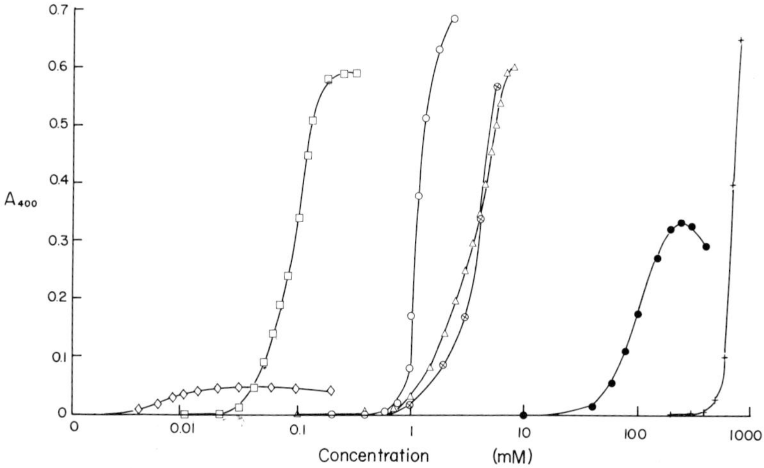

Figure 5. Turbidity of small unilamellar phosphatidylserine vesicles (0.2 μmoles lipid/ml of the NaCl buffer) for all cases except the La^{3+} case (0.02 μmoles lipid/ml of the NaCl buffer) suspended in 0.1 M NaCl buffer solution as a function of various metal cation and polyamine concentrations. (◇) La^{3+}; (○) Ca^{2+}; (⊕) H^+; (+) Na^+; (□) spermine^{4+}; (△) spermidine^{3+}; (●) putrescine^{2+}.

with spermine. The concentrations corresponding to these points were somewhat similar to those at which the vesicle mobility was reduced to about -1.5–-2.0 $\mu m \cdot sec^{-1}\ cm \cdot V^{-1}$, although there are slight deviations in these concentration values for the cases of spermidine, putrescine, and Na^+.

The qualitative agreement between the concentrations needed to reduce vesicle electrophoretic mobility to a certain value and to induce vesicle aggregation indicates that the charge densities, due to the adsorption of cations as well as polyamines on to the phosphatidylserine membrane surface, and vesicle aggregation are in fair correlation.

4. THEORY

As a simplest model representing a system of two interacting membranes, we consider two interacting large flat hydrocarbon plates of infinite thickness, which are separated by an aqueous medium of distance R, and these interacting plates possess a thin hydrophilic layer of thickness, *h,* on their surfaces. These surface layers may be different from those of the interior of the flat hydrocarbon plates as well as the aqueous phases in molecular nature and molecular density (Fig. 6).

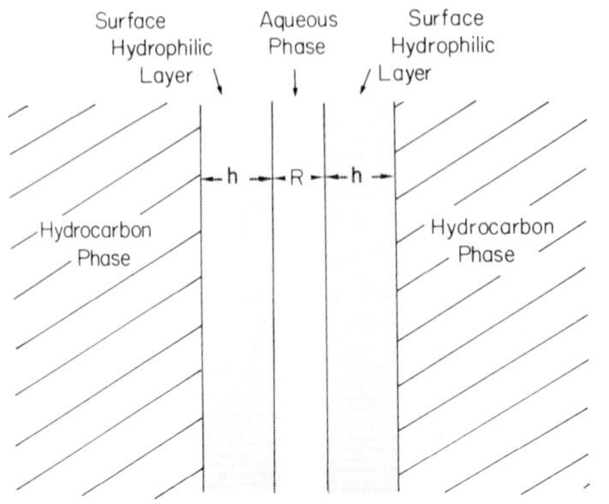

Figure 6. Schematic diagram of two interacting membranes. *R,* Separation distance between the two membranes; *h,* thickness of the surface hydrophilic layers.

The interaction energy, W, of two such flat plates can be expressed by the DLVO (Derjaguin, Landau, Verwey, and Overbeek) theory (Hiemenz, 1977; Verwey and Overbeek, 1948) as the sum of electrostatic repulsive interaction energy V, the van der Waals attractive interaction energy U, and the short-range attractive energy related to hydrogen bond interaction energy U^{SR}:

$$W = V + U + U^{SR} \tag{1}$$

The interaction energy profile expressed by Eq. (1) usually has an energy barrier (the primary maximum) as the distance between the two plates becomes small (see Fig. 7). In addition to the three terms in Eq. (1), a term giving a short-range repulsive force due to the interaction of electron clouds surrounding molecules should be included in the total interaction energy. However, because this term is not important for membrane fusion, it is not expressed in the equation here. Also, in some cases, the repulsive interaction energy term due to the structured water (LeNeveu *et al.,* 1976; Rand, 1981) should be considered in the expression [Eq. (1)]. However, in the present paper, we consider the case in which there is no structured water extended to a long distance from the membrane surface but only a few layers of hydrated water on the membrane surfaces. If the two interfaces can overcome the primary maximum barrier by reducing their surface charges and taking off any existing structured water in the interspace between the two membranes, they will come close to the point of the primary minimum, which we may call the position of *close adhesion.*

However, this close adhesion is not necessarily sufficient for the two interacting plates to fuse and become one. In order for the two plates to fuse and become one, it is necessary that the attractive interaction energy between the two plates become sufficiently larger than a certain value, the energy to dehydrate water from the two-membrane surface.

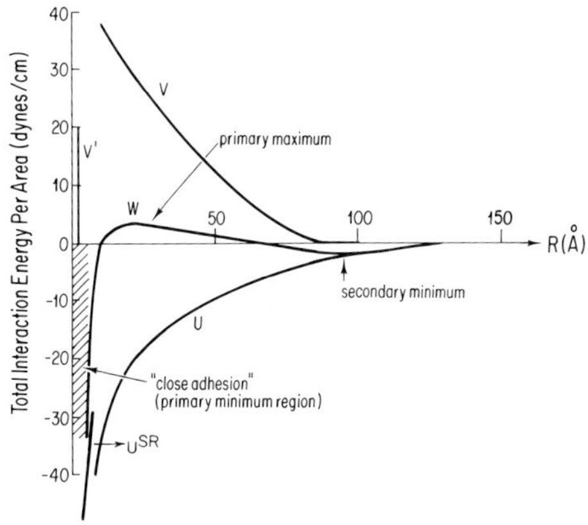

Figure 7. Schematic diagram of the interaction potential energy between the two membranes. *V*, Repulsive interaction energy; *U*, van der Waals attractive interaction energy, and U^{SR}: Short-range attractive energy. $W = V + U + U^{SR}$, total interaction energy, having a primary maximum. *R*, Interlayer separation distance. The scales of the ordinate and the abscissa are rather arbitrary.

When the separation distance is as small as in close adhesion, the electric field due to surface charges between the two layers is constant because of the Gauss theorem (Blum and Jancovici, 1984). Therefore, the electric free energy changes very little with respect to the separation distance, compared with the van der Waals interaction energy. From this we may assume the electrostatic energy term to be a constant contribution to the total interaction energy at the close adhesion separation distance. Also, since the magnitude of surface charge density is reduced enough, the electrostatic energy contribution is small compared to the other attractive energies. Therefore, we consider only the van der Waals and the short-range attractive terms to describe the above fusion, or *molecular cohesion.* The van der Waals interaction energy may consist of these parts: the London-dispersion, Debye, and Keesom terms. However, the latter two terms are, in general, small compared with the first term. Therefore, we consider only the London-dispersion term for van der Waals energy.

The van der Waals interaction energy between the two plates will be expressed as a function of the separation distance R, the thickness of the surface layer *h,* and the Hamaker constant A of two interacting phases [an interaction constant of the two interacting matters given in Eq. (3)] (Vincent, 1973; Vold, 1961):

$$U = -\sum_i \sum_j A_{ij} H(X_{ij}) \tag{2}$$

where A_{ij} is the Hamaker constant of interaction between the *i*th and *j*th phases and $H(X_{ij}) = 1/X_{ij}^2$, where X_{ij} is the distance between the two interfaces of either the surface hydrophilic layer or the bulk hydrocarbon phase—altogether four interfaces). The Hamaker constant (Hamaker, 1937) is approximately expressed in terms of the molecular polarization

α, ionization potential I, and the molecular density ρ, of each phase as follows (London, 1937):

$$A_{ij} = \frac{3\pi^2}{2} \frac{I_i I_j}{I_i + I_j} \alpha_i \alpha_j \rho_i \rho_j \tag{3}$$

Since the ionization potentials (*Handbook of Chemistry and Physics*, 1965) of the hydrocarbon unit C_2H_6 (12.8 eV), and some polar groups such as SO_2 (13.1 eV), and H_2O (12.6 eV) are similar quantities, Eq. (2) can be expressed approximately by

$$U = -\frac{\pi \cdot I}{16} \left\{ \left(\lambda_P - \lambda_W\right)^2 \frac{1}{R^2} + \left(\lambda_P - \lambda_W\right)\left(\lambda_H - \lambda_P\right) \frac{2}{(R+h)^2} + (\lambda_H - \lambda_P)^2 \frac{1}{(R+2h)^2} \right\} \tag{4}$$

Here, $\lambda_i \equiv \alpha_i \rho_i$, which we call the interaction coefficient, and suffixes W, P, and H refer to the aqueous, the polar hydrophilic layer and the hydrocarbon phases, respectively. In Eq. (4), the first term is mainly due to the interaction of the two hydrophilic surface layers separated by an aqueous phase. The third term is mainly due to the interaction between the two hydrocarbon plates (infinite thickness) through the hydrophilic interface as a medium. The second term is the cross-interaction term between the aqueous–hydrophilic interfacial layer interaction and the hydrocarbon body–hydrophilic interface layer interaction.

Since the interaction coefficients of water and hydrocarbon phases are 4.96×10^{-2} ($\alpha_{H_2O} = 1.48$ Å^3, $\rho_{H_2O} = 3.35 \times 10^{22}/cm^3$) and 7.72×10^{-2} ($\alpha_{C_2H_4} = 3.88$ Å^3, $\rho_{C_2H_4} = 1.99 \times 10^{22}/cm^3$), respectively (*Handbook of Chemistry and Physics*, 1965; Ohki and Fukuda, 1968), we can calculate approximately the van der Waals attractive energy between such two interacting membranes for the following three cases in which the thickness of the surface polar group layer of the membrane is 10 Å, 6 Å, and 3 Å, respectively, using the energy expression [Eq. (4)] for the above-mentioned model. Then, the energy can be expressed as a function of the interaction coefficient (λ) of the polar surface layer at a given separation distance (R = 2.5 Å). We consider this separation distance to be the thickness of one layer of water between the two membranes. The short-range interaction energy U^{SR} arising other terms than hydration energy may not be effective at this separation distance (2.5 Å) (C. J. van Oss, private communication). The results are shown as curves a, b, and c, respectively, in Figure 8.

We assume that a certain threshold energy of the attractive interaction must be exceeded for the two membranes to fuse. This energy is

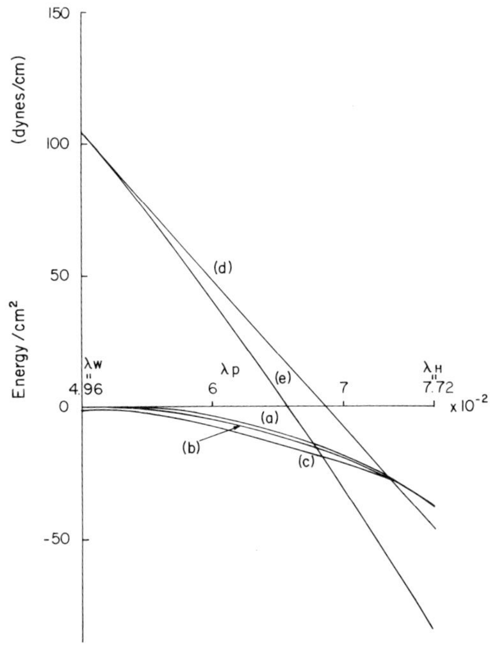

Figure 8. van der Waals interaction energy between the two membranes and the dehydration energy per unit membrane area. The ordinate refers to the calculated values of the van der Waals interaction energy between the two membranes separated by a distance R = 2.5 Å. (a) h = 10 Å. (b) h = 6 Å. (c) h = 3 Å. The abscissa refers to the interaction coefficient (λ_P) of the surface hydrophilic layer. λ_W (4.96×10^{-2}) and λ_H (7.72×10^{-2}) corresponds to those of the aqueous (water) and hydrocarbon phases, respectively. (d) Effective dehydration energy E_{dh}^{eff} to remove the last layer of water from the membrane surfaces. (e) Total energy of the van der Waals and dehydration energies, $U + E_{dh}^{eff}$.

assumed to be equal to the energy needed to dehydrate the last layer of water between the two interacting membranes. We may define such an energy E_{dh} in terms of the interfacial tensions of the membrane surface at two states:

$$E_{dh} = 2(\gamma_{m/A} - \gamma_{m/W}) \tag{5}$$

where $\gamma_{m/A}$ and $\gamma_{m/W}$ are the interfacial tension of the membrane–air and membrane–water phases, respectively (Davies and Rideal, 1961). The interfacial tension is defined as the excess energy of the interfacial region per unit area. If it is a positive value, in general, the interface exists, and if it is negative there will be no stable interface. Since we do not have sufficient information about $\gamma_{m/A}$ and $\gamma_{m/W}$, we assume that the nature of the membrane surface layer is expressed by the sum of water and hydrocarbon portions:

$$\lambda_m = p\lambda_H + q\lambda_W \tag{6}$$

where p and q are the fractions of each portion ($p + q = 1$).

In such a case, we can calculate the dehydration energy of the membrane using the interfacial tensions of the water/air (~72 dynes/cm), water/hydrocarbon (~50 dynes/cm), and hydrocarbon/air (~20 dynes/cm), respectively (*Handbook of Chemistry and Physics,* 1965). The results are shown as curve d in Figure 8. The effective dehydration energy E_{dh}^{eff} = 0.75 E_{dh}.

When the sum of the attractive energy and the dehydration energy ($U + E_{dh}^{eff}$) becomes negative, the two interfaces (membranes) would fuse if no molecular structural constraint exists at the membrane surfaces. From our calculation the interfacial tension of the membrane at this point is about 20 dynes/cm, which corresponds to the point of $\lambda_p = 6.5 \times 10^{-2}$.

From the graphic results of the energy calculation (Fig. 8) it is seen that for membranes having similar molecular components and structures, the two factors (interaction coefficient λ and thickness h of the surface hydrophilic layers) are involved in membrane fusion. The membrane fusion should be facilitated by either increasing the magnitude of the interaction coefficient or reducing the thickness of the surface layers. However, the latter factor is less effective judging from Figure 8 (curves a–c). The important factor is the magnitude of dehydration energy, which is reduced as the hydrophilic layer becomes more hydrophobic (see Fig. 8, curve d).

5. DISCUSSION

Experimental data on the divalent cation-induced fusion of acidic phospholipid membranes and the related physicochemical properties of the membrane indicate that the degree of fusion correlates well with the degree of increased surface tension or increased hydrophobicity of the membrane caused by cations (Ohki, 1977, 1982, 1984). A massive fusion of phosphatidylserine vesicles occurs at 1 mM Ca^{2+} in 100 mM NaCl, where the increase in interfacial tension of the membrane of about 8 dynes/cm from the reference state (0 mM Ca^{2+} and 0.1 mM NaCl). This is consistent with other cation cases: The increase in interfacial tension of the membrane is about 8 dynes/cm at the threshold concentration of each cation that induces the phosphatidylserine vesicle fusion. We may consider that this increase in interfacial tension is solely attributable to the change in free energy of the surface hydrophilic layer. Our recent studies suggest that this energy increase arises mainly from a conformational change in the interfacial polar groups of the membrane molecules upon divalent cation chelation binding, rendering the hydrocarbon portion of the membrane molecule open to the water phase, but not due to the change in the electrostatic energy of the surface due to the reduction of surface charges (Ohki and Ohshima, 1984; Ohshima and Ohki, 1985).

The interfacial tension at the reference state (0.1 M NaCl) for the phosphatidylserine monolayer membrane is approximately a few dynes (~ 2–4) dynes/cm (Ohki, 1977), and the small lipid vesicle (300 Å in diameter) is several (~ 5–7) dynes/cm higher in surface tension due to the expanded area per molecule (~85 Å^2/molecule) of the outer monolayer of the vesicle because of its vesicle curvature and molecular geometry (Ohki, 1982). Therefore, the total surface energy increase at the point corresponding to the threshold (where fusion occurs) for membrane fusion is about 15 dynes/cm. This value is slightly less than that (20–25 dynes) predicted in the present theory. It is not clear whether the short-range attractive force U^{SR} might contribute to the interaction energy or not (van Oss *et al.*, 1986; Pashley *et al.*, 1985).

Similar correlation between the interfacial tension increase and fusion of membranes is obtained for various monovalent cations and polyvalent cations including polyamines (see Table I). Obviously, the monovalent metal cations (except for H^+) do not alter the hydrophilic nature of the surface layer by their bindings, as they neither bind in the chelating manner with lipid polar groups, as divalent cations are known to do, nor dehydrate the hydrophilic polar surface layer as much as many other polyvalent cations and H^+ do (Ohki and Ohshima, 1984). Polyamines used in the experiments also do not cause an increase in interfacial tension of acidic phospholipid membranes. Although a considerable part of a polyamine is hydrophobic, the adsorption of many polyamines may effectively increase the thickness of the hydrophilic layer and therefore may not be effective in causing membrane fusion.

Another example is seen in the case of membrane fusion induced by increasing temperature or an osmotic pressure difference applied across the membrane. It has been shown (Breisblatt and Ohki, 1975; Chaudhury and Ohki, 1981; Ohki, 1984) that when the two phospholipid membranes are in close contact and are expanded by temperature or osmotic pressure change, the two membranes fuse after an increase of about 10 dynes/cm in interfacial tension of the membrane (Ohki, 1984). In this case, the lateral expansion of the membrane keeping its volume constant may result in thinning of the membrane, including the hydrophilic surface layer. This effectively increases the hydrophobicity in the hydrophilic surface layer region because the expansion of the hydrophilic surface layer in area would break up hydrogen bonds formed between the surface hydrophilic polar groups and water molecules. On an average, this would be expected to reveal more of the internal hydrophobic phase to the aqueous phase at the interface. This corresponds to the increase in interfacial tension of the membrane. The decrease in only the surface charge density of membranes

Table I. Properties of Phosphatidylserine Membrane in 0.1 M NaCl

Ions or molecules	Concn.	Increase in $\gamma_{m/W}$ of monolayer membranes (dynes/cm)	SUV fusion	SUV aggregation	Reference
Na^+	$\leqslant$1 M	No change	No	Yes	Ohki (1982)
K^+	$\leqslant$1 M	No change	No	Slight	S. Ohki, unpublished data; Ohki *et al.* (1982)
Li^+	0.8 M	A few dynes	No	Yes	S. Ohki, unpublished data; Ohki *et al.* (1982)
H^+	0.01 M	7–8	Yes	Yes	Ohki and Ohshima (1984)
Ca^{2+}	1.0 mM	7–8	Yes	Yes	Ohki (1982, 1984)
Mg^{2+}	8 mM	6–7	Yes	Yes	Ohki (1982, 1984)
Ba^{2+}	0.8 mM	7–8	Yes	Yes	Ohki (1982)
La^{3+}	20 μM	7–8	Yes	Yes	Ohki and Duax (1985)
Spermine 4+	$\leqslant$1 mM	A few dynes	No	Yes	Ohki and Duax (1985)
Spermidine 3+	$\leqslant$200 mM	No appreciable change	No	Yes	Ohki and Duax (1985)

is not a sufficient condition for two membranes to fuse. However, there is a good correlation between membrane aggregation and the decrease in the surface charge density (when vesicle electrophoretic mobility reaches −2 to −1.5 $\mu m \cdot sec^{-1}$ $cm \cdot V^{-1}$), as shown in Figures 4 and 5.

In divalent cation-induced fusion of acidic lipid membranes, the effective hydrophobicity of the hydrophilic surface layer of the membrane is increased by the chelation binding of divalent cation (Ohki and Ohshima, 1984; Ohshima and Ohki, 1985), while in the increased temperature or osmotic pressure-induced membrane fusion (Breisblatt and Ohki, 1975; Chaudhury and Ohki, 1981; Miller *et al.,* 1976; Ohki, 1984), the effective thickness of the hydrophilic surface layer of the membrane may be reduced by the expansion of the membrane and, consequently, the hydrocarbon phase reveals more to the aqueous phase. Both cases result in increase in the hydrophobic nature of the membrane surface.

The above argument was based on a physical quantity (i.e., the observed surface tension of the membrane) of the membrane averaged in time as well as space. However, some of the fusion processes could be due to a dynamic process or a local process: The separation distance between the two surfaces of the hydrocarbon phases at some portions of the membrane at a certain time becomes smaller than the average separation distance of the two membrane hydrocarbon phases by either or both thermal fluctuations and mechanical disturbances applied on membranes. Hence, the strong interaction may occur through small but closely interacting portions of the two membranes. In such a case, the interaction energy per unit area in highly local portions could be greater than the average interaction energy per unit area and exceed a threshold energy required for membrane fusion, even though other parts of the membrane surface remain more hydrophilic. Membrane fusion may be caused by the phase defect through a local surface energy increase, occurring within the membrane (1) upon mixture of different membrane molecular components (Hui *et al.,* 1981), (2) by lowering temperature below a certain phase-transition temperature (Schullery *et al.,* 1980), or (3) because of non-uniform distribution of membrane molecules (Hui *et al.,* 1981). In such a situation, the present proposed theory should be modified as a local event, and the physical quantities, which correlate to the membrane fusion, should be those of the local portions. However, the principal base for the theory presented here is unchanged. As previously proposed (Ohki, 1982, 1984), the membrane fusion events all have in common the following features: (1) the increase in surface free energy occurs at the site of membrane fusion; (2) the divalent or polyvalent cation-induced membrane fusion is directly related to the surface energy increase or to greater hydrophobicity of the membrane caused by binding of these ions with the membrane and resulting in the conformational change in the polar group region; (3) the increased temperature or osmotic pressure-induced membrane fusion is related to the membrane expansion whereby the energy of the surface hydrophilic layers increases; and (4) the boundary region between two different phases in a membrane caused by different membrane components or metal ion binding to the membrane or phase defect points in membrane could correspond to a local higher energy (more hydrophobic) state than the other parts of the homogeneous membrane phases.

Thus, membrane fusion can be explained by the theoretical interpretation presented here; this theory also appears to accommodate most of the existing membrane fusion modes proposed thus far.

ACKNOWLEDGMENTS. This work was supported by grant GM24840 from the National Institutes of Health.

REFERENCES

Blum, L., and Jancovici, B., 1984, Exactly soluble model for the interaction of two parallel charged plates in an ionic medium, *J. Phys. Chem.* **88**:2294–2297.

Breisblatt, W., and Ohki, S., 1975, Fusion in phospholipid spherical membrane. I. Effect of temperature and lysolecithin, *J. Membrane Biol.* **23**:385–401.

Chaudhury, M., and Ohki, S., 1981, Correlation between membrane expansion and temperature-induced membrane fusion, *Biochim. Biophys. Acta* **642**:365–374.

Chung, L., Kaloyanides, G., McDaniel, R., McLaughlin, A., and McLaughlin, S., 1985, Interaction of gentamicin and spermine with bilayer membrane containing negatively charged phospholipids, *Biochemistry* **24**:442–452.

Cohen, F. S., Zimmerberg, J., and Finkelstein, A., 1980, Fusion of phospholipid vesicles with planar phospholipid bilayer membranes. II. Incorporation of a vesicular membrane marker into the planar membrane, *J. Gen. Physiol.* **75**:251–270.

Davies, J. T., and Rideal, E. K., 1961, *Interfacial Phenomena,* Academic Press, New York.

Düzgüneş, N., and Ohki, S., 1977, Calcium-induced interaction of phospholipid vesicles and bilayer lipid membranes, *Biochim. Biophys. Acta* **467**:301–308.

Hammoudah, M. M., Nir, S., Isac, T., Kornhauser, R., Stewart, T. P., Hui, S. W., and Vaz, W. L. C., 1979, Interaction of La^{3+} with phosphatidylserine vesicle: Binding, phase transition, leakage and fusion, *Biochim. Biophys. Acta* **558**:338–343.

Herrman, A., Pratsch, L., Arnold, K., and Lassman, G., 1983, Effect of poly(ethylene glycol) on the polarity of aqueous solutions and on the structure of vesicle membranes, *Biochim. Biophys. Acta* **733**:87–94.

Hiemenz, P. C., 1977, *Principles of Colloid and Surface Chemistry,* Marcel Dekker, New York.

Hong, K., Schuber, F., and Papahadjopoulos, D., 1983, Biological modulators of membrane fusion, *Biochim. Biophys. Acta* **732**:469–472.

Hui, S. W., Stewart, T. P., Boni, L. T., and Yeagle, P. L., 1981, Membrane fusion through point defects in bilayers, *Science* **212**:921–923.

Knutton, S., 1979, Studies of membrane fusion. III. Fusion of erythrocytes with polyethylene glycol, *J. Cell Sci.* **36**:61–72.

LeNeveu, D. M., Rand, R. P., and Parsegian, V. A., 1976, Measurement of forces between lecithin bilayers, *Nature (Lond.)* **259**:601–603.

London, F., 1937, The general theory of molecular forces, *Trans. Faraday Soc.* **33**:8.

Lucy, J. A., 1970, The fusion of biological membranes, *Nature (Lond.)* **227**:814–817.

Miller, C., Arvan, P., Telford, J. N., and Racker, E., 1976, Ca^{2+}-induced fusion of proteoliposomes: Dependence on transmembrane osmotic gradient, *J. Membrane Biol.* **30**:271–282.

Ohki, S., and Fukuda, N., 1968, Interaction energy between water and hydrocarbon phases, *J. Colloid Interface Sci.* **27**:208–215.

Ohki, S., 1977, The oil–water interface, a model membrane system, in: *The Fifty-First Colloid and Surface Science Symposium,* pp. 134–135, Grand Island, New York.

Ohki, S., 1982, A mechanism of divalent ion-induced phosphatidylserine membrane fusion, *Biochim. Biophys. Acta* **689**:1–11.

Ohki, S., 1984, Effects of divalent cations, temperature, osmotic pressure gradient, and vesicle curvature on phosphatidylserine vesicle fusion, *J. Membrane Biol.* **77**:265–275.

Ohki, S., and Ohshima, H., 1984, Divalent cation-induced surface tension increase in acidic phospholipid membranes: ion binding and membrane fusion, *Biochim. Biophys. Acta* **776**:177–182.

Ohki, S., and Duax, J., 1986, Interaction of polyamines with phosphatidylserine membranes, *Biochim. Biophys. Acta* **861**:177–186.

Ohki, S., Düzgüneş, N., and Leonards, K., 1982, Phospholipid vesicle aggregation: Effect on monovalent and divalent ions, *Biochemistry* **21**:2127–2133.

Ohshima, H., and Ohki, S., 1985, Effects of divalent cations on the surface tension of a lipid monolayer-coated air/water interface, *J. Colloid Interface Sci.* **102**:85–94.

Pashley, R. M., McGuigganm, P. M., Ninhan, B. W., and Evans, D. F., 1985, Attractive forces between uncharged hydrophobic surfaces: Direct measurements in aqueous solution, *Science* **299**:1085–1089.

Papahadjopoulos, D., Vail, W. J., Newton, C., Nir, S., Jacobson, K., Poste, G., and Lazo, R., 1977, Studies on membrane fusion. III. The role of calcium-induced phase changes, *Biochim. Biophys. Acta* **465**:579–598.

Poste, G., and Allison, A. C., 1973, Membrane fusion, *Biochim. Biophys. Acta* **300**:421–465.

Poste, G., and Nicolson, G. L., eds., 1978, *Membrane Fusion,* Elsevier/North-Holland, Amsterdam.

Rand, P., 1981, Interacting phospholipid bilayers: Measured forces and induced structural charges, *Rev. Biophys. Bioeng.* **10**:277–314.

Rasmussen, H., 1970, Cell communication, calcium ion, and cyclic adenosine monophosphate, *Science* **170**:404–412.

Rouser, G., Bauman, A. J., Kritchevsky, G., Heller, D., and O'Brien, J. S., 1961, Quantitative chromatographic fractionation of complex lipid mixtures: Brain lipids, *J. Am. Oil Chem. Soc.* **38**:544–555.

Schuber, F., Hong, K., Düzgüneş, N., and Papahadjopoulos, D., 1983, Polyamines as modulators of membrane fusion: Aggregation and fusion of liposomes, *Biochemistry* **22**:6134–6140.

Schullery, S. E., Schmidt, C. F., Felgner, P., Tillack, T. W., and Thompson, T. E., 1980, Fusion of dipalmitoyl-phosphatidylcholine vesicles, *Biochemistry* **19**:3919–3923.

van Oss, C. J., Good, R. J., and Chaudhury, M. K., 1986, *J. Colloid Interface Sci.* (in press).

Verkleij, A. J., Van Echteld, C. J. A., Gerritsen, W. J., Cullis, P. R., and DeKruijff, B., 1980, The lipidic particle as an intermediate structure in membrane fusion processes and bilayer to hexagonal H_{II} transitions, *Biochim. Biophys. Acta* **600**:620–624.

Verwey, E. J. W., and Overbeek, J. Th. G., 1948, *Theory of the Stability of Lyophobic Colloids,* Elsevier, Amsterdam.

Vincent, B., 1973, The van der Waals attraction between colloid particles having adsorbed layers. II. Calculation of interaction curves, *J. Colloid Interface Sci.* **42**:270–285.

Vold, M. J., 1961, The effect of adsorption on the van der Waals interaction of spherical colloidal particles, *J. Colloid Sci.* **16**:1–12.

Weast, R. C., ed., 1965, *Handbook of Chemistry and Physics,* The Chemical Rubber Co., Cleveland, Ohio.

Wilschut, J., and Papahadjopoulos, D., 1979, Ca^{2+}-induced fusion of phospholipid vesicles monitored by mixing of aqueous contents, *Nature (Lond.)* **281**:690–692.

Wilschut, J., Düzgüneş, N., and Papahadjopoulos, D., 1980, Studies on the mechanism of membrane fusion: Kinetics of Ca ion induced fusion of phosphatidylserine vesicles followed by a new assay for mixing of aqueous vesicle contents, *Biochemistry* **19**:6011–6021.

Wilschut, J., Düzgüneş, N., and Papahadjopoulos, D., 1981, Calcium/magnesium specificity in membrane fusion: Kinetics of aggregation and fusion of phosphatidylserine vesicles and the role of bilayer curvature, *Biochemistry* **20**:3126–3133.

Zimmermann, U., 1982, Electric field-mediated fusion and related electrical phenomena, *Biochim. Biophys. Acta* **694**:227–277.

Chapter 16

Pressure as a Probe of Vesicle Fusion

E. L. Chang

1. INTRODUCTION

Fusion between two lipid membranes is superficially easy to understand: Two lipid bilayers come into contact and meld into one seamless membrane. But even at this basic level, one must distinguish whether fusion simply means the merging of the two membranes or whether the internal contents (of the liposomes or vesicles) must also mix. Certainly, for the case of vesicles, the requirement that "true" fusion involves the mixing of the internal volumes as well as lipid mixing is a reasonable and practical way of defining fusion. There are other situations, however, where the above definition is inadequate. For instance, in modeling exocytosis, where the internal contents of a liposome is excreted outward, there is no mixing per se of two internal volumes. In this case, the essence of membrane fusion has to be defined slightly differently. Leaving aside such interesting considerations concerning the definition of fusion, there is also much unknown about the process of fusion. Certainly there are many pathways for initiating vesicle membrane fusion; a partial listing would include induction by cations (Papahadjopoulos *et al.*, 1977), by pH (Ellens *et al.*, 1984), by proteins (Morero *et al.*, 1985), by temperature (Verkleij, 1980), by electric field (Zimmermann, 1982; Sowers, 1984), or even spon-

E. L. CHANG • Naval Research Laboratory, Bio/Molecular Engineering Branch, Code 6190, Washington, D.C. 20375.

taneous fusion between vesicles (Suurkuusk *et al.*, 1976; Schullery *et al.*, 1980; Chang *et al.*, 1982). It is not known whether there is a fundamental mode of fusion common to all these different types of initiations. It has been proposed, however, that lipids that can form an inverted hexagonal (H_{II}) phase (Luzzatti, 1968) are necessary in certain membranes for fusion to occur (Cullis and DeKruijff, 1979; Hope and Cullis, 1981; Nayar *et al.*, 1982). These H_{II}-forming lipids, such as the unsaturated phosphatidylethanolamines (PE) and cardiolipin, are also commonly found in biological membranes, leading to the hypothesis that nonbilayer phases may play an important function in fusion related biological events.

1.1. Role of Nonbilayer Structures in Fusion

The possible role of nonbilayer structures in membrane fusion has been hypothesized for some time and there are excellent reviews on this subject (Cullis and DeKruijff, 1979; Verkleij, 1984). A nonbilayer structure is thought to be important for membrane fusion because there is no way to piece together geometrically two bilayers without first destroying the bilayer structure at the point of contact in order to allow the two bilayers to merge into one (Hui *et al.*, 1981) (see Fig. 1). This disruption of the membrane would be transient in time and limited in spatial extent if the vesicular shape were to be preserved. Possible candidates for this localized perturbation of the bilayer include structures such as inverted H_{II} cylinders (of short length), lipidic particles, and inverted micellar intermediates (IMI) (see Fig. 2). One reason why the H_{II} phase has been intensely studied as a possible role player in fusion is that outside the lamellar phases, it is one of the few stable phases in excess water for many double-chained phospholipids. The H_{II} cylinders and other inverted structures (i.e., structures with the hydrophobic tails forming the outer surface) also satisfy a physical criteria: Bilayer structures can convert into the

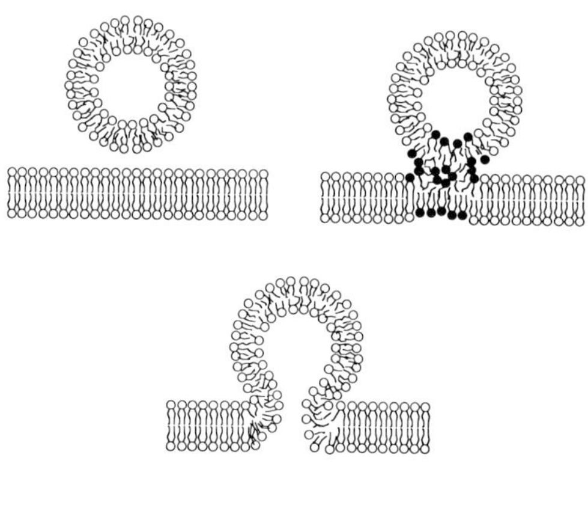

Figure 1. Fusion of two lipid bilayers involves the breakup of the bilayer structure at some point. The dark circles represent lipids that comprise the nonbilayer region of the fusing membrane.

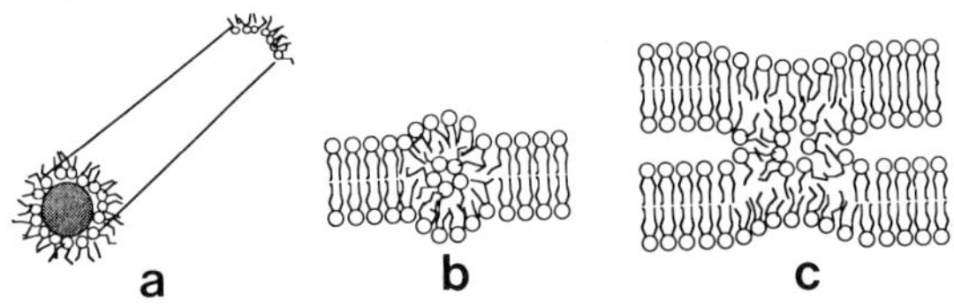

Figure 2. (a)Cylinders of lipids with the polar headgroups pointing inwards to the center of the water-filled cylinder while the tails splay out. The H_{II} phase consists of these cylinders packed in hexagonal arrays. (b) Inverted micelle inside a bilayer. (c) Inverted micelle between two bilayers. IMI, inverted micellar intermediate.

inverted structures as the average lipid parameters (e. g., head-group area, total molecular volume) change. Specifically, we show below that the bilayer structure can naturally fold into inverted structures such as the H_{II} cylinders or inverted micelles as the tail region becomes disordered and cone shaped. That is to say, if one imagines the molecular dimensions of the lipid molecule changing smoothly, such as the average volume occupied by the lipid tails increasing from an idealized straight cylindrical geometry to a cone shape, bilayers can "fold" into inverted structures and not into other lipid packing morphologies such as micelles or H_I cylinders (cylinders with a hydrophobic core and a polar outer surface). Let us define the following molecular parameters (see Fig. 3): v is the total

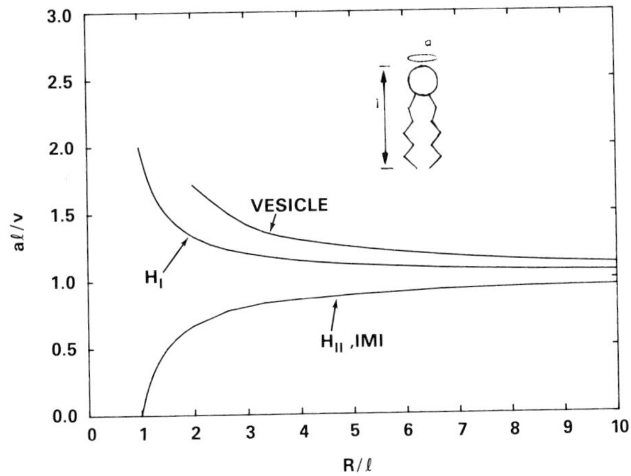

Figure 3. Plot of the variation of the of the ratio of the volume of an equivalent right cylinder (al) to the actual molecular volume as a function of R/L for different structures. Note that as the ratio al/v decreases, the overall shape changes from vesicular to inverted morphology (al/v = 1.0 corresponds to a flat bilayer structure). (Inset) Average total length and area of a lipid molecule.

molecular volume, a is the head-group area, l is the total lipid length, and N is the total number of lipids. Note that $al = v$ only for the straight cylindrical geometry. We can derive the relationship between the above molecular parameters and the principal radius, R, of the global morphology, e.g., the outer radius of vesicles. The following relations can be shown to follow from strictly geometrical arguments and the above definitions (Israelachvili *et al.*, 1976):

Vesicular outer monolayer

$$Nv = 4/3\pi[3R^2l - 3Rl^2 + l^3] \tag{1a}$$
$$Na = 4\pi R^2 \tag{1b}$$

H_{II} cylinders

$$Nv = \pi[2Rl - l^2]L \tag{2a}$$
$$Na = 2\pi(R - l)L \tag{2b}$$

H_I cylinders

$$Nv = \pi[2Rl - l^2]L \tag{3a}$$
$$Na = 2\pi RL \tag{3b}$$

Inverted micelles

$$Nv = 4/3\pi[3R^2l - 3Rl^2 + l^3] \tag{4a}$$
$$Na = 4\pi(R - l)^2 \tag{4b}$$

where L is the cylinder length.

A plot of al/v against $R/1$ (Fig. 3) shows that, as the average molecular shape changes from an upside-down cone, where the head area is larger than the tail area (as is the case for the outer monolayer of vesicles), to a right cylinder (flat bilayer case), and finally to a cone, where the head is smaller than the tail, the global shape changes from small vesicles to either inverted micelles or cylinders. Such molecular changes in conformation are indeed thought to occur when unsaturated phosphatidylcholines (PEs) undergo a bilayer–hexagonal phase transition (Mantsch *et al.*, 1981). The question of whether nonbilayer phases can coexist, even transiently, with the bilayer phase has been examined theoretically by Siegel (1984) for the cases of lipidic particles and inverted micellar particle with the conclusion that the inverted micellar particle has a good potential for forming in bilayer structures (see Chapter 9, this volume). These geometric and kinetic considerations are consistent with the hypothesis of nonbilayer structures participating in vesicle fusion.

In terms of experimental results, Cullis and co-workers, using nuclear magnetic resonance (NMR) and electron micrography methods, have evidence that certain lipid mixtures must include H_{II}-forming lipids before fusion can take place and that, for other lipid systems, the onset of the transition into the H_{II} phase is closely associated with massive fusion (Verkleij, 1980). The evidence that nonbilayer structures play a role in lipid membrane fusion is primarily inferential due to the transitory nature of the possible participation. In addition, the actual nonbilayer moiety involved in fusion could well be an intermediate that facilitates the bilayer–H_{II} transition rather than the equilibrium cylindrical structures themselves. For example, in the transition from lamellar (L) to hexagonal phase (H_{II}) there must be some intermediate structure to mediate the folding process (see Chapter 9, this volume). This intermediate structure in the L–H_{II} transition may also function to facilitate regular lipid membrane fusion which would account for the necessary presence of H_{II}-forming lipids in many fusing systems (Siegel, 1984; Caffrey, 1985).

1.2. Pressure and Fusion

The approach taken in our laboratory has been to look for the existence of transitory nonbilayer structures by exploiting the differences in the properties of the bilayer and nonbilayer phases in order to detect the nonbilayer structure. We hypothesized that, due to the drastic conformational changes that must accompany a bilayer–nonbilayer transition, there should be molar volume differences either between the initial and final equilibrium states (ΔV) or between a transition state and the equilibrium states ($\Delta V^{\ddagger}$). Thus if pressure were applied to a lipid system that could undergo this transition, the state of lower molar volume would be favored, which is the standard prediction from the Clapeyron equation (Callen, 1960):

$$dT/dP = T\Delta V/\Delta H$$

where T is transition temperature, P is pressure, ΔV is the molar volume change, and ΔH is the enthalpy change. Increasing pressure usually causes more restricted motions, closer packing, and more conformational order in molecules. In particular for lipids, high pressure decreases the orientational mobility, the number of *gauche* conformers, and increases interchain interactions (Yager and Peticolas, 1980; Wong and Mantsch, 1985). So, in a sense, temperature and pressure have opposite effects on the lipids. The effects of pressure on lipids listed above could modulate membrane fusion by affecting the lateral diffusion of lipids in the membrane, the disordering of the tails, and the rate of formation of intermediates with different molar volumes. In terms of models of nonbilayer structures (Siegel, 1984; see also Chapter 9, this volume), these pressure effects could affect the ability of lipids to achieve critical fluctuation size through the slowing down of diffusion and inhibition of disorder in the tails (see Siegel, 1984, for a definition of critical fluctuation), affect coalescence by impeding diffusion of small IMI into large tubules, affect the ability of individual lipids to reach the "right" dynamical geometrical shape for forming inverted structures, and affect the stability of the transient inverted structures, once formed. The balance of all these consequences of applying pressure against the opposing driving forces from temperature will, however, determine whether pressure can affect fusion and thus be useful as a physical method to elucidate the mechanism(s) of fusion.

2. THE EFFECT OF PRESSURE ON THE LAMELLAR–HEXAGONAL PHASE TRANSITION

We chose to study the lamellar–hexagonal II (L–H_{II}) phase transition for a PE in order to determine whether the lamellar–nonbilayer phase transition itself is affected by pressure and thereby establish whether pressure will stabilize or destabilize the nonbilayer phase. Unsaturated PEs in excess water have at least two phase transitions: a gel–liquid-crystal chain-melting transition at low temperature, T_m, and a L–H_{II} transition at a higher temperature, T_h (Boggs *et al.,* 1981). We have studied the T_h transition of egg PE made by transphosphatidylation of egg PE (Chang and Yager, 1983; Yager and Chang, 1983). The L–H_{II} transition, which occurs around 56°C, does not have a clear calorimetric or turbidimetric signature. For example, there is almost no abrupt change in the turbidity of a suspension of the transphosphatidylated egg PE (TPE) at the L–H_{II} phase transition temperature. Visual inspection of the sample cuvette shows that the lipid has aggregated into tiny particles that quickly settle. We devised a new, simple method for monitoring this L–H_{II} transition (Fig. 4). The lipid suspension, while in the lamellar phase, was pelleted inside a sealed capillary tube. The tube was placed inside a pressure cell (NOVA SWISS) with three sapphire windows. A temperature probe was inserted through a fourth port to sample the temperature inside the cell. A 50-mW He–Ne laser beam was focused onto the capillary through one of the sapphire windows. The depolarized, magnified image of the lipid pellet in the capillary was projected at zero angle onto the plane of a diaphragm which served to mask out all of the projected image of the capillary except for the spot where the laser beam penetrates the lipid sample. The pellet was quite opaque to the laser beam in the lamellar phase but turns translucent

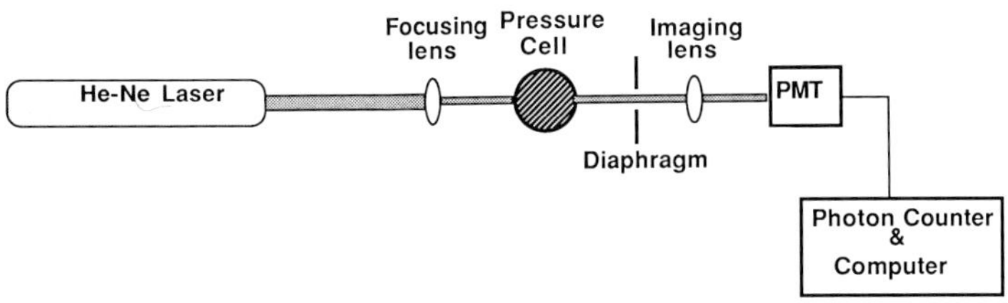

Figure 4. Schematic for measurement of the phase transitions of TPE. The sample cell sits inside the pressure cell; the sample's image is projected by the imaging lens onto the front face of the PMT.

above T_h. Photons were counted with standard photon-counting equipment as the temperature of the pressure cell was raised. In this manner, the level of photon counts rose dramatically at T_h, and the onset of the phase transition could be determined to 0.1°C.

We found pressure to increase both the T_m and T_h transition temperatures and that the increase in T_h as a function of pressure is linear to at least 300 bars (Yager and Chang, 1983). The slope of the heating line for the L–H_{II} is 47.1°/kbar, which, for comparison, is twice as steep as the slope for the chain melting transition of dipalmitoylphosphatidylcholine (Liu and Kay, 1977). The molar volume change associated with this phase transition, however, is small due to the low heat of transition (0.7 kcal/mole), so that we find, through the Clapeyron equation, that the molar increase in volume is only 4.1 cm^3/mole at atmospheric pressure—which is one-seventh the value for the chain-melting transition of DPPC.

In addition to the pressure sensitivity of the L–H_{II} transition, we find a large hysteresis of the heating and cooling curves that is independent of instrumental artifacts. This hysteresis, extrapolated to 0°C/min scanning rate, is 9°C (Fig. 5), whereas the chain-melting transition of the same lipid showed no such hysteresis. Since the coexistence region is only 2–3°C we are not looking at transitions that may occur at these boundaries. The hys-

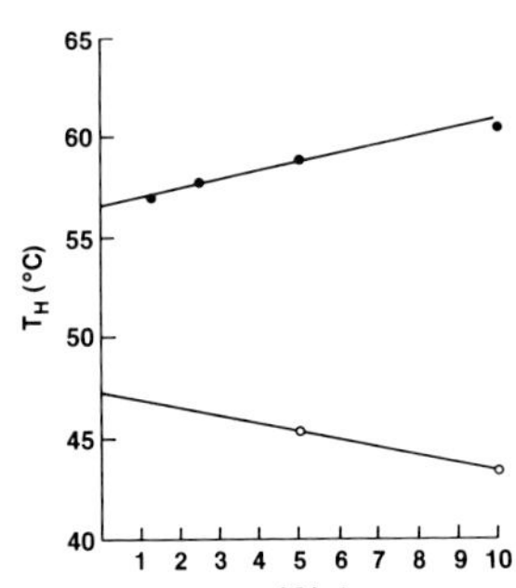

Figure 5. Hysteresis of the L–H_{II} phase transition of TPE. (Upper curve) Heating curve. (Lower curve) Cooling curve.

teresis is presumably due to a fairly large energy barrier of conversion from one state to another. Also, the differing water contents of the two phases may necessitate bulk water flow through the liposomes during the transition.

3. PRESSURE AND VESICLE FUSION

Since pressure was determined to inhibit the formation of the H_{II} phase, the next stop was to determine whether pressure will also delay the fusion of vesicles by inhibiting the onset of the non-bilayer phase. A mixture of dioleoyl phosphatidylethanolamine/dioleoyl phosphatidylcholine/cholesterol, in the molar ratio of 3 : 1 : 2, undergoes a L–H_{II} phase transition at 50°C. Vesicles formed from this mixture will fuse at the same temperature range as the L–H_{II} transition (Verkleij, 1980). It is of interest to determine whether the temperature at which fusion occurs for this system would also be offset upwards by pressure. To monitor the fusion process in real time, photon correlation spectroscopy (PCS) was used to determine the average diameter of the vesicles as a function of increasing temperature (Chang *et al.*, 1982). PCS is a light-scattering technique whereby the time-dependent fluctuations of light scattered by macromolecules in solution are analyzed by correlation techniques to determine an average diffusion coefficient (and thereby an average size) for the macromolecules (Berne and Pecora, 1976). In addition to simply determining the average vesicle size, we also mathematically inverted the correlation function generated by the photon correlator in order to produce the size distribution of the vesicles. The inverting method is a constrained, regularized algorithm (CONTIN) developed by Provencher (1982*a,b*) that can obtain a histogram of the vesicle sizes consistent within the uncertainties of the data and the smoothest curve.

The vesicle suspension of the lipid mixture was placed in a sealed, small, rectangular, four-side cuvette placed inside the NOVA pressure cell with a specially constructed cuvette holder. The cell was pressurized and the scattered light intensity at 90°C was collected as the temperature of the cell was raised to about 75°C. Each experiment took about half-hour of actual heating time and fresh sample was used for each experiment. Oxidation may affect some of the results at higher temperatures (>70°C), but the effects we observe start at much lower temperatures. The entire system was calibrated by determining the pressure dependence of dipalmitoyl phosphatidylcholine (DPPC). The rate of change of T_m for DPPC was found to be 23°/kbar, which is the accepted literature value. The scattered intensity was correlated with a Langley-Ford CM-64 correlator (Amherst, MA). The CONTIN program was implemented on a VAX 11/780.

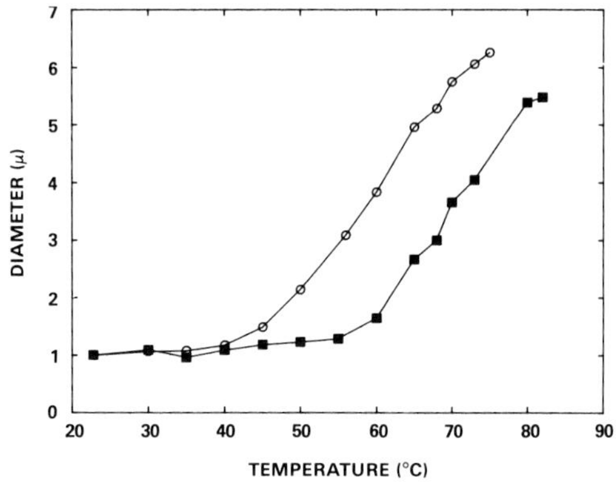

Figure 6. Increase in average vesicle diameter (in microns) as a function of temperature for 1 bar (○) and 100 bars (■).

Preliminary analysis of the data show that the fusion of vesicles, as determined by the increase in average vesicle size, is significantly inhibited by pressure. Pressure as low as 100 bars can dramatically delay the onset of fusion (Fig. 6). A rough estimation of the delay in the onset of fusion for the 100 bars gives about a 15°C offset for every 99 bars. This is a much higher value than would have been predicted from the pressure inhibition results for TPE. The results of the CONTIN analysis highlight the dramatic differences between the thermotropic behavior of the lipid vesicles at 1 bar (atmospheric) and 100 bars (Fig. 7). At 25°C both samples have nearly the same average size although the size distribution for the high pressure suspension appear to have narrowed with respect to the atmospheric one. At 65°C a bimodal distribution appears for the 1-bar suspension, while the distribution of the higher pressure vesicles has not significantly altered. At 75°C the 1-bar vesicle distribution shows that all the vesicles have converted into large structures (> 1 μm), while the sizes of the pressurized vesicle are just beginning to increase.

4. DISCUSSION

The results from modulating thermotropic properties of lipid membranes by pressure are promising. They show that pressure has a definite and significant effect on the ability of the lipids to rearrange themselves. The results do not distinguish whether the inhibition is a consequence of the difference between initial and final states of the lipid or whether it is some intermediate state that is inhibited by fusion. A study on the dynamics of the pressure inhibition should differentiate the two processes.

What is the physiological relevance of pressure on living organisms?

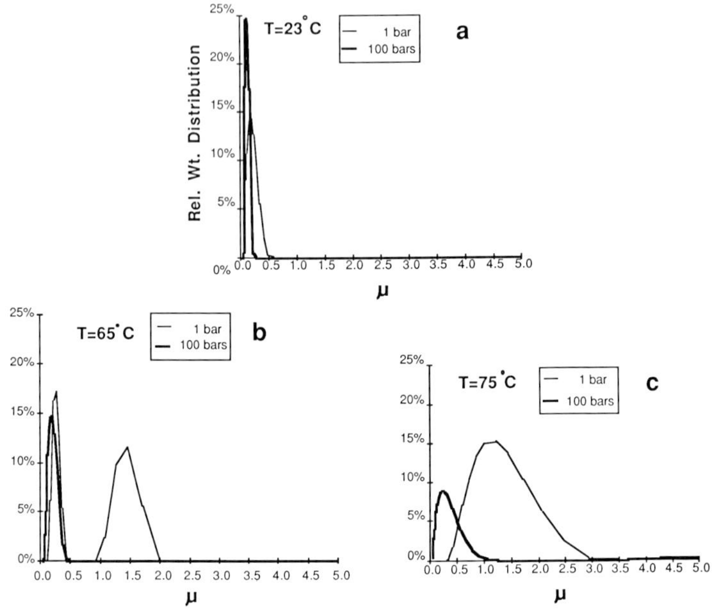

Figure 7. Results of the CONTIN analysis displaying the size distribution (microns) at different temperatures for 1 bar and 100 bars. The Y axis is a normalized, weight-averaged number.

Hydrostatic pressure on biological organisms can affect such vital properties as enzyme kinetics, receptor–agonist binding affinity, protein tertiary structure, and lipid conformations (Hochachka and Somero, 1973). Changing the normal thermodynamic state of these events will affect the smooth functioning of the organism. For example, high-pressure nervous syndrome (HPNS) is a life-threatening disorder suffered by humans free-diving below about 600 where the pressure is around 60 atm. HPNS is solely associated with the application of pressure and is a manifestation of malfunction in the central nervous system. One likely region where the breakdown could occur is in the neural synapse at which intercell communications occur through the release of chemicals called neurotransmitters. Release of these neurotransmitters is mediated by the fusion of the presynaptic membrane with synaptic vesicles containing the neurotransmitters. Since one of the main lipids of synaptic vesicles is unsaturated PE (Deutsch and Kelly, 1981), it may well be that nonbilayer structures play a role in the fusion of these vesicles to the presynaptic membrane, and, therefore, pressure could adversely affect the ability of the snyaptic vesicles to fuse. Because 100 bars is also within the onset of pressure-induced neural disorder, our results are consistent with the hypothesis that breakdown in synaptic vesicle fusion may be a major component of the molecular origins of HPNS.

ACKNOWLEDGMENTS. This study was funded by the Office of Naval Research, project no. RR041-01-02. I gratefully acknowledge Dr. P. Yager for his critical reading of the manuscript.

REFERENCES

Berne, B. J., and Pecora, R., 1976, *Dynamic Light Scattering,* Wiley, New York.

Boggs, J. M., Stamp, D., Hughes, D. W., and Deber, C. M., 1981, Influence of ether linkage on the lamellar to hexagonal phase transition of ethanolamine phospholipids, *Biochemistry* **20:**5728–5735.

Caffrey, M., 1985, Kinetics and mechanism of the lamellar gel/lamellar liquid crystal and lamellar/inverted hexagonal phase transition in phosphatidylethanolamine: A real-time X-ray diffraction study using synchrotron radiation, *Biochemistry* **24:**4826–4844.

Callen, H. B., 1960, *Thermodynamics,* Wiley, New York.

Chang, E. L., and Yager, Y., 1983, Effect of high pressure on a lipid non-bilayer phase, *Mol. Cryst. Liq. Cryst.* **98:**125–129.

Chang, E. L., Gaber, B. P., and Sheridan, J. P., 1982, Photon correlation spectroscopy study on the stability of small unilamellar DPPC vesicles, *Biophys. J.* **39:**197–201.

Cullis, P. R., and DeKruijff, B., 1979, Lipid polymorphism and the functional role of lipids in biological systems, *Biochim. Biophys. Acta* **559:**399–420.

Deutsch, J. W., and Kelly, R. B., 1981, Lipids of synaptic vesicles: Relevance to the mechanism of membrane fusion, *Biochemistry* **20:**378–385.

Ellens, H., Bentz, J., and Szoka, F. C., 1984, pH-induced destabilization of phosphatidylethanolamine-containing liposomes: Role of bilayer contact, *Biochemistry* **23:**1532–1538.

Hochachka, P. W., and Somero, G. N., 1973, *Strategies of Biochemical Adaptation,* W. B. Saunders, Philadelphia.

Hope, M. J., and Cullis, P. R., 1981, The role of nonbilayer lipid structures in the fusion of human erythrocytes induced by lipid fusogens, *Biochim. Biophys. Acta* **640:**82–90.

Hui, S. W., Stewart, T. P., Boni, L. T., and Yeagle, P. L., 1981, Membrane fusion through point defects in bilayers, *Science* **212:**921–923.

Israelachvili, J. N., Mitchell, D. J., and Ninham, B. W., 1976, Theory of self-assembly of hydrocarbon amphiphiles into micelles and bilayers, *J. Chem Soc. Faraday Trans. II* **72:**1525–1568.

Liu, N. I., and Kay, R. L., 1977, Redetermination of the pressure dependence of the lipid bilayer phase transition, *Biochemistry* **16:**3484–3486.

Luzzatti, V., 1968, X-ray diffraction studies of lipid–water systems, in: *Biological Membranes* (D. Chapman, ed.), pp. 71–123, Academic Press, New York.

Mantsch, H. H., Martin, A., and Cameron, D. G., 1981, An infrared spectroscopic characterization of the polymorphic phase behavior of human erythrocyte phosphatidylethanolamine, *Can. J. Spectrosc.* **26:**79–84.

Morero, R. D., Vinals, A. L., Bloj, B., and Farias, R. N., 1985, Fusion of phospholipid vesicles induced by muscle glyceraldehyde-3-phosphate dehydrogenase in the absence of calcium, *Biochemistry* **24:**1904–1909.

Nayar, R., Hope, M. J., and Cullis, P. R., 1982, Phospholipids as adjuncts for calcium ion stimulated release of chromaffin granule contents: Implications for mechanisms of exocytosis, *Biochemistry* **21**:4583–4589.

Papahadjopoulos, D., Vail, W. J., Newton, C., Nir, S., Jacobson, K., Poste, G., and Lazo, R., 1977, Studies on membrane fusion III. The role of calcium induced phase changes, *Biochim. Biophys. Acta* **465**:579–598.

Provencher, S. W., 1982, A constrained regularization method for inverting data represented by linear algebraic or integral equations, *Compt. Phys. Commun.* **27**:213–227.

Provencher, S. W., 1982, Contin: A general purpose constrained regularization program for inverting noisy linear algebraic and integral equations, *Compt. Phys. Commun.* **27**:229–242.

Schullery, S. E., Schmidt, C. F., Felgner, P., Tillack, T. W., and Thompson, T. E., 1980, Fusion of dipalmitoylphosphatidylcholine vesicles, *Biochemistry* **19**:3919–3923.

Siegel, D. P., 1984, Inverted micellar structures in bilayer membranes: Formation rates and half-lives, *Biophys. J.* **45**:399–420.

Sowers, A. E., 1984, Characterization of electric field-induced fusion in erythrocyte ghost membranes, *J. Cell. Biol.* **99**:1989–1996.

Suurkuusk, J., Lentz, B. R., Barenholz, Y., Biltonen, R. L., and Thompson, T. E., 1976, A calorimetric and fluorescent probe study of gel–liquid crystalline phase transition in small, single lamellar DPPC vesicles, *Biochemistry* **15**:1393–1401.

Tilcock, C. P. S., Bally, M. B., Farren, S. B., and Cullis, P. R., 1982, Influence of cholesterol on the structural preferences of DOPE-DOPC systems: A phosphorous-31 and deuterium NMR study, *Biochemistry* **21**:4595–4601.

Verkleij, A. J., 1984, Lipid intramembranous particles, *Biochim. Biophys. Acta* **779**:43–63.

Verkleij, A. J., van Echteld, C. J. A., Gerritsen, W. J., Cullis, P. R., and DeKruijff, B., 1980, The lipid particle as an intermediate structure in membrane fusion processes and bilayer to hexagonal HII transitions, *Biochim. Biophys. Acta* **600**:620–624.

Wong, P. T. T., and Mantsch, H. H., 1985, Effects of hydrostatic pressure on the molecular structure and endothermic phase transitions of phosphatidylcholine bilayers: A Raman scattering study, *Biochemistry* **24**:4091–4096.

Yager, P., and Chang, E. L., 1983, Destabilization of a lipid non-bilayer phase by high pressure, *Biochim. Biophys. Acta* **731**:491–494.

Zimmermann, U., 1982, Electric field-mediated fusion and related electrical phenomena, *Biochim. Biophys. Acta* **694**:227–277.

III

New Membrane Fusion Methods

Chapter 17

Electrofusion

Principles and Applications

George W. Bates, James A. Saunders, and Arthur E. Sowers

1. INTRODUCTION

Electric field-induced cell fusion (electrofusion) is emerging as a promising new tool in cell biology and somatic cell genetics. Compared with other cell-fusion techniques, electrofusion is rapid, simple, and highly efficient. What makes electrofusion truly unique, however, is its unusual physical basis. As a consequence, electrofusion creates some novel experimental opportunities. Although electrofusion has been the subject of several recent reviews (Pohl *et al.,* 1984; Zimmermann *et al.,* 1984*a,b*), these reviews have focused on the results of just a few laboratories; a wider review of the field may, therefore, be beneficial. This chapter has two components: (1) a discussion of the underlying physical principles of electrical fusion to elucidate the inherent strengths and limitations of the technique, and (2) a broad-based, critical account of progress in applying electrofusion to cell biology and somatic cell genetics. Wherever possible we have stressed the need for quantitative as opposed to qualitative observations

---

GEORGE W. BATES • Department of Biological Science and Institute of Molecular Biophysics, Florida State University, Tallahasee, Florida 32306. JAMES A. SAUNDERS • Germplasm Quality and Enhancement Laboratory, USDA Plant Genetics and Germplasm Institute, Beltsville, Maryland 20705. ARTHUR E. SOWERS • American National Red Cross, Jerome H. Holland Laboratory for the Biomedical Sciences, Rockville, Maryland 20855.

and have tried to indicate important areas requiring further research. Our goal is to provide both a theoretical and a practical basis from which workers interested in pursuing this technique, or those with a general interest in cell and membrane fusion, can compare electrofusion with other fusion methods.

One unique aspect of electrofusion is that it produces a high percentage of fused cells without the use of potentially cytotoxic chemical fusogens such as polyethylene glycol (PEG). In addition to their cytotoxicity, chemical fusogens cause general and nonspecific cell agglutination. Although adhesion is a prerequisite to fusion, general agglutination limits the formation of specific fusion products. Many chemical fusogens also cause cells to adhere to the fusion vessel, thus making subsequent isolation of the fusion products difficult. In electrofusion, cell adhesion and the induction of cell fusion can be controlled separately. Thus, specific cell–cell interactions can be generated prior to fusion, leading to the formation of specific fusion products. As demonstrated by Lo *et al.* (1984), this capability is particularly useful in monoclonal antibody production. During electrofusion individual fusion products can also be identified and readily isolated with micropipettes, a distinct advantage for plant cell fusion.

Electrofusion is also a valuable tool for studying general membrane properties, membrane interactions during cell fusion, and the mechanism of cell fusion itself. Cell–cell contacts induced by alternating current fields are restricted to a limited portion of the cell surface, and electrofusion is synchronous. As a consequence, fusion can be pinpointed in time and space and occurs under good optical conditions.

Although these are valuable attributes, electrofusion is not without problems. Electrofusion is ordinarily carried out in an osmotically balanced medium containing sucrose or mannitol but with little or no added salt. Most cells will tolerate these conditions, at least for the short period required to complete the fusion, but these same conditions will cause stress for some cell types. Moreover, the electric fields used to induce fusion can cause irreversible membrane damage if they are not carefully controlled. Thus, the viability of electrically fused cells has, until recently, been in question. Also, the small chambers in which electrofusion was originally carried out put some constraints on the number of recoverable fusion products. Large chambers are now available but need further evaluation. Although very high percentages of cells may be fused (up to 80% in some cases), the yields of recoverable fusion products are routinely in the range of 20–40%, and there can be considerable day-to-day variations in electrofusion yields. Finally, the high cost of commercially available electrofusion equipment has slowed the development and widespread use of this technique.

All these factors must be considered when electrofusion is compared with other fusion techniques. It is our opinion that once biologists have gained a realistic appreciation of the strengths and limitations of electrofusion, this technique will help make important contributions to the progress of cell biology and somatic cell genetics.

1.1. Historical Overview

Although the first publication on electrofusion is that of Senda *et al.* (1979), this technique was developed simultaneously by several laboratories and had its roots in more than 10 years of work on the properties of biological membranes in high-voltage electric fields. By 1979 it was already well established that high-voltage fields could induce (1) reversible breakdown of the electrical resistance of the plasma membrane, (2) transient loss of semipermeability, and (3) the exchange of intracellular and extracellular materials. The precise nature of the physical changes that take place during membrane breakdown is still unknown. However, it is widely believed that high-voltage electric pulses cause the formation of minute pores in the membrane (now termed electroporation; see Tsong, 1983, for review). Electroporation has been used to permeabilize the cell membrane for physiological studies (Baker and Knight, 1978; Adams, 1982; Shimmen and Tazawa, 1983), has been applied to the drug loading of erythrocytes (Zimmermann *et al.,* 1976, 1980), and is a popular procedure for DNA-transformation studies (Neumann *et al.,* 1982; Shivarova *et al.,* 1983; Wong and Neumann, 1982; Potter *et al.,* 1984).

However, it was not until Senda's work (Senda *et al.,* 1979) that it was shown that cells in contact could be induced to fuse by electrical stimulation. In this procedure, pairs of plant protoplasts were pushed together with micropipettes and then were fused by a single electric pulse. Following Senda's work, several laboratories demonstrated independently that animal (Neumann *et al.,* 1980; Scheurich and Zimmermann, 1980; Richter *et al.,* 1981), plant (Zimmermann and Scheurich, 1981) and yeast cells (Weber *et al.,* 1981*a,b*) could be fused by electric fields. Many of these studies used chemical agents or simply high cell densities to induce cell–cell contacts.

Zimmermann and colleagues provided a major advance by showing that dielectrophoresis could be used to induce cell contacts prior to the induction of cell fusion (Scheurich and Zimmermann, 1980; Zimmermann and Scheurich, 1981; Zimmerman *et al.,* 1981*a*).

This fusion procedure is a two-step process. The cells are introduced into a chamber in which parallel wires or plate electrodes are mounted (Fig. 1). A relatively low-intensity alternating current (AC) is applied to the electrodes, causing the cells to align into chains ("pearl chains") perpen-

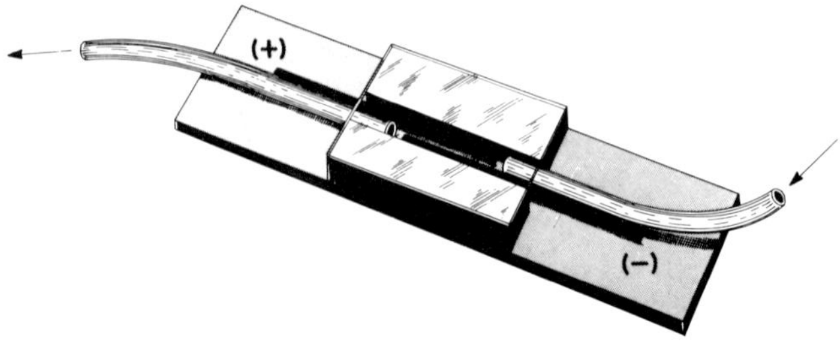

Figure 1. Drawing of a simple electrofusion chamber. Cells can be introduced into the chamber and removed from it by means of plastic tubing. The electrodes (indicated by (+) and (−) in the drawing) are two gold wires glued in parallel along the walls of the chamber. Distance between the electrodes is 0.5 mm. (From Bates *et al.,* 1983.)

dicular to the electrodes (Fig. 2). The alignment of cells into chains under the influence of AC fields is termed dielectrophoresis, and has a well understood physical basis (Pohl, 1978; Pethig, 1979; Sauer, 1983; Takashima and Schwan, 1985). Once the cells are in contact, they are fused by application of short direct current (DC) pulses of high intensity to the electrodes. The resulting DC field induces reversible membrane breakdown and fusion (Fig. 3). By means of these and similar procedures, it is now possible to fuse a wide range of cell types with high efficiency. Applications of electrofusion thus far include estimation of diffusion coefficients for membrane components (Sowers, 1985), generation of monoclonal antibody-secreting hybridomas (Lo *et al.,* 1984; Karsten *et al.,* 1985), and production of intraspecific and interspecific plant hybrids (Bates and Hasenkampf, 1985; Kohn *et al.,* 1985).

2. DIELECTROPHORESIS

The movement of particles under the influence of a nonuniform AC field has been termed dielectrophoresis (Pohl, 1978). This phenomenon can be demonstrated for any polarizable object, whether it is a cell (live or dead), a silicon bead, or even a metallic particle.

When placed in an electric field, neutral bodies become polarized as a result of the induction and reorientation of dipoles. In a uniform electrical field, a polarized body will not experience a net translational force. However, in nonuniform fields, such as the one diagrammed in Figure 4, a force develops that draws the polarized object toward regions of higher field strength. This movement occurs even in AC fields because the induced dipoles reorient in phase with the changing polarity of the electrodes. Dielectrophoresis is in contrast to electrophoresis, which involves

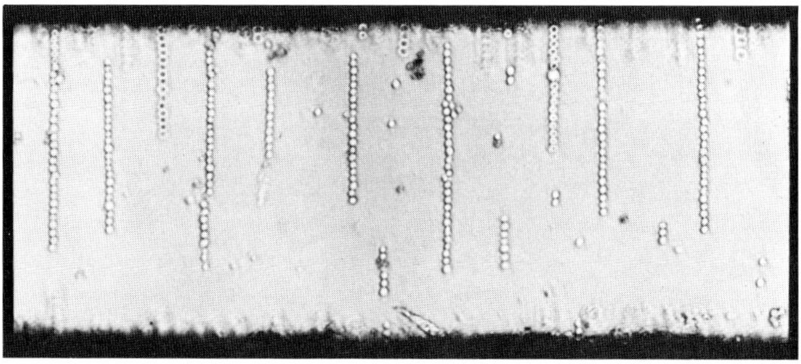

Figure 2. Pearl-chain formation by mouse myeloma cells in response to a sinusoidal AC field of (100 V · cm). The chains of cells, observed with an inverted microscope, radiate out from both electrodes.

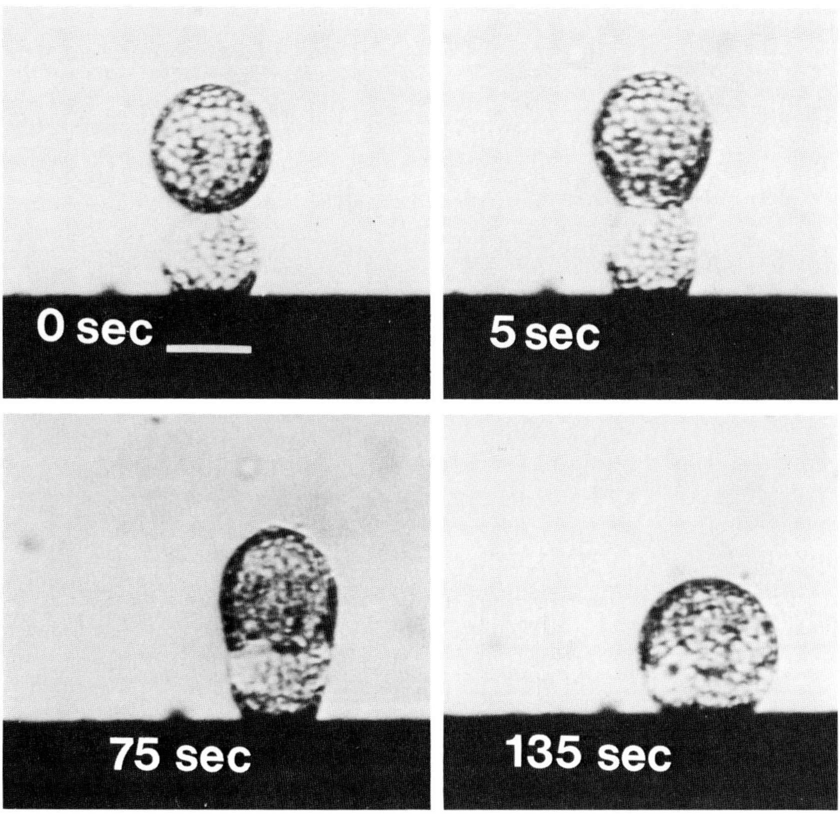

Figure 3. Fusion of two oat leaf protoplasts. The protoplasts were aligned by a 70-V · cm AC field and were fused by application of a single DC square-wave pulse (15 μsec, 800 V · cm) given just after the photograph in (a) was taken. Bar: 20 μm. (From Bates *et al.*, 1983.)

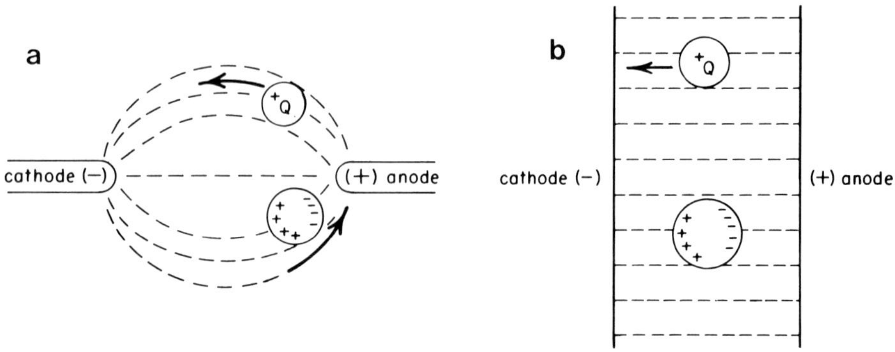

Figure 4. Movement of an electrically neutral (but polarizable) body and a charged body in (a) nonuniform and (b) uniform electric field. No movement of the neutral body is observed in the uniform field. (Adapted from Pethig, 1979.)

the movement of charged bodies in DC fields. Electrophoresis does not occur in AC fields.

In AC fields, chains of cells develop that radiate outward from the electrode surface like those in Figure 2. The cells create local distortions in the electric field, causing the ends of each cell to act as tiny electrodes. Consequently, when cells undergoing dielectrophoresis approach each other, they become mutually attracted (mutual dielectrophoresis) and finally line up in rows, end to end, parallel to the lines of the electric field.

The physical basis of dielectrophoresis lies in the difference between the permittivities (or complex dielectric constants) of the cell and its surrounding medium. A complete treatment of the subject is beyond the scope of this review (see Pohl, 1978; Pethig, 1979). However, several aspects of dielectrophoresis are particularly relevant to electrofusion: (1) dielectrophoresis and cell alignment are frequency dependent; (2) the rate of cell alignment depends on both the magnitude of the electric field and the radius of the cell (the smaller the cell, the larger the electric field that must be applied to achieve alignment); and (3) cell alignment is achieved most easily in media of low ionic strength.

Dielectrophoresis and cell alignment are frequency dependent because the permittivity of the cell relative to the medium changes with frequency. Effective alignment occurs principally in two frequency ranges—one between 0.1 and 10 MHz (centering around 0.5 MHz) and the other occurring at frequencies below 1 kHz. The results of Pohl and co-workers suggest that alignment at high AC frequencies is due to polarization of the plasma membrane, whereas alignment in the low frequency range probably results from reorientation of dipoles and mobile charges associated with the cell surface (Pohl, 1978).

Because the dielectrophoretic force depends on the radius of the cell, there is a size limit below which such high AC voltages are needed to overcome Brownian motion and permit pearl chain formation as to make the procedure impractical. This minimum size is in the range of 0.5 μm (Zimmermann, 1982) and therefore does not place restrictions on the electrofusion of cells. Mitochondria (Sowers, 1983), chloroplasts (Pohl, 1978), and bacterial spheroplasts (Ruthe and Adler, 1985) can all be aligned with AC fields. However, the 0.5-μm size limit does restrict the electrofusion of small liposomes.

Other cellular movements besides pearl chain formation are sometimes observed in AC fields. At AC frequencies of 1–200 kHz, a fraction of the cells in the chamber can be observed to rotate or spin. The speed of rotation increases with the AC voltage and can be very rapid at the voltages often used for cell alignment. Because the rotation of cells in pearl chains disrupts membrane contact, it interferes with fusion. Thus, frequencies causing rotation are to be avoided. Cell rotation was first reported by Teixeira-Pinto *et al.* (1960) and has been studied by Pohl and Crane (1971), Zimmermann *et al.* (1981*b*), and Saunders *et al.* (1986). The cause of this phenomenon is still unclear; however, it seems likely that dipole interaction between neighboring cells (or particles) is involved (Zimmermann *et al.,* 1981*b*).

Another movement often observed during dielectrophoretic collection of cells is turbulence of the medium due to electric current induced-heating (joule heating). Turbulence must be avoided during electrofusion experiments because it can interrupt the formation of stable cell–cell contacts. Heating and turbulence tend to be greatest near the electrodes, where field strengths and hence currents are greatest. Pohl (1978) showed that increases of even a few degrees at the electrode surface are sufficient to create convective currents that interfere with dielectrophoresis.

Turbulence due to heating can be prevented if the conductivity of the medium is kept low. Where necessary the osmolarity of the medium can be balanced with a nonionic solute. A medium of low ionic strength also facilitates cell alignment for two additional reasons: (1) at high AC frequencies, the permittivity difference between the cell and the medium is enhanced, and (2) cells create greater field distortions when in media of low ionic strength, thereby enhancing mutual dielectrophoresis.

In the high-frequency range (0.5–2 MHz), dielectrophoresis is optimal in media containing less than 1 mM in monovalent salts and preferably less than 0.5 mM. As demonstrated by Sowers (1984, 1985), however, cell alignment will occur at low AC frequencies in media of high ionic strength (up to 60 mM phosphate buffer). Heating and turbulence would be expected under these conditions, but are controlled if cells are kept a distance (1–2 mm) from the electrode surface.

3. ELECTROPORATION AND FUSION

Once cell–cell contact has been established, fusion is induced by application of a high-voltage DC pulse of microsecond duration. This brief pulse drives the membrane potential to such high values that the membrane undergoes physical breakdown, which apparently leads to the formation of minute pores. Cell fusion results when membrane breakdown occurs in areas of cell–cell contact. Membrane breakdown is reversible, and the cell remains viable, as long as the DC pulse is neither too large nor too long. However, excessive DC pulse strength (voltage or pulse length) can cause irreversible membrane damage and cell lysis.

3.1. Pore Formation

Cole (1968) described mathematically the membrane potential induced by an external electric field. In simplified form:

$$V = 3/2\, rE \cos\theta \tag{1}$$

where r is the cell radius, E is the strength of the imposed electric field, and θ is the angle with which any membrane patch intersects the field (Fig. 5).

The first application of this equation to cells in high strength electric fields was made by Sale and Hamilton (1968) in their study of the field-induced lysis of bacteria and red blood cells. These workers were able to rule out current flow and joule heating as causes of cell death and attributed it instead to a direct effect of the imposed field on the cell membrane. The field strength necessary to give an LD_{50} was found to be proportional to the cell radius. By applying Eq. (1), they determined that lysis occurred when membrane potentials reached a critical voltage V_{cr} of about 1 V for a variety of bacteria and erythrocytes.

Later, Coster and Zimmermann (1975) made direct measurements of membrane conductance during the electrical charging of giant algal cells by millisecond current pulses. They found that a dramatic increase in conductance occurred at membrane potentials of 0.85 V and that the membrane could not be charged further. This membrane breakdown was reversible; after a relaxation period of a few seconds, the membrane could be charged repeatedly to its breakdown voltage.

These electrical conductance changes are clearly related to changes in membrane permeabilities. In working with red blood cells, Zimmermann *et al.* (1976) and Kinosita and Tsong (1977*a,b*) showed that high-voltage electric pulses led to a breakdown of semipermeability and the diffusional exchange of Na^+ and K^+ across the membrane. This in turn caused cell swelling and eventual lysis (hemolysis) just as in Sale and Hamilton's experiments. However, both Zimmermann *et al.* (1976) and Kinosita and Tsong (1977*a,b*) found that hemolysis could be reduced or prevented by osmotic balancing of the medium. Using a range of carbohydrates from mannitol to the tetrasaccharides, stachyose, and inulin, these workers showed that osmotica with large molecular radii were most effective in preventing hemolysis. The interpretation was that semipermeability was lost because discrete pores were opened in the membrane by the electric pulse. As in Coster and Zimmermann's (1975) earlier work, these pores were found to reseal eventually. However, resealing could take from seconds to hours depending on conditions (e.g., temperature, pore diameter).

Based on electrical conductivity measurements (Benz *et al.*, 1979; Kinosita and Tsong, 1979), osmotic adjustment experiments (Kinosita and Tsong, 1977*b*), and leakage assays (Schwister and Deuticke, 1985), the pulse-induced pores have estimated diameters of 0.5–4 nm and number 1–1000 per membrane. These estimates involve assumptions that are difficult to verify and so must be considered preliminary. In view of their importance in understanding the mechanism of electrofusion and electroporation, these estimates of pore size and number should be reevaluated by different techniques.

The critical voltage (V_{cr}) for electroporation has been recorded for a wide range of animal and plant cells, as well as for liposomes (Teissie and Tsong, 1981) and black lipid membranes (Benz *et al.*, 1979). V_{cr} ranges from 0.5 to 1.5 V depending on the particular system and precise experimental conditions.

It seems reasonable that V_{cr} would vary for different cell types because of differences in membrane composition. However, the experimental conditions under whch V_{cr} has been measured have varied so much from laboratory to laboratory that little evidence can be cited for such a conclusion. Benz and Zimmermann (1981) showed that V_{cr} is substantially different for *Valonia* and *Halicystis*; however, as the authors point out, these algal species differ considerably in their osmotic properties. As will be shown below, membrane pressure resulting from osmotically driven turgor pressure does affect V_{cr}. Pilwat *et al.* (1975) and Mishra *et al.* (1981) found that membrane active substances (benzyl alcohol and chlorpromazine, respectively) can alter V_{cr} in erythrocytes.

This would appear to be good evidence that membrane composition can affect V_{cr}. However, these studies were based on hemolysis rather than measurement of the breakdown of the membrane's electrical properties. Thus, secondary effects of the substances on other cellular properties related to hemolysis, as for example on the cytoskeleton, cannot be ruled out.

Recently Tempelaar and Jones (1985*b*) reported that the threshold DC voltage for the electrofusion of plant cells varies for different cell types and for cells of different species. Although the cells they worked with differed somewhat in size, their results cannot be accounted for by size differences alone and so are likely to reflect inherent membrane properties as well.

Pulse length has a striking effect on V_{cr}. In *Halicystis,* for example, V_{cr} is 600 mV for pulses 100 µsec or longer, about 1 V for 20-µsec pulses, and almost 2 V for a 1-µsec pulse (Benz and Zimmermann, 1981). This same general phenomenon has been observed for *Fucus* eggs (Gauger and Bentrup, 1979), *Valerianella* protoplasts (Steinbiss, 1978), *Valonia* (Zimmermann and Benz, 1980), and erythrocytes (Riemann *et al.,* 1975; Kinosita and Tsong, 1979; Serpersu *et al.,* 1985). The pulse length dependence of V_{cr} is probably due to the time constant for charging the cell membrane. Charging times depend on cell size and experimental conditions (Sale and Hamilton, 1968; Cole, 1968) but generally fall in the range of 1–50 µsec. Clearly, if the pulse length is shorter than the charging time of the cell, a greater electric field must be applied to drive the membrane potential to V_{cr}.

What happens then if the membrane is charged to V_{cr} for periods longer than the membrane's charging time, say for 100 µsec or more? The answer has important practical implications for electrofusion. Because of the cosine function in Eq. (1), V_{cr} is reached first at the cell's poles, i.e., the points closest to the electrodes ($\cos \theta = 1$ for 0° and 180°). The voltage across the membrane decreases to zero at the cell equator ($\cos \theta = 0$ for 90°). Increasing the pulse length or voltage should result in poration over a large portion of the cell surface (Fig. 5).

Because mutual dielectrophoresis leads to cell contacts along the lines of field force (Figs. 2 and 4), DC pulses just sufficient to achieve V_{cr} should preferentially induce membrane breakdown at the sites of contact. Thus only a limited portion of the cell surface may be affected by the DC pulse, and it is precisely that area of the membrane where one hopes fusion will take place. Thus, within a chain of cells, fusion is highly favored over cell lysis. Increasing the pulse voltage or duration will induce more pores outside cell–cell contact areas, leading to loss of cytoplasmic contents and even lysis.

Interestingly, Mehrle *et al.* (1985), while studying the pulse-induced

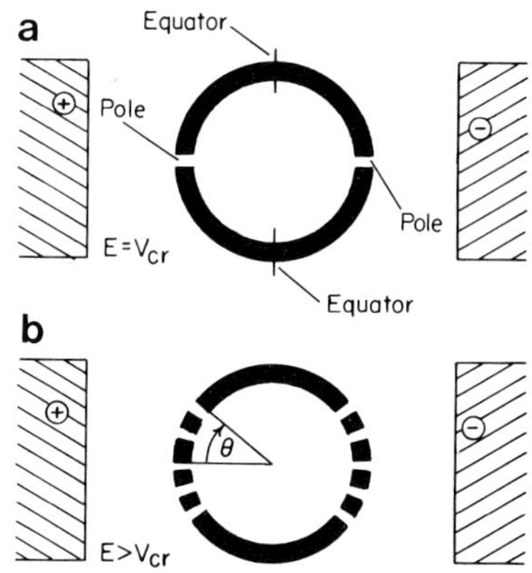

Figure 5. Effect of an applied electric field (E) on the extent of electroporation and the distribution of field-induced pores. (a) E equals the cell breakdown potential (V_{cr}), and pores form at the poles. (b) The larger applied electric field induces electroporation over more of the cell surface. (Adapted from Zimmermann *et al.,* 1984*a*.)

uptake of fluorescent dyes by plant protoplasts, observed that membrane breakdown occurs preferentially on the side of the cell facing the anode. One interpretation of this result is that the cell's own membrane potential also may have to be considered in Eq. (1) and thus may influence V_{cr} and the sites of electroporation.

A second aspect of Eq. (1) that has important practical implications for electrofusion is that the field-induced membrane potential is directly proportional to the cell radius. Large cells will reach V_{cr} at lower field strengths than small cells. Thus, fusion is likely to be nonrandom among a population of cells that varies in size. It may also be difficult to fuse large cells with very small ones without causing some large cells to lyse. Although Friend cells (radius 8 μm) have been fused with *Petunia* protoplasts (radius 28 μm) (Salhani *et al.,* 1985), electrical fusion is probably most efficient when cells are similar in size, and fusion between cells of widely different sizes may be impossible in practical terms. As an example, the electrofusion of submicroscopic liposomes with cells has never been reported, although giant liposomes have been electrically fused with each other (Büschl *et al.,* 1982).

3.2. Pore Resealing

The reversibility of pore formation is an important factor in determining the viability of electrically fused cells. Membrane breakdown is reversible, provided that the field strength and pulse duration are not excessive. However, defining the precise field values at which electroporation becomes irreversible is difficult. It is clear from the work of Kinosita and Tsong (1977*a,b*) that stabilizing cells osmotically makes them much less sensitive to field-induced lysis. However, strong electric pulses will still cause lysis or progressive cell death.

The plasma membrane is probably the primary cellular site affected by high-strength electric fields. Measurements of the electrical conductivity of pulsed cells (Pliquett and Wunderlich, 1980), liposomes (Teissie and Tsong, 1981), and black lipid membranes (Benz *et al.,* 1979) indicate that postpulse recovery of the normal low conductivity of the membrane is extremely rapid, being complete in a few seconds or less. However, when pulse-induced ion and solute leakage are measured with radiotracers, complete membrane recovery can take from minutes to hours. In red cells a 4-kV/cm, 2-μsec pulse was found to induce $^{86}Rb^+$ influx, and the membrane's normal impermeability to Rb^+ was restored in 40 min at 37°C (Serpersu *et al.,* 1985). Longer pulses (3.7 kV/cm, 20 μsec) induced the inward leakage of labeled sucrose as well as Rb^+ and, although the membrane's normal sucrose impermeability was restored after 10 min (at 37°C), inward leakage of Rb^+ continued for 20 hr. Resealing is also temperature dependent; the pores will remain open almost indefinitely at 4°C (Kinosita and Tsong, 1977*b*; Zimmermann *et al.,* 1980; Teissie and Tsong, 1981).

The discrepancy between the electrical conductivity and radiotracer measurements may arise from differences in experimental design. The larger and longer pulses used in radiotracer experiments are likely to create more pores, which take considerably longer to reseal. Electrical measurements of such cells might detect little or no membrane resistance for as long as several hours, suggesting that poration was irreversible when, in fact, the pores were slowly resealing.

Resealing of pure lipid vesicles (Teissie and Tsong, 1981) is orders of magnitude faster than for red blood cells (Zimmermann *et al.,* 1980; Serpersu *et al.,* 1985) and plant protoplasts (Steinbiss, 1978). Membrane proteins probably stabilize the pores and slow the resealing process.

3.3. Mechanism of Membrane Breakdown and Pore Formation

Several model mechanisms have been proposed to explain pore formation (Coster and Zimmermann, 1975; Abidor *et al.,* 1979; Berg *et al.,* 1979; Zimmermann, 1982; Berg *et al.,* 1984; Berg *et al.,* 1984; Dimitrov and Jain, 1984); however, none of them is entirely satisfactory. An accurate model would have to account for both the rapid formation of pores and their relatively long-term stability. Kinosita and Tsong (1977*a,b,* 1979) suggest that pore formation is a two-phase process. The first phase, pore opening, occurs in about 1 μsec and depends on pulse voltage and duration but is independent of medium ionic strength. The second phase, pore expansion, follows over the next 100 μsec. Pore expansion also depends on pulse voltage and duration but is promoted by media of low ionic strength.

Coster and Zimmermann (1975) proposed that the applied electric field exerts enough force on the membrane to compress it physically. Ordinarily, this compression is balanced by the membrane's elastic restoring forces. If compressed sufficiently, however, the restoring force is overwhelmed and the membrane collapses. There is considerable theoretical justification for this model (Crowley, 1973; White, 1980; White and Chang, 1981; Dimitrov and Jain, 1984). Moreover, Zimmermann *et al.* (1977) showed that increasing the turgor pressure (which is also believed to cause membrane compression) of giant algal cells decreases their critical breakdown voltage. However, according to Zimmermann's own calculations (Zimmermann, 1982), field strengths sufficient to induce breakdown would compress the membrane by only 10–20%—a 40% compression is believed necessary to cause membrane collapse.

Membrane compression is probably not the sole factor responsible for breakdown. The high field strengths present across the membrane at breakdown ($>10^6$ V/cm) may force ions into the lipid portion of the bilayer leading to structural defects (Zimmermann, 1982). Alternatively, Abidor *et al.* (1979) suggested that endogenous defects (minute hydrophobic pores) in bilayer structure may serve as nucleation sites for the formation of larger and hydrophilic pores during electroporation.

Berg *et al.* (1984) suggested that breakdown occurs through field-induced polarization and orientation of membrane proteins. Ensuing charge repulsions might then lead to opening of transmembrane channels. This mechanism has the advantage of taking membrane proteins into account, which is something most others fail to do. However, it cannot be a universal mechanism (if such exists), because both electroporation and electrofusion have been accomplished on protein-free liposomes and black lipid films (Benz *et al.,* 1979; Teissie and Tsong, 1981; Büschl *et al.,* 1982; Melikyan *et al.,* 1983). Nonetheless, field-induced changes in membrane protein orientation and conformation could play a role in pore expansion and in long-term pore stability. As observed by Neumann and Katchalsky (1972), high-voltage pulses can induce metastable conformational changes in macromolecules such as polynucleotides, so the pulses used in electrofusion may be sufficient to cause reorientation and conformational changes in membrane macromolecules. Moreover, the markedly shorter resealing times observed in liposomes (Teissie and Tsong, 1981) and black lipid membranes (Benz *et al.,* 1979) than in cells (Kinosita and Tsong, 1979; Zimmermann *et al.,* 1980; Serpersu *et al.,* 1985) suggest that membrane proteins are somehow involved in the postpulse stability of membrane pores.

3.4. Mechanisms of Electrofusion

Most workers assume that electroporation at the sites of membrane contact is responsible for electrofusion. Either poration occurs across both membranes simultaneously (Zimmermann and Vienken, 1982) or breakdown at one membrane leads to the charging and breakdown of a parallel site in the opposing membrane (Pohl *et al.*, 1984). Given that electroporation and electrofusion occur under the same conditions, it seems likely that electroporation has a role in fusion. However, Sowers (1984, 1985, 1986; also Chapter 18, this volume) indicates that pore formation may not be the sole process responsible for electrofusion. If simultaneous poration of opposing membrane sites were the cause of fusion, fusion would only be observed when cell–cell contact was established prior to application of the fusogenic DC pulse. Working with erythrocyte ghost membranes, Sowers found fusion nearly as efficient whether the DC pulses were given before or after cell contacts were established. Even as much as a 5-min interval between application of the DC pulses and the establishment of membrane–membrane contacts did not significantly reduce fusion. This observation has been confirmed by Bates (unpublished observations) for plant protoplasts and appears similar to an observation by Zimmermann *et al.* (1984*a*). The probability that small, randomly distributed pores on opposing cells would interact when the cells came together is very remote. Thus, Sowers's observation indicates that metastable pulse-induced defects in one membrane may induce similar defects in the other membrane on contact and stimulate fusion long after the DC pulse. The observations of Mehrle *et al.* (1985) are also consistent with this view. Plant protoplasts suspended in impermeant fluorescent dyes were given DC pulses sufficient to induce membrane breakdown. The dyes penetrated the cells through the membrane on the side of the cell facing the anode, but not on the side facing the cathode. Thus poration is asymmetric, which seems to preclude the possibility that fusion results from the simultaneous formation of opposed pairs of pores at cell–cell contact sites.

Furthermore, the metastable, pulse-induced membrane defects that lead to electrofusion might not be pores at all. For example, one speculative candidate for such a fusogenic membrane defect could be regions of Hex-II structured lipids. If Hex-II structures can be induced to form by high-strength electric fields, such regions might well initiate fusion at membrane contact sites through the formation of inverted micelles (DeKruijff *et al.*, 1985). Limited ultrastructural examination of membranes subjected to electrical breakdown (Sowers, 1983) has indicated the formation of inside-out vesicles during electrical fusion. However, additional studies are needed to determine whether electrofusion is associated with structural changes in the membrane.

Finally, some mention needs to be made of the effect of electric fields on the distribution of membrane proteins and their possible relationship to cell fusion. Based on the published electrophoretic migration rates of membrane proteins (Poo, 1981; McCloskey *et al.*, 1984), it seems unlikely that short DC pulses induce any more than local protein redistributions. Poo (1981) gives an approximate relationship between the diffusion coefficient (D) and electrophoretic mobility (m) of membrane proteins: D/m = 26 mV. Thus a membrane protein with a diffusion coefficient of 5×10^{-10} cm^2/sec moving in response to a 20-μsec, 1000-V/cm electric pulse would have a root-mean square displacement of only 0.1 nm.

The AC field used for dielectrophoretic cell alignment might exert a greater effect on the distribution of charged membrane components because the cells are exposed to it much longer than to the DC pulse. Zimmermann and Vienken (1982) have pointed out that membrane components could migrate in response to the magnitude of the electrical field component, which is tangential to the membrane surface. As this field component is greatest at the cell's equator (cell sites farthest from the electrodes) and decreases to zero at the cell's poles (areas of the cell facing the electrodes), charged membrane components could migrate away from the equator and toward the poles. Because the sites of cell contact (poles) are flattened, there would be no accumulation of membrane components at the contact sites themselves. In fact, within the zone of membrane contact there might be lateral electrostatic displacement of membrane proteins and charged lipids (Gingell and Ginsberg, 1978), leading to the emergence of protein-free membrane domains. Such protein-free domains have been postulated to have a role in chemically- (Hui *et al.*, 1985) and virally-induced cell fusion (Poste and Pasternak, 1978; Asano and Asano, 1982); Zimmermann (1982) has proposed a similar role for them in electrical fusion.

However, there is no direct evidence from either electron or light microscopic examination that membrane components undergo any redistributions prior to electrical fusion. Indeed, some evidence (Sowers, 1983) indicates that AC field-induced redistribution of membrane particles does not occur. Moreover, the observations of Sowers (1985, 1986) that fusion can result when cell contact is induced after the DC pulse, and that cells may be fused by DC pulses alone (Teissie *et al.*, 1982; Lo *et al.*, 1984), indicate that AC field-induced redistribution of membrane proteins probably does not have an important role in electrical fusion.

4. SOMATIC HYBRIDIZATION BY ELECTRICAL FUSION

4.1. Fusion of Animal Cells

Among the first demonstrations of electrofusion was that for *Dictyostelium* (Neumann *et al.*, 1980). Cell–cell contact was obtained by means of very high cell densities, rather than by dielectrophoresis, and fusion was initiated by treatment of the cells with a series of DC electric pulses. Variations on this protocol have been used with considerable success for several higher animal systems as well. Usually fibroblasts are the experimental material because of their natural tendency to grow in a monolayer with extensive cell–cell contacts. For example, 3T3 (Teissie *et al.*, 1982), mouse LM, Chinese hamster fibroblast (Finaz *et al.*, 1984), and Chinese hamster ovary (CHO) cells (Blangero and Teissie, 1983, 1985; Teissie and Blangero, 1984) have all been fused simply by application of DC pulses with fusion efficiencies ranging from 20 to 90%.

Fusion of cell monolayers is of limited usefulness. Consequently, cell adhesion is ordinarily generated before electrical fusion by dielectrophoresis, by addition of chemicals, or by modification of the cell surface. Scheurich and Zimmermann (1980) demonstrated that the adhesion and fusion of red blood cells were possible in high-voltage, nonuniform AC fields. AC fields strong enough to induce membrane breakdown also result in a massive dielectrophoretic force. Consequently this experiment resulted in the formation of long tubular fusion bodies from hundreds of individual cells. This approach was subsequently improved by use of weak AC fields to form discrete chains of cells followed by high-voltage DC pulses to induce fusion. With this modification, the electrofusion of Friend cells (Zimmermann *et al.*, 1981*a*), erythrocytes (Scheurich and Zimmermann, 1981*b*), and sea urchin eggs (Richter *et al.*, 1981) was achieved. By regulation of the density of the cell suspension and the AC field strengths, it was possible to control to some extent the size of the fusion products. At low cell densities and in weak AC fields, binucleate cells made up as much as 30% of the fusion products (Vienken and Zimmermann, 1982). Fusion at high cell densities can produce giant cells (Scheurich and Zimmermann, 1981*b*).

Zimmermann *et al.* (1981*a*) introduced the use of a pronase treatment prior to fusion—a procedure adopted by other workers as well (Podesta *et al.*, 1984; Karsten *et al.*, 1985). Although Zimmermann *et al.* (1981*a*) supplied no supporting data, they indicated that pronase improves fusion by stabilizing the plasma membrane against lysis during the DC pulse. They suggest that this process may reflect the creation of protein-free domains in the membrane, in which rapid resealing following electrical breakdown would stabilize the cell against colloid osmotic lysis.

A likely alternative is that proteases may improve fusion by removing cell surface determinants, promoting closer cell–cell contact. In the case of sea urchin eggs, with their extensive vitellin layer, protease treatment is a prerequisite to fusion. However, electrofusion of erythrocytes (Sowers, 1984), myeloma cells, and lymphocytes (Lo *et al.*, 1984) is possible without protease treatment. Vienken *et al.* (1983*a*) reported that myeloma cells could be fused without being bathed in pronase, provided that they were allowed to settle onto a polylysine-coated coverslip to which pronase had been attached. Because PMSF did not reduce the enhancement of fusion by pronase, Vienken *et al.* (1983*a*) concluded that the surface charge of pronase, rather than its enzymatic activities, was important. The authors overlooked the fact, however, that PMSF is not an effective inhibitor of pronase. Moreover, Ohno-Shosaku and Okada (1984) showed that dispase, pronase, and trypsin all enhanced the electrofusion of myeloma cells and that specific inhibitors of trypsin reversed its stimulatory effect. Thus, proteases probably enhance fusion through their proteolytic action. Whether this effect of proteases is due to the emergence of protein-free domains, improvement of cell–cell contact through the stripping away of surface glycoproteins, or the generation of fusogenic polypeptides (Lucy, 1984) remains to be determined.

The modulation of the electrofusion of mammalian culture cells by ions in the external medium has been examined by Teissie *et al.* (1982) and Blangero and Teissie (1983, 1985). Increasing concentrations of Na^+, K^+, Li^+, and Ca^{2+} inhibited electrofusion, whereas the pH and anion composition of the medium were less important. Electrofusion was completely blocked by 50 mM Na^+ or 1 mM Ca^{2+}, and the authors proposed that the ions exert their effects by reducing the formation of cell–cell contacts. It should be borne in mind that fusion in these experiments depends on the naturally occurring cell–cell contacts in a high-density cell culture, so the results may have no bearing on electrofusion when cell–cell contact is generated by dielectrophoresis or PEG. Indeed, dielectrophoresis and electrofusion are certainly possible in 60 mM Na-phosphate buffer (Chapter 18, this volume).

The viability of animal cells following electrical fusion has been questioned until quite recently. Richter *et al.* (1981) showed that fused sea urchin eggs could be fertilized and would even undergo a first division but developed no further. In some cases, such as the fusion of mouse blastomeres by 5-msec, 4-kV/cm pulses (Berg, 1982), the viability of the fusion products is quite doubtful. Still, the extensive documentation of the resealing of electroporated membranes makes it clear that, under the proper conditions, electrically fused cells should be viable. Now that appropriate equipment and culture techniques have been developed, several types of viable hybridomas have been produced. For example, Finaz *et al.* (1984) fused mouse LM cells and Chinese hamster fibroblasts and recovered hybrids by selection on HAT. Podesta *et al.* (1984) demonstrated normal hormonal responses by fusion products of Leydig and adrenal cells. Even the electrical fusion of Friend cells and *Petunia* protoplasts to form interkingdom hybrids has yielded cells that, although unable to divide, did synthesize both cellulosic walls and hemoglobin (Salhani *et al.*, 1985).

4.1.1. Hybridomas for Monoclonal Antibody Production

The application of electrofusion to monoclonal antibody production has been a major goal of several laboratories. Possible advantages are that (1) high fusion rates could reduce the dependency on rigid postfusion selection schemes and improve the production of hybridomas from human–human cell fusions, and (2) specific cell fusion might be obtained (rather than random fusion in a mixed population of cells), leading to the formation of high-affinity monoclonal antibodies. Both goals have now been realized to some extent.

In initial work on hybridoma formation Vienken and Zimmermann (1982) demonstrated the fusion of mouse myeloma cells and mouse lymphocytes. A layer of myeloma cells was first built up on the electrodes followed by a layer of lymphocytes; thus, specific myeloma–lymphocytes could be generated. The authors state that M/L combinations were achieved with a 60–80% efficiency (fraction of aligned cells that are fused). However, they gave no information on cell viability, cell recovery from the chamber, how fusion was actually assessed, or the total yield of fusion products, yield being the number of fusion products among the total number of cells recovered from the chamber.

Working with plant cells, but with similar equipment and procedures, Bates (1985) has suggested that actual yields of recoverable fusion products are likely to be low because of difficulties in removing cells from the chamber. Nevertheless, Vienken and Zimmermann's (1982) work indicates the possibility of producing specific fusions with electric fields. Bischoff *et al.* (1982) used a simpler chamber and procedure to form human cell hybridomas. Fusion was directed by mixing B-lymphocytes and myeloma cells in a 5 : 1 ratio. After fusion, the cells were cultured on macrophage feeder cells in microtiter plates. Three weeks later, the presence of antibodies was detected by use of anti-Ia sera. Because anti-Ia is directed toward a general lymphocyte surface antigen, this experiment indicates the formation of viable hybridomas but does not provide evidence of antibody secretion. The authors also did not supply information on the number of clones recovered per cells fused or the long-term viability of these clones.

Using procedures similar to those of Bischoff *et al.* (1982), Karsten *et al.* (1985) reported the production of monoclonal antibodies to cytokeratin. Electrofusion gave an average colony formation frequency of 3×10^{-4}, which was more than 10-fold better than obtained with PEG.

In one of the most interesting and innovative studies employing electrofusion, Lo *et al.* (1984) reported a "directed" method of fusing myeloma cells and B-lymphocytes. The hybridomas that were produced all secreted antibodies directed against the same specific antigen, (angiotensin-converting enzyme or ACE). Lymphocytes were isolated from mice that had been previously injected with ACE and were reacted with an avidin–antigen complex. Myeloma cells were labeled with NHS–biotin. When the two cell types were mixed together, specific cell–cell contact resulted from interaction between the NHS–biotin-labeled myeloma cells and those lymphocytes bearing the antigen–avidin complex. Fusion was induced by four 5-μsec, 4-kV/cm DC pulses. Because B cells express surface antigen receptors of the same antigenic specificity as the antibodies they will eventually secrete, all of the clones recovered from this fusion would be expected to secrete anti-ACE antibodies. Starting with 3×10^6 cells, 41 clones were obtained. Not only did all the clones secrete anti-ACE, but these were high-affinity antibodies (10^{-8}–10^{-10}M) presumably because cell–cell contact and hence successful fusion depends on a strong interaction between the avidin–antigen complex and the antigen receptors on the lymphocytes.

4.2. Plant Protoplasts

The fusion of naked plant cells (protoplasts) is an active field because of its implications for plant genetic engineering. In contrast to the situation with animal cells, it is possible, in culture, to stimulate the development of plant cells and their fusion products and thereby to recover intact plants and interspecific somatic hybrids (Gleba and Sytnik, 1984). Plant protoplasts lack sialic acid residues and so cannot be fused by Sendai virus but are readily fused by PEG. However, PEG-induced fusion efficiencies vary greatly (1–20%), and certain cell types show marked cytotoxic responses to this fusogen (Kartha *et al.*, 1974; Kao, 1981; Kao and Michayluk, 1974). Thus a simple, efficient, physical fusion procedure such as electrofusion should be a valuable tool for plant cell biologists. Progress in plant somatic hybridization has also been slowed by a need for biochemical markers suitable for hybrid selection in culture. Thus, the manual isolation of fusion products and their culture as single cells or on feeder layers has become an important procedure in plant somatic hybridization (Gleba and Sytnik, 1984). Chemical fusogens such as PEG induce extensive agglutination and adhesion of cells to the fusion vessel thus making cell isolation difficult. Here again electrofusion may offer distinct advantages over chemical fusion, as it facilitates the manual isolation of heterokaryons (Bates and Hasenkampf, 1985) and even allows the fusion of single pairs of cells (Koop *et al.*, 1983).

The first morphological demonstration of electrofusion was the fusion of barley and *Rauwolfia* protoplasts (Senda *et al.*, 1979). The protoplasts were pushed together with a pair of microelectrodes and fused with a 5-msec 12-μA pulse. Protoplasts were observed to coalesce into a single body in 10–30 min, but no cell culture work was reported.

Zimmermann and Scheurich (1981) demonstrated the electrofusion of *Vicia faba* leaf protoplasts using the combined approach of dielectrophoresis by AC fields and fusion by DC pulses. Using modified versions of this procedure, many other workers have fused a wide variety of plant protoplasts, indicating that electrofusion is both general and effective.

The reported electrofusion efficiencies range from 100% (Zimmermann and Scheurich, 1981) down to 20% (Zachrisson and Bornman, 1984). This range probably reflects the perspective of the experimenters more than the actual biology of fusion (Bates, 1985; see also Chapter 17, this volume). If one ignores any unaligned cells in the fusion chamber and just scores the percentage of cells in pearl chains that fuse, fusion efficiencies can be very high. By contrast, the culture of fusion products requires recovering the cells from the chamber. Given that most fusion chamber designs leave substantial numbers of unaligned protoplasts either above or below the electrodes, it is not surprising that the percentage of fused cells among protoplasts recovered from the chamber is substantially lower. Realistic values probably range from 20 to 40% (Zachrisson and Bornman, 1984; Bates, 1985; Saunders *et al.,* 1986) and are still considerably better than those obtained by other fusion procedures. There is also evidence that different cell types may display different fusion sensitivities (Tempelaar and Jones, 1985*a,b;* Chapter 24, this volume). Still, it is unclear to what extent this situation is due to differences in how the protoplasts are prepared and handled, rather than in their inherent properties. Much in the literature on electrofusion is still observational, rather than quantitative; thus, assessing the field is sometimes difficult. For example, Watts and King (1984) indicate that small amounts of Ca^{2+} improve the electrofusion of plant protoplasts, whereas Chapel *et al.* (1984) maintain that Ca^{2+} inhibits, but that Mg^{2+} is beneficial. Most other workers obtain satisfactory results in the absence of any salts during fusion. Still, great progress has been made recently in providing quantitative data on plant cell electrofusion; undoubtedly, such apparently contradictory observations as these will be resolved in the near future.

4.2.1. Viability of Electrically Fused Plant Protoplasts

As with the electrofusion of animal cells, the viability of electrically fused plant protoplasts was initially questioned. Resolution of this issue was hindered by the lack of good selection systems for the recovery of plant somatic hybrids. Viable hybrids have now been recovered following electrofusion (Kohn *et al.,* 1985; Bates and Hasenkampf, 1985; J. A. Saunders and B. F. Mathews, unpublished observations).

Despite the obvious importance of the question of cell viability, early protoplast electrofusion studies (Zimmermann and Scheurich, 1981; Scheurich and Zimmermann, 1981*a;* Vienken *et al.*, 1981, 1983*b;* Verhoek-Köhler *et al.,* 1983) were not complemented by any cell culture work. Later, Bates *et al.* (1983), Zachrisson and Bornman (1984), and Watts and King (1984) were able to show that electrically fused protoplasts were able to grow and divide in culture. Saunders *et al.* (1986) and Tempelaar and Jones (1985*b*) have also demonstrated the viability of electrically fused protoplasts using the vital dye fluorescein diacetate. However, none of these studies documented the viability of actual fusion products.

This problem has now been resolved by two different approaches. Using two complementary nitrate reductase-deficient mutants of *Nicotiana tabacum,* Kohn *et al.* (1985) reported the recovery, after electrofusion, of nitrate reductase-competent calli by selection on media lacking reduced nitrogen. These mutants are known to have extremely low reversion rates and to be incapable of complementing each other except following cell fusion or sexual crossing. Thus, the evidence for somatic hybridization here is quite strong. However, these lines are genetically identical, except for the nitrate reductase mutations, so confirmation of somatic hybridization by a second test was not possible.

Bates and Hasenkampf (1985) recovered intraspecific somatic hybrids following the fusion of *N. tabacum* mesophyll protoplasts and *N. plumbaginifolia* suspension cell protoplasts. Heterokaryons were identified by their distinctive morphology, manually isolated with micropipettes, and cultured in small groups or as single cells on conditioned media. Individual calli recovered from these cultures were regenerated and their esterase isozyme profiles examined. The isozyme analysis showed that almost 70% of the lines were somatic hybrids.

These studies clearly document the viability of electrically fused plant protoplasts. Bates and Hasenkampf (1985) reported heterokaryon plating efficiencies of 30%, which compare favorably with unfused controls (50%). Although Kohn *et al.* (1985) did not calculate plating efficiencies, they recovered hybrid calli among their fusion products at frequencies of 10^{-2}. This frequency is two orders of magnitude higher than the frequencies obtained with PEG (Gleba and Sytnik, 1984). It should be pointed out that the frequency with which hybrids are recovered is only partially related to the efficiency of cell fusion. Two further events, nuclear fusion and the formation of viable colonies, are also involved. These events occur in only a fraction of the fused cells (Gleba and Sytnik, 1984). Consequently, large increases in the efficiency of fusion do not *a priori* lead to equally large increases in the recovery of somatic hybrids.

Many other laboratories are now involved in plant protoplast electrofusion and will undoubtedly provide more examples of somatic hybridization by electrofusion in the near future. For example, Tempelaar and Jones (1985*b*) have reported the culture and regeneration of *Nicotiana* and *Datura* electrofusion products but have not yet provided biochemical evidence of somatic hybridization. Conversely, J. A. Saunders and B. F. Mathews (unpublished observations) have recovered somatic hybrids in *Daucus* following the electrofusion of two separate amino acid analogue-resistant lines. The resulting calli have isozyme profiles intermediate between the two parent lines, but plants have not yet been regenerated.

4.4.2. Chamber Design for Cell Culture Work

Culture of both plant and animal cells is facilitated by high cell densities. Unfortunately, early electrofusion chambers held only a few microliters of cell suspension. Thus, considerable emphasis has been placed on developing chambers that hold large numbers of cells (Watts and King, 1984; Zachrisson and Bornman, 1984; Bates, 1985; Tempelaar and Jones, 1985*a*).

Several promising designs have emerged. Bates (1985, also this volume) described the use of a chamber consisting of an interdigitated network of parallel wires laid across a plastic slide and covered by a plastic collar (which forms a well to hold the cells). This chamber holds 0.5 ml of cell suspension and allows the fusion of batches of 2.5×10^5 protoplasts with efficiencies approaching 40%. Watts and King (1984) report using a chamber consisting of a series of parallel plates separated by plastic spacers. This chamber held volumes of about 1 ml and gave fusion efficiencies as high as 80–90%. However, optimal fusion was obtained at lower cell densities than used by Bates (1985), so the yield of fusion products was similar in the two designs. A third approach (Zachrisson and Bornman, 1984) involved the use of a flat plate chamber (internal volume 110 μl) through which the protoplasts were pumped in batches. Fusion efficiencies were in the range of 20%; however, the authors do not give any data on total yields. The commercial suppliers of electrofusion equipment are now providing larger fusion chambers as well; no published data on their effectiveness is yet available.

Somatic hybridization also benefits from the formation of as many binucleate (as opposed to multinucleate) heterokaryons as possible. Bates (1985) found that approximately 15% of the cells recovered after fusion were binucleate and about one-half of these were binucleate heterokaryons. However, using a chamber similar to that of Watts and King (1984), Tempelaar and Jones (1985*a*) report efficiencies for the recovery of binucleate heterokaryons similar to those found by Bates (i.e., 5–9%). More quantitative analyses like these are clearly needed to clarify the effect of chamber design on the yields of different types of fusion products. Still, one conclusion emerges; even with the high overall fusion efficiencies offered by electrofusion, a good postfusion selection system remains necessary for the recovery of somatic hybrids.

4.3. Microorganisms: Bacteria and Yeasts

Two approaches have been taken in the electrofusion of bacteria. Shivarova and Grigorova (1983) used PEG to induce cell–cell contact between protoplasts of different strains of *Bacillus thuringiensis,* which were then fused by application of three successive DC pulses. Fusion products were recovered on the basis of genetic complementation between the strains, one of which had a distinctive colony morphology and the other of which was kanamycin resistant. Hybrid colonies were recovered with a frequency of 10^{-3} if given the DC pulses; no hybrid colonies resulted from the use of PEG alone.

In a second study, Ruthe and Adler (1985) reported the fusion of spheroplasts of *Escherichia coli* and *Salmonella typhimurinum* using the combined electrofusion protocol of dielectrophoretic cell alignment followed by DC pulses. Giant spheroplasts (2–10 μm in diameter) were prepared by growing the bacteria in a cell division inhibitor (cephalexim) for several replication cycles prior to cell wall digestion. Fusion of these giant spheroplasts was monitored by light microscopy, which allowed optimization of fusion parameters. Hybrids were recovered by complementation between tetracycline- and kanamycin-resistant strains at frequencies up to 3×10^{-4}. Fusion was not successful between spheroplasts of normal size (0.5–1-μm diameter) or between cells retaining their peptidoglycan cell wall.

Weber *et al.* (1981*a,b*) investigated the viability and fusion of protoplasts of auxotrophic strains of several yeast species following electrical stimulation. Agglutination was induced by PEG, after which fusion was induced by a single DC pulse. Fusion products were recovered by genetic complementation and growth on selective media. The efficiency with which prototrophic colonies were recovered ranged from $2-7 \times 10^{-4}$, depending on the particular species and strain combination. Fusion induced by PEG alone was 10–100-fold less.

Halfmann *et al.* (1982, 1983) investigated the fusion of compatible and incompatible *Saccharomyces* strains that had been aggregated by dielectrophoresis. The strains fused were haploid lines carrying either nuclear or mitochondrial deficiencies. Because of their small size, these cells required extremely strong AC fields (1 kV/cm) for cell alignment. DC pulses led to visible fusion, and prototrophic colonies were recovered on selective media. Fusion of incompatible strains led to restoration of mitochondrial deficiencies, but not nuclear ones. Thus plasmogamy, but not karyogamy, had occurred. Fusion of compatible strains resulted in both plasmogamy and karyogamy. These results are qualitatively the same as those obtained with PEG. However, electrofusion gave colony recovery efficiencies of 5×10^{-2} for plasmogamy and 5×10^{-3} for karyogamy, again about 100-fold greater than ordinarily obtained with PEG.

ACKNOWLEDGMENTS. This is contribution No. 698 from the American Red Cross Laboratories.

REFERENCES

Abidor, I. G., Arakelyan, V. G., Chernomordik, L. V., Chizmadzhev, Yu. A., Pastushenko, V. F., and Tarasevich, M. R., 1979, Electric breakdown of bilayer lipid membranes, *Bioelectrochem. Bioenerg.* **6:**37–52.

Adams, R. J., 1982, Organelle movement in axons depends on ATP, *Nature (Lond.)* **297:**327–329.

Asano, A., and Asano, K., 1982, Mechanism of HVJ-induced cell fusion, in: *Structure, Dynamics, and Biogenesis of Biomembranes* (R. Sato and S. Ohinishi, eds.), pp. 57–77, Japan Scientific Societies Press, Tokyo, and Plenum Press, New York.

Baker, P. F., and Knight, D. E., 1978, Calcium dependent exocytosis in bovine adrenal medullary cells with leaky plasma membranes, *Nature (Lond.)* **276:**620–622.

Bates, G. W., 1985, Electrical fusion for the optimal formation of protoplast heterokaryons in *Nicotiana, Planta* **165:**217–224.

Bates, G. W., and Hasenkampf, C. A., 1985, Culture of plant somatic hybrids following electrical fusion, *Theor. Appl. Genet.* **70:**227–233.

Bates, G. W., Gaynor, J. J., and Shekhawat, N. S., 1983, The fusion of plant protoplasts by electric fields, *Plant Physiol.* **72:**1110–1113.

Benz, R., and Zimmermann, U., 1981, High electric field effects on the cell membranes of *Halicystis parvula.* A charge pulse study, *Planta* **152:**314–318.

Benz, R., Beckers, F., and Zimmermann, U., 1979, Reversible electrical breakdown of lipid bilayer membranes: A charge-pulse relaxation study, *J. Membrane Biol.* **48:**181–204.

Berg, H., 1982, Biological implications of electric field effects. Part V. Fusion of blastomeres and blastocysts of mouse embryos, *Bioelectrochem. Bioenerg.* **9:**223–228.

Berg, H., Augsten, K., Bauer, E., Forster, W., Jacob, H. -E., Muhlig, P., and Weber, H., 1984, Possibilities of cell fusion and transformation by electrostimulation, *Bioelectrochem. Bioenerget.* **12:**119–133.

Bischoff, R., Eisert, R. M., Schedel, I., Vienken, J., and Zimmermann, U., 1982, Human hybridoma cells produced by electro-fusion, *FEBS Lett.* **147:**64–68.

Blangero, C., and Teissie, J., 1983, Homokaryon production by electrofusion: A convenient way to produce a large number of viable mammalian fused cells, *Biochem. Biophys. Res. Commun.* **114:**663–669.

Blangero, C., and Teissie, J., 1985, Ionic modulation of electrically induced fusion of mammalian cells, *J. Membrane Biol.* **86:**247–253.

Büschl, R., Ringsdorf, H., and Zimmermann, U., 1982, Electric field-induced fusion of large liposomes from natural and polymerizable lipids, *FEBS Lett.* **150:**38–42.

Chandler, D. E., Heuser, J. E., 1980, Arrest of membrane fusion events in mast cells by quick-

freezing, *J. Cell Biol.* **86**:666–674.
Chapel, M., Teissie, J., and Alibert, G., 1984, Electrofusion of spermine-treated plant protoplasts, *FEBS Lett.* **173**:331–336.
Cole, K. S., 1968, *Membranes, Ions and Impulses. A Chapter of Classical Biophysics,* University of California Press, Berkeley.
Coster, H. G. L., and Zimmermann, U., 1975, Dielectric breakdown in the membranes of *Valonia utricularis, Biochim. Biophys. Acta* **382**:410–418.
Crowley, J. M., 1973, Electrical breakdown of bimolecular lipid membranes as an electromechanical instability, *Biophys. J.* **13**:711–724.
de Kruijff, B., Cullis, P. R., Verkleij, A. J., Hope, M. J., van Echteld, C. J. A., and Taraschi, T. F., 1985, Lipid polymorphism and membrane function, in: *The Enzymes of Biological Membranes,* Vol. 1: *Membrane Structure and Dynamics* (A. N. Martonosi, ed.), pp. 131–204, Plenum Press, New York.
Dimitrov, D. S., and Jain, R. K., 1984, Membrane stability, *Biochim. Biophys. Acta* **779**:437–468.
Finaz, C., Lefevre, A., and Teissie, J., 1984, A new, highly efficient technique for generating somatic cell hybrids, *Exp. Cell Res.* **150**:477–482.
Gauger, B., and Bentrup, F. W., 1979, A study of dielectric membrane breakdown in the *Fucus* egg, *J. Membrane Biol.* **48**:249–264.
Gingell, D., and Ginsberg, L., 1978, Problems in the physical interpretation of membrane interaction and fusion, in: *Membrane Fusion* (G. Poste and G. L. Nicolson, eds.), pp. 792–833, North-Holland, Amsterdam.
Gleba, Y. Y., and Sytnik, K. M., 1984, Protoplast Fusion. Genetic Engineering in Higher Plants, in: *Monographs on Theoretical and Applied Genetics,* Vol. 8 (R. Frankel, ed.), Springer-Verlag, Berlin.
Halfmann, H. J., Rocken, W., Emeis, C. C., and Zimmermann, U., 1982, Transfer of mitochondrial function into a cytoplasmic respiratory-deficient mutant of *Saccharomyces* yeast by electro-fusion, *Curr. Genet.* **6**:25–28.
Halfmann, H. J., Emeis, C. C., and Zimmermann, U., 1983, Electro-fusion of haploid *Saccharomyces* yeast cells of identical mating type, *Arch. Microbiol.* **134**:1–4.
Hui, S. W., Isac, T., Boni, L. T., and Sen, A., 1985, Action of polyethylene glycol on the fusion of human erythrocyte membranes, *J. Membrane Biol.* **84**:137–146.
Jacob, H. -E., Siegemund, F., and Bauer, E., 1984, Fusion von pflanzlichen Protoplasten durch elektrischen Feldimpuls nach Dielectrophorese, *Biol. Zentralbl.* **103**:77–82.
Kao, K. N., 1981, Plant protoplast fusion and somatic hybrids, in: *Plant Tissue Culture* (H. Han, ed.), pp. 331–339, The Pitman International Series in Applied Biology, Proceedings of Beijing Symposium, London.
Kao, K. N., and Michayluk, M. R., 1974, A method for high frequency of intergeneric fusion of plant protoplasts, *Planta* **115**:355–367.
Karsten, U., Papsdorf, G., Roloff, G., Stolley, P., Abel, H., Walther, I., and Weiss, H., 1985, Monoclonal anti-cytokeratin antibody from a hybridoma clone generated by electrofusion, *Eur. J. Cancer Clin. Oncol.* **21**:733–740.
Kartha, K. K., Gamborg, O. L., Constabel, F., and Kao, K. N., 1974, Fusion of rapeseed and soybean protoplasts and subsequent division of heterokaryocytes, *Can. J. Bot.* **52**:2435–2436.
Kinosita, K., and Tsong, T. Y., 1977*a,* Hemolysis of human erythrocytes by a transient electric field, *Proc. Natl. Acad. Sci. U.S.A.* **74**:1923–1927.
Kinosita, K., and Tsong, T. Y., 1977*b,* Formation and resealing of pores of controlled sizes in human erythrocyte membrane, *Nature (Lond.)* **268**:438–441.
Kinosita, K., and Tsong, T. Y., 1979, Voltage-induced conductance in human erythrocyte membranes, *Biochim. Biophys. Acta* **554**:479–497.
Kohn, H., Schieder, R., and Schieder, O., 1985, Somatic hybrids in tobacco mediated by elec-

trofusion, *Plant Sci.* **38:**121–128.

Koop, H. U., Dirk, J., Wolff, D., and Scheiger, H. G., 1983, Somatic hybridization of two selected single cells, *Cell Biol. Int. Rep.* **7:**1123–1128.

Lo, M. M. S., Tsong, T. Y., Conrad, M. D., Strittmatter, S. M., Hester, L. D., and Snyder, S. H., 1984, Monoclonal antibody production by receptor-mediated electrically induced cell fusion, *Nature (Lond.)* **310:**792–794.

Lucy, J. A., 1984, Fusogenic mechanisms, in: *Cell Fusion,* Ciba Foundation Symposium 103 (J. Evered and J. Whelan, eds.), pp. 28–39, Pitman, London.

McCloskey, M. A., Liu, Z. Y., and Poo, M. M., 1984, Lateral electromigration and diffusion of Fce receptors on rat basophilic leukemia cells: Effects of IgE binding, *J. Cell Biol.* **99:**778–787.

Mehrle, W., Zimmermann, U., and Hampp, R., 1985, Evidence for asymmetrical uptake of fluorescent dyes through electro-permeabilized membranes of *Avena* mesophyll protoplasts, *FEBS Lett.* **185:**89–94.

Melikyan, G. B., Abidor, I. G., Chernomordik, L. V., and Chailakhyan, L. M., 1983, Electrostimulated fusion and fission of bilayer lipid membranes, *Biochim. Biophys. Acta* **730:**395.

Mishra, K. P., Binh, L. D., and Singh, B. B., 1981, Effect of dielectric discharge on drug treated mammalian cells, *Ind. J. Exp. Biol.* **19:**520–523.

Neumann, E., and Katchalsky, A., 1972, Long-lived conformation changes induced by electric impulses in biopolymers, *Proc. Natl. Sci. U.S.A.* **69:**993–997.

Neumann, E., Gerisch, G., and Opatz, K., 1980, Cell fusion induced by high electric impulses applied to *Dictyostelium, Naturwissenschaften* **67:**414–415.

Neumann, E., Schaefer-Ridder, M., Wang, Y., and Hofschneider, P. H., 1982, Gene transfer into mouse lymphoma cells by electroporation in high electric fields, *EMBO J.* **1:**841–845.

Ohno-Shosaku, T., and Okada, Y., 1984, Facilitation of electrofusion of mouse lymphoma cells by the proteolytic action of proteases, *Biochem. Biophys. Res. Commun.* **120:**138–143.

Pethig, R., 1979, *Dielectric and Electronic Properties of Biological Materials,* Wiley, New York.

Pilwat, G., Zimmermann, U., and Riemann, F., 1975, Dielectric breakdown measurements of human and bovine erythrocyte membranes using benzyl alcohol as a probe molecule, *Biochim. Biophys. Acta* **406:**424–432.

Pliquett, F., and Wunderlich, S., 1980, Relationship between cell parameters and pulse deformation due to these cells as well as its change after electrically induced membrane breakdown, *Bioelectrochem. Bioenerg.* **10:**467–475.

Podesta, E. J., Solano, A. R., Molina, Y., Vedia, L., Paladini, A., Sanchez, M. L., and Torres, H. N., 1984, Production of steroid hormone and cyclic AMP in hybrids of adrenal and Leydig cells generated by electrofusion, *Eur. J. Biochem.* **145:**329–332.

Pohl, H. A., 1978, *Dielectrophoresis,* Cambridge University Press, London.

Pohl, H. A., and Crane, J. S., 1971, Dielectrophoresis of cells, *Biophys. J.* **11:**711–728.

Pohl, H. A., Pollock, K., and Rivera, H., 1984, The electrofusion of cells, *Int. J. Quant. Chem. Quant. Biol. Symp.* **11:**327–345.

Poo, M., 1981, *In situ* electrophoresis of membrane components, *Annu. Rev. Biophys. Bioeng.* **10:**245–276.

Poste, G., and Pasternak, C. A., 1978, Virus-induced cell fusion, in: *Membrane Fusion* (G. Poste and G. L. Nicolson, eds.), pp. 306–367, North-Holland, Amsterdam.

Potter, H., Weir, L., and Leder, P., 1984, Enhancer-dependent expression of human k immunoglobulin genes introduced into mouse pre-B lymphocytes by electroporation, *Proc. Natl. Acad. Sci. U.S.A.* **81:**7161–7165.

Richter, H. -P., Scheurich, P., and Zimmermann, U., 1981, Electric field-induced fusion of sea urchin eggs, *Dev. Growth Diff.* **23:**479–486.

Riemann, F., Zimmermann, U., and Pilwat, G., 1975, Release and uptake of haemoglobin and ions in red blood cells induced by dielectric breakdown, *Biochim. Biophys. Acta* **394**:449–462.

Ruthe, H. -J., and Adler, J., 1985, Fusion of bacterial spheroplasts by electric fields, *Biochim. Biophys. Acta* **819**:105–113.

Sale, A. J. H., and Hamilton, W. A., 1968, Effects of high electric fields on micro-organisms. III. Lysis of erythrocytes and protoplasts, *Biochim. Biophys. Acta* **163**:37–43.

Salhani, N., Vienken, J., Zimmermann, U., Ward, M., Davey, M. R., Clothier, R. H., Balls, M., Cocking, E. C., and Lucy, J. A., 1985, Haemoglobin synthesis and cell wall regeneration by electric field-induced interkingdom heterokaryons, *Protoplasma* **126**:30–35.

Sauer, F. A., 1983, Forces on suspended particles in the electromagnetic field, in: *Coherent Excitations in Biological Systems* (H. Fröhlich and F. Kremer, eds.), pp. 134–144, Springer-Verlag, Berlin.

Saunders, J. A., Roskos, L. A., Mischke, S., Aly, M., and Owens, L. D., 1986, Behavior and viability of tobacco protoplasts in response to electrofusion parameters, *Plant Physiol.* **80**:117–121.

Scheurich, P., and Zimmermann, U., 1980, Membrane fusion and deformation of red blood cells by electric fields, *Z. Naturforsch.* **35c**:1081–1085.

Scheurich, P., and Zimmermann, U., 1981*a*, Electrically stimulated fusion of different plant cell protoplasts, *Plant Physiol.* **67**:849–853.

Scheurich, P., and Zimmermann, U., 1981*b*, Giant human erythrocytes by electric-field-induced cell-to-cell fusion, *Naturwissenschaften* **68**:45–47.

Schwister, K., and Deuticke, B., 1985, Formation and properties of aqueous leaks induced in human erythrocytes by electrical breakdown, *Biochim. Biophys. Acta* **816**:332–348.

Senda, M., Takeda, J., Abe, S., and Nakamura, T., 1979, Induction of cell fusion of plant protoplasts by electrical stimulation, *Plant Cell Physiol.* **20**:1441–1443.

Serpersu, E. H., Kinosita, K., and Tsong, T. Y., 1985, Reversible and irreversible modification of erythrocyte membrane permeability by electric field, *Biochim. Biophys. Acta* **812**:779–785.

Shimmen, T., and Tazawa, M., 1983, Permeabilization of *Nitella* internodal cell with electrical pulses, *Protoplasma* **117**:93–96.

Shivarova, N., and Grigorova, R., 1983, Microbiological implications of electric field effects. Part VIII. Fusion of *Bacillus thuringiensis* protoplasts by high electric field phase, *Bioelectrochem. Bioenerget.* **11**:181–185.

Shivarova, N., Forster, W., Jacob, H. -E., and Grigorova, R., 1983, Microbiological implications of electric field effects. VII. Stimulation of plasmid transformation of *Bacillus cereus* protoplasts by electric field pulses, *Allg. Mikrobiol.* **23**:595–599.

Sowers, A. E., 1983, Fusion of mitochondrial inner membranes by electric fields produces inside-out vesicles. Visualization by freeze-fracture electron microscopy, *Biochem. Biophys. Acta* **735**:426–428.

Sowers, A. E., 1984, Characterization of electric field-induced fusion in erythrocyte ghost membranes, *J. Cell Biol.* **99**:1989–1996.

Sowers, A. E., 1985, Movement of a fluorescent lipid label from a labeled erythrocyte membrane to an unlabeled erythrocyte membrane following electric-field-induced fusion, *Biophys. J.* **47**:519–525.

Sowers, A. E., 1986, A long-lived fusogenic state is induced in erythrocyte ghosts by electric pulses, *J. Cell Biol.* **102**:1358–1362.

Steinbiss, H. -H., 1978, Dielektrischer Durchbruch des Plasmalemmas von *Valerianella locusta*-Protoplasten, *Z. Pflanzenphysiol.* **88**:95–102.

Takashima, S., and Schwan, H. P., 1985, Alignment of microscopic particles in electric fields and its biological implications, *Biophys. J.* **47**:513–518.

Teissie, J., and Blangero, C., 1984, Direct experimental evidence of the vectorial character of the interaction between electric pulses and cells in cell electrofusion, *Biochim. Biophys. Acta* **775**:446–448.

Teissie, J., and Tsong, T. Y., 1981, Electric field induced transient pores in phospholipid bilayer vesicles, *Biochemistry* **20**:1548–1554.

Teissie, J., Knutson, V. P., Tsong, T. Y., and Lane, M. D., 1982, Electric pulse-induced fusion of 3T3 cells in monolayer culture, *Science* **216**:537–538.

Teixeira-Pinto, A. A., Nejelski, L. L., Jr., Cutler, J. L., and Heller, J. H., 1960, The behavior of unicellular organisms in an electromagnetic field, *Exp. Cell Res.* **20**:548–564.

Tempelaar, M. J., and Jones, M. G. K., 1985*a*, Directed electrofusion between protoplasts with different responses in a mass fusion system, *Plant Cell Rep.* **4**:92–95.

Tempelaar, M. J., and Jones, M. G. K., 1985*b*, Fusion characteristics of plant protoplasts in electric fields, *Planta* **165**:205–216.

Tsong, T. Y., 1983, Voltage modulation of membrane permeability and energy utilization in cells, *Biosci. Rep.* **3**:487–505.

Verhoek-Köhler, B., Hampp, R., Ziegler, H., and Zimmermann, U., 1983, Electro-fusion of mesophyll protoplasts of *Avena sativa*, *Planta* **158**:199–204.

Vienken, J., and Zimmermann, U., 1982, Electric field-induced fusion: electro-hydraulic procedure for production of heterokaryon cells in high yield, *FEBS Lett.* **137**:11–13.

Vienken, J., Ganser, R., Hampp, R., and Zimmermann, U., 1981, Electric field-induced fusion of isolated vacuoles and protoplasts of different developmental and metabolic provenience, *Physiol. Plant* **53**:64–70.

Vienken, J., Zimmermann, U., Fouchard, M., and Zagury, D., 1983*a*, Electrofusion of myeloma cells on the single cell level; fusion under sterile conditions without proteolytic enzyme treatment, *FEBS Lett.* **163**:54–56.

Vienken, J., Zimmermann, U., Ganser, R., and Hampp, R., 1983*b*, Vesicle formation during electro-fusion of mesophyll protoplasts of *Kalanchoe diagremontiana*, *Planta* **157**:331–335.

Watts, J. W., and King, J. M., 1984, A simple method for large-scale electrofusion and culture of plant protoplasts, *Biosci. Rep.* **4**:335–342.

Weber, H., Forster, W., Berg, H., and Jacob, H. -E., 1981*a*, Parasexual hybridization of yeasts by electric field stimulated fusion of protoplasts, *Curr. Genet.* **4**:165–166.

Weber, H., Forster, W., Jacob, H. -E., and Berg, H., 1981*b*, Microbiological implications of electric field effects. III. Stimulation of yeast protoplast fusion by electric field pulses, *Allg. Mikrobiol.* **21**:555–562.

White, S. H., 1980, How electric fields modify alkane solubility in lipid bilayers, *Science* **207**:1075–1077.

White, S. H., and Chang, W., 1981, Voltage dependence of the capacitance and area of black lipid membranes, *Biophys. J.* **36**:449–453.

Wong, T. -K., and Neumann, E., 1982, Electric field mediated gene transfer, *Biochem. Biophys. Res. Commun.* **107**:584–587.

Zachrisson, A., and Bornman, C. H., 1984, Application of electric field fusion in plant tissue culture, *Physiol. Plant* **61**:314–320.

Zimmermann, U., 1982, Electric field-mediated fusion and related electrical phenomena, *Biochem. Biophys. Acta* **694**:227–277.

Zimmermann, U., and Benz, R., 1980, Dependence of electrical breakdown voltage on the charging time in *Valonia utricularis*, *J. Membrane Biol.* **53**:33–43.

Zimmermann, U., and Scheurich, P., 1981, High frequency fusion of plant protoplasts by electric fields, *Planta* **151**:26–32.

Zimmermann, U., and Vienken, J., 1982, Electric field-induced cell-to-cell fusion, *J. Membrane Biol.* **67**:165–182.

Zimmermann, U., Pilwat, G., Beckers, F., and Riemann, F., 1976, Effects of external electrical fields on cell membranes, *Bioelectrochem. Bioenerget.* **3**:58–83.

Zimmermann, U., Beckers, F., and Coster, H. G. L., 1977, The effect of pressure on the electrical breakdown in the membranes of *Valonia utricularis, Biochim. Biophys. Acta* **464**:399–416.

Zimmermann, U., Vienken, J., and Pilwat, G., 1980, Development of drug carrier systems: Electrical field induced effects in cell membranes, *Bioelectrochem. Bioenerget.* **7**:553–574.

Zimmermann, U., Pilwat, G., and Richter, H. -P., 1981*a,* Electric-field-stimulated fusion: Increased field stability of cells induced by pronase, *Naturwissenschaften* **68**:577–579.

Zimmermann, U., Vienken, J., and Pilwat, G., 1981*b,* Rotation of cells in an alternating electric field: The occurrence of a resonance frequency, *Z. Naturforsch.* **36c**:173–177.

Zimmermann, U., Buchner, K. H., and Arnold, W. M., 1984*a,* Electrofusion of cells: Recent developments and relevance for evolution, in: *Charge and Field Effects in Biosystems* (M. J. Allen and P. N. R. Usherwood, eds.), pp. 293–318, Abacus Press, Normal, Ill.

Zimmermann, U., Vienken, J., Pilwat, G., and Arnold, W. M., 1984*b,* Electrofusion of cells: principles and potential for the future, in: *Cell Fusion,* CIBA Foundation Symposium 103, pp. 60–73, Pitman, London.

Chapter 18

The Electrofusion Mechanism in Erythrocyte Ghosts

Arthur E. Sowers and Veena Kapoor

1. INTRODUCTION

Electrofusion is a new method of inducing membrane fusion that uses electric field pulses. A general review of the method will be found elsewhere in this volume (Chapter 17). This chapter reviews what is known about the mechanism of electrofusion and will include a discussion of the results from our laboratory on studies using erythrocyte ghosts as a model membrane.

2. THE PRESENT HYPOTHESIS

All previously reported membrane electrofusion protocols and their variations have called for membrane–membrane contact to be established and maintained while one or more strong but short-duration fusion-inducing direct current pulses is passed through the medium. How electrofusion works was first proposed by Pilwat *et al.* (1981). Their explanation involves the phenomenon of electroporation but, in the context of their experiments, utilized alternating current-induced dielec-

ARTHUR E. SOWERS and VEENA KAPOOR • American National Red Cross, Jerome H. Holland Laboratory for the Biomedical Sciences, Rockville, Maryland 20855.

trophoresis (Pohl, 1978; Pohl *et al.,* 1984) as the means by which membrane–membrane contact was established before the pulses were applied. Presumably the same explanation would hold for any membrane system as long as close membrane–membrane contact existed before and during the application of the pulses. Moreover, recently published reviews (e.g., Arnold and Zimmermann, 1984; Zimmermann *et al.,* 1984, 1985) still refer to this mechanism. To date no alternative hypothetical mechanisms have been published.

Electroporation (Fig. 1), the process by which pores are induced in membranes treated with pulses, was well studied in the laboratories of Tsong, Newmann, and Zimmermann during the late 1970s and early 1980s (reviewed by Tsong, 1983). Electroporation, also referred to as membrane breakdown, can be reversible or irreversible. If it is reversible, the pores reseal. The resealing rate depends on conditions (reviewed by Tsong, 1983; Serpersu *et al.,* 1985; Schwister and Deuticke, 1985).

According to the mechanism proposed by Pilwat *et al.* (1981), breakdown occurs at one or more locations in the contact area shared by both membranes in response to a step increase in transmembrane voltage (Fig. 2). This breakdown results in the simultaneous induction of a pore in each of the two closely-spaced membranes and each pore is concentric with the other. Alternatively, breakdown could occur at a given location in only one of the two membranes, causing a rapid incremental increase in the transmembrane voltage across the other membrane. This would lead to breakdown in the second membrane immediately afterward. Either of these breakdown scenarios is presumed to result in a set of two concentric pores: a pore in one membrane in concentric juxtaposition to a pore in the other membrane. Presumably more than one of these concentric pore pairs can exist in different locations in the contact area. Hereafter, this explanation of electrofusion is referred to as the concentric pore-pair (CPP) hypothesis. The cylindrical space between the concentric pairs of pores is also simultaneously and transiently occupied by a cloud composed of individually dispersed lipid molecules (Fig. 2). The fused membrane geometry then condenses from this cloud. A recently published variation (Dimitrov and Jain, 1984) of the hypothetical CPP mechanism however does not involve a cloud of lipids between the concentric pair of pores in originally separate (prefusion) membranes before membrane–membrane fusion occurs (Fig. 3). Instead, the shape of the membrane in the immediate vicinity of the pore develops an outward flare that bridges the gap between the originally separate membranes. In both versions of the CPP hypothesis the pores are viewed as having a diameter about equal to the thickness of the lipid bilayer. Neither version addresses the role of integral proteins, cytoskeletal elements, or the glycocalyx.

If the constriction of the now fused bilayer is stable, it will determine

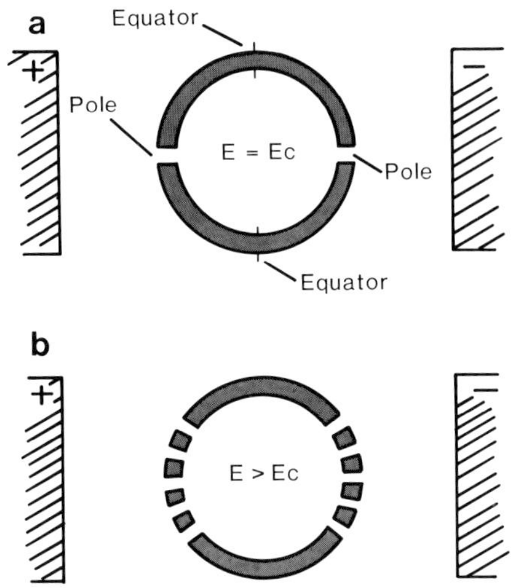

Figure 1. Electropore formation in a spherical membrane in a pulsed electric field. (a) Electric field strength, E, induces a transmembrane electric field strength, equal to the breakdown field strength E_C. Note pore at each pole of membrane. (b) Electric field strength is greater than the breakdown field strength. Note more pores appear at locations farther from the poles. (Adapted from Zimmermann, 1983.)

the upper limit for the diameter of molecules, which can pass from one originally separate cytoplasmic compartment to the other (Figs. 2 and 3). If the constriction is unstable, the inside diameter of this constriction may increase sufficiently to form a lumen that will eventually become large enough to be visible by phase-contrast optics. Both possibilities have been shown to occur in erythrocyte ghosts (Sowers, 1984). Continued increase in the lumen diameter would eventually convert the hourglass morphology to the morphology of a single large sphere.

3. PULSES INDUCE A LONG-LIVED FUSOGENIC STATE

We discovered that fusion would also occur if membranes were treated with pulses before being brought into membrane–membrane contact by dielectrophoresis (Sowers, 1983*b*, 1984). Further studies showed that fusion would occur even if significant time intervals were to separate the treatment of the membranes with pulses and the moment when membrane–membrane contact was induced (Sowers, 1985*a*, 1985*c*, 1986*a*). This new fundamental observation showed that a structural alteration was induced in the membrane and that this alteration was both fusogenic and long-lived. To distinguish between the original method of inducing electrofusion and the method of electrofusion which involves treating membranes with pulses before membrane contact, we introduce the designations contact first (CF) and pulse first (PF) to describe the protocol used to achieve electrofusion.

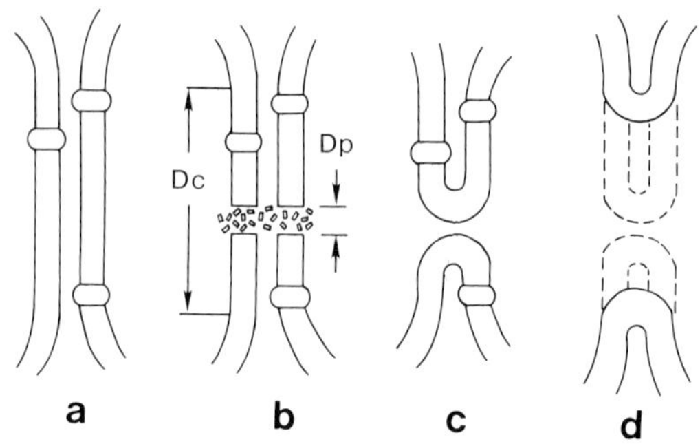

Figure 2. Electrofusion mechanism as proposed by Pilwat *et al.* (1981). (a) Closely-spaced membranes (contact), (b) concentric pairs of pores simultaneously induced by the pulse treatment (note cloud of dispersed lipid molecules), (c) subsequent connection of originally separate membranes (fusion), (d) increase in lumen diameter (needed for identification by phase optics). Dc: diameter of contact area; Dp: diameter of pore. (See Sowers, 1984.)

Long-lived structural alterations induced in membranes by electric fields have been detected by other means in other laboratories (Stulen, 1981; Chernomordik *et al.,* 1985). Indeed, residual electropores with very small diameters can be detected up to minutes or hours after pulse treatments and themselves are one known form of pulse-induced membrane structural alteration (Serpersu *et al.,* 1985).

The long-lived fusogenic state is not predicted by the hypothetical CPP mechanism. Furthermore, fusion by the use of the PF protocol suggests that fusion might not occur according to this hypothesis because (1) the pores must already exist on the membranes before the membranes come into contact; and (2) at the time membrane–membrane contact exists, no pulses are applied to induce pores.

4. CAN AN ELECTROPORE BE A FUSOGENIC SITE?

Concentric pore pairs could be involved in the electrofusion mechanism when the PF protocol is used if they are formed upon membrane–membrane contact as a pore on one membrane precisely collides with a pore on another membrane. In this view of the electrofusion mechanism, a pore is a fusogenic site. Two fusogenic sites on each of two membranes, therefore, must be assembled into a concentric pore pair for fusion to occur.

A first-approximation estimate of the probability of such a collision should be represented by the ratio of total pore area to membrane area in the contact zone (Fig. 2). Previously published estimates of the total area of electropores in erythrocyte membranes (Kinosita and Tsong, 1979; Schwister and Deuticke, 1985) are 10^{-5} of the total cell area and are lower by orders of magnitude than would be needed to account for the observed fusion yields. If the pores were concentrated within a contact area that is only 1% of the total membrane area, this would raise the probability to

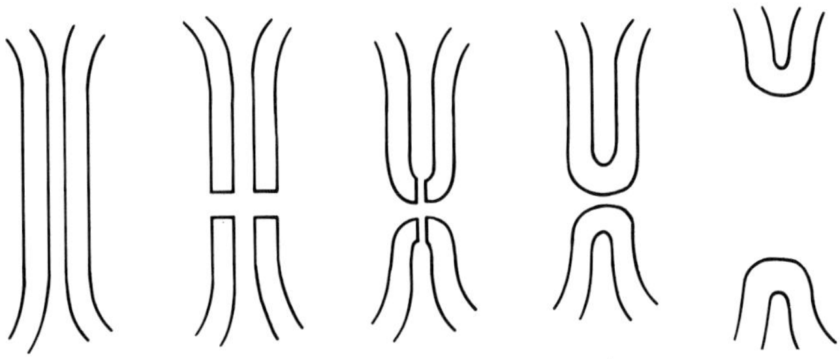

Figure 3. Electrofusion mechanism as proposed by Dimitrov and Jain (1984). (a) Closely-spaced membranes (contact), (b) concentric pairs of pores induced by the pulse treatment, (c) bending of membrane in vicinity of both pores so that the hydrophobic edges move towards each other, (d) connection of originally separate membranes (fusion), (e) increase in lumen diameter. (See Fig. 2.)

only 0.001. This is still much lower than observed yields. The same papers estimate that the numbers of pulse-induced electropores range from 1 to 200 or more per membrane but their location in the membrane is obscure. Hence collision probability may not be a simple calculation if large numbers of pores are present. It should be pointed out, however, that no direct evidence has yet been obtained for electropore number, size, heterogeneity, lifetime, and location on single membranes.

Alternatively, the electropores could be laterally mobile and could thus move to locations where they were not induced. Concentric pore pairs could then be formed after (instead of upon) membrane–membrane contact. If electropores were laterally mobile and electropores were fusogenic sites, fusion yields would continue to increase for some time after membranes came into contact. This is because the probability of two pores, each laterally diffusing in each of two membranes in close contact, encountering each other would increase with time. However, as monitored by the instantaneous movement of FITC dextran to unlabeled cytoplasmic compartments all observed fusion events were simultaneous, within human perception, with the application of pulses if the CF protocol was used, or upon membrane–membrane contact if the PF protocol was used (Sowers, 1984). This is indirect evidence that the pores either diffuse laterally in the membrane at immeasureably slow rates or are laterally immobile (possibly due to cytoskeleton-membrane restraints on lateral mobility).

A time-dependent acceleration in loss of FITC dextran from random-positioned ghosts after any pulse treatment was also never observed. This observation suggests that if pores diffuse laterally, they did not collide with themselves on the same membrane and coalesce into larger pores.

5. FUSOGENIC SITES ARE NOT LATERALLY MOBILE

Previously published papers consider an electropore a hole in the lipid bilayer. The edge of the hole is composed of an annulus of lipids surrounded by a sea of lipids that extends to nearby proteins (Pilwat *et al.*, 1981; Dimitrov and Jain, 1984). Except for experimental evidence in one case (Teissie and Tsong, 1980), it has been generally assumed that integral proteins do not play a role in electroporation or electrofusion. This may have come about because it has been reported that artificial lipid vesicles can undergo electroporation (Teissie and Tsong, 1981; Benz and Zimmermann, 1981) and electrofusion (Buschel *et al.*, 1982). As fusion can occur in the absence of integral proteins, it was thought that they were neither involved nor necessary. From what is commonly accepted about membrane fluidity, however, it can be predicted that small pores should diffuse laterally in the plane of the membrane at rates that should be similar to lipids (Sowers, 1985*b*), if they are indeed composed exclusively of lipids and are no larger in diameter than the largest known integral proteins.

The hypothesis that the fusogenic sites are laterally mobile electropores was tested by using a special four electrode chamber in which any angle of 0–90° could be generated between the orientation of the pearl-chain formation and the orientation of the DC-pulsed electric field (a complete account of these experiments will be published elsewhere). Thus the two different electrode sets could be used to align the membranes so that membrane–membrane contact would take place either at the poles or at the equator or at any location between these limits. Using the PF protocol, the data showed that (1) high yields of fusion occurred for membrane–membrane contacts made at the poles, (2) no fusion above background fusion occurred for contacts made at or near the equator, and (3) intermediate fusion yields occurred for contacts made between these limits (Sowers, 1986*b*). These results implied that (1) only a fraction of the total erythrocyte membrane area became fusogenic; (2) the fusogenic property was induced in the area, which coincided with the point at which the highest transmembrane field strengths were expected; and (3) the fusogenicity was a continuously variable function of location on the membrane. Passage of greater time intervals between the moment of pulse application and the moment of membrane–membrane contact showed that fusion yields went down with time when membrane–membrane contact was established at the poles, but the fusion yields remained at background levels for membrane contact at or near the equator. If fusogenic sites laterally diffused in the plane of the membrane at rates comparable to lipid molecules, fusion yields at some location away from the poles should have increased measureably with time. Our more detailed studies showed that this did not happen (Sowers, 1986*b*). It would thus appear that the long-lived fusogenic structural alterations in the fusogenic area are laterally immobile. This is inconsistent with the concept of a fusogenic site as an electropore composed exclusively of lipids.

6. ELECTROPORE DIAMETER AND THE TWO ELECTROFUSION PROTOCOLS

Our studies of electropore formation used FITC-dextran (10 kD = 3-nm effective diameter) as a soluble fluorescent label (Sowers, 1984). This molecular probe is loaded into cytoplasmic compartments of erythrocyte ghosts through the low-ionic-strength-induced hemolytic pore (Lieber and Steck, 1982*a,b*) held open at low temperatures (0–4°C) followed by resealing at room temperature (several washes remove the excess label from the medium). The pulse-induced movement of the label from a cytoplasmic compartment to the medium or to an adjacent unlabeled cytoplasmic compartment upon fusion can be positively and instantly identified visually by using fluorescence microscopy. This provides rigorous evidence of electropore formation or fusion, respectively, and does not depend on postfusion macroscopic morphological changes for the fusion event to be scored or evaluated. Indeed, we showed that use of phase contrast optics alone may seriously underestimate fusion yields (Sowers, 1984).

We found (Sowers, 1984, 1986*a*) that the pulse-induced loss of FITC-dextran from cytoplasmic compartments of labeled erythrocytes through electric field pulse-induced pores was always directly related to pulse strength (pulse voltage and pulse decay half-time), and total number of applied pulses and inversely related to ionic strength (over the concentration range of sodium phosphate buffer: 20–60 mM). These results are consistent with previous studies of electropore formation in erythrocyte membranes (reviewed by Tsong, 1983).

More recent studies in our laboratory (Sowers, 1986*a*) utilized 6-deoxy-*N*-(7-nitrobenz-2-oxa-1,3-diazol-4-yl) aminoglucose (NBD-G) (Speizer *et al.*, 1985), Lucifer Yellow (Stewart, 1981), and R-phycoerythrin (Oi *et al.*, 1982). These fluorescent soluble molecules have effective molecular diameters of about 0.9, 1.2, and 8.8 nm, respectively. The first two are much smaller than 10 kD FITC-dextran and leaked out of labeled cytoplasmic compartments, with fewer pulses through electropores induced by the use of the same pulses used in the FITC-dextran efflux study (Sowers, 1984). However, R-phycoerythrin did not come out regardless of the pulses used. Significant loss of FITC-dextran and Lucifer Yellow did not occur between pulses. However, as little as one pulse resulted in a noticeable and continuous loss of NBD-G from these membranes. Thus it

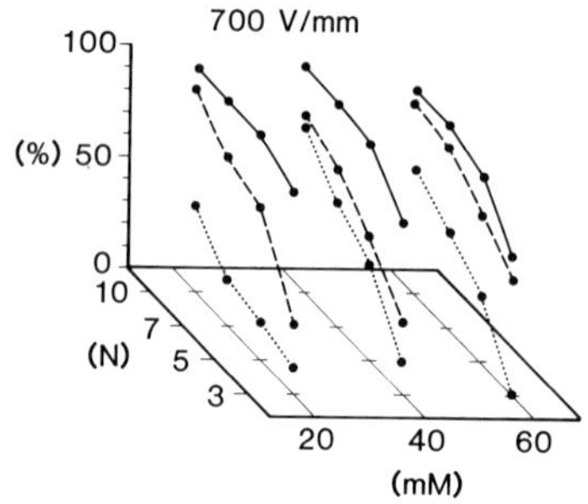

Figure 4. Fusion yield (%) in erythrocyte ghosts held in pearl chains by dielectrophoresis during the pulse treatment. Pulses generated a field strength of 700 V/mm in the chamber. Pulse decay half-times were 0.2 msec (. . .), 0.6 msec (---), and 1.2 msec (—). *N*, total number of applied pulses. Ionic strength of NaPi suspension buffer (pH 8.5) in millimeters. (See Sowers, 1984, 1986*a,b,* for further details.)

is concluded that few if any of the electropores reach a size near or larger than 8.8 nm during a pulse, but that after a pulse the retention of FITC-dextran and Lucifer Yellow between pulses indicated that the pores reclosed to very small sizes (e.g., ~1 nm) very quickly.

If it can be assumed that electropores induced in membranes while they are in random positions in suspension are comparable to those induced in dielectrophoretically aligned membranes, the fusion yields obtained from the use of the CF and PF protocols should be predictable in the following way. When the PF protocol is used, the electropores will have a residual diameter of about 1 nm or less at the time of membrane–membrane contact. When the CF protocol is used, the electropores will have a peak diameter during the pulse of $\geqslant 4$ nm while close membrane–membrane contact is maintained. Thus, the difference between the two electropore areas will be about $4^2/1^2 = 16$. If pore collisions are a factor in the mechanism of fusion, fusion yields should be about 16 times higher using the CF protocol than using the PF protocol. Experimental results indicate for $N = 3$ pulses with a decay half-time of 1.2 msec a fusion yield of 60–90% (Fig. 4) using the CF protocol and a yield of 15–25% when using the PF protocol (Fig. 5). The ratio of CF to PF fusion is lower than predicted, but the trend is consistent in this case with predictions.

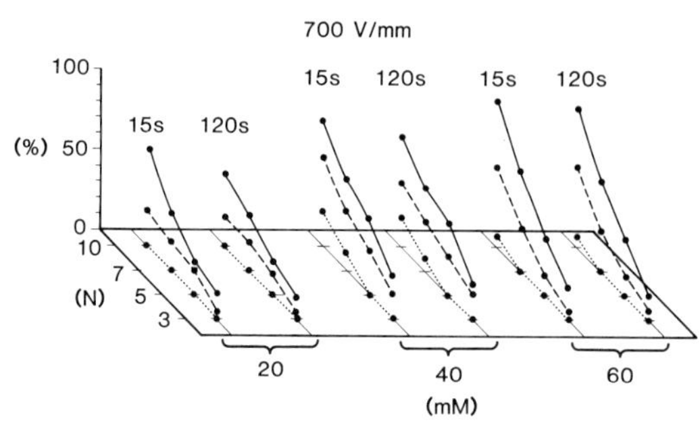

Figure 5. Fusion yield (%) in erythrocyte ghosts obtained by use of the PF protocol (see text). The complete pulse treatment preceded the moment of membrane–membrane contact by either 15 sec or 120 sec. Pulse parameters and suspension medium same as in Figure 4. (See Sowers, 1984, 1986*a,b,* for further details.)

Pohl, H. A., Pollock, K., and Rivera, H., 1984, The electrofusion of cells, *Int. J. Quantum Chem.: Quantum Biol. Symposium* **11**:327–345.

Schwister, K., and Deuticke, B., 1985, Formation and properties of aqueous leaks induced in human erythrocytes by electrical breakdown, *Biochim. Biophys. Acta* **816**:332–348.

Serpersu, E. H., Kinosita, K., Jr., and Tsong, T. Y., 1985, Reversible and irreversible modifications of erythrocyte membrane permeability by electric field, *Biochim. Biophys. Acta* **812**:779–785.

Sowers, A. E., 1983*a*, Fusion of mitochondrial inner membranes by electric fields produces inside out vesicles: Visualization by freeze-fracture electron microscopy, *Biochim. Biophys. Acta* **735**:426–428.

Sowers, A. E., 1983*b*, Red cell and red cell ghost membrane shape changes accompanying the application of electric fields for inducing fusion, *J. Cell Biol.* **97**:179a.

Sowers, A. E., 1984, Characterization of electric field-induced fusion in erythrocyte ghost membranes, *J. Cell Biol.* **99**:1989–1996.

Sowers, A. E., 1985*a*, Mechanism of membrane electrofusion in erythrocyte ghosts, *J. Cell Biol.* **101**:309a.

Sowers, A. E., 1985*b*, Movement of a fluorescent lipid label from a labeled erythrocyte membrane to an unlabeled erythrocyte membrane following electric-field-induced fusion, *Biophys. J.* **47**:519–525.

Sowers, A. E., 1985*c*, Electric field-induced membrane fusion in erythrocyte ghosts: Evidence that pulses induce a long-lived fusogenic state and that fusion may not involve pore formation, *Biophys. J.* **47**:171a.

Sowers, A. E., 1986*a*, A long-lived fusogenic state is induced in erythrocyte ghosts by electric pulses, *J. Cell Biol.* **102**:1358–1362.

Sowers, A. E., 1986*b*, Long-lived fusogenic membrane sites induced by electric field pulses are not free to diffuse laterally in the plane of the membrane, *Biophys. J.* **49**:102a.

Sowers, A. E., and Hackenbrock, C. R., 1981, Rates of lateral diffusion of intramembrane particles: Measurement by electrophoretic displacement and rerandomization, *Proc. Natl. Acad. Sci. U.S.A.* **78**:6246–6250.

Sowers, A. E., and Lieber, M. R., 1986, Electropore diameters, lifetimes, numbers, and locations in individual erythrocyte ghosts, *FEBS Lett.* **205**:179–189.

Speizer, L., Haugland, R., and Kutchai, H., 1985, Asymmetric transport of fluorescent glucose analogue by human erythrocytes, *Biochim. Biochem. Acta* **815**:75–84.

Stenger, D. A., and Hui, S. W., 1986, Kinetics of ultrastructural changes during electrically-induced fusion of human erythrocytes, *J. Membr. Biol.*, in press.

Stewart, W. W., 1981, Lucifer dyes—highly fluorescent dyes for biological tracing, *Nature (Lond.)* **292**:17–21.

Stulen, G., 1981, Electric field effects on lipid membrane structure, *Biochim. Biophys. Acta* **640**:621–627.

Teissie, J., and Tsong, T. Y., 1980, Evidence of voltage-induced channel opening in Na/K ATP-ase of human erythrocyte membrane, *J. Membrane Biol.* **55**:133–140.

Teissie, J., and Tsong, T. Y., 1981, Electric field induced transient pores in phospholipid bilayer vesicles, *Biochemistry* **20**:1548–1554.

Tsong, T. Y., 1983, Voltage modulation of membrane permeability and energy utilization in cells, *Biosci. Rep.* **3**:487–505.

Zimmermann, U., 1982, Electric field-mediated fusion and related electrical phenomena, *Biochim. Biophys. Acta* **694**:227–277.

Zimmermann, U., 1983, Electrofusion of cells: Principles and industrial potential, *Trends Biotechnol.* **1**:149–155.

Zimmermann, U., Vienken, J., and Pilwat, G., 1984, Electrofusion of cells, in: *Investigative Microtechniques in Medicine and Biology* (J. Chayen and L. Bitensky, eds.), pp. 89–167, Dekker, New York.

Zimmermann, U., Vienken, J., Halfman, J., and Emeis, C. C., 1985, Electrofusion: A novel hybridization technique, in: *Advances in Biotechnological Processes,* Vol. 4 (A. Mizrahi and A. L. van Wezel, eds.), pp. 79–150, Liss, New York.

Chapter 19

Laser-Induced Cell Fusion

Einhard Schierenberg

1. INTRODUCTION

In recent years a number of different techniques have been developed to fuse cells of various origin in order to obtain hybrids with qualities different from each of the original cells. In many cases, one wants to fuse large numbers of cells (e.g., to generate hybridomas), and fusion occurs more or less randomly (e.g., Köhler and Milstein, 1975). With other methods, individual cells have been fused under controlled conditions (e.g., McGrath and Solter, 1984). So far a prerequisite for fusing two preselected cells has been that they must be isolated away from other cells.

This chapter reports a technique that permits exclusive fusion of two preselected adjacent cells within an aggregate of cells at precisely chosen points of development and cell cycle phase. The system for which this method has been established is the embryo of the well-studied nematode *Caenorhabditis elegans*. The embryo of this organism is surrounded by a transparent, impermeable eggshell. The microenvironment inside this egg is crucial for normal embryonic development. Therefore, the embryo is hardly accessible for mechanical manipulation from outside. Instead, with a laser microbeam coupled to a microscope, cells of different

Einhard Schierenberg • Max-Planck-Institut für Experimentelle Medizin, Abteilung Chemie, D-3400 Göttingen, Federal Republic of Germany.

developmental potential can be fused. As fusion is performed under continuous visual control, the success of the operation and consequential abnormal cell behavior can be readily observed.

In summary, this method has proved useful for small numbers of selected fusions under specific conditions in which other currently available techniques are not suitable.

2. MATERIAL AND METHODS

2.1. Laser-Microscope Assembly

The different parts of the laser microbeam unit are shown in Figure 1a. The vacuum pump generates a permanent N_2 gas flow (pressure 45–180 mm Hg) in the laser tube of a nitrogen laser (6). At high voltage (10–16 kV), very short light pulses are emitted with a wavelength of 337 nm. These in turn trigger the emission of light from a dye laser (7). The dye laser contains a sealed interchangeable cuvette in which dye is circulated (5). Depending on the dye, the laser beam can vary in wavelength overlap range from approximately 365 to 870 nm. Parellel to the beam of the dye laser runs that of a red He–Ne pilot laser (4), which is used to visualize the target area that will be hit by the dye laser. Along an optical bench, the beams pass through apertures, mirrors, prisms, and lenses (Fig. 1b), collectively enter the microscope, and are focused by a sliding intermediary optic (12—a collecting lens with a focal length of 10 cm) in the image plane of the microscope. After passing through an interchangeable beam splitter (15—with barrier filters for specific wavelengths), the beams enter the 100 × objective to form a focused spot on the specimen of less than 1 μm in diameter. Located in the optical pathway of each beam is a selector wheel (8) containing a set of different gray filters by which the intensity of the light can be decreased stepwise down to 0.01% of the original value.

The control unit (20) permits monitoring and regulation of gas flow rate, gas pressure, voltage, and repetition rate of laser pulses (up to 100 pulses/sec). Via a photocell in the pathway of the laser beam and an oscilloscope (16), the intensity of the flash can be visualized and compared with the predetermined standard specific to that dye. Laser-induced manipulation of the specimen under the microscope is observed via a TV camera (14) on a monitor (19) and can be recorded on a video recorder.

The unit as described above is no longer commercially available. As I am not aware of any company that now offers a laser microbeam coupled to a microscope, the interested reader may have to assemble such a unit himself. Different microbeam designs have been described and may serve

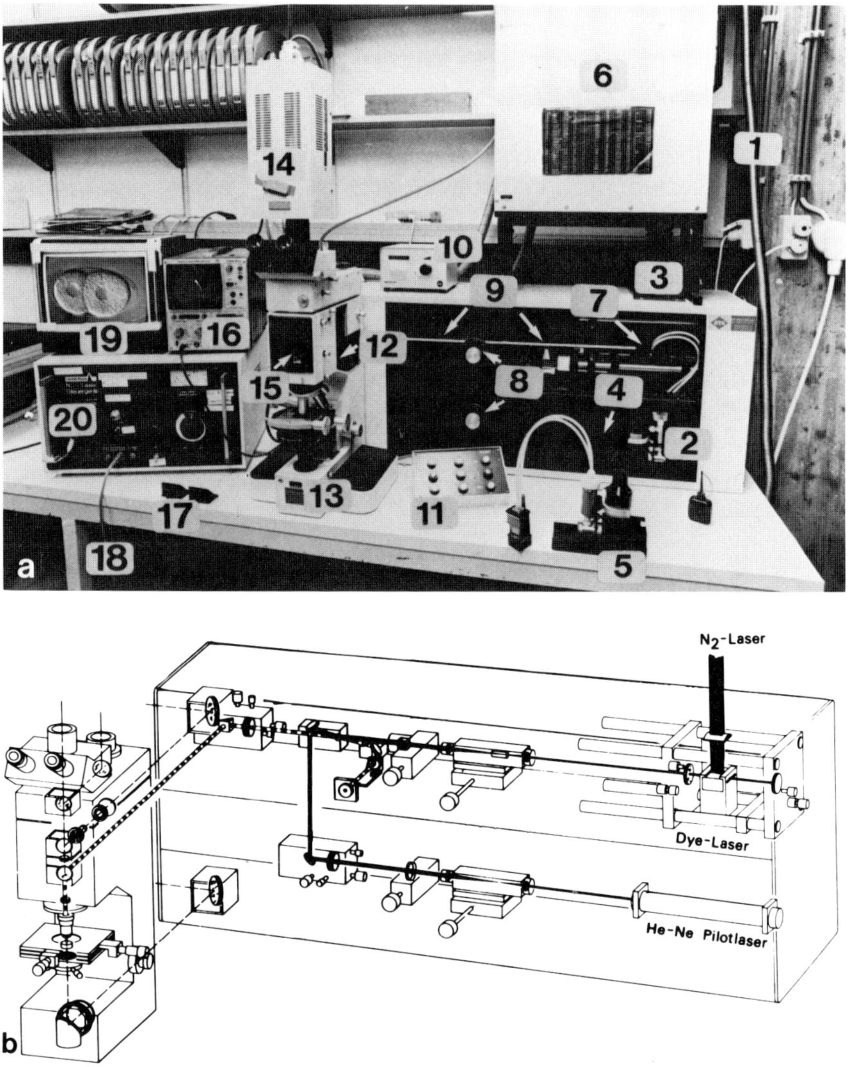

Figure 1. (a) Laser–microscope assembly. 1, Connecting tube between N_2 gas laser and vacuum pump; 2, laser dye reservoir bottle with circulation pump in working position; 3, junction between gas laser and dye laser; 4, position of He–Ne pilot laser; 5, extra laser dye unit with cuvette, tube, filter, reservoir bottle, and circulation pump; 6, N_2 gas laser; 7, dye cuvette in working position; 8, selectors for gray filters; 9, pathway of dye laser beam; 10, transformer for microscope illumination; 11, on–off console for microscope and laser; 12, position of intermediary optics; 13, modified Orthoplan microscope; 14, TV camera; 15, position of beam splitter; 16, oscilloscope; 17, extra beam splitters with various barrier filters; 18, cable connection to foot pedal; 19, TV monitor; and 20, laser control unit. (The lasers are from Lambda Physik, Göttingen, F.R.G.; the microscope is from Leitz, Wetzlar, F.R.G.. The unit was assembled by BTG, Biotechnik GmbH, Munich, F.R.G.) (b) Optical pathways of laser beams. (Courtesy of BTG, Biotechnik GmbH, Munich, F.R.G.)

as a reference (e.g., Moreno *et al.*, 1969; Berns, 1972; Isenberg *et al.*, 1976; Pataki *et al.*, 1978; White and Horvitz, 1979). Zeiss, (D-7082 Oberkochen, F. R. G.) is developing a laser–microscope interface, which should soon be available.

The simplest way to introduce the laser beam into the microscope is from the side through an epicondensor such as the one used for epifluorescence, in which a beam splitter conducts the laser beam down through the objective (see, e.g., Bereiter-Hahn, 1972).

For the purpose of cell fusion, one requires the following for the laser:

1. Enough power to do the operation with a single pulse (our laser has a pulse energy of approx. 250 μJ using the laser dye Rhodamine 6G)
2. Short pulse duration for minimal heat damage (a few nanoseconds with our laser)
3. The option of fast repetition rate (for cell fusion up to 20 pulses/sec, see below)
4. Interchangeable dye cells (variable wavelengths for different operations; not necessary if only cell fusion is intended)
5. A sufficient number of filters to regulate the intensity of the laser beam

For the microscope, one requires the following:

1. An objective with low magnification (6.3× or 10×) for searching and one objective with high magnification (100×)
2. Differential interference contrast (Nomarski) optics for easy focusing on the cell surface
3. Enlarged knobs on the microscope mechanical stage for precise X–Y movement of the specimen (with regular-sized handles, it is quite difficult to position the thin cell membrane exactly on the target spot)

If a TV camera is used, a pilot laser is not essential as the target area where the power laser will hit can be marked on the screen of the video monitor. This makes the laser design simpler and saves time-consuming realignments between the two lasers.

A new generation of nitrogen-pumped dye lasers are on the market which need neither a vacuum pump (normally needed for the nitrogen laser) nor the circulation of the dye in the cuvette. Their small size and simple construction make them particularly suitable for adaptation to a microscope (PRA, Photochemical Research Associates Inc., London, Ontario, Canada, with distributors in several other countries).

2.2. Technique of Cell Fusion

Various laser dyes have been tested in my laboratory for their suitability to fuse cells. In my hands the dye BiBuQ (4,4-bis-(2-butyloctyloxy)-*p*-quaterphenyl), Lambda Physics, Göttingen (main U.S. office: Bedford, MA) with a wavelength of 386 nm works best. With dyes of other wavelengths (Coumarin, Rhodamine) fusion is also possible but the rate of success was found to be lower and the extent of undesired damage higher.

Early embryos are dissected out of the mother and transferred to a microscope slide in a drop of liquid. A coverglass sealed on the edges with vaseline is placed on top (Deppe *et al.,* 1978; Schierenberg, 1982). To obtain a good image and to allow cell fusion with a minimum of laser energy, the embryos can be slightly flattened under the coverslip. To avoid damage they are best placed on a thin agar pad as a cushion (Sulston and Horvitz, 1977). But if manipulations other than cell fusion (such as pressure-induced removal of portions of the embryo) are to be performed, an agar pad is not suitable.

After transfer to the laser microscope stage, a specimen is selected under low magnification. Good candidates for cell fusion are adjacent cells that have broad contact with each other. Preferably, they should lie in the top layer of cells and the membrane to be disrupted should run parallel to the direction of the laser beam. Membranes that run obliquely are difficult to hit.

Just prior to the experiment, the spot marked on the video screen (or generated by the pilot laser) is reconfirmed to still be identical to the target area of the power laser. Then, with the help of intermediary optics, the laser beam must be precisely prefocused in the image plane of the microscope. This can be done by firing on a microscope slide coated with a thin layer of dye complementary to the color of the laser beam (a coverslip covers the dye). The intermediary optics are adjusted until a position is found where with a minimum of energy a small, clear-cut hole can still be burned into the dye (the position is different for each dye). Fusion is performed at high magnification (100 × objective) either with a single pulse to the membrane close to the surface or with 3–10 pulses at a 15–20-Hz frequency and reduced intensity (10–50% of that for single pulse fusion). The fusions performed with a single pulse yield a somewhat higher survival rate but require more practice. Correctly applied, undesired visible damage to cells is rare. It can be reliably detected by retardation (or absence) of subsequent mitoses. The proper selection of intensity which is just sufficient to puncture the membrane (with our laser around one-tenth the maximal power for single pulse fusion) is crucial. Because of its shallow depth of focus Nomarski optics allows a precise focusing on the membrane and observation of any cytoplasmic streaming after fusion. Therefore it is ideally suited for this type of micromanipulation.

In embryonic cells fused just prior to mitosis the membrane starts to retract and disappear within a few seconds (Fig. 2e–h), while in interphase cells the membrane may disappear more slowly (Fig. 2a–d) so that certainty about the success of the operation may be delayed. In many cases, fusion requires more than one attempt (if, however, the membrane is visibly disrupted, no further irradiation is necessary to cause its retraction). In this case, it is advisable to wait a minute or so to avoid accumulation of heat in the specimen which can lead to developmental arrest or even rupture. It is most convenient to trigger the laser with a foot pedal connected to the control unit. In small doses, embryos can tolerate without damage more than 10-fold of the amount of laser irradiation, which is necessary for fusion (Schierenberg, 1984; unpublished observations).

The method developed to fuse embryonic nematode cells also appears to be suitable for fusing cells of other origin. Figure 2i–l shows the fusion of two transformed human epithelial cells. Here visible fusion takes longer to complete than in the nematode cells. The fused pair has not been observed to assume a round shape, but development has not been followed for more than 2 hr. A technical problem is that the cells need the environment of the incubator but must be sealed under a coverslip for manipulation and observation. Preliminary tests indicate that laser-induced fusion between cultured cells of different type is also feasible.

3. EXPERIMENTS

No extensive account is given here of all the results obtained after laser-induced cell fusion in embryos of *C. elegans.* Rather, a few examples are briefly reported and some additional applications for a laser microbeam are outlined. This approach is intended to give the reader an idea of the kind of questions that have been addressed and it is hoped, will facilitate decisions as to whether such an instrument could be a useful experimental tool for other areas of research.

Early embryonic development is largely controlled by maternal gene products that are deposited in the unfertilized egg (cf. Davidson, 1976).

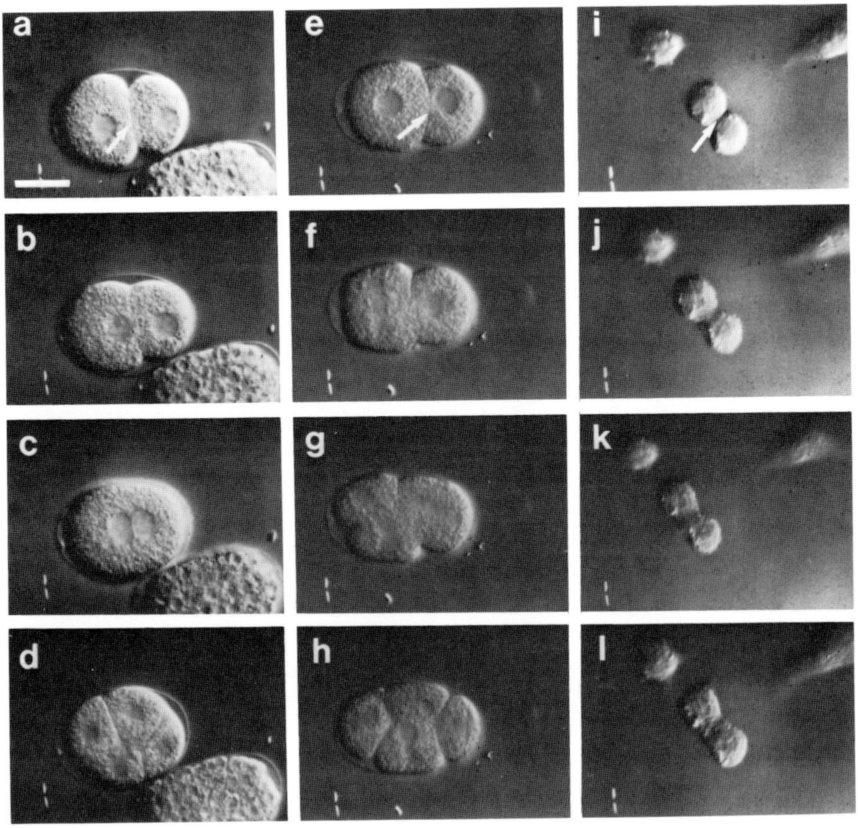

Figure 2. Laser microbeam-induced cell fusion. (a–d) Fusion of interphase cells in *Caenorhabditis.* (a) 2-cell embryo. Arrow points to laser target area. Time: 1 min before laser-induced fusion (b.l.f.) (b) Division membrane partially retracted. Migration of nuclei towards the center. Time: 2 min after laser-induced fusion (a.l.f.) (c) Division membrane completely disappeared. Nuclei have met in the center. Time: 3 min a.l.f. (d) 4 cells after simultaneous tetrapolar cleavage. Time: 10 min a.l.f. (e–h):Fusion of mitotic cells in *C. elegans.* (e) 2-cell embryo. Arrow points to laser target area. Time: 1 min b.l.f. (f) Division membrane partially retracted. Both dividing nuclei stay in original position. At time of fusion left cell was in metaphase, right cell was in late prophase. Differences in cell cycle position are retained. Time: 2 min a.l.f. (g) Both nuclei are in division. Formation of 4 cells. Time: 2 min a.l.f. (h) 4-cell stage. Time: 5 min a.l.f. (i–l) Fusion of transformed human epithelial cells. (i) Two adjacent cells with broad contact. Arrow points to laser target area. Time: 1 min b.l.f. (j) Time: 5 min a.l.f. (k) Time: 15 min a.l.f. (l) Time 60 min a.l.f. No visible signs of damage besides ruptured membrane. All pictures are taken from video tape. Nomarski optics. Bar: 20 μm.

In order to study the ensuing early cell determination, embryos of *C. elegans* have been manipulated using a laser microbeam.

During early embryogenesis of the nematode, a number of somatic cells are generated that serve as founder cells for different cell lineages. The members of each cell lineage divide synchronously and regularly with cell cycle periods different from that of other lineages (Sulston *et al.*, 1983). This results in a specific order of cleaving cell groups, which seems to be crucial for correct positioning and thus for normal morphogenesis (Schierenberg *et al.*, 1980). Fusion of two cells with different cell cycle periods leads to cytoplasmic mixing. Synchronous mitoses of the two nuclei result in the direct formation of four cells simultaneously. Subsequently an acceleration of the cell cycle rhythms in certain (but not all) descendants of the fused cells is observed (Schierenberg, 1984).

The combination of nuclear extrusion and cell fusion enables one to test further the influence of cytoplasmic components on cell behavior. Cytoplasmic transfer between members of different cell lineages can considerably alter subsequent cell cycle periods, while a mere increase in cellular volume by the addition of cytoplasm from the same lineage has only marginal effects (Schierenberg and Wood, 1985).

Typical for embryogenesis in nematodes is the early separation of germ line and soma (Boveri, 1899; reviewed by Chitwood and Chitwood, 1974). While in the young embryo the germ-line cells visibly cleave unequally, the somatic cells divide in an equal fashion and more rapidly. To test whether this might be due to a cytoplasmic asymmetry in the uncleaved egg, the zygote nucleus was extruded (with a major portion of cytoplasm) from the anterior pole. For this experiment, the eggshell was punctured using the laser dye rhodamine 6G (Laufer and von Ehrenstein, 1981). The cleavage pattern of the extruded partial embryo was found to be somalike (i.e., equal). Cell fusion between one descendant of the extruded part and the uncleaved remainder of the egg inside the shell results in the migration of a nucleus back (through the hole in the eggshell) into the latter region. This way a new cell is formed which cleaves in a fashion characteristic of the early germ line (i.e., unequal) (Schierenberg, 1985). The result not only demonstrates non-nuclear determination of cleavage pattern but also shows that the nucleus (or centriole) does not receive any imprint which binds it irreversibly to follow a once-chosen alternative.

Cytoplasmic transfer can also induce descendants of the recipient cell to express characteristics typical for the donor cell. This has been shown using autofluorescence specific to intestinal cells as a marker for cell specific differentiation (Wood *et al.*, 1984; Schierenberg, 1985).

Taken all together, laser microbeam-induced manipulations indicate that determination of cells involves differential segregation of cytoplasmic components. In addition a mechanism that counts the number of cell cycles independent of cell division seems to play a role in determination (Schierenberg, 1985).

In addition to inducing cell fusion and extrusion, a laser microbeam has been used for cell ablation to test for cell autonomy in embryonic and larval development in *C. elegans* (e.g., Kimble and White, 1981; Sulston *et al.,* 1983).

In general, lasers have found a wide range of applications in biological research. They have been used for microsurgery of cells and organelles (e.g., Bereiter-Hahn, 1972; Berns, 1976) as well as for tissue microdissection (Meier-Ruge *et al.,* 1976), analysis of mobility as fluorescence redistribution after photobleaching (Jacobson *et al.,* 1983; Salmon *et al.,* 1984), and studies of ultrafast processes on the molecular level (Hochstrasser and Johnson, 1985).

ACKNOWLEDGMENTS. I would like to thank Kati Vehmeyer for discussion and the supply of cultured cells, Randy Cassada for critical reading of the manuscript, and Mechthild Ziemer for expert photographic work.

REFERENCES

Bereiter-Hahn, J., 1972, Laser micro-radiation as a tool in biology and medicine. Part 2: Method and Applications, *Microsc. Acta* **72:**1–33.

Berns, M. W., 1972, Partial cell irradiation with a tuneable organic dye laser, *Nature (Lond.)* **240:**483–484.

Berns, M. W., 1976, *Biological Microirradiation,* Prentice-Hall, Englewood Cliffs, New Jersey.

Boveri, T., 1899, Die Entwicklung von *Ascaris megalocephala* mit besonderer Rücksicht auf die Kernverhältnisse, *Festschrift für C. v. Kupffer,* Fischer, Jena, pp. 383–430.

Chitwood, B. G., and Chitwood, M. B., 1974, *Introduction to Nematology,* University Park Press, Baltimore.

Davidson, E. H., 1976, *Gene Activity in Early Development,* 2nd ed., Academic Press, New York.

Deppe, U., Schierenberg, E., Cole, T., Krieg, C., Schmitt, D., Yoder, B., and von Ehrenstein, G., 1978, Cell lineages of the embryo of the nematode *Caenorhabditis elegans, Proc. Natl. Acad. Sci. U.S.A.* **75:**376–380.

Hochstrasser, R. M., and Johnson, C. K., 1985, Lasers in biology, *Laser Focus/ElectroOpt.* **21:**100–118.

Isenberg, G., Bielser, W., Meier-Ruge, W., and Remy, E., 1976, Cell surgery by laser microdissection: a preparative method, *J. Microsc.* **107:**19–24.

Jacobson, K., Elson, E., Koppel, E., and Webb, W., 1983, International workshop on the application of fluorescence photobleaching techniques to problems in cell biology, *Fed. Proc.* **42:**72–79.

Laufer, J. S., and von Ehrenstein, G., 1981, Nematode development after removal of egg cytoplasm: Absence of localized unbound determinants, *Science* **211:**402–405.

Kimble, J. E., and White, J. G., 1981, On the control of germ cell development in *Caenorhabditis elegans, Dev. Biol.* **81:**208–219.

Köhler, G., and Milstein, C., 1975, Continuous culture of fused cells secreting antibody of predefined specificity, *Nature (Lond.)* **256:**495–497.

McGrath, J., and Solter, D., 1984, Nuclear transplantation in the mouse embryo by microsurgery and cell fusion, *Science* **220:**1300–1302.

Meier-Ruge, W., Bielser, W., Remy, E., Hillenkamp, F., Nitsche, R., and Unsöld, E., 1976, The laser in the Lowry technique for microdissection of freeze-dried tissue slices, *Histochem. J.* **8:**387–401.

Moreno, G., Lutz, M., and Bessis, M., 1969, Partial cell irradiation by ultraviolet and visible light: conventional and laser sources, *Int. Rev. Exp. Pathol.* **7:**99–137.

Pataki, A., Meier-Ruge, W., Wiederhold, K. -H., and Remy, E., 1978, The laser micro beam as a tool for tissue sampling in biochemistry, *Laser ElektroOptik* **1:**14–16.

Salmon, E. D., Saxton, W. M., Leslie, R. J., Karow, M. L., and McIntosh, J. R., 1984, Diffusion coefficient of fluorescein-labeled tubulin in the cytoplasm of embryonic cells of a sea urchin: Video image analysis of fluorescence redistribution after photobleaching, *J. Cell Biol.* **99:**2157–2164.

Schierenberg, E., 1982, Development of the nematode *Caenorhabditis elegans,* in: *Developmental Biology of Freshwater Invertebrates* (F. W. Harrison and R. R. Cowden, eds.), pp. 249–281, Liss, New York.

Schierenberg, E., 1984, Altered cell-division rates after laser-induced cell fusion in nematode embryos, *Dev. Biol.* **101:**240–245.

Schierenberg, E., 1985, Cell determination during early embryogenesis of the nematode *Caenorhabditis elegans, Cold Spring Harbor Symp. Quant. Biol.* **50:**59–68.

Schierenberg. E., and Wood, W. B., 1985, Control of cell-cycle timing in early embryos of *Caenorhabditis elegans, Dev. Biol.* **107:**337–354.

Schierenberg, E., Miwa, J., and von Ehrenstein, G., 1980, Cell lineages and developmental defects of temperature-sensitive embryonic arrest mutants in *Caenorhabditis elegans, Dev. Biol.* **76:**141–159.

Sulston, J. E., and Horvitz, H. R., 1977, Post-embryonic cell lineages of the nematode *Caenorhabditis elegans, Dev. Biol.* **56:**110–156.

Sulston, J. E., Schierenberg, E., White, J., and Thomson, N., 1983, The embryonic cell lineage of the nematode *Carnorhabditis elegans, Dev. Biol.* **100:**64–119.

White, J. G., and Horvitz, H. R., 1979, Laser microbeam techniques in biological research, *Electro-Opt. Syst. Design* **August:**23–24.

Wood, W. B., Schierenberg, E., and Strome, S., 1984, Localization and determination in early embryos of *Caenorhabditis elegans,* in: *Molecular Biology of Development* (E. H. Davidson and R. A. Firtel, eds.), pp. 37–49, Liss., New York.

IV

Applications of Membrane Fusion

Chapter 20

The Early History of Mammalian Somatic Cell Fusion

John W. Littlefield

Probably the primary value of recounting the early history of any subject is to preserve for some future scholar a chronology and a bibliography that otherwise might be difficult to reconstruct. Accordingly, this brief review of the way in which the technique of mammalian cell fusion or hybridization originated includes both the key original papers and all the significant early reviews concerning this topic known to this author, from 1960, when experiments on fusion began, until about 1975. During the subsequent 10 years up to the present, the use of cell fusion techniques has been extensive and diverse, often in directions that were not apparent in 1975.

For many years, multinucleated cells have been observed by histologists and pathologists, particularly as a consequence of virus infections, and in this case probably the result of fusion of two separate cells rather than the failure of two daughter cells to separate. But it was only 25 years ago that the experimental production of hybrid cells began when Barski *et al.* (1960) reported that occasionally cultures containing a mixture of two sublines of heteroploid mouse cells were overgrown by a new cell line, the karyotype of which was the sum of the two parental

John W. Littlefield • Department of Physiology, Johns Hopkins University School of Medicine, Baltimore, Maryland 21205.

karyotypes. This first method of experimental fusion was slow and relied upon the fact that occasionally spontaneous hybrids grew faster than either parental line. These initial results received little attention, or at best skepticism, but Ephrussi, who had been looking for a method to begin the genetic analysis of animal cells, as had Lederberg (1958) and Pontecorvo (1961), sensed the potential of this new approach. Sorieul and Ephrussi (1961) confirmed the validity of the observation by Barski's group, and then Ephrussi and co-workers went on to study a number of such intraspecific hybrids in regard to the codominance of genes specifying histocompatibility antigens and certain enzymes. They also obtained the first hybrid involving a diploid mouse cell (see Ephrussi, 1972, and references cited therein). Similar confirmatory studies were also reported by Gershon and Sachs (1963). At that time, following the pioneering work of Szybalski, who was unsuccessful in his own attempt to fuse established cell lines (Szybalski *et al.,* 1962; Kit *et al.,* 1963), and others (see Littlefield, 1976, and references cited therein), we were working with mutant sublines of mouse L cells which were resistant to either 8-azaguanine or 5-bromodeoxyuridine because they lacked hypoxanthine-guanine phosphoribosyltransferase or thymidine kinase respectively. Szybalski had designed a selective culture medium called hypoxanthine-aminopterin-thymidine (HAT) to isolate revertants of mutants of his permanent cell line (later shown probably to be HeLa), which were deficient in the phosphoribosyltransferase. We were using this selective medium to study reversion in both our L cell mutants, and it occurred to us that these two cell types should complement each other in a hybrid. Neither parental cell should survive in HAT because aminopterin blocked the endogenous pathways to both guanylic acid and thymidylic acid, and the mutants lacked the ability to salvage hypoxanthine or thymidine from the medium. But if each mutant contributed its remaining salvage enzyme to the hybrid, the hybrid cell should survive. This selection for hybrid cells was immediately successful and it allowed the quantitation and the isolation of hybrid cells much earlier than had been possible previously (Littlefield, 1964, 1966). HAT medium was the first, and has remained the best known, of the media to select hybrid cells, although several others have been described (Choy *et al.,* 1982; Hooper, 1985). [I have had to deny that I originated HAT medium (Littlefield, 1973), just as Ephrussi had to deny that he was the first to discover experimental hybridization (Ephrussi, 1972).]

The next advance was the development by Harris and Watkins (1965) of a technique to increase the frequency of hybrid cells, using ultraviolet (UV) light-inactivated hemagglutinating virus of Japan (HVJ), now called Sendai virus, which had been shown earlier by Okada (1958) to produce the fusion of Ehrlich ascites cells into homokaryons and heterokaryons.

At first it seemed that cells fused in this way were just multinucleated, but did not propagate. Indeed Harris and co-workers subsequently carried out striking experiments on the reactivation of the chick erythrocyte nucleus following fusion to a cell which was metabolically active (H. Harris, 1968). But soon it was clear that at least some virus-stimulated cell fusions were able to grow and divide (Yerganian and Nell, 1966), and the Sendai method became important for hybridization experiments in which the rate of spontaneous fusion was low.

Meanwhile Davidson and Ephrussi (1965) modified the HAT-selective system to make it more generally applicable by using a slowly growing cell as one partner. No selective marker was needed in this cell, because the other marked partner could be killed with HAT as previously, while any hybrid cells quickly outgrew the slowly growing partner. This scheme plus Sendai virus was used by Ephrussi and Weiss (1965) to isolate the first viable interspecific hybrid, which combined an established mouse cell with a selective marker and a freshly explanted diploid rat cell. It was now possible to exploit the differences in chromosome morphology of two species, as well as the differences in electrophoretic mobilities of proteins. Then Weiss and Green (1967) made the remarkable discovery that, in contrast to what occurred in intraspecific fusions, hybrids between a mouse cell line and diploid human fibroblasts progressively lost human chromosomes. Subsequent studies have shown that the nature of the cells used, that is, whether they are primary (diploid) or permanent (heterploid), may be more important than the species of origin (Ephrussi, 1972; Ringertz and Savage, 1976; Hooper, 1985). In any case, this quite unexpected observation opened up the field of gene mapping, which has advanced so extensively and rapidly ever since (Ruddle, 1974). The first studies with either mouse–human (Weiss and Green, 1967) or hamster–human (Kao and Puck, 1970) hybrids focused on the genes in the few human chromosomes remaining after extensive segregation, but much more important subsequently have been analyses of the coincident loss of chromosomes and gene markers of all sorts, as chromosome loss progresses.

Another topic that received early attention concerned the regulation of differentiated properties in hybrid cells. This field began with the studies of Davidson *et al.* (1966) on pigment synthesis in hybrids of mouse fibroblasts and hamster melanoma cells. These experiments suggested that the differentiated or luxury functions of a cell are extinguished in hybrids, in contrast to the household functions essential to maintenance and growth. Subsequently examples were found in which luxury functions were not extinguished, and even in which dormant differentiated functions were activated, so that this subject has become complex (reviewed by Ephrussi, 1972; Davis and Adelberg, 1973; Davidson, 1974).

Another area of considerable controversy has concerned the outcome of fusions between malignant cells and normal cells (see H. Harris, 1970, and Ephrussi, 1972, for different views), a topic that remains incompletely resolved (H. Harris, 1979).

Fifteen years after the discovery of experimental hybridization, this technique, considerably refined and improved, was also beginning to contribute to studies in virology (reviewed by Davidson and de la Cruz, 1974), immunology (Kohler and Milstein, 1975), and the analysis of gene expression in heterkaryons (Ringertz and Savage, 1976). Hybrids between two human diploid cells had been isolated (Migeon *et al.,* 1974); polyethylene glycol, a more easily prepared and standardized fusogen than Sendai virus, had been discovered (Pontecorvo, 1976); new selective systems to isolate hybrids had been devised (Hooper, 1985); and fusions of cell components, such as cytoplasm or nuclei or chromosomes, were underway (Ringertz and Savage, 1976). So this early history blends imperceptibly into the more recent history of cell fusion described elsewhere in this book.

Experimental mammalian somatic cell fusion, like other aspects of somatic cell genetics, could not have begun without the original work of Puck (1956), Eagle (1959), and others who developed the techniques necessary to culture single mammalian cells. In this regard, reviews of the origins of somatic cell genetics are relevant to this volume (Puck, 1974; Green and Todaro, 1967; Krooth *et al.,* 1968). Reviews of cell–cell variation in culture, which were necessary for the initial hybridization experiments as well as related topics, may also be of interest (M. Harris, 1964; Littlefield, 1976). The views of Ephrussi (1972) and H. Harris (1968, 1970) are personal, sometimes contrasting, and very readable, and the popular article by Ephrussi and Weiss (1969) should be included for historical interest. The comprehensive review by Ringertz and Savage (1976) was published at the end of the early era. At the 1973 meeting in Winter Park, Florida (Davidson and de la Cruz, 1974) were presented interesting individual reports from that early era, as at the previous meetings of somatic cell geneticists in Philadelphia, Paris, and Amalfi, of which only the Philadelphia meeting was published (Defendi, 1969). Would that financial constraints had not prevented the continuation of these delightful affairs!

REFERENCES

Barski, G., Sorieul, S., and Cornefert, F., 1960, Production dans des cultures in vitro de deux souches cellulaires en association de cellules de caractére "hybride," *C. R. Acad. Sci. (Paris)* **251:**1825–1827.

Choy, W. N., Gopalakrishnan, T. V., and Littlefield, J. W., 1982, Techniques for using HAT selection in somatic cell genetics, in: *Techniques in Somatic Cell Genetics* (J. W. Shay, ed.), pp. 11–21, Plenum Press, New York.

Davidson, R. L., 1974, Control of expression of differentiated functions in somatic cell hybrids, in: *Somatic Cell Hybridization* (R. L. Davidson and F. de la Cruz, eds.), pp. 131–150, Raven Press, New York.

Davidson, R. L., and de la Cruz, F., 1974, *Somatic Cell Hybridization,* Raven Press, New York.

Davidson, R. L., and Ephrussi, B., 1965, A selective system for the isolation of hybrids between L cells and normal cells, *Nature (Lond.)* **205:**1170.

Davidson, R. L., Ephrussi, B., and Yamamoto, K., 1966, Regulation of pigment synthesis in mammalian cells, as studied by somatic hybridization, *Proc. Natl. Acad. Sci. U.S.A.* **56:**1437–1440.

Davis, F. M., and Adelberg, E. A., 1973, Use of somatic cell hybrids for analysis of the differentiated state, *Bacteriol. Rev.* **37:**197–214.

Defendi, V., 1969, *Heterospecific Genome Interaction,* Wistar Institute Symposium Monograph No. 9, Wistar Institute Press, Philadelphia.

Eagle, H., 1959, Amino acid metabolism in mammalian cell cultures, *Science* **130:**432–437.

Ephrussi, B., 1982, *Hybridization of Somatic Cells,* Princeton University Press, Princeton, New Jersey.

Ephrussi, B., and Weiss, M. C., 1965, Interspecific hybridization of somatic cells, *Proc. Natl. Acad. Sci. U.S.A.* **53:**1040–1042.

Ephrussi, B., and Weiss, M. C., 1969, Hybrid somatic cells, *Sci. Am.* **220:**26–35.

Gershon, D., and Sachs, L., 1963, Properties of a somatic hybrid between mouse cells with different genotypes, *Nature (Lond.)* **198:**912–913.

Green, H., and Todaro, G. J., 1967, The mammalian cell as a differentiated microorganism, *Annu. Rev. Microbiol.* **21:**573–600.

Harris, H., 1968, *Nucleus and Cytoplasm,* Clarendon Press, Oxford.

Harris, H., 1970, *Cell Fusion,* Harvard University Press, Cambridge, Massachusetts.

Harris, H., 1979, Some thoughts about genetics, differentiation, and malignancy, *Somatic Cell Genet.* **5:**923–930.

Harris, H., and Watkins, J. F., 1965, Hybrid cells derived from mouse and man: Artificial heterokaryons of mammalian cells from different species, *Nature (Lond.)* **205:**640–646.

Harris, M., 1964, *Cell Culture and Somatic Variation,* Holt, Rinehart and Winston, New York.

Hooper, M. L., 1985, *Mammalian Cell Genetics,* John Wiley, New York.

Kao, F. T., and Puck, T. T., 1970, Genetics of somatic mammalian cells: Linkage studies with human–Chinese hamster cell hybrids, *Nature(Lond.)* **228:**329–332.

Kit, S., Dubbs, D. R., Piekarski, L. J., and Hsu, T. C., 1963, Deletion of thymidine kinase activity from L cells resistant to bromodeoxyuridine, *Exp. Cell Res.* **31:**297–312.

Kohler, G., and Milstein, C., 1975, Continuous cultures of fused cells secreting antibody of predefined specificity, *Nature (Lond.)* **256:**495–497.

Krooth, R. S., Darlington, G. A., and Velazquez, A. A., 1968, The genetics of cultured mammalian cells, *Annu. Rev. Genet.* **2:**141–164.

Lederberg, J., 1958, Genetic approaches to somatic cell variation: Summary comment, *J. Cell Comp. Physiol. (Suppl.)* **1:**383–392.

Littlefield, J. W., 1964, Selection of hybrids from matings of fibroblasts in vitro and their presumed recombinants, *Science* **145:**709–710.

Littlefield, J. W., 1966, The use of drug-resistant makers to study the hybridization of mouse fibroblasts, *Exp. Cell Res.* **41:**190–196.

Littlefield, J. W., 1973, Medium for hybrid selection (letter), *Science* **180:**255.

Littlefield, J. W., 1976, *Variation, Senescence, and Neoplasia in Cultured Somatic Cells,* Harvard University Press, Cambridge, Massachusetts.

Migeon, B. R., Norum, R. A., and Corsaro, C. M., 1974, Isolation and analysis of somatic hybrids derived from two human diploid cells, *Proc. Natl. Acad. Sci. U.S.A.* **71:**937–941.

Okada, Y., 1958, The fusion of Ehrlich's tumor cells caused by HVJ virus in vitro, *Biken's J.* **1:**103–110.

Pontecorvo, G., 1962, Methods of microbial genetics in an approach to human genetics, *Br. Med. Bull.* **18:**81–84.

Pontecorvo, G., 1975, Production of mammalian somatic cell hybrids by means of polyethylene glycol treatment, *Somatic Cell Genet.* **1:**397–400.

Puck, T. T., 1972, *The Mammalian Cell as a Microorganism: Genetic and Biochemical Studies in vitro,* Holden-Day, San Francisco.

Puck, T. T., Marcus, P. I., and Cieciura, S. J., 1956, Clonal growth of mammalian cells in vitro: Growth characteristics of colonies from single HeLa cells with and without a "feeder" layer, *J. Exp. Med.* **103:**273–284.

Ringertz, N. R., and Savage, R. E., 1976, *Cell Hybrids,* Academic Press, New York.

Ruddle, F. H., 1974, Human genetic linkage and gene mapping by somatic cell genetics, in: *Somatic Cell Hybridization* (R. L. Davidson and F. de la Cruz, eds.), pp. 1–13, Raven Press, New York.

Sorieul, S., and Ephrussi, B., 1961, Karyological demonstration of hybridization of mammalian cells in vitro, *Nature (Lond.)* **190:**653–654.

Szybalski, W., Szybalska, E. H., and Ragni, G., 1962, Genetic studies with human cell lines, *Natl. Cancer Inst. Monog.* **7:**75–89.

Weiss, M. C., and Green, H., 1967, Human–mouse hybrid cell lines containing partial complements of human chromosomes and functioning human genes, *Proc. Natl. Acad. Sci. U.S.A.* **58:**1104–1111.

Yerganian, G., and Nell, M., 1966, Hybridization of dwarf hamster cells by UV-inactivated Sendai virus, *Proc. Natl. Acad. Sci. U.S.A.* **55:**1066–1073.

Chapter 21

Bioselective Cell–Cell Fusion for Antibody Production

Mary K. Conrad, Mathew M. S. Lo, Tian Y. Tsong, and Solomon H. Snyder

1. INTRODUCTION

1.1. Background

For the past 10 years, monoclonal antibodies have been produced in many laboratories around the world by hybridoma cells formed during the random fusion, in the presence of polyethylene glycol, of hyperimmunized mouse spleen cells and selected myeloma cells (Kohler and Milstein, 1975; Galfre *et al.*, 1977; Galfre and Milstein, 1981). This technology has provided major advancements in the biomedical sciences. The methodology, however, has two inherent disadvantages. The first is related to the random nature of fusion of cells and the high probability of destroying, losing or simply failing to fuse the relatively rare B cell producing the desired antibody. The second disadvantage is the fact that

MARY K. CONRAD and MATHEW M. S. LO • National Institute on Drug Abuse, Addiction Research Center, Baltimore, Maryland 21224. TIAN Y. TSONG • Department of Biological Chemistry, Johns Hopkins University School of Medicine, Baltimore, Maryland 21205. SOLOMON H. SNYDER • Departments of Neuroscience, Pharmacology and Experimental Therapeutics, and Psychiatry and Behavioral Sciences, Johns Hopkins University School of Medicine, Baltimore, Maryland 21205.

chemically induced fusions result in very large numbers of growing colonies that need to be screened for antibody production. In cases in which the antigen in question is very rare or is in a crude form, or both, this screening process poses insurmountable problems at times. In addition, chronic hyperimmunization of the animals prior to fusion has been the standard practice, in order to stimulate and expand the appropriate subpopulations of B cells and therefore increase the chance of including these cells of interest in the group ultimately fused. This practice necessitates the use of large amounts of antigen. Furthermore, in cases in which only a few B cells can be obtained (such as in the production of human hybridomas), or when the antibodies desired must be of very high affinity, conventional fusion methods may be ineffective.

In recent years, much attention has been given to the fusion of a variety of cell types with intense, short-duration electric fields (Neumann *et al.,* 1980; Tsong, 1983; Zimmermann *et al.,* 1985). Cell fusion induced electrically can be stringently controlled and is typically 1000–10,000 times more efficient than fusion induced by chemicals. In addition, this method has a general advantage over chemically induced fusions in the fact that exogenous chemicals need not be used. Instead, minute pores are generated in membranes, and those membranes in contact at the time of electric pulse subsequently merge in the process of spontaneous membrane repair (Kinosita and Tsong, 1977, 1978), and a new hybrid cell is born. The key to successful fusion under these circumstances is cell–cell contact at the time of, or immediately following (Sowers, 1984, 1986), electric field pulse application.

1.2. Difficulties with Conventional Methods

Many innovative techniques have been used in the past to achieve contact between cells for the purpose of fusion in an electric field. One of these is dielectrophoresis (Pohl, 1978; Zimmermann *et al.,* 1985), a process during which an alternating current is used to polarize and align cells electrically before the intense direct current pulse is applied for fusion. This is done with the aid of direct visualization under the light microscope, and cells are fused pair by pair or group by group (Bischoff *et al.,* 1982; Vienken and Zimmermann, 1985). Other methods involve the use of substances such as pronase (Karsten *et al.,* 1985) and concanavalin A (Con A) (see Table I) which cause cells in suspension to adhere to each other. Aliquots of such suspensions are then subjected to high-voltage pulses, and any cells in contact at that time fuse. However, all these techniques have in common their nonspecificity. In one case, all cells present in the fusion chamber have an equal probability of being brought into contact, and therefore fused. In the other case, cells to be fused must first be care-

Table I. Electric Field-Induced Fusion: Efficiency

	Spleen cell count	Hybridomas after HAT selection		
		Cell count	[^{3}H]Thymidine	Frequency
Experiment 1				
Pronase/Con A	1.6×10^7		4.0×10^5	2.5×10^{-2}
	6.1×10^6		1.5×10^5	2.4×10^{-2}
	4.6×10^6	4.0×10^5	4.6×10^4	9.9×10^{-3}
	4.9×10^6		1.7×10^5	3.5×10^{-2}
Experiment 2				
Con A	2.0×10^7		1.9×10^5	9.5×10^{-3}
Con A	2.0×10^7		1.7×10^5	8.4×10^{-3}
Pronase	2.0×10^7		3.7×10^4	1.8×10^{-3}
Pronase	2.0×10^7		2.7×10^4	1.3×10^{-3}
Pronase/Con A	2.0×10^7		1.3×10^5	6.3×10^{-3}
Experiment 3				
Pronase/Con A	2.5×10^7	2.3×10^4	2.0×10^4	0.8×10^{-3}
Con A	2.5×10^7	2.4×10^4	2.0×10^4	0.8×10^{-3}

Fusion efficiency using various methods to promote cell–cell contact. Myeloma cells were grown in the presence of [^{3}H]thymidine. Specific activity of the cells was approximately 0.6 cpm per cell. Fusions were carried out using either pronase, concanavalin A (Con A), or both, to elicit adhesion between cells. Hybridomas were then selected in aminopterin and either counted visually or harvested to determine the amount of stably incorporated tritium. This gives an estimate of the fusion frequency between myeloma and B cells, as any other fused cell pairs will be nonviable. Untreated cells and treated cells not subjected to high-voltage pulse gave fusion frequencies of less than 1 in 100 million cells. Chemical fusions gave fusion frequencies of 1 in 0.1–1.0 million cells.

fully preselected in order to prevent the formation of undesired hybridomas. In the attempt to create monoclonal antibodies, such advance isolation of pertinent B cells (typically fewer than 1% of cells in a hyperimmunized spleen), although theoretically possible, may, for most practical purposes, be safely called not feasible. The problem of nonspecific fusion is, of course, shared with chemical methods of fusion; however, since electrically induced fusion in bulk suspension is up to three to four orders of magnitude more efficient than chemically induced fusion (Zimmermann *et al.,* 1985), the resultant millions of growing colonies pose a considerable problem for screening. To avoid these impossible situations, one must devise a method of very selectively and specifically placing into contact only those cells of interest in a heterogeneous bulk cell suspension.

1.3. Bioselective Targeting

Fortunately for those interested in making and using monoclonal antibodies, nature has provided, on the surface of B lymphocytes, antigen-receptor immunoglobulins of the same antigenic specificity as the immunoglobulins they are destined ultimately to secrete (Siskind and Benaceraff, 1969; Tonegawa, 1983). Binding of the antigen to these surface receptors *in vivo* stimulates proliferation of the B cell and differentiation into the mature plasma cells that actually produce the antibody for subsequent secretion and binding to the initiator antigen. We have devised a means (Lo *et al.,* 1984) of utilizing this high affinity antigen-surface immunogobulin interaction for the purpose of achieving intimate and highly selective cell–cell contact in heterogeneous bulk cell suspensions (Fig. 1). In designing this methodology, we have used another well known phenomenon, that is the tight binding of avidin to biotin (Green, 1975; Wormmeester *et al.,* 1984) as well as a number of techniques by which these substances may be manipulated and bound to a wide variety of materials, including living cells, to effect the bringing together of diverse pairs of biochemical or biological entities (Green, 1975). Using these techniques in combination, we have been successful in consistently producing a manageable number of hybridomas, a very substantial percentage of which secrete very high affinity monoclonal antibodies. Furthermore, this system of bioselection in a heterogeneous suspension does away with the need to hyperimmunize the animal prior to fusion: We have

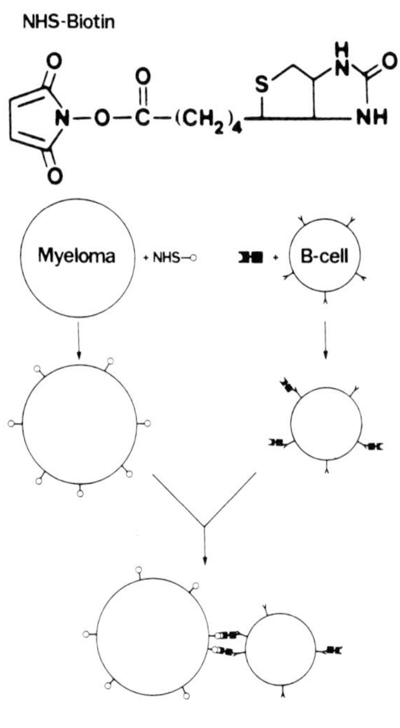

Figure 1. Bioselective targeting. Cell–cell contact is achieved when biotin (-○), covalently attached to myeloma cells, binds avidin (■) linked to antigen (◧), which in turn is tightly held to receptor immunoglobulins on stimulated B cells.

found that two immunizations one week apart with minute amounts of antigen are adequate for very high yields of a variety of subclasses of high-affinity antibodies. [However, it has been shown that chronic immunization has the effect of generating specific antibodies with higher affinity than those formed in response to acute immunization (Berek *et al.,* 1985). Therefore, chronic immunization increases the probability of producing monoclonal antibodies of higher affinity.] The remainder of this chapter is devoted to a detailed explanation of this methodology; in conclusion, we discuss some applications of the method for studying a variety of problems in novel ways.

2. CONSTRUCTION OF CELL CROSS-LINKING SYSTEMS

Linkage between proliferating myeloma cells and pertinent B cells is achieved by the attaching biotin to the surface of the myeloma cell, and covalently attaching avidin to the antigen that binds to the surface receptor on these B cells. The high-affinity interaction between biotin and avidin creates specific adherence between treated B cells and biotinylated myeloma cells when they are mixed together.

Myeloma cells are reacted with an N-hydroxysuccinimide (NHS) derivative of D-biotin (Berman and Basch, 1980; Orr, 1981). This procedure is gentle and cell viability, along with the ability to proliferate and form hybridomas, is preserved.

Antigens are covalently attached to avidin by a variety of chemical cross-linking procedures. Briefly, these are as follows;

1. Attachment of the antigen to avidin with homobifunctional cross-linkers, e.g., 1,5-difluoro-2,4-dinitrobenzene (DFDNB) (Golds and Braun, 1978; Harris and Stahl, 1980) immobilized on an iminobiotin matrix (Lo *et al.,* 1984)
2. Attachment of the antigen to avidin with heterobifunctional cross-linkers, e.g., maleimidobenzoyl-*N*-hydroxysuccinimide ester (MBS) (Lui *et al.,* 1979; Youle and Neville, 1980) and *N*-succinimidyl 3-(2-pyridyldithio)propionate (SPDP) (Carlsson *et al.,* 1978) to avidin in solution or on solid phase (H. Tomita, M. M. S. Lo, T. Y. Tsong, and S. H. Snyder, in preparation)
3. Chemical activation of the antigen to its *N*-hydroxysuccinimide derivative and subsequent reaction with avidin (J. F. Schwartz, M. M. S. Lo, T. Y. Tsong, and S. H. Snyder, in preparation)
4. Biotinylation of the antigen with NHS-biotin, and subsequent binding to avidin (M. M. S. Lo, T. Y. Tsong, and S. H. Snyder, in preparation)

Large peptides and proteins are commonly derivatized by methods 1, 2, and 4. Smaller peptides (less than 50 to 100 residues) are often derivatized with methods 2 and 4. Many haptens (particularly carbohydrates and steroids) have been successfully derivatized with method 3. Membrane-bound antigens, such as those found on tumor cells or bacterial cell walls, may be biotinylated directly *in situ* and then solubilized with detergents. When necessary the chemical reactions may be performed on a small scale using minute amounts (micrograms) of the antigen. In several instances, hybridomas were formed from fusions using only partially purified antigens. DFDNB is a useful substance because of its relative simplicity and its reactivity with amino groups, which are found in abundance in avidin and other proteins. There are, however, several disadvantages in using DFDNB as a cross-linker: (1) when used to couple an antigen covalently to a large molecule like avidin, its small size may contribute to steric hindrance; (2) it may not be efficient for crosslinking minute amounts of antigens; (3) it requires rigorous techniques to ensure control over the way molecules couple together in a mixture; and (4) it renders molecules less soluble by reacting with and neutralizing negatively charged amino groups. The last factor mentioned is not a problem when one uses a heterobifunctional cross-linker such as MBS or SPDP. These react with sulfhydryl groups as well as amino groups, thereby sparing some of the molecule's negative charge and solubility. Chemical activation of the antigen to its NHS derivative is the simplest and most direct method of attaching antigen to avidin. This attachment requires that one must maintain a low ratio of antigen to avidin in order to preserve negative charge and solubility. Finally, when avidinylation is very difficult or impossible, one may directly biotinylate the antigen coat the biotinylated antigen with avidin, and subsequently bind the complex directly to biotinylated myeloma cells. Again, to maintain solubility, one must avoid excessive chemical modification of the antigen with NHS–biotin. Antigens conjugated to either biotin or avidin are chemically characterized and, when possible, assayed to determine their biological activities while bound to avidin or biotin, using a reference polyclonal antibody raised to the antigen (Fig. 2).

3. ELECTRIC FIELD-INDUCED CELL FUSION

Treated myeloma cells and B cells are brought into selective contact by suspension in an isosmotic sucrose solution. Fusion between selected cell pairs is induced with a transient electric field produced by a high-voltage pulse generator connected to the fusion chamber (Fig. 3). To avoid

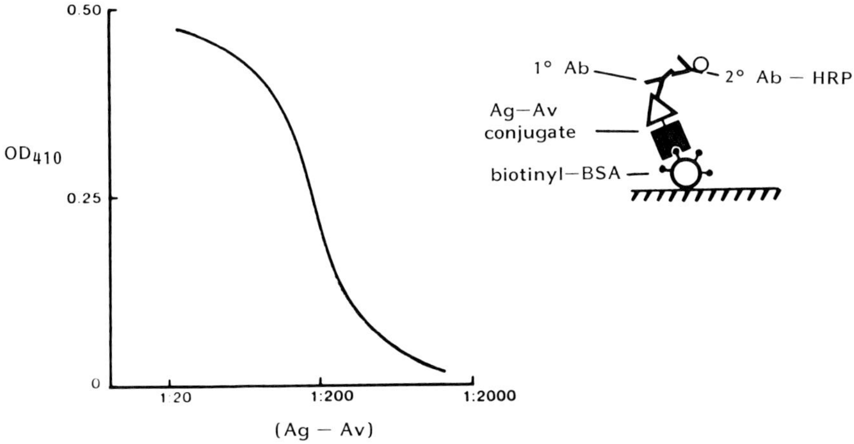

Figure 2. Characterization of activity of antigen–avidin conjugate. Avidin conjugated to a peptide is first bound to plastic plates (Dynatek) coated with biotinylated bovine serum albumin (BSA). Plates are washed after incubation, and a polyclonal antiserum raised against the peptide is added. After washing, a second antibody covalently attached to horseradish peroxidase (HRP) is added to the plate, which is subsequently washed and developed with substrate. The curve shows that the amount of bound conjugate, as measured by the optical density, can be detected to at least a 1 : 200 dilution of the original conjugate preparation.

cell damage due to heating, isosmotic sucrose (which has a low electrical conductance) is used. Typically, a 0.5-ml suspension containing about 50 million cells is exposed to 2 to 4 pulses of 5 μsec duration and a field intensity of 3 kV/cm.

We have determined the efficiency of electrically induced cell fusion in bulk suspension to be about 0.1–1%, a value 1000–10,000 times higher than that attained during fusion using conventional methods (Table I). In these experiments, myeloma and B cells were paired together nonspecifically using lectins, pronase, or an avidinylated antibody raised against mouse IgM (present on B cells).

Figure 3. Fusion chamber. A circular well is drilled out of a square piece of Plexiglas (thickness 3 mm), to hold a final volume of 0.5 ml. Two stainless steel discs are sandwiched around the Plexiglas and clamped between brass cubes that serve as electrodes connected to a high-voltage pulse generator. The cell suspension is introduced into the chamber through a 22-gauge port. A similar port serves as an air vent.

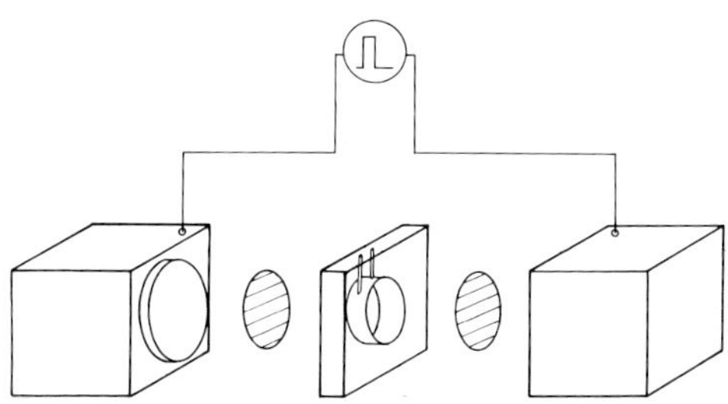

4. FUSION PROCEDURES

4.1. Immunization

We have found that immunizing animals with minute (microgram) quantities of antigen twice, 1 week apart before fusion suffices for production of numerous high-affinity antibodies of various subclasses and high affinities for the antigen. In the typical case of first injection consists of antigen emulsified in Freund's adjuvant, while the second injection is antigen alone. Both injections are intraperitoneal. These acute low doses have given rise to antibodies of mainly the IgG type, rather than the IgM type, as one may predict. We hypothesize that our bioselective system, which utilizes binding of antigen to surface antibody, selects out those B cells stimulated to produce high-affinity immunoglobulins, leaving those producing weaker antibodies behind. It has been shown, however, that somatic mutation and shifts to alternative germ-line combinations that occur during the process of chronic immunization give rise to a greater variety of even higher-affinity antibodies than those which may be seen following acute exposure to antigen (Berek *et al.,* 1985). We have been especially interested in producing antibodies to rare or hard-to-isolate entities and were therefore pleasantly surprised at the predominance of high-affinity IgG antibodies obtained following the above immunization regimen. When chronically immunizing animals we have found that using Bentonite rather than Freund's adjuvant results in fewer problems with intraperitoneal adhesions.

4.2. Preparation of Cells for Fusion

B cells from the appropriately immunized source are reacted with the antigen–avidin or antigen–biotin–avidin conjugate. The reaction is carried out at 4°C to prevent capping and internalization of the conjugate. The B cells are then washed to remove unbound conjugate. Myeloma cells are reacted with 10 μM NHS–biotin for 15 min at room temperature and then washed to remove unreacted biotin. NHS–biotin is dissolved in dimethylformamide (DMF). The latter is a solvent, which even at low concentrations (greater than 0.5%), causes cells to become sensitive to mechanical manipulations. Release of intracellular components into a cell suspension, following cell lysis, causes aggregation of cells. In order to avoid this situation, NHS–biotin with DMF is made fresh in a large volume of phosphate buffered saline (PBS) before addition to the myeloma cell pellet. The treated myeloma and B cells are then incubated together at 4°C, permitting adhesion of the biotinylated myeloma cells with those B cells having the antigen–avidin conjugate bound to their surface immunoglobulins. At this step, it is important to avoid any condition that promotes the formation of large aggregates of cells, as such clumps are destroyed in the intense electrical field. Therefore, the suspension is kept dilute, the tube is kept shaking gently during incubation, and excessive manipulation of the cells is avoided. In addition, a small amount of DNase I (10–50 μg/ml) is used in the initial stages of preparation (to degrade cellular DNA released from lysed cells, which may promote cell aggregation); in later stages, however, this is removed by washing, as DNase, if taken up by cells during fusion, will kill them. Another precaution taken is the omission of all possible divalent cations, as these also promote nonspecific cell–cell adhesion. Cells are fused in an electric field. Following fusion, cells are maintained at 37°C for 30 min in Dulbecco's modified Eagle's medium (DMEM) before plating into 96-well tissue culture plates with DMEM containing aminopterin, which kills all unfused myeloma cells (as these cells are chosen for use because they lack the enzyme hypoxanthine guanine ribosyltransferase, thereby rendering them susceptible to aminopterin. Fused cells are protected by the enzyme provided by the B-cell component of the hybrid). We have found that a wide range of cell-type ratios may be used in this system and that this factor does not seem to be of such critical importance as it is when fusing with polyethylene glycol.

4.3. Screening Hybridomas for Antibody Production

Cell colonies are typically detected from about two weeks up to eight weeks following fusion. It appears to us that hybridomas formed in electric fields require a longer period of latency before growth as compared to those formed during chemical fusions. We have observed that colonies detected early have less likelihood of producing antibodies than those detected at later times. However, we suspect that this is partly dependent on the nature of the antigen used and the immunization schedule followed. Initially, supernatants are screened for mouse Ig using solid phase enzyme-linked immunosorbent assays (ELISA) (Mishell and Shiigi, 1980). Subsequently, the more concentrated supernatants (greater than 50 ng/ml) are studied for the presence of specific antibody. The assay used depends on the antigen involved; where possible, we use ELISA. However, this is dependent on the availability of the pure antigen, as substantial amounts of antigen are required for adsorption onto 96-well plastic plates for the procedure. When this is not feasible, a small amount of antigen is radiolabeled with iodine-125 for use in filtration radioimmunoassays (RIAs). Using either screening method, antibodies having affinities with dissociation constants above about 50 nM cannot be detected.

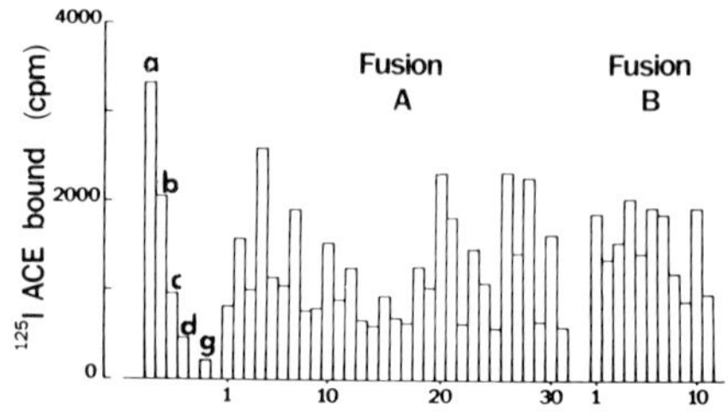

Figure 4. Detection of antibody by radioimmunoassay. Two separate fusions resulted in a total of 42 growing colonies. Supernatants were assayed with [^{125}I]labeled antigen. Columns a,b,c, and d represent four dilutions (3000-, 10,000-, 30,000-, and 100,000-fold, respectively) of a polyclonal serum raised against the antigen. Column g represents a control monoclonal antibody raised against an unrelated antigen. It can be seen that all 42 colonies showed significant levels of immunoreactivity with labeled antigen. (From Lo *et al.,* 1984.)

To illustrate, we have included some data obtained during initial experiments (Fig. 4). The antigen used in these studies was angiotensin converting enzyme (ACE). A total of 42 growing colonies resulted from two separate fusions, all of which produced unique monoclonal antibodies directed against ACE. Supernatants were assayed with radiolabeled ACE by the filtration method. Figure 5 shows that the antibodies had different epitope specifities; i.e., the antibodies formed recognized a variety of antigenic sites. Subsequent fusions have not always yielded a 100% return of purified antigen-specific antibodies. This is partly due to the fact that not all hybridomas are genetically stable entities. We have found this dependent in part on the myeloma cell line used. In our hands, the P3X63/Ag8.653 line provides the most stable hybrids. Another factor involved is the physical nature of the antigen used; i.e., the size, the charge, and the chemical composition of the molecule. Another factor may be the recognition by stimulated B cells of a variety of sites in the antigen–avidin conjugate, resulting in some antibodies that are not subsequently reactive with the pure antigen alone and therefore not detectable by assay with the same. Such immunoglobulins may have their uses for some applications of this bioselective fusion methodology.

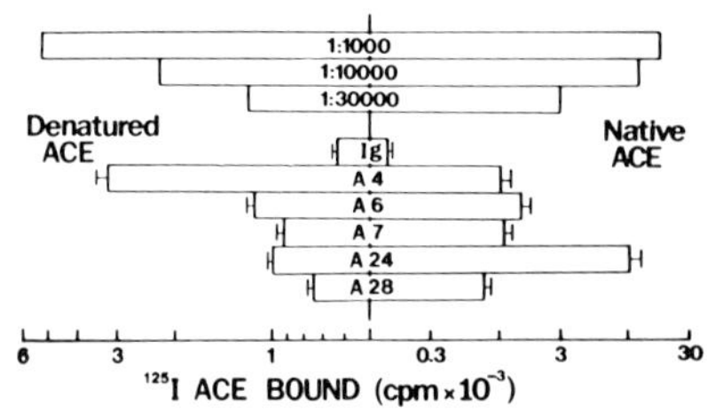

Figure 5. Antibodies display differences in immunoreactivity. Five monoclonal antibodies shown in Figure 4 were reacted with either native or denatured antigen in the same radioimmunoassay. Note the log scale of the abscissa. This illustrates the fact that the various antibodies recognize somewhat different antigenic sites of the immunogen.

5. CONCLUSION

The advantages of bioselection in bulk cell suspension, followed by fusion in an electric field, promise to greatly improve the overall efficiency of specific hybridoma production. Antibodies may be preselected by regulating cell–cell contact obtained with the appropriate antigen conjugate prior to fusion. Specificity of cell–cell adhesion may also be controlled by the amount of antigen conjugate added, the incubation time and conditions used, and the presence of competing, related or cross-reacting antigens.

This approach may help overcome a number of problems often encountered in monoclonal antibody research. This methodology preselects for the production of very high affinity antibodies, thereby eliminating the need for costly and time-consuming screening of unproductive hybridomas, or those producing weaker, and therefore less useful immunoglobulins. The number of growing colonies following a fusion is small, reflecting the population of B cells actually producing antibodies of interest, again limiting the amount of screening required after fusion, and sparing valuable antigen. Producing antibodies to unknown antigens, such as tumor-specific membrane components, may be facilitated. Here, crude membrane fragments could simply be conjugated to avidin, and a library of potentially useful antibodies would be generated from spleen cells immunized with these fragments. Finally, in cases in which stimulated B cells are rare, such as with human hybridomas, this technology offers a means of maximum conservation and utilization of the resource.

In conclusion, our method embodies two techniques, bioselective targeting and electrically induced fusion. Random fusion of cells in electric fields is useful when cell types can be isolated in advance or when multinucleated giant cells are needed for a particular purpose. However, without a method of selecting specific cells in bulk heterogeneous cell suspensions, the technology does not facilitate monoclonal antibody production; rather, it raises the specter of screening millions of hybridoma colonies for a potential few useful antibodies. In addition, bioselection alone without a highly efficient fusion technique confers no advantage over existing methodologies. We believe that combining the two approaches provides a new and more rigorous way of generating monoclonal antibodies.

REFERENCES

Berek, C., Griffiths, G. M., and Milstein, C., 1985, Molecular events during maturation of the immune response to oxazolone, *Nature (Lond.)* **316:**412–418.

Berman, J. W., and Basch, R. S., 1980, Amplification of the biotin–avidin immunofluorescence technique, *J. Immunol. Methods* **36:**335–338.

Bischoff, R., Eisert, R. M., Schedel, I., Vienken, J., and Zimmermann, U., 1982, Human hybridoma cells produced by electro-fusion, *FEBS Lett.* **143:**64–68.

Carlsson, J., Drevin, H., and Axen, R., 1978, Protein thiolation and reversible protein–protein conjugation. N-Succinimidyl 3-(2-pyridyldithio) propionate, a new heterobifunctional reagent, *Biochem. J.* **173:**723–737.

Galfre, G., and Milstein, C., 1981, Preparation of monoclonal antibodies: Strategies and procedures, *Methods Enzymol.* **73:**1–46.

Galfre, G., Howe, S. C., Milstein, C., Butcher, G. W., and Howard, J. C., 1977, Antibodies to major histocompatibility antigens produced by hybrid cell lines, *Nature (Lond.)* **266:**550–552.

Golds, E. E., and Braun, P. E., 1978, Protein associations and basic protein conformation in the myelin membrane. The use of difluorodinitrobenzene as a cross-linking reagent, *J. Biol. Chem.* **253:**8162–8170.

Green, N. M., 1975, Avidin, *Adv. Protein Chem.* **29:**85–133.

Harris, W. E., and Stahl, W. L., 1980, Organization of thiol groups of electric-eel electric-organ sodium-plus-potassium adenosine triphosphatase studied with bifunctional reagents, *Biochem. J.* **185:**787–790.

Karsten, U., Papsdorf, G., Roloff, G., Stolley, P., Abel, G., Walther, I., and Weiss, H., 1985, Monoclonal anti-cytokeratin antibody from a hybridoma clone generated by electrofusion, *Eur. J. Cancer Clin. Oncol.* **21:**733–740.

Kinosita, K., and Tsong, T. Y., 1977, Formation and resealing of pores of controlled sizes in human erythrocyte membrane, *Nature (Lond.)* **268:**438–441.

Kinosita, K., and Tsong, T. Y., 1978, Survival of sucrose-loaded erythrocytes in the circulation, *Nature (Lond.)* **272:**258–260.

Kohler, G., and Milstein, C., 1975, Continuous cultures of fused cells secreting antibody of predefined specificity, *Nature (Lond.)* **256:**495–497.

Lo, M. M. S., Tsong, T. Y., Conrad, M. K., Strittmatter, S. M., Hester, L. D., and Snyder, S. H., 1984, Monoclonal antibody production by receptor-mediated electrically induced cell fusion, *Nature (Lond.)* **310:**792–794.

Lui, F. T., Zinnecker, M., Hamaoka, T., and Katz, D., 1979, New procedures for preparation and isolation of conjugates of proteins and a synthetic copolymer of D-amino acids and immunochemical characterization of such polymer, *Biochemistry* **18:**690–697.

Mishell, B. B., and Shiigi, S. M., eds., 1980, *Selected Methods in Cellular Immunology,* W. A. Freeman, San Francisco.

Neumann, E., Gerisch, G., and Opatz, K., 1980, Cell fusion induced by high electric impulses applied to *Dictyostelium, Naturwissenschaften* **67:**414–415.

Orr, G. A., 1981, The use of the 2-iminobiotin-avidin interaction for the selective retrieval of labeled plasma membrane components, *J. Biol. Chem.* **256:**761–766.

Pohl, H. A., 1978, *Dielectrophoresis,* Cambridge University Press, Cambridge.

Siskind, G. W., and Benaceraff, B., 1969, Cell selection by antigen in the immune response, *Adv. Immunol.* **10:**1–50.

Sowers, A., 1984, Characterization of electric field induced fusion in erythrocyte ghost membrane, *J. Cell Biol.* **99:**1989–1996.

Sowers, A., 1986, A long-lived fusogenic state is induced in erythrocyte ghost by electric pulses, *J. Cell Biol.* **102:**1358–1362.

Tonegawa, S., 1983, Somatic generation of antibody diversity, *Nature (Lond.)* **302:**575–581.

Tsong, T. Y., 1983, Voltage modulation of membrane permeability and energy utilization in cells, *Biosci. Rep.* **3:**487–505.

Vienken, J., and Zimmermann, U., 1985, An improved electrofusion technique for production of mouse hybridoma cells, *FEBS Lett.* **182:**278–280.

Wormmeester, J., Stiekema, F., and De Groot, K., 1984, A simple method for immunoselective cell separation with the avidin–biotin system, *J. Immunol. Methods* **67**:389–394.

Youle, R. J., and Neville, D. M., Jr., 1980, Anti-Thy 1.2 monoclonal antibody linked to ricin is a potent cell-type-specific toxin, *Proc. Natl. Acad. Sci. U.S.A.* **77**:5483–5486.

Zimmermann, U., Vienken, J., Halfmann, J., and Emeis, C. C., 1985, Electrofusion: A novel hybridization technique, *Adv. Biotech. Proc.* **4**:79–150.

Chapter 22

Micromanipulation and Elastic Response of Electrically Fused Red Cells

Donna M. Miles and Robert M. Hochmuth

1. INTRODUCTION

Fusion of human erythrocytes may be obtained by several methods such as incubation with uranyl ions or some rare earth metals (Majumdar *et al.*, 1980), treatment with calcium and phosphate ions (Baker and Kalra, 1979), incubation with viral fusogens such as the Sendai virus (Knutton, 1979*b*), and treatment with polyethylene glycol (Knutton, 1979*a*). The method of electrofusion is relatively new in comparison and in its early stages of investigation and application. According to Zimmermann and Vienken (1982), electrofusion has several advantages compared with other methods of fusion: (1) the number of cells to be fused may be selected, (2) the loss of intracellular contents is decreased, and (3) the membrane properties are not changed by the addition of other substances.

The method of electric field fusion for the human erythrocyte and other biological cells has evolved with the use of an alternating current (AC) field to bring the cells into close contact and a pulsed direct current (DC) field to initiate the process of fusion (Zimmermann and Vienken, 1982). A red cell may be modeled electrically as a conductive core surrounded by an insulating membrane (Sale and Hamilton, 1968). Factors

DONNA M. MILES and ROBERT M. HOCHMUTH • Department of Biomedical Engineering, Duke University, Durham, North Carolina 27706. *Present address for* R. M. H.: Department of Mechanical Engineering, Duke University, Durham, North Carolina 27706.

such as Brownian motion, the net negative charge on the membrane, and the water molecules associated with the hydrophilic membrane components usually prevent cells from coming into close membrane contact naturally. A slightly nonuniform AC field polarizes the cells, causing them to realign with their longest axis parallel to the field lines. Cells aligned on the same field lines and in close proximity experience an attractive force due to their polarization. Because the strength of this force at short distances is greater than the opposing electrostatic force and the hydration force (which actually becomes quite large at distances less than 2 nm (Rand, 1981), the cells are forced into very close membrane contact. This motion in an electric field has been termed dielectrophoresis (Pohl, 1978). Therefore, an AC field with a field strength above the threshold required to polarize and align the cells establishes the membrane contact between cells necessary for the fusion process. The cells form chains of varying lengths that depend on the number of cells in the solution and the strength and divergence of the field.

The application of DC pulses to a suspension with cell chains causes some of the cells to fuse (Zimmermann, 1982). It is known that a rapid (less than 100 μsec) increase in membrane potential to a value above a certain threshold dramatically increases the permeability and conductivity of the membrane resulting in the breakdown of the selective permeability of the barrier. Two types of breakdown have been characterized (Benz *et al.*, 1979). Reversible electrical breakdown is due to local membrane compression by the high field strength intensity. The compression produces defects, or pores, in the membrane that begin to reseal naturally after the removal of the field due in part to the liquid character of the lipid components. Thus, the original membrane properties are restored. Mechanical breakdown (lysis) occurs when the critical threshold is greatly exceeded or is slightly exceeded for long periods of time thereby resulting in the complete and irreversible breakdown of the permeability barrier and possible destruction of the membrane. Either the pores created are too large to reseal naturally before all the inner contents of the cell are lost and ghosts are produced or they do not reseal at all, resulting in the breakup of the membrane.

During the fusion process, the maximum membrane potential occurs in those sections of the membrane that are perpendicular to the field lines (Sale and Hamilton, 1968). If the voltage applied to a cell chain is just above the value necessary for reversible breakdown, pores are formed only in those areas of the membranes that are in contact. It has been suggested by Zimmermann and Vienken (1982) that the apposing pores form channels between the cells, and that the lipid edges join to reform their bilayer structure. This results in the sealing of the pores and the fusion of the cells.

Our electrofusion experiments are intended to gently fuse two or more red cells in such a way that the fused membranes stabilize and reseal before losing their internal components (ions and hemoglobin). The cell doublet that results is either flaccid with each half joined by a membrane tunnel or prestressed and joined by a circular lumen or hole. In either case, aspiration of the fused-cell doublet into a micropipette allows us to measure its total surface area and volume and its elastic response to the aspiration pressure. We observe that a large aspiration pressure expands the lumen which changes the cell doublet into a single sphere, and that the doublet quickly recovers its original shape when released from the micropipette.

2. MATERIALS AND METHODS

Since a continuous AC field may result in local heating which leads to budding and lysis of the cells, a pulsed AC field is used to align cells and produce the close contact necessary for fusion (Takashima and Schwan, 1985). To produce a variable pulsed AC signal for our experiments, we have developed circuitry consisting of two complementary metal oxide semiconductor (CMOS) multivibrators (4047), as shown in Figure 1a. The first is wired in the astable form (oscillator) and the second in the monostable form (rising edge-triggered one-shot). When power is applied, the oscillator produces a square wave output with a frequency set by an external resistor and capacitor. These square waves periodically trigger the one-shot, which produces single square pulses of variable width. For each square pulse, the gated function generator produces an equal-width AC signal at the set frequency. This signal is then amplified and sent through a relay system to the chamber containing the cells.

Several DC pulses are needed to initiate the fusion process once the cells have been brought into close contact by the AC field. To generate DC pulses, a circuit has been developed that uses a power field effect transistor (FET) as a fast switching device and large batteries as the voltage source, V_L (Fig. 1a). A one-shot, triggered by a signal from a manual switch, produces a pulse the width of which depends on the external components. This pulse opens the transistor, thereby enabling the charge on the large capacitor, C4, to be delivered to the chamber. The transistor is held open for the length of the pulse by the diode, D2. A resistance is placed in parallel with the chamber to decrease the amount of current flow through the chamber. The transformer and diode, D1, separate and protect the low-voltage side from the high-voltage side of the circuit.

The AC and DC circuits are joined and controlled by a relay system (Fig. 1b) with two electromagnetic mercury-wetted relays (Sowers, 1984).

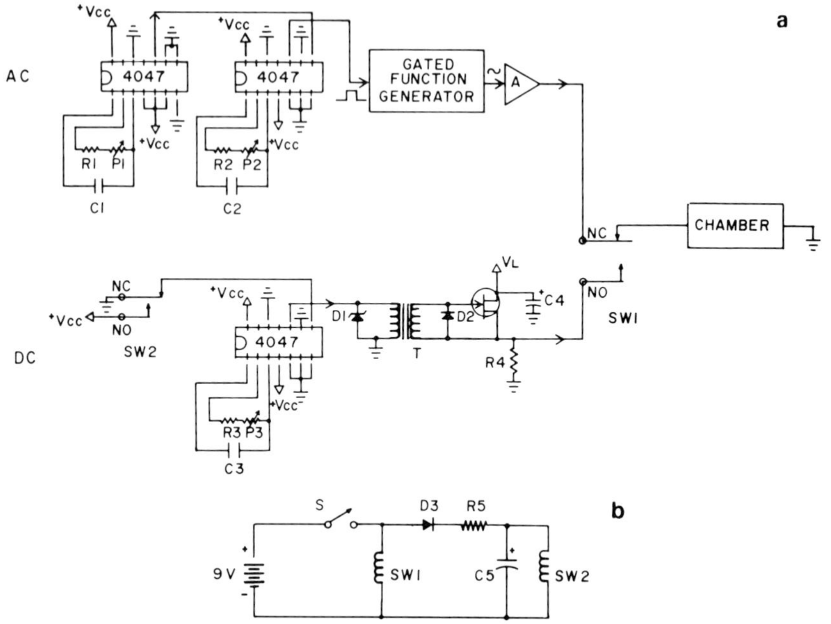

Figure 1. (a) Schematic drawing of the AC and DC circuits joined by a relay, SW1, to the chamber containing the cell suspension. The oscillator (first 4047) and the one-shot (second 4047) trigger the gated function generator (Wavetek, Model 182A, Beechgrove, IN), and the output signal is amplified (Hewlett Packard amplifier, model 6824 A, Palo Alto, CA) and sent to the chamber. High-voltage V_L for the DC pulses is provided by ±225V and ±90V Eveready batteries. The power FET transistor (Siemens, type BUZ50A, Broomfield, CO) acts as a high-voltage switch to produce the DC pulses, and the 1 : 1 transformer *T* (Radio Shack, Fort Worth, TX) separates the high- and low-voltage sides of the circuit. A power supply (Heath, model 2718, St. Joseph, MI) is used to supply V_{CC} (5V) to the circuit, and an oscilloscope (Tektronix, model 5223, Beaverton, OR) is used to monitor the signals. The components have the following values: R1 = 474 kΩ, R2 = 10 kΩ, R3 = 510 Ω, R4 = 22 kΩ; potentiometers, P1 = 25 kΩ, P2 = 500 kΩ, P3 = 200 kΩ; capacitors, C1 = 3900 pf, C2 = 3900 pf, C3 = 500 μf, C4 = 0.1 μf; diodes, D1 = 9.1 V Zener, D2 = 5.1 V Zener. (The 4047 pins are numbered according to convention with #1 in the lower left corner and #14 in the upper left corner.) (b) Control circuit for the mercury-wetted relays (Clare, type HGSM-5009, Chicago, IL), shown as SW1 and SW2 in parts a and b. The values for the components are R5 = 51 Ω and C5 = 500 μf. Component D3 may be any silicon diode. The closure of the mechanical switch, S, begins the triggering sequence that leads to the delivery of the high-voltage pulse to the chamber.

When the manual switch S is pressed, the first relay disconnects the chamber from the AC circuit and connects it to the DC circuit. The second relay after a finite delay time, connects the DC circuit to a positive voltage, and this begins the triggering sequence that leads to the delivery of the high-voltage pulse to the chamber.

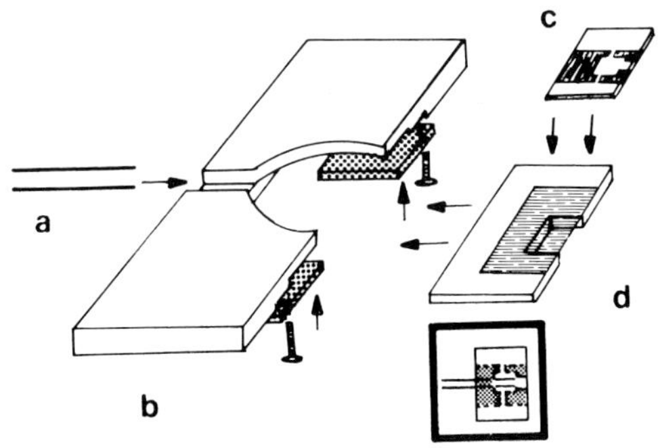

Figure 2. Assembly of chamber for use with an inverted microscope. (a) Platinum-wire (32-gauge, Fisher Scientific, Pittsburgh, PA) electrodes are inserted through the channel as shown. (b) Plexiglas holder (same size as a standard 25 × 75 mm microscope slide) and aluminum plates form slot for support of part d. (c) Cut piece of microscope slide with parafilm gasket heat sealed on bottom side. (d) Cut piece of microscope slide with #1½ coverslip glued on top over open slot. Part d is placed in holder such that electrodes lay flat on top, 2 mm apart, and visible through slot. Part c is placed on top of d such that the electrodes are centered in the inner chamber formed by the gasket. The assembly is heat sealed, which secures the electrodes and seals the chamber. The cell suspension is injected into the chamber through the channel at the open end of part c (formed by gasket).

In summary, this apparatus delivers a pulsed AC signal to the chamber containing the cell suspension until a manual switch triggers the delivery of a high-voltage DC pulse. When the switch is released, the AC signal is restored to the chamber and continues until the switch is pressed again. This method allows us to observe the effects of the DC pulses on the cells and to apply only the number of pulses necessary to initiate the fusion process.

The chamber shown in Figure 2 is designed to yield a maximum number of fused cells with a minimum amount of cell trauma. The bottom piece (d) is inserted as shown into the Plexiglas holder, and the platinum wire electrodes are inserted through (b), as shown, and positioned on top of (d) so that they are parallel, are approximately 2 mm apart, and span the length of the chamber. The piece with the gasket on the underside (Fig. 2c) is placed on top, and the assembly is heat sealed, securing the electrodes and sealing the chamber. The volume of the inner chamber is approximately 7 mm^3, while the volume V between the electrodes is approximately 3 mm^3. The small chamber and closely spaced electrodes reduce the current produced and the voltage needed for the fusion process but maximize the effects of evaporation and changing surface tension. These effects are controlled by the use of a large filled outer chamber connected to the inner chamber via the small side channels and by exposing the chamber assembly to 98% saturated air during an experiment.

The micromanipulation system consists of a glass micropipette mounted in a De Fonbrune type C micromanipulator and a pressure apparatus that allows manual control of the aspiration pressure and gives a differential pressure measurement. The micropipettes are made from Kimax 51 borosilicate glass (Preiser Scientific, Evanston, IL) by pulling a closed-tip pipette on a Kopf Vertical Pipette Puller (Tujunga, CA) and creating an open tip by a modified version of the Merrill and Ainsworth technique (Nash *et al.,* 1984). The pipette, which is filled with the same solution used in the cell suspension, is connected by water-filled tubes to a partially filled reservoir and a Validyne pressure transducer (Northridge, CA) with one port open to air. A syringe, attached to the top of the reservoir, is used to change the pressure in the reservoir and, subsequently, in the pipette and transducer. These controlled pressure changes allow the red cell to be aspirated into or released from the pipette.

As shown in Figure 3, a video display of the fusion and manipulation processes is obtained by a Dage-MTI video camera (Michigan City, IN) mounted on the eyepiece of the trinocular head of an inverted Leitz microscope (Rockleigh, NJ). The camera signal is sent through a Vista Electronics mixer (La Mesa, CA) and a ¾-inch Sony videocassette recorder (model VO-2800) to a video monitor. The differential pressure measured by the transducer is shown on the video display by the electronic mixer.

In order to decrease the amount of current flow to the cells, the solution used in this experiment is a low-ionic-strength isotonic phosphate-buffered sucrose solution. The ingredients in grams per liter of solvent (distilled water) are as follows: KH_2PO_4, 0.653; $Na_2HPO_4 \cdot 12\ H_2O$, 10.743; NaCl, 0.880; sucrose, 65.0; albumin, 20.0; streptomycin, 0.133; penicillin, 0.063. The result is approximately 30 mM phosphate, 45 mM sodium, 4.8 mM potassium, 15 mM chloride, and 200 mM sucrose. The measured pH

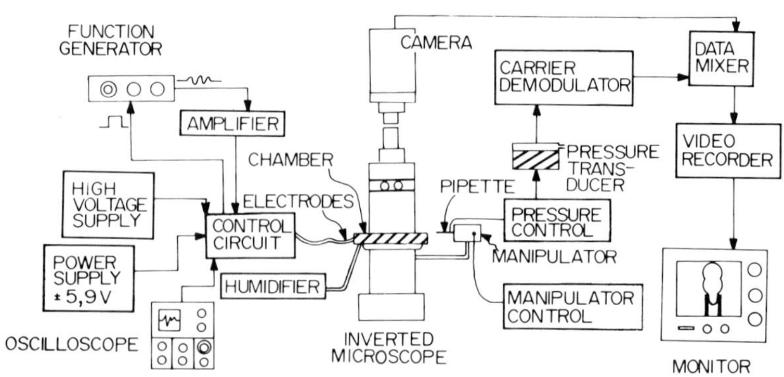

Figure 3. Diagram of experimental setup.

is 7.55, and tonicity is 300 mOsm. Many other solutions can be used that result in viable fusion products. Ionic strength and tonicity should be based on size of the chamber, electrode material, distance between electrodes, and desired fusion products (i.e., cells or ghosts). As found by Donlon and Rothstein (1969), an isotonic solution must contain a minimum of 30 mM sodium to prevent a rapid outflow of potassium from the cell. The large quantity of albumin is required to produce viable fused cells and to aid in preventing cells from sticking to the chamber and to the pipette. The pipettes are filled by boiling them under a vacuum in the phosphate–sucrose solution (excluding the albumin) for 20 min and allowing them to cool for several hours while the vacuum is maintained. A small amount of the solution is then removed and used in the preparation of the cell suspension to ensure a match in osmotic pressure between the solution in the pipette and the solution in the chamber. A drop of blood is obtained by fingerprick from a healthy donor, immediately placed in 10 ml of the isotonic phosphate–sucrose solution with albumin, washed once, and resuspended in the same solution. When the solution is injected into the chamber, there are approximately 50 cells in a 100 μm × 100 μm field of view. Experiments are conducted immediately at room temperature.

To begin the fusion process, the cells are aligned and brought into close contact by an AC signal from the circuit. The signal used in these experiments has a frequency of 5 kHz and an amplitude of 70 V/cm rms. The field has a pulse frequency of 125 Hz and a pulse length of 4 msec (50% duty cycle). The AC signal is turned on immediately following the injection of the cell suspension into the sealed chamber so that the cells begin to line up before contacting and possibly adhering to the bottom of the chamber. Depending on the quantity of cells in the suspension and the desired length of the chains, the time required for chain formation may be 1–5 min.

A critical potential of 1 V across a red cell membrane, corresponding to approximately 2.5 ± 0.5 kV/cm (Kinosita and Tsong, 1979), causes that portion of the membrane to experience electrical breakdown. A membrane defect is formed, and the resulting increase in membrane cation permeability leads to an exchange of sodium and potassium (Riemann *et al.,* 1975). Several DC pulses (5–10) of approximately 2 kV/cm and 50 μsec duration are applied at a rate of 2 sec^{-1} to initiate the fusion process and are terminated when cells begin to crenate. To ensure continuous cell contact, the AC field is left on until fusion is first observed.

After allowing 15 min for the completion of the fusion process, a filled micropipette with an inside tip diameter between 0.8 and 1.2 μm is secured in the manipulator and inserted into the chamber between the electrodes. When a fused cell doublet is to be aspirated, the pipette tip is placed very

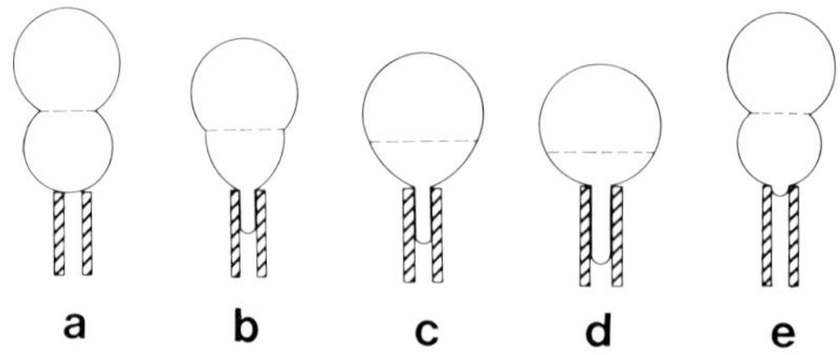

Figure 4. Aspiration of stressed fused cell into a pipette. Aspiration pressure is zero in (a) and increases (b,c) until cell doublet is sphered (d). When pressure is released, the doublet returns to its original shape (e).

close to one end of the doublet directly opposite the lumen. The pipette is then brought into contact with the cell, and a slight negative pressure is manually applied, resulting in the aspiration of a small portion of the cell into the pipette (Fig. 4). The aspiration pressure is slowly increased stepwise, and the corresponding increases in membrane tongue length and lumen diameter are measured. Also measured are the cell diameters, lengths of sphere segments, and contact angles. Eventually, the lumen expands until the outer portion of the doublet is completely sphered and the two cells are indistinguishable (Fig. 4d). The doublet is then released from the pipette by decreasing the aspiration pressure. Each cell is aspirated several times to check for possible volume losses and confirm experimental repeatability.

3. ANALYSIS

The aspiration into a pipette of a fused cell doublet is illustrated in Figure 4. Before aspiration, the doublet consists of either two spherical, prestressed cells joined by a relatively large lumen (Fig. 5a) or two flaccid stress-free cells joined by a small membrane tunnel (Fig. 5b). The initial aspiration of a prestressed doublet results in the formation of a hemispherical bump (of radius R_p) and a small increase in the lumen radius R_L. These dimensions are shown in Figure 6a. For two fused but flaccid cells, the initial aspiration pressure is smaller than for the sphered cells and results in a substantial increase in the lumen radius and a membrane tongue length L_p in the pipette. For either type of fusion product, further stepwise pressure increases result in L_p and R_L increasing until the doublet is transformed into a single sphere (Fig. 4d) with radius Rc_2 and tongue length L_p. The doublet returns to its original shape when the aspiration pressure is released (Fig. 4e).

The force resultants per unit length or tensions in the membrane of a cell doublet are calculated from measurements of the aspiration pressure, $P_o - P_p$, where P_o is the pressure in the reservoir and P_p is the pressure in the

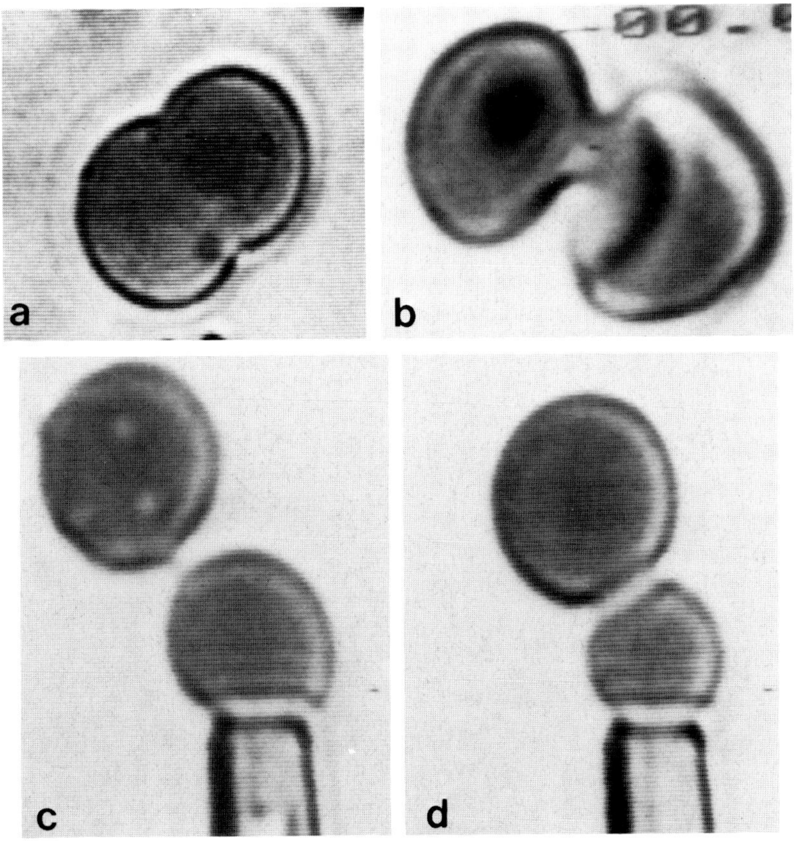

Figure 5. (a) Prestressed cell doublet with large lumen. (b) Flaccid cell doublets joined by a membrane tunnel. (c) Fused cells joined by a thin membrane filament or tether. (d) Expansion of tether into a lumen as one cell is aspirated into the pipette. Note the increase of tongue length L_p and the further sphering of cell 2, which was initially in a slightly crenated state.

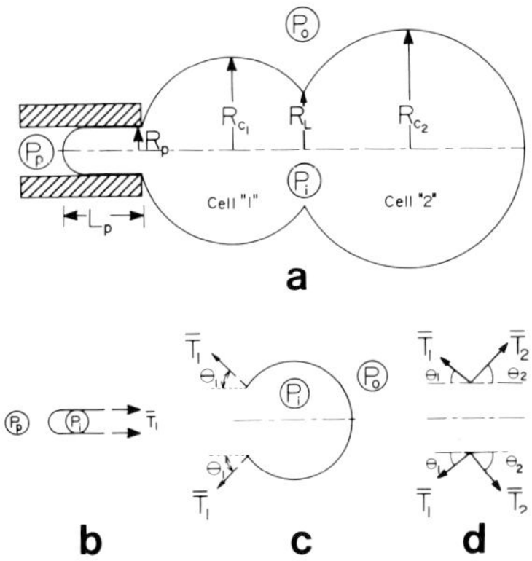

Figure 6. (a) Line drawing of fused cell aspirated into a pipette. P_o, reservoir pressure; P_p, pressure in pipette; P_i, pressure inside cell; L_p, length of aspirated membrane tongue; R_p, radius of pipette; Rc_1, radius of cell 1 (aspirated cell); Rc_2, radius of cell 2; R_L, radius of lumen. (b) Free-body diagram of that portion of membrane aspirated into pipette. (c) Free-body diagram of cell 2 and small portion of cell 1. (d) Drawing showing the contact angles, θ_1 and θ_2, at the lumen. The isotropic tensions in the membrane of cells 1 and 2 are $\bar{T}_1$ and $\bar{T}_2$.

pipette, lumen radius R_L, and contact angles, θ_1 and θ_2 (Fig. 6). The equations for such calculations are developed from force balances with the free body diagrams shown in Figure 6b–d. A force balance on that portion of the cell aspirated into the pipette (Fig. 6b) gives an expression for the tension in the membrane of cell 1:

$$\bar{T}_1 \cdot 2\pi R_p = (P_i - P_p)\, \pi R_p^2$$

or

$$\frac{2\bar{T}_1}{R_p} = P_i - P_p \tag{1}$$

The tension $\bar{T}_1$ is isotropic and constant as long as the membrane does not stick to the pipette wall and the pressure difference across the membrane is large enough to create spherical surfaces (Evans *et al.*, 1976; Waugh and Evans, 1979). A force balance on the other half of the cell doublet (cell 2 in Fig. 6c) in which a small portion of the membrane from cell 1 is included gives another expression for the tension in the membrane of cell 1:

$$\bar{T}_1(\cos\theta_1) \cdot 2\pi R_L = (P_i - P_o)\, \pi R_L^2$$

or

$$\frac{2\bar{T}_1 \cos\theta_1}{R_L} = (P_i - P_o) \tag{2}$$

Note that Eq. (1) gives $\bar{T}_1$ in terms of the pressure in the interior relative to the pressure in the pipette, while Eq. (2) is in terms of the interior pressure relative to the outside or reservoir pressure. The subtraction of Eq. (2) from Eq. (1) eliminates the unknown interior pressure P_i, and results in the following expression:

$$2\bar{T}_1\left[\frac{1}{R_p} - \frac{\cos\theta_1}{R_L}\right] = P_o - P_p \tag{3}$$

The measurement of θ_i, R_L, and $P_o - P_p$ (the aspiration pressure) permits the calculation of a value for $\bar{T}$. For the case in which cell 1 has a distinct radius Rc_1, such that $Rc_1 > R_L$, Eq. (3) simplifies to

$$2\bar{T}_1\left[\frac{1}{R_p} - \frac{1}{Rc_1}\right] = P_o - P_p \tag{4}$$

since for a sphere, $R_L = Rc_1 \cos\theta_1$. To obtain an expression for the tension in cell 2, an axial force balance at the point of contact between the two cells (Fig. 6d) yields

$$\bar{T}_2 \cos\theta_2 = \bar{T}_1 \cos\theta_1 \tag{5}$$

Also, for a sphere, $R_L = Rc_2 \cos\theta_2$; therefore,

$$\bar{T}_2 = Rc_2 \left[\frac{\cos\theta_1}{R_L} \right] \bar{T}_1 \tag{6}$$

For the special case in which cell 1 has a radius $Rc_1 > R_L$, Eq. (6) can be written as

$$\bar{T}_2 = \frac{Rc_2}{Rc_1} \bar{T}_1 \tag{7}$$

Thus, for a given aspiration pressure, the tension in cell 1 can be calculated from either Eq. (3) or (4) while that in cell 2 can be calculated from either Eq. (5), (6), or (7).

4. RESULTS

Cells are fused according to the technique discussed in Section 2. The number of cells in the suspension is kept low, approximately 50 cells in a 100 μm × 100 μm field, for facilitation of the manipulation process. After the application of the DC pulses, all cells become completely crenated within 2 min. About 75% of the cells sphere slowly (10 min), while the others either swell very quickly and become ghosts or do not swell at all. Within approximately 15 min, complete fusion, defined as the cessation of relative motion of the two cells, is observed in almost all of the sphered cells that have flat areas of membrane contact. Thus, the fusion yield is approximately 35% of the original cells, but an average of 20% actually form doublets suitable for this experiment. The others either continue to swell and eventually become ghosts or fuse in groups of three or more. Most of the fused pairs are completely sphered and have a large lumen radius R_L. Because of a slow resealing process in which the selective permeability of the membrane is restored, approximately one-half of these pairs become flaccid after 20–30 min (Fig. 5b). The others maintain their original sphered state (Fig. 5a) even though they are sealed as shown during manipulation. In most cases, the structure joining the flaccid cells is a large membrane cylinder or tunnel, however a few are very thin membrane tethers (Fig. 5c). The tethered cells are completely fused as shown by the expansion of the thin structure into a lumen when suction is applied (compare Fig. 5c with 5d). The viable, fused-cell yield and final state of membrane tension are both very dependent on several factors other than the amount of applied potential. These include: pulse length, number of pulses, rate of pulsing, and the ionic strength and tonicity of the solution.

The surface areas and volumes for six fused cells (one of which was initially flaccid while five were prestressed) are monitored during the aspiration process. Since the membrane surface area of an aspirated cell is composed of a cylinder and spherical segments (Fig. 4), its surface area and volume are readily calculated from the measured values for L_p, R_p, θ_1 (or Rc_1), R_L, and Rc_2 (Fig. 6). Linderkamp and Meiselman (1982) use a similar technique to calculate the surface area and volume accurately for single red cells. The results of these calculations for two typical cells are given in Table I. For five prestressed cells, the average surface area is 244 μm^2 and the average volume is 302 μm^3. We note that when these cells are expelled from the pipette following the aspiration process, they return to their original shape, illustrating that the lumen has an elastic memory.

For a given aspiration pressure and cell size, the tension in the membrane of cell 1, $\bar{T}_1$, can be calculated with either Eq. (3) or (4). The change in $\bar{T}_1$ with lumen size R_L is shown in Figure 7 for a prestressed cell and for an initially flaccid cell. The point at which the radius of cell 1 disappears (i.e., $Rc_1 = R_L$ or $\theta_1 = 0$) is shown as a dashed line. The tension required to expand the lumen to this point is similar for the two cells.

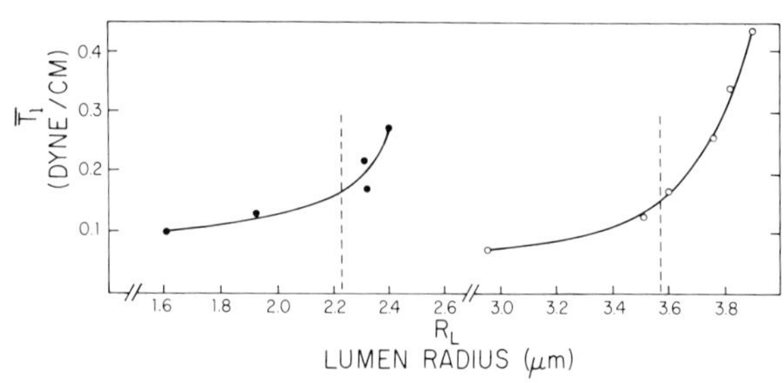

Figure 7. Plot of membrane tension in cell 1, $\bar{T}_1$, versus lumen radius R_L for a prestressed cell doublet (O) and for an originally flaccid cell joined by a tunnel (●). Dashed lines show the points at which $Rc_1 = R_L$ or $\theta_1 = 0$ for both cells. (The initial radius of the tunnel when $T_1 = 0$ is $R_L = 0.82$ μm. This value is not plotted.)

Table I. Surface Area, Volume, and Lumen Size for Two Fused Cells[a]

Flaccid				Prestressed			
ΔP ($\times 10^4$ dynes/cm^2)	R_L (μm)	S (μm^2)	V (μm^3)	ΔP ($\times 10^4$ dynes/cm^2)	R_L (μm)	S (μm^2)	V (μm^3)
0.23	1.66	226	238	0.22	2.96	213	262
0.30	1.93	211	219	0.38	3.51	203	254
0.40	2.33	219	235	0.50	3.60	219	285
0.50	2.32	235	266	0.76	3.77	213	285
0.61	2.40	237	263	1.00	3.82	218	282
0.75	Sphered	238	261	1.25	3.90	221	287
				1.50	Sphered	223	292

[a]Same cells as in Figure 7.

5. DISCUSSION

During the aspiration process, the surface area and volume stay essentially constant, as indicated by the data in Table I. The volume of the flaccid cell is slightly smaller than that of the prestressed cell, which probably indicates that the latter has expanded by gaining water through a "leaky" membrane. The total membrane surface area of two fused cells (244 μm^2) is slightly smaller than twice the average value of 134 μm^2 reported by Linderkamp and Meiselman (1982) for pooled red cells. This suggests that a small amount of membrane surface may be lost during the fusion process.

Significant pressures and, therefore, significant isotropic tensions are required to increase the radius of the lumen. For a given value of R_L, approximately the same tension is required regardless of whether the cell is flaccid or prestressed (Fig. 7). Thus, a flaccid cell is in the reference or stress-free state and becomes stressed by gaining water through a leaky membrane. Conversely, a prestressed cell becomes flaccid after its membrane reseals and water is lost during restoration of osmotic and ionic equilibrium in an isotonic medium.

If the expansion of the lumen between two fused cells is resisted solely by the shear elasticity of the membrane, then much smaller membrane tensions than those shown in Figure 7 would be required to dilate the lumen. This is shown by analyzing the elastic expansion of a flat, membrane "disk" with a hole of initial radius R_o. Far from the hole, an isotropic tension $\bar{T}$ acts around the circumference of the disk and expands the hole to a radius R_h. We assume that this expansion is resisted only by the shear elasticity of the membrane. For this case, we obtain the following expression from a solution of the equation of membrane equilibrium for a disk in which a constitutive equation that relates the shear deformation in the membrane to the shear-force resultant (Evans, 1973) is used:

$$\bar{T} = \frac{\mu}{2} \left[1 + \ln\left[\frac{R_h}{R_o}\right]^2 - \left[\frac{R_h}{R_o}\right]^{-2}\right] \tag{8}$$

In Eq. (8), μ is the shear modulus of elasticity for the membrane surface. Typically, $\mu = 0.007$ dyne/cm (Waugh and Evans, 1979). The value for $\bar{T}$ is only about 0.04 dyne/cm when $R_h = 1$ μm, even if we assume a minimum value for R_o of 100 Å. Increasing the value for R_h to 2 μm only increases the value of $\bar{T}$ by 14%, since it has a weak logarithmic dependence on R_h, as shown by Eq. (8). Since $\bar{T}$ and R_h are strongly related (Fig. 7), we assume that membrane shear elasticity does not play a major role in resisting the elastic dilation of the lumen between two fused red cells.

Since the curvature is quite large where two cells fuse, bending or curvature elasticity could possibly contribute to the stiff elastic response of the lumen. If we assume that the radius of curvature at the point of fusion is on the order of the total thickness (lipid plus cytoskeleton) of the red cell membrane, then the isotropic tension in the membrane will be on the order of

$$\bar{T} \approx \frac{B}{t^2}$$

where t is the membrane thickness and B is the curvature elastic modulus or bending modulus (Evans, 1985). For $B = 10^{-12}$ dynes · cm (Evans, 1983) and $t = 100$ Å, $\bar{T} \approx 1$ dyne/cm, and therefore, the high degree of bending where two cells fuse could account for the stiffness of the lumen.

Although curvature elasticity could explain the stiffness of the lumen, it does not explain its "memory." We do not know why a fused-cell doublet has an elastic memory of its original shape, which it recovers rapidly when expelled from a pipette. Somehow the membranes of the two cells become permanently "crimped" when they fuse. Thus, when unstressed, a tunnel exists between the two cells (Fig. 5b) and when stressed, the cells are joined by a larger lumen (Fig. 5a). We conclude from our observations and measurements that electrofusion is not a perfect process. It is possible that the lipids intermix completely, whereas the structural proteins of the cytoskeleton do not.

ACKNOWLEDGMENTS. This work was supported by Grant HL 23728, from the National Heart, Lung, and Blood Institute. We would like to thank David Berk for his derivation of Eq. (8), Dr. Robert Frank for his assistance with the development of the electronic circuitry, Dr. Shiro Takashima for his suggestions concerning the chamber and AC field signal, and Dr. Arthur Sowers for introducing us to cell fusion and for his continued support.

REFERENCES

Baker, R. F., and Kalra, V. K., 1979, Chemically induced fusion of human erythrocytes, *Biochem. Biophys. Res. Commun.* **86**:920–928.

Benz, R., Beckers, F., and Zimmermann, U., 1979, Reversible electrical breakdown of lipid bilayer membranes: A charge-pulse relaxation study, *J. Membrane Biol.* **48**:181–204.

Donlon, J. A., and Rothstein, A., 1969, The cation permeability of erythrocytes in low ionic strength media of various tonicities, *J. Membrane Biol.* **1**:37–52.

Evans, E. A., 1973, A new material concept for the red cell membrane, *Biophys. J.* **13**:926–939.

Evans, E. A., 1983, Bending elastic modulus of red blood cell membrane derived from buckling instability in micropipet aspiration tests, *Biophys. J.* **43:**27–30.

Evans, E. A., 1985, Detailed mechanics of membrane–membrane adhesion and separation. I. Continuum of molecular cross-bridges, *Biophys. J.* **48:**175–183.

Evans, E. A., Waugh, R., and Melnik, L., 1976, Elastic area compressibility modulus of red cell membrane, *Biophys. J.* **16:**585–595.

Kinosita, K., Jr., and Tsong, T. Y., 1977, Voltage-induced pore formation and hemolysis of human erythrocytes, *Biochim. Biophys. Acta* **471:**227–242.

Knutton, S., 1979*a*, Studies of membrane fusion. III. Fusion of erythrocytes with polyethylene glycol, *J. Cell Sci.* **36:**61–72.

Knutton, S., 1979*b*, Studies of membrane fusion. V. Fusion of erythrocytes with non-haemolytic Sendai virus, *J. Cell Sci.* **36:**85–96.

Linderkamp, O., and Meiselman, H. J., 1982, Geometric, osmotic, and membrane mechanical properties of density-separated human red cells, *Blood* **59:**1121–1127.

Majumdar, S., Baker, R. F., and Kalra, V. K., 1980, Fusion of human erythrocytes induced by uranyl acetate and rare earth metals, *Biochim. Biophys. Acta* **598:**411–416.

Nash, G. B., Johnson, C. S., and Meiselman, H. J., 1984, Mechanical properties of oxygenated red blood cells in sickle cell disease, *Blood* **63:**73–82.

Pohl, H. A., 1978, *Dielectrophoresis,* Cambridge University Press, Cambridge.

Rand, R. P., 1981, Interacting phospholipid bilayers: Measured forces and induced structural changes, *Annu. Rev. Biophys. Bioeng.* **10:**277–314.

Riemann, F., Zimmermann, U., and Pilwat, G., 1975, Release and uptake of haemoglobin and ions in red blood cells induced by dielectric breakdown, *Biochim. Biophys. Acta* **394:**449–462.

Sale, A. J. H., and Hamilton, W. A., 1968, Effects of high electric fields on micro-organisms. III. Lysis of erythrocytes and protoplasts, *Biochim. Biophys. Acta* **163:**37–43.

Sowers, A. E., 1984, Characterization of electric field-induced fusion in erythrocyte ghost membranes, *J. Cell Biol.* **99:**1989–1996.

Takashima, S., and Schwan, H. P., 1985, Alignment of microscopic particles in electric fields and its biological implications, *Biophys. J.* **47:**513–518.

Waugh, R., and Evans, E. A., 1979, Thermoelasticity of red blood cell membrane, *Biophys. J.* **26:**115–132.

Zimmermann, U., 1982, Electric field-mediated fusion and related electrical phenomena, *Biochim. Biophys. Acta* **694:**227–277.

Zimmermann, U., and Vienken, J., 1982, Electric field-induced cell-to-cell fusion, *J. Membrane Biol.* **67:**165–182.

Chapter 23

Microinjection of Culture Cells via Fusion with Loaded Erythrocytes

Robert A. Schlegel and Michael R. Lieber

1. INTRODUCTION

Because of its relative simplicity, the non-nucleated mammalian erythrocyte has often served as a paradigm for study of more complex eukaryotic cells. About 10 years ago, studies of the erythrocyte membrane and investigations employing erythrocytes as models for understanding cell fusion converged, resulting in the technique referred to as erythrocyte-mediated microinjection. The seminal studies of Seeman (1967) indicated that macromolecules could enter and be trapped inside erythrocytes whose membrane integrity was reversibly interrupted by a hypotonic shock treatment. Loyter and colleagues (Peretz *et al.*, 1974) then described conditions under which erythrocytes could be fused with one another by Sendai virus without losing their membrane integrity, so that intracellular macromolecules were retained. Under these same sorts of conditions hemoglobin could be transferred from erythrocytes to nucleated eukaryotic cells to which they were fused (Zakai *et al.*, 1974; Furusawa *et al.*, 1974).

ROBERT A. SCHLEGEL • Department of Molecular and Cell Biology, Pennsylvania State University, University Park, Pennsylania 16802. MICHAEL R. LIEBER • Laboratory of Pathology, National Institutes of Health, Bethesda, Maryland 20205. *Present address for* M. R. L.: Laboratory of Molecular Biology, NIDDK, National Institutes of Health, Bethesda, Maryland 20892.

Finally, these findings on loading of erythrocytes with exogenous macromolecules and fusion of such carriers to recipient cells were combined to afford an efficient method for introducing macromolecules into eukaryotic cells (Schlegel and Rechsteiner, 1975; Loyter *et al.,* 1975; Nishimura *et al.,* 1976). The advantages and disadvantages of this and other techniques (e.g., glass microcapillaries, liposomes, osmotic lysis, etc.) for introduction of macromolecules were recently presented by Celis (1984).

As numerous reviews on the methodology and applications of erythrocyte-mediated microinjection are available (Furusawa *et al.,* 1976; Schlegel and Rechsteiner, 1978; Kulka and Loyter, 1979; Furusawa, 1980; Rechsteiner, 1982; McElligott and Dice, 1984; Schlegel and Rechsteiner, 1986; Rechsteiner and Schlegel, 1986), it is not our intention to provide an exhaustive review of the literature. Rather we hark back to the origins of the technique and place emphasis on the usefulness of erythrocytes as models for the study of membrane structure and dynamics and in the elucidation of mechanisms of cellular fusion.

2. ERYTHROCYTE LOADING

Human erythrocytes have been used most often for erythrocyte-mediated microinjection, although rabbit (Kaltoft *et al.,* 1976), guinea pig (Heumann *et al.,* 1981), and commercially available sheep erythrocytes (Kriegler *et al.,* 1978) have also been successfully employed. Outdated as well as fresh blood can be used (Wasserman *et al.,* 1976), without regard to human ABO blood type (Yamaizumi, 1978*a*).

When human erythrocytes are swollen from their biconcave shape (87 μm^3) to their spherical form (approximately 150 μm^3) in a hypotonic solution of less than 160 mOsm, their plasma membrane ruptures. The membrane ruptures at a localized region of the cell surface, presumably a statistical weak point, releasing cytoplasmic molecules and equalizing the osmotic gradient (Danon, 1961; Baker, 1967; Yee and Mel, 1978). The size of the rupture transiently may reach several hundred nanometers during the seconds in which the cytoplasmic molecules are effluxing (Seeman, 1967; Huhn *et al.,* 1970; Seeman *et al.,* 1973; Seeman, 1974). However, the residual membrane defect, the hemolytic hole, closes within minutes to a smaller radius, typically between 0.5 and 20 nm, the exact size being determined by the electrostatic charge repulsion among anionic membrane molecules in the vicinity of the membrane hole (Lieber and Steck, 1982*a,b*).

Following lysis, one is left with an erythrocyte ghost. The ghost consists of a plasma membrane, including its underlying cytoskeleton. Although a few percent of the cytoplasmic contents may remain, the internal solution of the ghost is nearly the same ionic composition as the extracorpuscular solution at this point. (These generalizations assume lysis in the absence of extracellular macromolecule concentrations higher than 5–10 mg/ml.) The osmotic gradient across the membrane has been dissipated to a low or negligible level. The membrane bilayer and cytoskeletal components which impart the biconcave shape to the parent erythrocyte are present in the erythrocyte ghost, which is typically mono- or biconcave. Erythrocyte ghost membrane shape and, therefore, volume are responsive to the ionic composition of the solution in which they are suspended (Johnson *et al.*, 1980; Lange *et al.*, 1982*a,b*). The erythrocyte ghost internal volume varies with the cation concentration of the solution from about 30 μm^3 at 300 mM monovalent cation to 90–130 μm^3 at 10 mM.

2.1. Conventional Techniques

The dialysis and preswell methods of loading have been most commonly used in erythrocyte-mediated microinjection, although exact protocols differ somewhat among investigators. Recently, a new methodology in which erythrocytes are loaded by one or more freeze-thaw cycles has been introduced.

2.1.1. Dialysis

In all dialysis protocols, erythrocytes are placed in a dialysis bag containing a solution of the molecule to be loaded. The suspension is dialyzed against hypotonic buffer until lysis and loading are complete. Then tonicity is restored by dialysis against isotonic buffer or by addition of concentrated salt solution, and resealing is afforded by incubation at room temperature or 37°C (Yamaizumi *et al.*, 1978*a;* Wasserman *et al.*, 1976).

2.1.2. Preswell

In the preswell method of loading (Rechsteiner, 1975; Schlegel and Rechsteiner, 1978; Rechsteiner, 1982; Schlegel and McEvoy, 1986), erythrocytes are placed in a solution whose tonicity has been adjusted such that cells swell but do not rupture. The addition of only a small volume of lysis buffer (typically 10 mM Tris) to a pellet of these preswollen erythrocytes affords lysis. Concentrated saline is added to restore isotonicity and cells are incubated at 37°C to reseal. The complete preswell protocol is schematically depicted in Figure 1.

Several investigators have found it necessary to clear erythrocytes of endogenous molecules which might interfere upon injection into recipient cells. This goal has been accomplished by one (Capecchi *et al.*, 1977) or

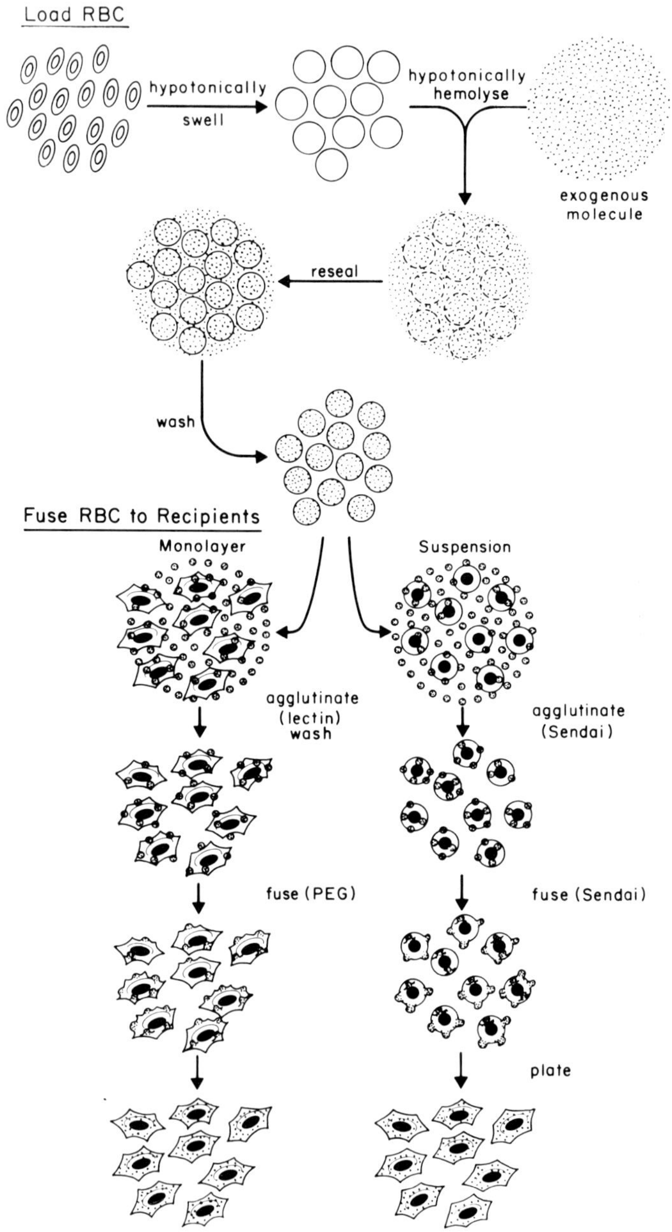

Figure 1. Loading of erythrocytes and their fusion with recipient culture cells. The preswell method of loading erythrocytes (red bood cells, RBC) is depicted in the upper portion of the diagram. In the lower left portion of the diagram loaded erythrocytes are fused to monolayers of culture cells using a lectin to enhance fusion induced by PEG. In the lower right portion of the diagram, loaded erythrocytes are fused with culture cells in suspension using Sendai virus as fusogen. (From Schlegel, 1984.)

multiple (Boogaard and Dixon, 1983*a*) rounds of lysis and resealing, or by multiple washings of unsealed erythrocytes (Kaltoft and Celis, 1978), before the actual loading cycle.

2.1.3. Freeze–Thaw Loading

Because neither the dialysis nor the preswell method is effective at loading high-molecular-weight DNA, freeze–thaw loading protocols have been devised to overcome this shortcoming. DNA (or protein) is simply added to a suspension of erythrocytes (Iino *et al.,* 1983) or unsealed, washed erythrocytes (Wiberg *et al.,* 1983), which are then frozen and thawed once or several times, followed by resealing at 37°C. Transformation or transfection of cells to which loaded erythrocytes were fused indicated successful transfer of DNA.

2.1.4. Characteristics of Loading

Under most conditions, erythrocyte lysis is an all-or-none process with respect to each individual erythrocyte. This situation holds true regardless of whether lysis is mediated by osmotic shock, freeze–thawing, or detergent and regardless of how rapidly the erythrocyte lyses (Davies *et al.,* 1968; Zade-Oppen *et al.,* 1979). The erythrocyte departs from a discrete all-or-none lysis, however, when macromolecules are present in the extracorpuscular solution (Davies *et al.,* 1968; Zade-Oppen *et al.,* 1979). In the presence of macromolecules of at least 2000 M_r at a weight concentration greater than 10 mg/ml, fractional lysis of individual erythrocytes is observed such that each cell may release 20–90% of its hemoglobin rather than all or none of it. Such concentrations of macromolecule are not unusual when one is lysing erythrocytes in dialysis bags or in relatively small solution volumes after preswelling because the hemoglobin itself is a macromolecule. Therefore, hemoglobin released by the portion of the population of erythrocytes lysing first may cause fractional lysis in the remaining cells of the population as the volume ratio of lysing solution to erythrocytes approaches three to one (Yamaizumi *et al.,* 1978*a;* Kulka and Loyter, 1979). As the concentration of macromolecule approaches 100 mg/ml, each erythrocyte may release as little as 20% of its hemoglobin. The basis for this fractional cell lysis when extracellular macromolecules are present may be solvent exclusion (Seeman, 1973).

The kinetics of about 80% of the hemoglobin released through the hemolytic hole of an osmotically lysing erythrocyte can be accounted for on the basis of diffusion (Heedman, 1958). However, the initial 20% of hemoglobin released from each cell appears to escape faster, probably due to convective outflow occurring to relieve the osmotic pressure.

Because of the convective flow, it is difficult to load molecules into the cell during this period. Therefore, release of greater than 20% of the hemoglobin is correlated with a marked increase in loading ability (Davies *et al.,* 1968).

The major problem of the hypotonic loading methods (dialysis and preswell) is the progressive exclusion of higher molecular weight materials from entry into the erythrocyte by the seiving effect of a hemolytic hole of finite dimensions (see Section 2.2.1). Empirically, whereas 100% of the initial concentration of a protein of 22,000 M_r can be loaded, only 50% of a 68,000-M_r protein and 30% of a 160,000-M_r protein can be loaded (cited in Mekada *et al.,* 1978*b*), which translates to nearly 10^7 copies of a smaller protein loaded into each erythrocyte, but only about 10^5 copies of a protein with a molecular weight of roughly 500,000. Available evidence indicates a similar size dependence for loading of RNA (Schlegel and Rechsteiner, 1986) or DNA (Rechsteiner *et al.,* 1984), the latter being efficiently loaded only if below 800 base pairs, although transfection of KB cells fused with erythrocytes hypotonically loaded with adenovirus type 2 DNA has been reported (Straus and Raskas, 1980). Because of this limitation, the freeze–thaw methods of loading mentioned above were introduced in order to entrap high-molecular-weight DNA. Whether these protocols are effective with proteins or RNA of high molecular weight has yet to be determined.

2.2. The Hemolytic Hole

The conventional loading techniques just described have empirical evidence for their success and utility under defined conditions. However, individual applications may require deviation from standard protocols in order to preserve the integrity or activity of a molecule during loading, thus necessitating loading under prescribed ionic strength, osmolarity, temperature, concentration of the loaded molecule, erythrocyte density, or other parameters. In order to design alternate conditions rationally under which successful and efficient loading might be accomplished, further understanding of erythrocyte lysis and the characteristics of the resulting hemolytic hole is required.

2.2.1. Factors Controlling the Size of the Hemolytic Hole

Following lysis of the erythrocyte, the membrane appears to be unchanged over more than 99% of its surface. However, a single hole has been blown in the membrane bilayer of each erythrocyte by the osmotic rupture at the time of lysis (Lieber and Steck, 1982*a,b*). That residual hemolytic hole can be maintained stably in the membrane for the life of the ghost. The hemolytic hole is a right circular cylinder, and its radius can be increased or decreased by increasing or decreasing, respectively, the electrostatic charge repulsion among the anionic membrane phospholipids, glycolipids, and proteins that constitute the erythrocyte membrane. The hole can be varied in size to include from 10^{-8} (0.6-nm radius) to 10^{-2} (1000-nm radius) of the ghost membrane surface area. It is reversible in size for a period of time determined by the history of ionic environments and temperatures to which the erythrocyte ghost has been exposed.

Varying the ionic strength of the solution is the simplest method of varying the size of the hemolytic hole (Lieber and Steck, 1982*b*). Cations appear to bind to the anionic membrane components, reducing the net negative charge in the vicinity of the hemolytic hole. This allows the channel to decrease in radius. All monovalent cations appear to be of similar potency. The calcium ion is 60–120 times more potent than the monovalent cations and four times more potent than the magnesium ion. Approximate hole sizes at various concentrations of sodium and calcium ions are given in Table I. The channel size can be reversibly changed to larger and smaller sizes with changes in the solution cation concentration. However, with time the channel loses its reversibility (Lieber and Steck, 1982*b*).

As with other cations, the size of the hemolytic hole varies inversely with the hydrogen ion concentration. Between pH 7.1 and 7.8, the estimates in Table I apply. At pH 5.5–7.1, the equilibrium radius is smaller. Above pH 7.8, the equilibrium radius may be larger.

Membrane fluidity influences the rate at which the hole size can be changed (Lieber and Steck, 1982*b*). Below 15°C, changes in the radius of the hole with cation concentration are very slow (less than 0.05 nm/min). Above 15°C, the rate at which the radius of the hole reaches equilibrium rises markedly and at 37°C it is 0.8 nm/min. Temperatures between 20° and 40°C are practical.

Although the hemolytic hole can be closed to the size of a few phospholipid molecules, complete closure of the hole is not attained by a large fraction of the ghost population, even under vigorous sealing conditions (e.g., incubation in 150 mM NaCl for 1 hr at 37°C). Under such conditions, about 50% of the ghosts have a hemolytic hole of radius 0.6 to 1.0 nm (Lieber and Steck, 1982*a,b*). The other 50% have a hole radius less than 0.5 nm, and this group is nearly as impermeable to cations as the native erythrocyte (Bodeman and Passow, 1972; Bjerrum, 1979; Lieber and Steck, 1982*a,b*).

Factors relevant to controlling hemolytic hole size have been surveyed, and are covered in more detailed accounts (Lieber, 1981; Lieber and Steck, 1982*a,b*).

Table I. Hemolytic Hole Size as a Function of Solution Cation Concentration[a]

$[Na^+]$(mM)	$[Ca^{2+}]$(mM)[b]	radius(nm)
<5		>15.0
5–9		10.0–15.0
10–14	0.04–0.10	5.0–10.0
15–20	0.10–0.25	3.0–5.0
20–35	0.25–0.50	2.0–3.0
35–50	0.50–0.80	1.5–2.0
>50	>0.80	0.6–1.5[c]

[a] Human erythrocyte ghosts were made as described in Section 2.2.2 and were washed twice after the lysis step in 40 vol of 5 mM NaP_i (pH 8) at 4°C. The erythrocyte ghosts were incubated at 20°C in 24 vol of various concentrations of NaP_i (pH 7.2) or 2 mM imidazole (pH 7.2) and various concentrations of $CaCl_2$. Hemolytic hole radii were determined using dextran density barriers standardized with kinetic permeability measurements (Lieber and Steck, 1982*a*). The radii were measured as a function of time until the hole size reached a plateau (typically 15–45 min). The plateau radius is given as a function of cation concentration.

[b] Below 0.040 mM Ca^{2+}, the hole size is greater than 10 nm radius.

[c] Closing the hemolytic hole below 1.5 nm typically requires that the ghosts be treated with sealing conditions (37°C for 20 min) within 1 hr of lysis. One wash of the ghosts instead of two was used here in order to reach the sealing step within 1 hr of lysis.

2.2.2. Loading Molecules through the Hemolytic Hole

2.2.2a. Small Molecules. Most molecules of less than 2000 daltons have radii of less than 1.5 nm. One method for loading such molecules into ghosts is described as follows. (All procedures are done at 0–4°C.) After three washings of one volume of fresh or blood bank erythrocytes in 20 vol of phosphate-buffered saline, 1 vol of packed cell pellet is lysed in 40 vol of 5 mM NaP_i (pH 8), incubated 5 min to allow hemoglobin efflux and then centrifuged. These ghosts contain about 2–4% of their original hemoglobin load. If hemoglobin contamination is a problem, one additional wash in 40 vol 5 mM NaP_i (pH 8) prior to adding the molecule to be loaded will reduce the hemoglobin content another 5–10-fold. If hemoglobin contamination is not a problem, one can proceed directly to adding the molecule to be loaded to the ghost pellet. Equilibration of small molecules with the internal volume of the ghost typically takes less than two minutes. [Equations for calculating equilibration time across the hemolytic hole are described elsewhere (Lieber and Steck, 1982*a*).] In order to close the hemolytic channel to the smallest size possible after loading, the pellet is made to greater than 50 mM in NaCl and incubated at 37°C for 15–45 min. The hemolytic hole will remain small as long as the cation concentration is maintained at greater than 50 mM and the cells are not osmotically reshocked.

2.2.2b. Proteins. Most proteins are less than 20 nm in maximal diameter. In order to attain the highest efficiency of loading, the lysis described for loading small molecules should be followed by a wash in 5 mM NaP_i (pH 8), the pellet from which is mixed with the molecule to be loaded. The equilibration time prior to sealing should be 5–20 min, the longer times being required as larger molecules are loaded.

2.2.2c. DNA. Typical DNA constructs and expression vectors are 5000–10,000 base pairs in size. Supercoiled DNA molecules in this size range are about 700–1400 nm in length and less than 40 nm in width. These long, narrow molecules may be loaded into erythrocyte ghosts as follows. Ghosts are made as described for loading proteins. At time zero, 1 vol of pelleted ghosts is diluted into 15–30 vol of 0.5 mM NaP_i (pH 7.2), 0.05 mM EDTA on ice, and then pelleted. This step dilates the hemolytic hole to several hundred nanometers in diameter. At 20 min, the DNA construct (typically in a low-ionic-strength buffer such as 10 mM Tris (pH 7.5), 1 mM EDTA) is added to give a final concentration of 1–100 μg/ml from a DNA stock sufficiently concentrated so as not to increase the monovalent cation concentration above 1 mM in the ghost pellet. The DNA molecules are allowed to equilibrate with the ghost internal volume for 15 min. Finally, the pellet is made 150 mM in NaCl and 1 mM in EDTA and incubated at 37°C for 20 min to close the hole as much as possible.

This procedure allows the DNA molecules to achieve more than 75% equilibration with the ghost internal volume (M. R. Lieber, M. Gellert, and K. Mizuuchi, unpublished observations). However, dilation of the hemolytic hole to this extent causes loss of resealability (Lieber and Steck, 1982*b*). Therefore, the hole closes to only 3–8 nm radius after the sealing step. Nevertheless, the DNA is retained inside the ghosts. However, a residual hemolytic hole may result in leakage after fusion, decreasing viability of recipient cells. Despite this problem, subsequent fusion of the loaded ghosts with relatively large cells such as fibroblasts does result in successful transfection, though the efficiency of transfection may be lower than that achieved with more conventional transfection methods such as those employing DEAE-dextran and calcium phosphate–DNA precipitates (M. R. Lieber, M. Gellert, and K. Mizuuchi, unpublished observations). The advantage to this method is, however, its applicability to cell types that are not easily transfected by conventional methodology and in allowing transfer of both DNA and potential regulatory proteins into the cell at the same time.

3. FUSION

Although Sendai virus was used as fusogen in the original description of erythrocyte-mediated microinjection, poly(ethylene glycol) (PEG) is now used almost as often. Each fusogen offers its own advantages. Sendai virus is most often employed when large numbers of cells are injected in suspension; PEG is most often used for injections of cells in monolayer, when numerous variables and timepoints are to be examined, since as few as 10^3–10^4 cells are sufficient for an experimental point. General schemes for both types of protocols are presented in Figure 1 and elaborated upon in this section.

3.1. Sendai Virus

Fusions in suspension are performed by mixing 10^6–10^7 culture cells, a 100-fold excess of loaded erythrocytes and 400–1000 HAU of UV-inactivated Sendai virus (Schlegel and Rechsteiner, 1978; Kulka and Loyter, 1979; Furusawa, 1980). After 10–15 min on ice to allow cells to agglutinate, the mixture is incubated at 37°C while gently shaking for 15–30 min. Cultures are then returned to ice and/or diluted with buffer to terminate fusion. Routinely, 60–90% of culture cells fuse with one to three erythrocytes following this simple procedure, and viability of injected cells is 90–100% of uninjected controls as judged by cloning efficiency (Mekada *et al.*, 1978*a,b*). The only real variable in the protocol is the selection of divalent (or trivalent) cation, which is included during fusion. Specific requirements appear to vary from cell type to cell type, as discussed by Schlegel and Rechsteiner (1986).

The vast majority of erythrocytes do not fuse to culture cells and must be removed from the injected population. Removal is accomplished in a number of ways. The simplest method is to plate the fusion mixture. After the culture cells have adhered, the nonadherent erythrocytes can be washed away. Differential centrifugation (Yamaizumi *et al.*, 1978*d*; Rechsteiner, 1982) and sedimentation through a shelf of calf serum (Wasserman *et al.*, 1976; Yamaizumi *et al.*, 1979; Bennett *et al.*, 1983) have also been successfully employed, as has "panning" for culture cells using dishes coated with antisera specific for determinants on the culture cell surface (Ohara and Watanabe, 1982). Following any one of these primary procedures, residual erythrocytes can be preferentially lysed by two 10-min treatments with 0.83% NH_4Cl (Kaltoft *et al.*, 1976) or by a 10-sec treatment with 10mM buffer (Kruse *et al.*, 1981).

When microinjections are performed in monolayer, virus alone is first added to the adherent culture cells. Following incubation at 4°C, erythrocytes are added, and cultures are incubated at 37°C to induce fusion. Several investigators find that no more than 5–20% of cells in monolayer cultures can be successfully injected (Mercer, 1980; Kulka and Loyter, 1979; Yamaizumi *et al.*, 1978*b;* Iino *et al.*, 1983), but efficiencies of injection as high as 84% have been reported when the above protocol is followed (Yamaizumi *et al.,* 1978*a*). Replacing the virus particle with the HN (hemagglutinin) and F (fusion) glycoproteins isolated from it results in microinjection efficiencies comparable to those induced by the intact whole virus (Kruse *et al.,* 1981, 1984).

3.2. Polyethylene Glycol

No underlying formula applies in the selection of the brand or molecular weight of PEG best suited to a particular microinjection protocol. Beyond some general guidelines for the selection of the most fusogenic and least toxic PEG (Schlegel and McEvoy, 1986), the type actually used is for the most part empirically determined. It is therefore advisable when adopting a particular protocol to also adopt the PEG used in that protocol, at least in initial trials.

In the simplest procedures for injection in suspension (Boogaard and Dixon, 1983*b;* Wiberg *et al.,* 1983) culture cells and erythrocytes in excess are combined in a very small volume or in a pellet, 1 ml of 50–55% PEG is added; after 1 min at 37°C, the fusion mixture is slowly diluted with buffer or media. To enhance fusion, Ohara and Watanabe (1982) used poly-L-arginine in their PEG solution to bring erythrocytes and culture cells in closer apposition and reported a microinjection efficiency of 80%. Similar efficiencies can be achieved by immunospecific targeting of erythrocytes to recipients (Godfrey *et al.,* 1983) with the added advantage that selected cells within a population can be specifically targeted for injection.

In the simplest monolayer protocols (Neff *et al.,* 1981; Dice, 1982; Backer *et al.,* 1983; McElligott and Dice, 1983) a 100-fold excess of loaded erythrocytes in a small volume is added to a confluent culture. After swirling to distribute the erythrocytes evenly, a small volume of PEG is added and also evenly distributed by swirling. After 1 min at room temperature, medium is added and the culture is incubated at 37°C for 30 min, resulting in microinjection frequencies of 50–60%. In a variation of this procedure, loaded erythrocytes and PEG are mixed together and then applied to the monolayer (Kriegler and Livingston, 1977). Microinjection efficiencies of 90% are reported.

As was the case with fusions in suspension, promotion of contact between erythrocytes and cells in monolayer culture leads to enhancement of fusion. Erythrocytes can be centrifuged onto coverslips of culture cells to bring the two types into closer apposition (Kaltoft and Celis, 1978; Jonak and Mora, 1980). If cells which are nonadherent are to be injected, they and erythrocytes can be mixed and centrifuged together onto coverslips (Celis *et al.,* 1980). In a variation of this protocol, erythrocytes and culture cells are centrifuged onto coverslips previously coated with concanavalin A to promote attachment of cells to the surface of the coverslip (Brown *et al.,* 1985).

Schlegel and colleagues (Mercer *et al.,* 1979; Schlegel and Mercer, 1980; Schlegel and McEvoy, 1986) agglutinated erythrocytes to culture cells by application of the lectin phytohemagglutinin (PHA). Just before the addition of erythrocytes to monolayer cultures, PHA is added to the erythrocyte suspension. After allowing time for contact and attachment of erythrocytes to culture cells, nonattached erythrocytes are washed from the cultures and PEG is added to induce fusion. Microinjection efficiencies in the 60–90% range are achieved. Substitution of concanavalin A or polylysine for PHA has the advantage that attachment of erythrocytes to culture cells by these crosslinkers is reversible (by chelation of divalent cations or thorough washing, respectively), so that nonfused erythrocytes are easily removed from fused cultures (Schlegel and McEvoy, 1986). Heumann *et al.* (1981) reported better viability of cells following injection if cultures are fused by a five minute exposure to ice-cold PEG rather than the usual one minute exposure at room temperature. Finally, by substituting specific antibodies for nonspecific agglutination by lectins or polyanions targeting of erythrocytes to selected types of cells in a mixed population can be accomplished (Godfrey *et al.,* 1983).

In contrast to fusions induced by Sendai virus where Ca^{2+} is sometimes found to enhance fusion, several groups caution that Ca^{2+} should be specifically excluded from fusion and postfusion buffers for at least 15 min following fusion with PEG in order to preserve cell viability (Schneiderman *et al.,* 1979; Schlegel and Mercer, 1980). The extent of the toxic effects of Ca^{2+} is apparently dependent on the brand of PEG which is used (Mercer and Baserga, 1982). Whether this caveat applies to all cell types or just those tested in trials is unknown.

3.3. Influenza Virus Hemagglutinin

In each of the examples given in Sections 3.1 and 3.2 the fusogen was exogenously added. Recently, a system was devised in which the fusogen is endogenous to one of the fusing partners but is maintained in a latent state until required. The fusogenic protein of influenza virus is the hemagglutinin (HA) protein. Unlike the F protein of Sendai virus, HA is activated only at reduced pH (Maeda and Ohnishi, 1980; White *et al.,* 1981). The gene coding for HA has been cloned and introduced into fibroblasts resulting in a stably transformed line of cells which constitutively synthesize and efficiently express HA on their surface (Sambrook *et al.,* 1985). When erythrocytes are added to monolayers of these cells, they become tightly bound. When the pH of the medium is briefly dropped to 5.0, fusion is triggered and injection of erythrocyte contents is accomplished (Doxsey *et al.*, 1985). The utility of such a self-contained system is quite evident.

4. MECHANISM OF PEG-INDUCED FUSION OF ERYTHROCYTES WITH CULTURE CELLS

Fluorescent membrane and cytoplasmic probes have been used to study the mechanism by which PEG induces fusion of erythrocytes to culture cells (Wojcieszyn *et al.*, 1983). The system used was the one described above where PHA is utilized to bring erythrocytes into close apposition with culture cells attached in monolayer. In these studies, erythrocytes were loaded with fluoresceinated bovine serum albumin (fl-BSA), which then served as a marker for determining transfer of cytoplasmic contents; the plasma membranes of erythrocytes were labeled with one of several fluorescent lipid probes before fusion.

PHA enhances the fusion of erythrocytes to culture cells by causing more erythrocytes to bind to target cells (Schlegel and Mercer, 1980). The action of the lectin itself does not produce fusion since in its presence alone neither cytoplasmic nor membrane probes were transferred to recipient culture cells (Wojcieszyn *et al.*, 1983). When PEG was added, transfer of lipid probes to recipients was observed in <30 sec, and by 1–2 min fluorescence had spread over the entire cell surface. Fl-BSA was not transferred, however, as long as PEG was present. Upon dilution of PEG, the cytoplasmic marker began to enter culture cells within 1–3 min presenting the picture seen in Figure 2.

These results raise the question of whether fusion occurs when PEG is present, with delay of transfer of cytoplasmic probe, or whether fusion occurs only sometime after dilution of the polymer. To address this question fluorescence recovery after photobleaching (FRAP) was used to determine the diffusion coefficients of fl-BSA within erythrocytes and within fibroblasts injected with the cytoplasmic probe. In both types of cells fl-BSA was found to be operationally immobile ($D \lesssim 10^{-12}$ cm^2/sec) in the presence of 44% PEG. When PEG was removed by washing, fl-BSA diffused freely in both types of cells with a D value of $\simeq 10^{-8}$ cm^2/sec (Wojcieszyn *et al.*, 1983). Therefore, the delay in transfer of fl-BSA upon dilution of PEG can most likely be attributed to the immobilization of the probe produced by the dehydrating effects of PEG which are relieved upon dilution of the polymer.

Honda *et al.* (1981) reported that when commercial grade PEG was recrystallized from organic solvents, it lost its ability to fuse cells. Although this purified PEG is apparently not nonfusogenic in all fusion systems (Smith *et al.*, 1982; Hui *et al.*, 1985), it was unable to induce the

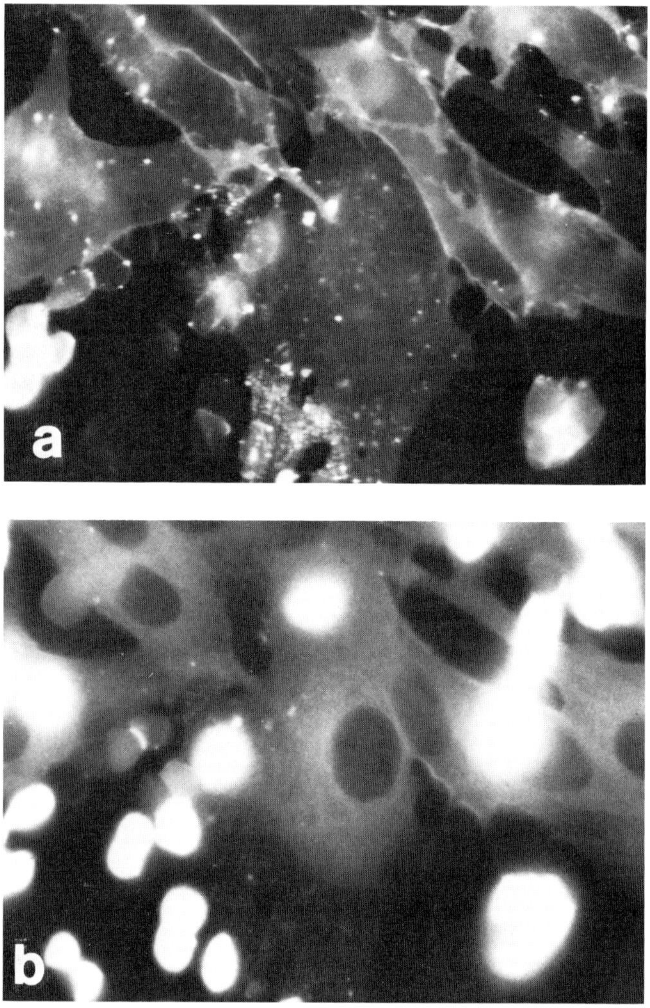

Figure 2. Transfer of fluorescent lipid probes and fl-BSA from erythrocytes to culture cells. Erythrocytes were loaded with fl-BSA, labeled with the lipid probe diI-C_{18} and fused to BHK fibroblasts in monolayer using phytohemagglutinin and PEG. Live cells were photographed at 30 min after fusion to reveal diI (a) or fl-BSA (b) fluorescence. Highly fluorescent areas are unfused erythrocytes.

transfer of fl-BSA from erythrocytes to culture cells indicative of fusion. Yet fluorescent lipid probes were still transferred from the membranes of erythrocytes when nonfusogenic PEG was applied (Wojcieszyn *et al.*, 1983). These contrasting results with cytoplasmic and membrane probes point to a nonfusogenic mechanism of exchange of lipid probe between erythrocytes and culture cells. Freeze-fracture electron micrographs of erythrocytes treated with either fusogenic or nonfusogenic PEG support the view that PEG, by virtue of its dehydrating ability, promotes a unique membrane apposition that permits rapid transfer of lipid probes irrespective of its fusing ability (Wojcieszyn *et al.,* 1983).

This unexpected behavior of lipid probes led to experiments designed to ask whether membrane proteins are transferred from erythrocytes to culture cells upon fusion. Although membrane proteins intermix when mouse and human culture cells are fused (Frye and Edidin, 1970) and when erythrocytes are fused with each other (Fowler and Branton, 1977), this does not appear to be the case when erythrocytes are fused to culture cells. Membrane proteins of erythrocytes loaded with fl-BSA were covalently labeled with fluorescent tags and fused with culture cells. Although transfer of fl-BSA indicated that fusion had occurred, fluorescence attributable to the membrane markers was limited to erythrocytes (Wojcieszyn *et al.,* 1983). Similar results have been reported using the influenza HA system described in the preceding section (Doxsey *et al.,* 1985).

To counter the argument that labeling was responsible for immobilization of the proteins, erythrocytes loaded with fl-BSA, but whose membranes were not labeled, were fused with culture cells, and erythrocyte membrane components were then detected by antibodies directed against them. All culture cells that exhibited cytoplasmic fl-BSA indicative of fusion were also found to exhibit at least one localized area of surface fluorescence when indirect immunofluorescence was used to detect erythrocyte membrane components. Such patches were seen on fewer than 1 in 500 culture cells that did not contain fl-BSA, arguing strongly that the patches were not attached unfused erythrocytes or debris (Schlegel *et al.,* 1985*a*). These localized areas of fluorescence, presumably sites at which erythrocytes had fused, persisted for up to 24 hr at 37°C following fusion (Wojcieszyn *et al.,* 1983; Schlegel *et al.,* 1985*a;* Doxsey *et al.,* 1985). There is as yet insufficient information to permit a distinction of various mechanisms which might explain the lack of diffusion of erythrocyte membrane components. However, the absence of an intact cytoskeletal network of spectrin does not remove this inhibition (Wojcieszyn *et al.,* 1983), leading to speculation that perhaps lipid–protein interactions are involved in the mechanism which maintains segregation (Schlegel *et al.,* 1985*a*).

These collected observations led to the following scheme of how fusion might occur in this system. In the presence of either fusogenic or nonfusogenic PEG, membranes come extremely close to one another, possibly allowing transfer of lipid probes at these regions of ultraclose apposition. Actual fusion, as detected by transfer of cytoplasmic contents, only occurs, however, in the presence of ingredients of PEG, which can be removed upon recrystallization. Likely candidates are the antioxidants and polymerization agents added to PEG during production, since their readdition to nonfusogenic PEG restores fusion competence (Honda *et al.,* 1981; unpublished observations cited in Wojcieszyn *et al.,* 1983). Therefore, at least in this system, PEG appears to perform a dual role. First, the polymer itself acts as a dehydrant to promote close contact of membranes. Second, these close contact areas are destabilized, not by the polymer itself but by contaminants therein, leading to disruption and subsequent coalescence of adjacent bilayers, forming intercellular pores of communication.

5. ENUMERATION AND SEPARATION OF INJECTED CELLS

In order to determine the fraction of culture cells in an injected population which have fused with at least one erythrocyte, a method for identifying injected cells is required. One of the most common methods, the detection by fluorescence microscopy of transfer of a fluorescent cytoplasmic tracer molecule, was illustrated in the preceding section. Besides fl-BSA, fluoresceinated IgG is also popular (Yamaizumi *et al.,* 1979; Ohara and Watanabe, 1982). Transfer of a cytoplasmic molecule endogenous to erythrocytes, hemoglobin, can be detected by benzidine staining (Furusawa *et al.,* 1974; Yamaizumi *et al.,* 1978*a;* Rechsteiner, 1982). Increased sensitivity of hemoglobin detection is provided by using antihemoglobin antibodies and a second fluoresceinated antibody (Hendil, 1980). In fact, almost any molecule which can be loaded and for which an antibody is available can be used to identify injected cells, e.g., IgG and fluoresceinated anti-IgG (Antman and Livingston, 1980; Petrie *et al.,* 1982; Schlegel, 1984; Schlegel *et al.,* 1985*b;* Doxsey *et al.,* 1985). If a fluorescence microscope is not available, cytological detection by second antibodies conjugated with, for instance, horseradish peroxidase can be used. In fact, horseradish peroxidase itself can be used as a cytoplasmic marker (Heumann *et al.,* 1981; Rechsteiner, 1982; Doxsey *et al.,* 1985). Finally, radioactively labeled molecules can be injected as tracers (see, e.g., Rechsteiner and Kuehl, 1979; Neff *et al.,* 1981). However, since injected cells must be identified by autoradiography, a lag of several days is usually required for exposure of emulsions.

As also illustrated in the preceding section, the incorporation of endogenous erythrocyte membrane components into culture cells can be used to detect and enumerate injected cells (Wasserman *et al.,* 1976; Yamaizumi *et al.,* 1978*a*).

Separation of injected from uninjected cells can be accomplished by fluorescence-activated cell sorting following fluorescent labeling of injected cells by one of the methods described. Mekada *et al.* (1978*a*) used the erythrocyte membrane detection system to obtain a population that contained 95% injected cells (confirmed by benzidine staining) from a starting population which contained only 50% injected cells, while maintaining cloning efficiency at 90%. More elegantly, culture cells fused with one, two or three erythrocytes can be sorted from each other when fl-BSA is used as a marker (Mekada *et al.,* 1978*b*; Yamaizumi *et al.,* 1978*c*) with no apparent reduction in cloning efficiency among the sorted populations.

Several investigators have used the ability of an injected substance to rescue a cell as a means for selecting only the injected cells in a population. Yamaizumi *et al.* (1978*b,* 1979) injected cells with antibody to fragment A of diphtheria toxin. When the injected population was subsequently exposed to toxin, only injected cells survived, the toxic fragment A being neutralized intracellularly by the antibody. Because several days of culture are required for this selection, during which a co-loaded molecule of primary interest might be degraded, Rogers and Rechsteiner (personal communication) have injected anti-ricin and challenged with ricin, which killed uninjected cells within 24 hr. Both the toxin and its antibody are available commercially.

6. APPLICATIONS

The main advantage of erythrocyte-mediated microinjection over injection of cells by microneedles is that far greater numbers of cells (10^7 vs. 10^3) can be injected. Therefore, erythrocyte-mediated microinjection is well suited for studies in which injected molecules must be recovered and examined biochemically or where the response elicited by the injected molecule is only measurable with large numbers of injected cells. For example, the technique has been used extensively in studies of intracellular proteolysis since proteins labeled radioactively prior to injection can be recovered from recipient cells in quantities sufficient to follow readily their degradation biochemically. Likewise, intracellular inactivation or neutralization of specific molecules by antibodies introduced into cells by erythrocyte-mediated microinjection is being used increasingly to probe a variety of cellular functions. These and other applications of erythrocyte-mediated microinjection have been recently reviewed by Rechsteiner and Schlegel (1986).

An apparent limitation of erythrocyte-mediated microinjection is that molecules cannot be directly introduced into the nucleus. Even though proteins of nuclear origin rapidly accumulate in the nucleus following injection into the cytoplasm (Yamaizumi *et al.,* 1978*d*; Rechsteiner and Kuehl, 1979; Wu *et al.,* 1981; Rechsteiner *et al.,* 1984), most other proteins remain in the cytoplasm following injection. This problem can be indirectly overcome, however, if recipients are synchronized in mitosis when injected. As injected cells leave mitosis, the cytoplasmically dispersed nuclear membrane reforms around injected molecules, enclosing even those normally excluded (Schlegel, 1984; Schlegel *et al.,* 1985*b*). Thus the effects of injected molecules on nuclear as well as cytoplasmic functions can be studied.

7. CONCLUDING REMARKS

It has been a decade since the introduction of erythrocyte-mediated microinjection as a method for transferring macromolecules into culture cells. Only in the last five years, however, has the technique gained acceptance and been put to fruitful use. The ability to inject as many as 10^7 protein molecules or 10^8 small RNA molecules into each of 10^7 culture cells in a matter of hours promises that the method will find even wider application in the future as a tool for deciphering the intricacies of cellular structure and function.

ACKNOWLEDGMENTS. We thank Katy Lumley-Sapanski for the photographs in Figure 2. R. A. S. is an established investigator of the American Heart Association.

REFERENCES

Antman, K. H., and Livingston, D. M., 1980, Intracellular neutralization of SV40 tumor antigens following microinjection of specific antibody, *Cell* **19:**627–635.

Backer, J. M., Bourret, L., and Dice, J. F., 1983, Regulation of catabolism of microinjected ribonuclease A requires the amino-terminal 20 amino acids, *Proc. Natl. Acad. Sci. U.S.A.* **80:**2166–2170.

Baker, R. F., 1967, Ultrastructure of the red blood cell, *Fed. Proc.* **26:**1785–1801.

Bennett, F., Busch, H., Lischwe, M. A., and Yeoman, L. C., 1983, Antibodies to a nucleolar protein are localized in the nucleolus after red cell-mediated microinjection, *J. Cell Biol.* **97:**1556–1572.

Bjerrum, P. J., 1979, Hemoglobin-depleted human erythrocyte ghosts: Characterization of morphology and transport functions, *J. Membrane Biol.* **48:**43–46.

Bodeman, H., and Passow, H., 1972, Factors controlling the resealing of the membrane of human erythrocyte ghosts after hypotonic hemolysis, *J. Membrane Biol.* **8:**1–26.

Boogaard, C., and Dixon, G. H., 1983*a*, Red cell ghost-mediated microinjection of RNA into HeLa cells. I. A comparison of two techniques for the entrapment and microinjection of tRNA and mRNA, *Exp. Cell Res.* **143:**175–190.

Boogaard, C., and Dixon, G. H., 1983*b*, Red cell ghost-mediated microinjection of RNA into HeLa cells: II. Cellular translation of protamine mRNA; post-translational modifications and nuclear binding of newly-synthesized protamine, *Exp. Cell Res.* **43**:191–205.

Brown, D. B., Hanks, S. K., Murphy, E. C., and Rao, P. N., 1985, Early initiation of DNA synthesis in Gl phase HeLa cells following fusion with red cell ghosts loaded with S-phase extracts, *Exp. Cell Res.* **156**:251–259.

Capecchi, M. R., Von der Haar, R. A., Capecchi, N. E., and Sveda, M. M., 1977, The isolation of a suppressable nonsense mutant in mammalian cells, *Cell* **12**:371–381.

Celis, J. E., 1984, Microinjection of somatic cells with micropipets: Comparison with other transfer techniques, *Biochem. J.* **223**:281–291.

Celis, J. E., Kaltoft, K., and Bravo, R., 1980, Microinjection of somatic cells, in: *Introduction of Macromolecules into Viable Mammalian Cells* (R. Baserga, C. Croce, and G. Rovera, eds.), pp. 99–123, Alan Liss, New York.

Danon, D., 1961, Osmotic hemolysis by gradual decrease in the ionic strength of the surrounding medium, *J. Cell Comp. Physiol.* **57**:111–117.

Davies, H. G., Marsden, N. V. B., Ostling, S. G., and Zade-Oppen, A. M. M., 1968, The effect of some neutral macromolecules on the pattern of hypotonic hemolysis, *Acta Physio. Scand.* **74**:577–593.

Dice, J. F., 1982, Altered degradation of proteins microinjected into senescent human fibroblasts, *J. Biol. Chem.* **257**:14624–14627.

Fowler, V., and Branton, D., 1981, Lateral mobility of human erythrocyte integral membrane proteins, *Nature (Lond.)* **268**:23–26.

Frye, L. D., and Edidin, M., 1970, Rapid mixing of cell surface antigens after formation of mouse-human heterokaryons, *J. Cell Sci.* **7**:319–335.

Furusawa, M., 1980, Cellular microinjection by cell fusion: Technique and applications in biology and medicine, *Int. Rev. Cytol.* **62**:29–67.

Furusawa, M., Nishimura, T., Yamaizumi, M., and Okada, Y., 1974, Injection of foreign substances into animal cells by fusion, *Nature (Lond.)* **249**:449–450.

Furusawa, M., Yamaizumi, M., Nishimura, T., Uchida, T., and Okada, Y., 1976, Use of erythrocyte ghosts for injection of substances into animal cells by cell fusion, *Methods Cell Biol.* **14**:73–80.

Godfrey, W., Doe, B., and Wofsy, L., 1983, Immunospecific vesicle targeting facilitates microinjection into lymphocytes, *Proc. Natl. Acad. Sci. U.S.A.* **80**:2267–2271.

Heedman, P. A., 1958, Hemolysis of individual red blood cells, *Exp. Cell Res.* **14**:9–22.

Hendil, K. B., 1980, Intracellular degradation of hemoglobin transferred into fibroblasts by fusion with red blood cells, *J. Cell. Physiol.* **105**:449–460.

Heumann, R., Schwab, M., and Theonen, H., 1981, A second messenger required for nerve growth factor biological activity?, *Nature (Lond.)* **292**:838–840.

Honda, K., Maeda, Y., Sasakawa, J., Ohno, H., and Tsuchida, E., 1981, Activities of cell fusion and lysis of the hybrid type of chemical fusogens. I. Structure and function of the promoter of cell fusion, *Biochem. Biophys. Res. Commun.* **100**:442–448.

Huhn, D., Pauli, G. D., and Grassmann, D., 1970, Die Erythrocytenmembran: Feinstruktur der gefrierfeatzten Membran nach Einwirkung von hyptonen Lösungen und Saponin, *Klin. Wochenschr.* **48**:939–943.

Hui, S. W., Isac, T., Boni, L. T., and Sen, A., 1985, Action of polyethylene glycol on the fusion of human erythrocyte membranes, *J. Membrane Biol.* **84**:137–146.

Iino, T., Furusawa, I., and Obinata, M., 1983, Transformation of L cells with virus thymidine kinase genes introduced by red cell-mediated microinjection, *Exp. Cell Res.* **148**:475–480.

Johnson, R. M., Taylor, G., and Meyer, D. B., 1980, Shape and volume changes in erythrocyte ghosts and spectrin-actin networks, *J. Cell Biol.* **86**:371–376.

Jonak, G. J., and Mora, M., 1980, Rapid identification of microinjected mammalian cells by fusion with loaded chicken erythrocytes, in: *Introduction of Macromolecules into Viable Mammalian Cells* (R. Baserga, C. Croce, and G. Rovera, eds.), pp. 157–167, Alan Liss, New York.

Kaltoft, K., and Celis, J. E., 1978, Ghost mediated transfer of human hypoxanthineguanine phosphoribosyl transferase into deficient Chinese hamster ovary cells by means of polyethylene glycol-induced fusion, *Exp. Cell Res.* **115**:423–428.

Kaltoft, K., Zeuthen, J., Engberg, F., Piper, P. W., and Celis, J. E., 1976, Transfer of tRNAs to somatic cells mediated by Sendai virus-induced fusion, *Proc. Natl. Acad. Sci. U.S.A.* **73**:2793–2797.

Kriegler, M. P., and Livingston, D. M., 1977, Chemically facilitated microinjection of protein into intact monolayers of tissue culture cells, *Somat. Cell Genet.* **3**:603–610.

Kriegler, M. P., Griffen, J. D., and Livingston, D. M., 1978, Phenotypic complementation of the SV40 tsA mutant defect in viral DNA synthesis following microinjection of SV40 T antigen, *Cell* **14**:983–994.

Kruse, C. A., Spector, E. B., Cederbaum, S. D., Wisnieski, B. J., and Popjack, G., 1981, Microinjection of arginase into enzyme deficient cells with the isolated glycoproteins of Sendai virus as fusogen, *Biochem. Biophys. Acta* **645**:339–345.

Kruse, C. A., Wisnieski, B. J., and Popjack, G., 1984, Characterization of a glycoprotein fusogen isolated from Sendai virus, *Biochem. Biophys. Acta* **797**:40–50.

Kulka, R. G., and Loyter, A., 1979, The use of fusion methods for the microinjection of animal cells, *Curr. Topics Membrane Trans.* **12**:365–430.

Lange, Y., Hadesman, R. A., Steck, T. L., 1982*a*, Role of the reticulum in stability of the isolated human erythrocyte membrane, *J. Cell Biol.* **92**:714–721.

Lange, Y., Gough, A., and Steck, T. L., 1982*b*, Role of the bilayer in the shape of the isolated erythrocyte membrane, *J. Membrane Biol.* **69**:113–123.

Lieber, M. R., 1981, Characterization of the hemolytic holes in human erythrocyte membranes, Ph.D. dissertation, University of Chicago.

Lieber, M. R., and Steck, T. L., 1982*a*, A description of the holes in human erythrocyte membrane ghosts, *J. Biol. Chem.* **257**:11651–11659.

Lieber, M. R., and Steck, T. L., 1982*b*, Dynamics of the holes in human erythrocyte membrane ghosts, *J. Biol. Chem.* **257**:11660–11666.

Loyter, A., Zakai., N., and Kulka, R. G., 1975, "Ultramicroinjection" of macromolecules or small particles into animal cells, *J. Cell Biol.* **66**:292–304.

Maeda, T., and Ohnishi, S., 1980, Activation of influenza virus by acidic media causes hemolysis and fusion of erythrocytes, *FEBS Lett.* **122**:283–287.

McElligott, M. A., and Dice, J. F., 1983, Erythrocyte-mediated microinjection: A technique to study protein degradation in muscle cells, *Biochem. J.* **216**:559–566.

McElligott, M. A., and Dice, J. F., 1984, Microinjection of cultured cells using red-cell-mediated fusion and osmotic lysis of pinosomes: A review of methods and applications, *Biosci. Rep.* **4**:451–466.

Mekada, E., Yamaizumi, M., and Okada, Y., 1978*a*, An attempt to separate mononuclear cells fused with human red blood cell-ghosts from a cell mixture treated with HVJ (Sendai virus) using a fluorescence activated cell sorter (FACS II), *J. Histochem. Cytochem.* **26**:62–67.

Mekada, E., Yamaizumi, M., Uchida, T., and Okada, Y., 1978*b*, Quantitative introduction of a given macromolecule into cells by fusion with erythrocyte ghosts using a fluorescence activated cell sorter, *J. Histochem. Cytochem.* **26**:1067–1073.

Mercer, W. E., 1980, Intracellular factor(s) involved in the control of mammalian cell proliferation, Ph.D. dissertation, Pennsylvania State University.

Mercer, W. E., and Baserga, R., 1982, Techniques for decreasing the toxicity of polyethylene glycol, in: *Techniques in Somatic Cell Genetics* (J. W. Shay, ed.), pp. 23–34, Plenum Press, New York.

Mercer, W. E., Terefinko, D. J., and Schlegel, R. A., 1979, Red cell-mediated microinjection of macromolecules into monolayer cultures of mammalian cells, *Cell Biol. Int. Rep.* **3**:265–271.

Neff, N. J., Bourret, L., Maio, P., and Dice, J. F., 1981, Degradation of proteins microinjected into IMR-90 human diploid fibroblasts, *J. Cell Biol.* **91**:184–194.

Nishimura, T., Furusawa, M., Yamaizumi, M., and Okada, Y., 1976, Method for intracellular injection by cell fusion using erythrocyte ghosts, *Cell Struct. Funct.* **1**:197–200.

Ohara, J., and Watanabe, T., 1982, Microinjection of macromolecules into normal murine lymphocytes by cell fusion technique, *J. Immunol.* **128**:1090–1096.

Peretz, H., Toister, Z., Laster, Y., and Loyter, A., 1974, Fusion of intact human erythrocytes and erythrocyte ghosts, *J. Cell Biol.* **63**:1–11.

Petrie, H. T., and McDonel, J. L., and Schlegel, R. A., 1982, Intracellular antibody to *Clostridium perfringens* enterotoxin fails to counteract toxin-induced damage, *Cell Biol. Int. Rep.* **6**:705–711.

Rechsteiner, M., 1975, Uptake of proteins by red blood cells, *Exp. Cell Res.* **93**:487–492.

Rechsteiner, M. C., 1982, Transfer of macromolecules using erythrocyte ghosts, in: *Techniques in Somatic Cell Genetics* (J. W. Shay, ed.), pp. 385–398, Plenum Press, New York.

Rechsteiner, M., and Kuehl, L., 1979, Microinjection of the nonhistone chromosomal protein HMG1 into bovine fibroblasts and HeLa cells, *Cell* **16**:901–908.

Rechsteiner, M. C., and Schlegel, R. A., 1986, Erythrocyte-mediated transfer: Applications, in: *Microinjection and Organelle Transplantation Techniques: Methods and Applications* (J. E. Celis, A. Graessman, and A. Loyter, eds.), pp. 89–116, Academic Press, New York.

Rechsteiner, M., Wu, L. H., and Miller, A. O. A., 1984, RBC-mediated microinjection of chromatin components into cultured mammalian cells, in: *Red Blood Cells as Carriers for Drugs* (J. R. DeLoach and U. Sprandel, eds.), pp. 142–149, Karger, Basel.

Sambrook, J., Rogers, L., White, J., and Gething, M. J., 1985, Lines of murine cells that constituitively express influenza virus hemagglutinin, *EMBO J.* **4**:91–103.

Schlegel, R. A., 1984, Red cell-mediated microinjection of antibodies, in: *Red Blood Cells as Carriers for Drugs* (J. R. DeLoach and U. Sprandel, eds.), pp. 134–141, Karger, Basel.

Schlegel, R. A., and McEvoy, L., 1986, Red cell-mediated microinjection of proteins and nucleic acids, *Methods Enzymol.* (in press).

Schlegel, R. A., and Mercer, W. E., 1980, Red cell-mediated microinjection of quiescent fibroblasts, in: *Introduction of Macromolecules into Viable Mammalian Cells* (R. Baserga, C. Croce, and G. Rovera, eds.), pp. 145–155, Alan Liss, New York.

Schlegel, R. A., and Rechsteiner M. C., 1975, Microinjection of thymidine kinase and bovine serum albumin into mammalian cells by fusion with red blood cells, *Cell* **5**:371–379.

Schlegel, R. A., and Rechsteiner, M. C., 1978, Red cell-mediated microinjection of macromolecules into mammalian cells, *Methods Cell Biol.* **20**:341–354.

Schlegel, R. A., and Rechsteiner, M. C., 1986, Erythrocyte-mediated transfer: Methods, in: *Microinjection and Organelle Transplantation Techniques: Methods and Applications* (J. E. Celis, A. Graessman, and A. Loyter, eds.), pp. 67–87, Academic Press, New York.

Schlegel, R. A., and Lumley-Sapanski, K., and Williamson, P., 1985*a*, Insertion of lipid domains into plasma membranes by fusion with erythrocytes, *Biochem. Biophys. Acta* **846**:234–241.

Schlegel, R. A., Miller, L. S., and Rose, K. M., 1985*b*, Reduction in RNA synthesis following red cell-mediated microinjection of antibodies to RNA polymerase I, *Cell Biol. Intl. Rep.* **9**:341–350.

Schneiderman, S., Farber, J. L., and Baserga, R., 1979, Techniques for decreasing the toxicity of polyethylene glycol, *Somat. Cell Genet.* **5**:263–269.

Seeman, P., 1967, Transient holes in the erythrocyte membrane during hypotonic hemolysis and stable holes in the membrane after lysis by saponin and lysolecithin, *J. Cell Biol.* **32:**55–70.

Seeman, P., 1973, Macromolecules may inhibit diffusion of hemoglobin from lysing erythrocytes by exclusion of solvent, *Can. J. Physiol. Pharmacol.* **51:**226–229.

Seeman, P., 1974, Ultrastructure of membrane lesions in immune lysis, osmotic lysis, and drug-induced lysis, *Fed. Proc.* **33**(10)**:**2116–2134.

Seeman, P., Cheng, D., and Iles, G. H., 1973, Structure of membrane holes in osmotic and saponin hemolysis, *J. Cell Biol.* **56:**519–527.

Smith, C. L., Ahkong, Q. F., Fisher, D., Lucy, J. A., 1982, Is purified polyethylene glycol able to induce cell fusion?, *Biochem. Biophys. Acta* **692:**109–114.

Straus, S. E., and Raskas, H. J., 1980, Transfection of KB cells by polyethylene glycol-induced fusion with erythrocyte ghosts containing adenovirus type 2 DNA, *J. Gen. Virol.* **48:**241–245.

Wasserman, M., Zakai, N., Loyter, A., and Kulka, R. G., 1976, A quantitative study of ultramicroinjection of macromolecules into animal cells, *Cell* **7:**551–556.

White, J., Helenius, A., and Kartenbeck, J., 1982, Membrane fusion activity of influenza, *EMBO J.* **1:**217–222.

Wiberg, F. C., Sunnerhagen, P., Kaltoft, K., Zeuthen, J., and Bjursell, G., 1983, Replication and expression in mammalian cells of transfected DNA: Description of an improved erythrocyte ghost fusion technique, *Nucl. Acids Res.* **11:**7287–7302.

Wojcieszyn, J. W., Schlegel, R. A., Lumley-Sapanski, K., and Jacobson, K. A., 1983, Studies of the mechanism of polyethylene glycol-mediated cell fusion using fluorescent membrane and cytoplasmic probes, *J. Cell Biol.* **96:**151–159.

Wu, L., Kuehl, L., and Rechsteiner, M., 1981, Comparative studies on microinjected high mobility group chromosomal proteins, HMG1 and HMG2, *J. Cell Biol.* **91:**488–496.

Yamaizumi, M., Furusawa, M., Uchida, T., Nishimura, T., and Okada, Y., 1978*a*, Characterization of the ghost fusion method: A method for introducing exogenous substances into cultured cells, *Cell Struct. Funct.* **3:**293–304.

Yamaizumi, M., Uchida, T., Okada, Y., and Furusawa, M., 1978*b*, Neutralization of Diptheria toxin in living cells by microinjection of antifragment A contained within resealed erythrocyte ghosts, *Cell* **13:**227–232.

Yamaizumi, M., Mekada, E., Uchida, T., and Okada, Y., 1978*c*, One molecule of Diptheria toxin fragment A introduced into a cell can kill the cell, *Cell* **15:**245–250.

Yamaizumi, M., Uchida, T., Okada, Y., Furusawa, M., and Mitsui, H., 1978*d*, Rapid transfer of non-histone chromosomal proteins to the nucleus of living cells, *Nature (Lond.)* **273:**783–784.

Yamaizumi, M., Uchida, T., Mekada, E., and Okada, Y., 1979, Antibodies introduced into living cells by red cell ghosts are functionally stable in the cytoplasm of the cells, *Cell* **18:**1009–1014.

Yee, J. P., and Mel, H. C., 1978, Cell-membrane and rheological mechanisms: Dynamic osmotic hemolysis of human erythrocytes and repair of ghosts, as studied by resistive pulse spectroscopy, *Biorheology* **15:**321–339.

Zade-Oppen, A. M. M., Ostling, S. G., and Marsden, N. V. B., 1979, On how macromolecules reduce hemoglobin loss in hypotonic hemolysis, *Upsula J. Med. Sci.* **84:**155–161.

Zakai, N., Loyter, A., and Kulka, R., 1974, Fusion of erythrocytes and other cells with retention of erythrocyte cytoplasm: Nuclear activation in chicken erythrocyte–melanoma heterokaryons, *J. Cell Biol.* **61:**241–248.

Chapter 24

Electrofusion and Plant Somatic Hybridization

George W. Bates, Lawrence J. Nea, and Clare A. Hasenkampf

1. INTRODUCTION

The transfer of genetic information between sexually incompatible species creates unique experimental opportunities for geneticists and cell biologists. This is especially true of plant somatic cell fusion because the totipotency of plant cells makes it possible to grow fusion products into mature plants (reviewed by Gleba and Sytnik, 1984; Vasil and Vasil, 1980). Although Carlson *et al.* (1972) reported recovering the first interspecific plant hybrid through protoplast fusion in 1972, subsequent progress in this field has been slower than originally hoped. This problem is due to several biological and technical barriers. This chapter describes our progress in the application of electrofusion to plant somatic hybridization (see Bates *et al.*, 1983; Bates, 1985; Bates and Hasenkampf, 1985) and suggests solutions to some of the problems in this field.

There are two principal difficulties in plant somatic hybridization;

GEORGE W. BATES, LAWRENCE J. NEA, and CLARE A. HASENKAMPF • Department of Biological Science and Institute of Molecular Biophysics, Florida State University, Tallahassee, Florida 32306. *Present address for* C. A. H.: Department of Biology, University of California at San Diego, La Jolla, California 92093.

they are interwoven with each other and with the method chosen for inducing fusion.

Protoplast culture, although routine in some species, is difficult or impossible in other species. Furthermore, successful protoplast culture and subsequent plant regeneration from these cultures often depends on the type of tissue from which the protoplasts were isolated. Those tissues that most readily yield protoplasts (leaf mesophyll and cell suspension cultures) are sometimes not a suitable source of protoplasts for culture or plant regeneration.

A second general difficulty in plant somatic cell hybridization is hybrid selection or screening. Often the availability of a selection method forces one to work with protoplasts from a tissue less than ideal in terms of cell culture or regeneration. Few auxotrophs or toxin-resistant mutants are available for the selection of plant somatic hybrids, and those mutants that do exist are restricted to specific varieties or cell lines in a limited number of species (Davey and Kumar, 1983). Consequently, plant biologists have generally relied on morphological markers to screen for somatic hybrids rather than biochemical markers to select for hybrids. Such screens often take the form of fusing a chlorophyll-containing mesophyll protoplast with an achlorophyllous tissue culture cell. Fusion products, identified visually by the presence of distinct green and nongreen sectors in the same cell, can then be isolated by means of micropipettes and cultured at low cell densities. However, the number of species amenable to low density culture is limited. Thus, the need for additional general selection methods is clear.

Although the fusion of plant protoplasts can be induced by a number of chemicals (see Saunders and Bates, this volume), polyethylene glycol (PEG), the most efficient treatment (Kao, 1977), results in the fusion of 1 to 10 percent of the protoplasts (Power *et al.,* 1977; Maliga *et al.,* 1977; Glimelius *et al.,* 1978). Fusion efficiency can be increased by an increase in the PEG concentration, but this also decreases cell viability and shifts the balance among the fusion products from binucleate to multinucleate cells. Furthermore, mesophyll cells of some species are so sensitive to cytolysis in PEG (Kao and Michayluk, 1974; Kartha *et al.,* 1974; Constabel *et al.,* 1975*b*; Kao, 1981) as to make PEG-induced fusion impractical. This problem is particularly unfortunate because mesophyll protoplasts are a component of the most frequently used hybrid selection scheme. PEG also induces general cell agglutination and causes the protoplasts to stick to the fusion vessel; this makes heterokaryon identification and manual isolation difficult.

With the introduction of electrofusion (Zimmermann and Scheurich, 1981), it was suggested that the high fusion efficiencies obtained might make selection unnecessary (Zimmermann, 1982). Further advantages of electrofusion are that it eliminates the use of cytotoxic, fusogenic chemicals; it is effective with an impressive range of species and cell types; and it facilitates the manual isolation of fusion products with micropipettes. It has since been pointed out that increasing the efficiency of fusion may not greatly increase the yield of hybrids (Gleba and Sytnik, 1984) because development depends on the successful completion of a series of additional transitions (nuclear fusion, sustained cell divisions, and regeneration). With PEG, fusion frequencies range from 10^{-2} to 10^{-1}; however, hybrid plants are recovered at much lower frequencies, 10^{-5}–10^{-4} (Gleba and Sytnik, 1984). Thus a 10-fold increase in fusion frequency would still leave hybrid recovery in the range of 10^{-3} or less. Clearly, such recoveries would not eliminate the need for a good postfusion selection system. It is also apparent that the electric fields used in electrofusion may in themselves be toxic; thus, one source of toxicity (PEG) may simply be exchanged for another (electricity).

With these issues in mind, our laboratory set out to evaluate electrofusion as a practical method for producing plant somatic hybrids. The first objective was to recover viable hybrids with this technique. During work toward this goal, it became clear that the equipment originally described for electrofusion (Zimmermann and Scheurich, 1981) was poorly designed for cell culture work. Consequently, we designed better fusion chambers. With these chambers, we have characterized the types of fusion products recovered under different experimental conditions in order to maximize the efficiency of fusion and to obtain a quantitative comparison between electrofusion and fusion with PEG.

Our results indicate that viable somatic hybrids can be recovered after electrofusion. Fusion is simple and is achieved in minutes. Batches of several hundred thousand protoplasts can be fused with efficiencies in the range of 20–40%. Because electrofusion does not lead to persistent cell agglutination, heterokaryons are easily isolated with micropipettes. Although mesophyll protoplasts are more sensitive to damage during electrofusion than protoplasts from tissue culture cells, they are nonetheless fused with high efficiency, and many remain viable after fusion. Using a biochemical selection system, which should be applicable to most dicotyledenous plants, we have recovered hybrids with efficiencies of 10^{-3} or greater, which is at least an order of magnitude better than obtained with PEG.

2. OPTIMIZATION OF ELECTROFUSION FOR HETEROKARYON FORMATION

PEG produces a range of binucleate and multinucleate fusion products (Constable *et al.*, 1975*b*; Kao, 1977). Only those fusion products containing two or at most three nuclei are viable (Gleba and Sytnik, 1984). We have conducted a similar analysis for electrofusion.

Mesophyll protoplasts from leaves of *Nicotiana tabacum* and *N. plumbaginifolia* suspension-cell protoplasts were resuspended in 0.4 M mannitol at densities of 5×10^5 cells/ml and mixed together in equal volumes. A droplet of the suspension was pipetted into the fusion chamber (DEP Systems Inc., Metamora, MI, Open Fusion Slide 2050S), and appropriate fusogenic electric fields were applied. Then, the cells were removed from the chamber and mixed with an equal volume of K3 culture medium (Nagy and Maliga, 1976). The electronics package used in all the work described here is the GCA (Chicago, IL) Zimmermann cell fusion power supply. We have also successfully used the homemade electrical system described by Bates *et al.* (1983) but find the GCA equipment easier to use and capable of greater electrical output.

Following fusion, mesophyll–suspension cell heterokaryons can be easily identified by the presence of green and nongreen sectors within the same cell (Fig. 1). Cytoplasmic mixing is slow. Thus, distinct cytoplasmic sectors can be seen in some heterokaryons for several hours or more. However, after 24 hr, the chloroplasts in all heterokaryons have moved to a position surrounding the nuclear area (see the heterokaryon in the upper right of Fig. 1). This morphology is distinctly different from that of the mesophyll protoplasts, where the chloroplasts are evenly distributed around the edge of the cell. Thus, visual inspection permits reliable identification of suspension cell–mesophyll fusion products during the first 24 hr in culture.

Although visual inspection of living fusion products permits determination of the number of heterokaryons, the number of nuclei per cell can only be determined by fixation and staining because, in living cells, the nuclei are obscured by the chloroplasts. Staining with modified carbol–fuchsin (Kao, 1975*a*) facilitates determination of the number of nuclei per cell and gives sufficient differential staining to allow identification of heterokaryons (Bates and Hasenkampf, 1985). Thus, we were able to describe in quantitative terms the effects of varying electric field strengths and other procedural details.

The data in Figure 2 were collected by fusion of batches of protoplasts with DC pulses of different voltage, while the AC field was held constant (150 V/cm, 600 kHz). Fusion occurs only when a DC voltage threshold (somewhere between 500 and 1000 V/cm) has been exceeded (Fig. 2a). Increasing the voltage beyond this threshold does not increase the percentage of cells fused but does shift the balance of cells fused from binucleate to multinucleate fusion products (Fig. 2b). The fusion observed at 0 V (no DC pulses given) is due to the spontaneous fusion of suspension-cell protoplasts that occurs during their isolation (Miller *et al.*, 1971) and not to the induction of fusion by the AC field alone.

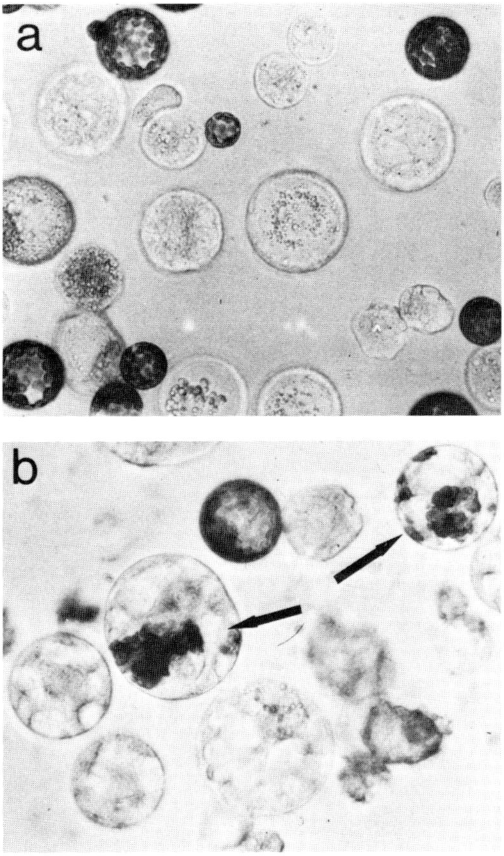

Figure 1. Mixture of mesophyll and suspension-cell protoplasts before and after electrofusion. The clear, granular cells are the suspension-cell protoplasts, whereas the mesophyll protoplasts are much darker because of the presence of chloroplasts. (a) Protoplasts before fusion. (b) Protoplasts in culture several hours after fusion. Two heterokaryons (arrows) are in this field as well as several suspension-cell protoplasts and a single mesophyll protoplast. (From Bates and Hasenkampf, 1985.)

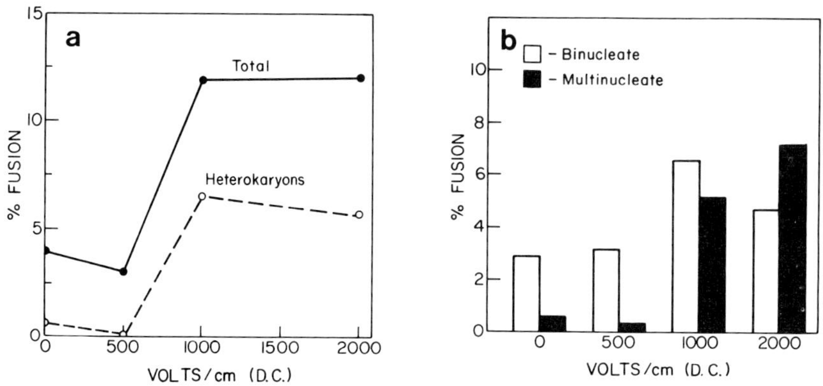

Figure 2. Types of fusions observed as a function of DC voltage. A mixture of *N. plumbaginifolia* and *N. tabacum* protoplasts was aligned by the AC field. After 90 sec, two DC pulses (each 50 μsec long) of the voltage indicated were applied. The protoplasts were removed from the chamber 60 sec later and placed in culture medium. One hour later, they were fixed and stained with modified carbol–fuchsin (Kao, 1975*a*), and the extent and types of fusions were determined. (a) Overall fusion and heterokaryon production. (b) Proportions of bi- and multinucleate cells. (From Bates, 1985.)

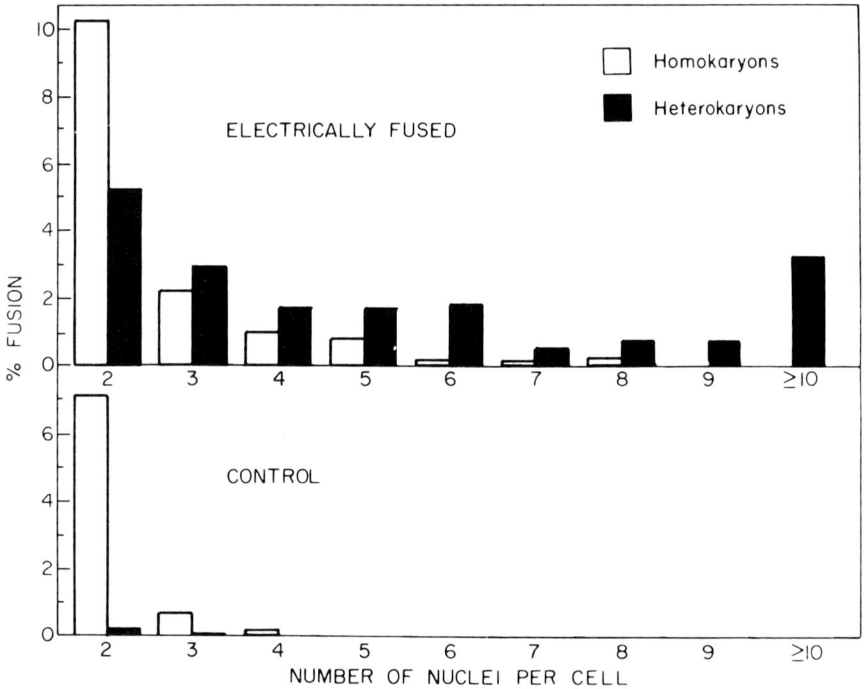

Figure 3. Distribution of fusion products recovered after a typical experiment. *N. tabacum* mesophyll protoplasts were fused with *N. plumbaginifolia* suspension-cell protoplasts. The types of fusion products were determined by means of carbol-fuchsin staining as described in Figure 2. (From Bates, 1985.)

In a parallel experiment, the AC voltage was varied while the DC voltage was held constant. In this case, there was no threshold; any AC field strong enough to induce some cell–cell contacts led to fusion. Up to 150 V/cm, increasing the AC field strength increased the extent of fusion. AC voltages greater than 150 V/cm resulted in a decrease in fusion due to increased cell lysis.

Figure 3 shows the types of fusions produced by optimal electric fields (150 V/cm, AC; 1000 V/cm, 50 μsec, DC). Most of the fusion products (60%) are bi- or trinucleate; however, a significant number contain four or more nuclei per cell. Some giant cells (nine to ten nuclei per cell) are also produced. Aside from these giant cells, the distribution of fusion products resulting from electrofusion is roughly similar to that seen with PEG-induced fusion (Constabel *et al.,* 1975*a*; Kao, 1977).

Because high cell densities have been reported to favor the formation of giant cells (Scheurich and Zimmermann, 1981), we believed that reducing the cell density might increase the recovery of binucleate fusion products. Reducing the cell density by a factor of 10 did greatly reduce the production of giant cells, but it also reduced the fusion yield (total number of cells fused) 20-fold. Apparently, at lower cell densities, many of the protoplasts fail to make any cell–cell contacts, and, therefore, fail to fuse. Thus, yields of the bi- and trinucleate heterokaryons are optimal when fusion is carried out at high cell densities.

3. CHAMBER DESIGN

Protoplast culture is density dependent. Ordinarily, 5×10^4–10^5 protoplasts/ml is optimal; culture usually fails completely if the cell density falls below 10^4 protoplasts/ml. The DEP fusion slide pictured in Figure 4a holds about 10 µl of cell suspension between its electrodes. Even when fusion is carried out at 5×10^5 cells/ml, the total number of cells contained by this chamber is too low for routine cell culture. Therefore, we sought to design a fusion chamber that would permit the high efficiency fusion of larger numbers of protoplasts.

We have tried three different types of chambers; examples are shown in Figure 4. The chamber in Figure 4d is our most successful version. This

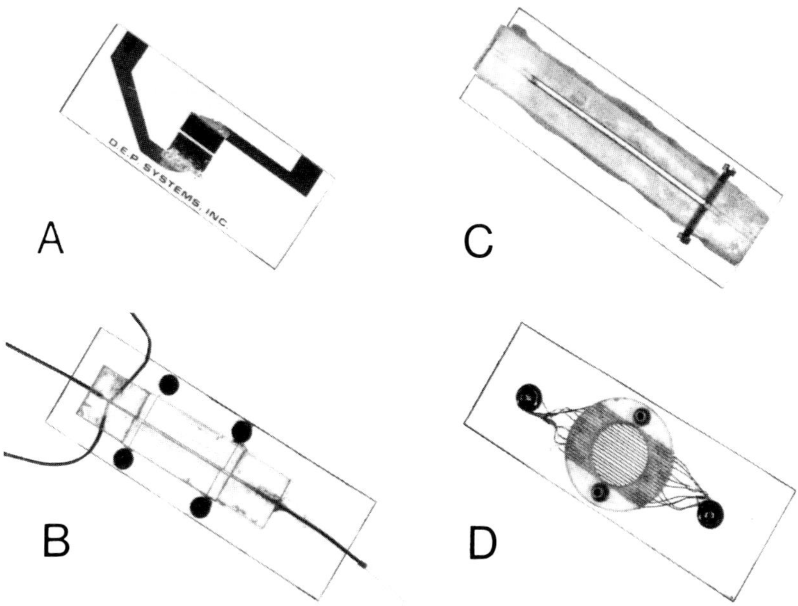

Figure 4. Photographs of four different chambers tested for their efficiency in fusing plant protoplasts. All the chambers are shown at the same magnification. (a) DEP Systems, Inc., Open Fusion Slide 2050S. (b) Flow-through chamber with three inlet ports and one outlet. (c) Rectangular slot chamber. (d) "Mass" fusion chamber, 2.8 cm in diameter.

"mass" chamber will accommodate up to 0.5 ml of cell suspension (about 2.5×10^5 protoplasts) and consists of wires (80 : 20 Au–Pd) laid parallel to each other, 0.5 mm apart, across a polycarbonate slide. Each wire is connected to the posts at the ends of the chamber such that its electrical polarity is opposite to that of its neighbors. A polycarbonate ring glued down over the wires forms a well that holds the protoplasts.

Fusion is accomplished as follows. The two cell types are mixed together (5×10^5 protoplasts/ml) and pipetted into the chamber. The AC field is immediately applied and, within 2–3 min, the protoplasts have settled onto the wires and have aligned into pearl chains. Then, one or two DC pulses are applied and, in 60 sec, the protoplasts can be removed from the chamber with a pipette and mixed with culture medium. The entire procedure is completed within 3–4 min.

Table I compares the results obtained with the mass chamber, with the DEP chamber (Fig. 4a) and with PEG-induced fusion. The mass chamber gives greater fusion for all classes of fusion products. In addition, the total number of cells fused is not only much larger than is possible with the DEP chamber, but also large enough to make protoplast culture a simple, straightforward matter. Although overall fusion approaches 40% with this chamber, binucleate heterokaryon production is only in the range of 5–6%.

The chamber in Figure 4b is a flowthrough chamber designed to allow batches of protoplasts to be pumped into the chamber, fused, and then pumped out into a collecting vessel. The model shown in Figure 4b has multiple inlet ports so each type of proptoplast can be introduced into

Table I. Percentages and Types of Fusions Observed by Different Techniques[a,b]

	Fusions (%)					
	Total		Heterokaryons		Binucleate heterokaryons	
Technique[c]	%	SE	%	SE	%	SE
Mass chamber (n = 4)	36.8 ± 2.1		19.5 ± 1.0		5.4 ± 0.8	
DEP chamber[d] (n = 6)	15.4 ± 2.3		6.2 ± 0.1		2.9 ± 0.9	
PEG treated (n = 2)	17.0 ± 1.0		10.0 ± 0.5		3.1 ± 1.0	
Control[e]	6.8 ± 1.3		0.2 ± 0.1		0.2 ± 0.1	

[a]From Bates (1985).
[b]Values are given as the percentage of cells that contain more than one nucleus.
[c]n is the number of trials.
[d]The PEG fusion procedure is that of Kao (1975*b*).
[e]*N. tabacum* mesophyll and *N. plumbaginifolia* suspension-cell protoplasts were mixed together and then fixed and stained with modified carbol-fuchsin without first being subjected to electric fields or PEG treatment.

the chamber separately. It was possible to induce fusion in this chamber; however, recoveries of fusion products were consistently low—in the range of a few percent or less. It was very difficult to control the flow of cells into and through the chamber; many cells were pushed directly through the chamber without ever being aligned by the AC field. Flushing the chamber also takes a large volume of medium, thereby diluting the protoplasts and necessitating their reisolation. Additionally, the tendency of the protoplasts to clog and settle in the syringes and tubing used with flowthrough chambers makes this type of chamber cumbersome.

Vienken and Zimmerman (1982) reported that flowthrough chambers could be used to "target" fusion. In principal, one type of protoplast could be pumped into the chamber to form a monolayer of cells on the electrodes. Then, the second cell type would be pumped in. Fusion would then lead to the preferential formation of binucleate heterokaryons. We have been unable to obtain "targeted" fusion in the flowthrough chamber in Figure 4b or in any of our other chambers. Although it may be possible to reproduce the observations of Vienken and Zimmerman (1982), it seems unlikely that this will become a routine, practical procedure.

The interior of the chamber in Figure 4c is a rectangular slot (vol 0.25 ml) with long walls lined by two stainless steel plates 1 mm apart. Protoplasts are easily pipetted into and out of this chamber. The electric fields produced by flat-plate electrodes are uniform rather than nonuniform (in contrast to all the other chambers shown in Fig. 4). Dielectrophoresis is theoretically impossible in uniform fields. However, some chain formation still occurs because each cell distorts the electric field, creating local field inhomogeneities (so-called mutual dielectrophoresis, see Chapter 17, this volume, and Pohl, 1978). One advantage of chambers having flat-plate electrodes is that the chains of cells do not form on the electrode surface; instead, they form out in the medium (Watts and King, 1984). This result eliminates the adherence of fusion products to the electrode surface as is sometimes observed with wire electrodes (Bates *et al.*, 1983). A second potential advantage is that the uniform electric fields should lead to relatively uniform fusion throughout the chamber. In chambers with wire electrodes, fusion is generally limited to areas near the electrode surface where the electric field strengths are greatest.

Despite these considerations, the chamber shown in Figure 4c was unsuccessful. Good cell alignment was only observed when the AC field was made quite strong. These strong fields led to heat generation and severe convective currents in the chamber. Under such conditions, fusion is not possible.

Watts and King (1984) and Tempelaar and Jones (1985*a,b*) report electrofusion using chambers similar to that in Figure 4c. It is not yet clear why their flat-plate electrode chambers are more successful than ours.

However, one possibility is that our chamber may have been too large. Making the electrodes smaller or placing them closer together might improve performance; it would reduce the number of protoplasts that could be fused.

The results of Watts and King (1984) and Tempelaar and Jones (1985*b*) differ quantitatively and qualitatively even though similar equipment and procedures were used for fusion. Watts and King report that 90% of their protoplasts were fused, and one-half of these were two-cell fusions. In a more detailed study, Tempelaar and Jones (1985*b*) report fusion efficiencies (considering only protoplasts aligned in chains) of 50% or less. One-half of these (20–30%) were two-cell fusions. However, 50% of the protoplasts in the chamber never aligned into chains. Thus, only 10–15% of the cells introduced into the chamber could be recovered as binucleate fusion products, and only a fraction of these would be heterokaryons. Given these considerations, the results Tempelaar and Jones (1985*b*) obtained with their chamber having flat-plate electrodes are similar to ours obtained with the mass fusion chamber.

4. CULTURE AND SELECTION OF FUSION PRODUCTS

Cell culture and hybrid selection are widely believed to be the limiting steps in the recovery of plant somatic hybrids. These problems are often interrelated. The availability of a specific selection method can dictate the cell culture approach and also the types of cells that may be fused. Conversely, the researcher's interest in producing hybrids between a specific pair of species often dictates the choice of cell types and the culture system. These, in turn, can place severe constraints on the selection system. Because of these considerations and the fact that few reliable biochemical selection systems are available to plant biologists, considerable effort has been devoted to developing widely applicable procedures for identifying and screening plant heterokaryons and somatic hybrids.

We have taken two different approaches to recovering somatic hybrids following electrofusion. Both approaches offer some potential for being readily adapted for use with a wide range of plant species.

The first approach is to fuse mesophyll protoplasts with suspension cell protoplasts. Heterokaryons are identified by visual inspection, manually isolated with micropipettes, and cultured in small groups at low density on a conditioned medium. Low-density culture such as this is possible only with a few plant species. However, the recent introduction of feeder-layer culture techniques (Horsch and Jones, 1980; Aviv and Galun, 1984) suggests that visual identification and manual isolation of heterokaryons can now be applied to a much broader range of species.

The second approach uses a biochemical selection based on resistance to the antibiotic kanamycin. The bacterial gene conferring kanamycin resistance has been introduced into higher plants by use of *Agrobacterium tumefaciens* as the vector (Fraley *et al.,* 1983). Resistant plants have now been recovered for several species (Horsch *et al.,* 1984, 1985; Potrykus *et al.,* 1985*a*) and could ultimately include all plant species susceptible to infection by *Agrobacterium* (dicots). Recent advances in the use of free-DNA transformation procedures will probably make kanamycin-resistant plants (or cell lines) available for monocot species as well (Potrykus *et al.,* 1985*b;* Fromm *et al.,* 1985). Fusing kanamycin-resistant protoplasts with sensitive protoplasts provides only one-half of a selection scheme. Some method must also be used to select against growth of any unfused kanamycin-resistant protoplast. This selection can be achieved by mitotic inactivation of the resistant protoplasts before fusion with either iodoacetate (Sidorov *et al.,* 1981) or X-ray or γ-irradiation (Menczel *et al.,* 1982; Fluhr *et al.,* 1984). After this treatment, the only cells capable of growth on kanamycin would be somatic hybrids.

4.1. Mesophyll–Suspension-Cell Fusions and Manual Isolation

Mesophyll protoplasts of *N. tabacum* were fused with suspension-cell protoplasts of *N. plumbaginifolia.* Following fusion, live heterokaryons are readily identified by visual inspection (see Fig. 1b). However, when the unfused mesophyll protoplasts enter their first division (two to three days of culture), their chloroplasts cluster around the nucleus and so are difficult to distinguish from heterokaryons.

So that the development of the heterokaryons could be followed further, they were manually isolated, with micropipettes, 24–48 hr after fusion. These heterokaryons were cultured as groups of 20–40 cells in 50–100 ml of conditioned medium (Bates and Hasenkampf, 1985). We found that manual isolation could be routinely performed with hand-held micropipettes; the use of micromanipulators appears unnecessary. Because of the natural tendency of protoplasts to clump during culture, manual isolation does not prevent the carryover of some unfused parental cells. However, when isolated with care, heterokaryons usually comprised 75% or more of the cells. Conditioned medium was prepared by growth of *N. tabacum* mesophyll protoplasts in Nagy and Maliga's (1976) K3 medium (Bates and Hasenkampf, 1985). After 5–6 days of culture, the cells were removed by centrifugation, and the supernatant was saved as a source of conditioned medium.

When placed in this conditioned medium, 30% of the heterokaryons were capable of sustained divisions. This figure compares favorably with the 50% plating efficiency observed in unfused controls, especially when one considers that many of the heterokaryons were undoubtedly multinucleate (see Fig. 3).

As the heterokaryons grew, they were diluted with fresh medium and separated to ensure growth as individual calli. We recovered shoots from several of these calli; some shoots were capable of limited root formation. However, none of these plants were vigorous. Cytogenetic analysis revealed that all these fusion products are aneuploid, as is the original *N. plumbaginifolia* suspension culture line. This cell culture line is incapable of regeneration.

Esterase isozyme profiles confirmed the recovery of somatic hybrids. As is evident from the polyacrylamide gel shown in Figure 5, the parental plant species can be distinguished from each other whether esterases are examined in leaf or in undifferentiated callus tissues. The somatic hybrids show a banding pattern that is largely intermediate between the two

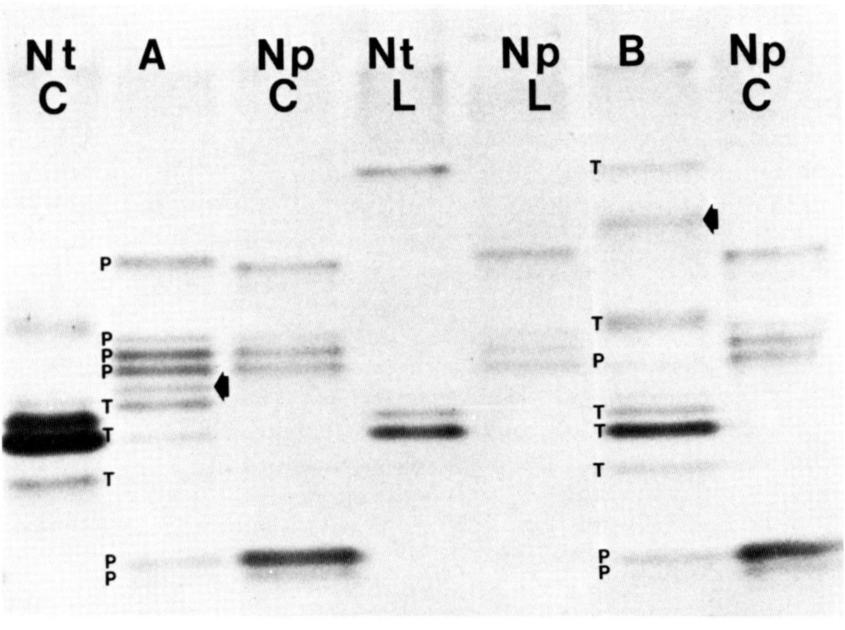

Figure 5. Esterase isozyme profiles for callus and leaves of *N. plumbaginifolia, N. tabacum,* and hybrids. Soluble protein extracts were separated on a nondenaturing 10% polyacrylamide slab gel as previously described (Bates and Hasenkampf, 1985). Esterase isozymes were visualized according to the procedure of Brewbaker *et al.* (1968). C, callus; L, leaf; Nt, *N. tabacum;* Np, *N. plumbaginifolia;* A and B, hybrids. Band assignments for A and B are indicated by the small letters to the left of these lanes (P for *N. plumbaginifolia* and T for *N. tabacum*). The arrows indicate unique bands. (From Bates and Hasenkampf, 1985.)

parental profiles. However, some unique bands appear that are not seen in either parent (arrows), providing evidence for the recovery of somatic hybrids rather than stable chimeras.

4.2. Mesophyll-Mesophyll Fusions, Kanamycin Selection

In this series of experiments, *N. tabacum* mesophyll protoplasts were fused with mesophyll protoplasts of a kanamycin-resistant seed line of *N. plumbaginifolia.* We had two initial objectives: (1) to demonstrate the utility of kanamycin resistance for recovering somatic hybrids, and (2) to see whether hybrid regeneration is improved when both parental cell types are diploid prior to fusion.

Kanamycin-resistant *N. plumbaginifolia* plants were obtained from Dr. R. Horsch (Monsanto Company, St. Louis, MO). These plants had been transformed to kanamycin resistance by means of the split-end vector pMON 200 (Horsch *et al.,* 1984, 1985), which had been engineered into *Agrobacterium tumefaciens.* The neomycin phosphotransferase gene that was transferred to *N. plumbaginifolia* by this procedure is stably inherited as a single dominant nuclear gene and confers resistance to kanamycin levels exceeding 400 mg/ml. The growth of wild type *N. plumbaginifolia* or *N. tabacum* cells is completely inhibited by 50 μg/ml kanamycin.

Protoplasts were isolated and fused as described earlier but with the following modifications: *N. plumbaginifolia* and *N. tabacum* cells were mixed together in a 1 : 3 ratio. The protoplasts were aligned by a 2–3-min application of a 125-V/cm, 600-kHz AC field. Then the AC field was reduced to 50 V/cm, and two DC pulses (875 V/cm, 25-μsec duration) were used to induce fusion. One minute later, the AC field was turned off, and the cells were removed from the chamber and placed in culture. Reduction of the electric fields used for alignment and fusion was necessary to prevent the lysis of the *N. plumbaginifolia* mesophyll protoplasts.

Following fusion, the protoplasts were cultured for 14 days without kanamycin. Then, 100 μg/ml kanamycin was added to the culture. One week later, the developing colonies were solidified into an agar-containing medium, and selection was continued by the agarose bead technique (Shillito *et al.,* 1983). After 4 weeks, kanamycin-resistant colonies are easily seen growing among a background of dying, sensitive cells. Two approaches were taken for mitotic inactivation of the *N. plumbaginifolia* protoplasts prior to fusion: iodoacetate treatment and ^{37}Cs γ-irradiation. Iodoacetate proved difficult to use. Apparently, the protoplasts exhibit day-to-day variation in their sensitivity to the treatment. Irradiation, however, proved highly reliable. A 4-krad dose completely prevents colony formation.

In these experiments, kanamycin-resistant colonies were recovered at a frequency of about 10^{-3}. This is 10–100-fold greater than ordinarily obtained in other systems following fusion with PEG. We have never recovered a single resistant colony from control cultures (*N. tabacum* and irradiated *N. plumbaginifolia* protoplasts mixed together but not fused), so the selection appears extremely efficient and reliable. Therefore, all resistant colonies recovered after fusion are assumed to be somatic hybrids. However, we have not yet completed the biochemical analysis necessary to verify this assumption.

Kanamycin-resistant calli recovered from these fusions exhibit good regeneration, vigorous growth, and good root formation. None of the plants recovered from these fusions has yet been cytogenetically analyzed. However, their vigorous growth and normal morphology argues that they are diploid or nearly so.

5. COMPARISON OF THE ELECTROFUSION OF DIFFERENT CELL TYPES

Tempelaar and Jones (1985*b*) suggested that results obtained with electrofusion may vary depending on species and cell type. Some of our data appear consistent with this suggestion. For example, fusions between tobacco suspension cell and mesophyll protoplasts occurred with an overall efficiency of 37.5%, while tobacco mesophyll–mesophyll fusions had a maximum overall fusion of 24%. In other experiments, we found that protoplasts of a carrot suspension culture have a maximum fusion efficiency of 30%.

However, we consider it premature to conclude that there are real species and/or cell type differences in the efficiency of electrofusion. The data shown in Table II indicate that differences in protoplast preparation can have dramatic effects on electrofusion.

A battery of enzyme preparations are available for protoplast production. These preparations differ in their polysaccharide hydrolyzing capacities as well as in the presence of contaminating proteases and lipases (see Chapter 17, this volume). Ordinarily, one seeks an enzyme that releases large numbers of uniformly round, viable protoplasts. Protoplast formation does not, however, mean *a priori* that the cell surface has been thoroughly stripped. Cell wall fragments or integral membrane glycoproteins may remain. These cell surface factors could alter electrofusion efficiency and clearly seem to be doing so in the experiment shown in Table II. Protoplasts were released from a carrot suspension cultured by a 5-hr digestion with 2% Driselase and then were electrically fused. Addition of 0.5% Cellulysin, or a 5-min treatment with 1 mg/ml Pronase at the end of the digestion, enhanced fusion substantially. In order to make this

Table II. Enhancement of the Electrofusion of Carrot Protoplasts by Treatments with Pronase or Cellulysin[a]

Enzyme treatment[e]	Fusion (%) % ± SE
Driselase alone[b] (n = 19)	7.3 ± 1.5
+ Pronase[c] (n = 10)	13.8 ± 0.9
+ Cellulysin[d] (n = 9)	28.7 ± 3.0

[a]Protoplasts were prepared from a nonembryogenic suspension culture of *Daucus carota* by a 5-hr digestion with 2% Driselase. The protoplasts were then washed with 0.4 M mannitol. When aligned using a 125-V/cm AC field and stimulated by application of two 1000-V/cm DC pulses (50 μs long), 30.3 ± 1.0% of the protoplasts underwent fusion.
[b]Although Driselase alone produces very clean protoplasts without any suggestion of residual cell wall, we noticed that addition of other digestive enzymes to the Driselase significantly enhanced fusion. The data shown here were taken from protoplasts that were fused using 600-V/cm rather than 1000-V/cm DC pulses. This procedure serves to reduce the level of fusion in the control (protoplasts isolated with Driselase alone) and more clearly reveals the effects of the other digestive enzymes.
[c]Pronase (1 mg/ml) was added to the protoplasts for 5 min at the end of the Driselase treatment.
[d]Cellulysin was given as 0.5% in combination with 1.5% Driselase and was present during the entire protoplast digestion period (5 hr).
[e]n is the number of trials.

effect clear, the protoplasts were fused with suboptimal electric fields. Under these conditions, the control (Driselase alone) value drops to 7% fusion. Addition of Pronase or Cellulysin doubles or quadruples the fusion efficiency, respectively. The effect obtained with Cellulysin can be eliminated by boiling this enzyme preparation but not by dialysis. This result suggests that some enzyme activity in Cellulysin is responsible for the enhancement of fusion. Because Cellulysin contains proteases (Flores *et al.,* 1981) and Pronase partially mimics the effect of Cellulysin, it is tempting to speculate that removal of membrane proteins facilitates electrofusion. We have as yet no additional evidence to support such a conclusion; still, it is clear that the method of protoplast preparation can strongly affect electrofusion efficiency.

NOTE ADDED IN PROOF

Since the writing of this work a number of other groups have documented formation of viable plant somatic hybrids by electrofusion

(Koop, 1985; Puite *et al.*, 1985; Chapel *et al.*, 1986; Morikawa *et al.*, 1986). In the context of fusion methodology, the recent work of Negrutiu *et al.* (1986) is of particular interest. In a side by side comparison they found that electrofusion gave greater recoveries of hybrid colonies than did fusion using several PEG fusion procedures. They attributed this effect to a reduction of cell lysis when using electrofusion as compared with PEG-induced fusion.

ACKNOWLEDGMENTS. This work was supported by Competitive Research Grant 83-CRCR-1-1257 from the U.S. Department of Agriculture. The authors would also like to thank Drs. Fraley and Horsch of Monsanto Corp., St. Louis, Missouri, for their generous gifts of kanamycin-resistant *Nicotiana plumbaginifolia* seeds.

REFERENCES

Aviv, D., and Galun, E., 1984, The feeder layer technique, in: *Cell Culture and Somatic Cell Genetics of Plants,* Vol. 1 (I. K. Vasil, ed.), pp. 199–203, Academic Press, New York.

Bates, G. W., 1985, Electrical fusion for optimal formation of protoplast heterokaryons in *Nicotiana*, *Planta* **165:**217–224.

Bates, G. W., and Hasenkampf, C. A., 1985, Culture of plant somatic hybrids following electrical fusion, *Theor. Appl. Genet.* **70:**277–233.

Bates, G. W., Gaynor, J. J., and Shekhawat, N. S., 1983, Fusion of plant protoplasts by electric fusion, *Plant Physiol.* **72:**1110–1113.

Brewbaker, J. L., Upadhya, M. D., Makinen, Y., and MacDonald, T., 1968, Isoenzyme polymorphism in flowering plants. 3. Gel electrophoretic methods and applications, *Physiol. Plant.* **121:**930–940.

Carlson, P. S., Smith, H. H., and Dearing, R. D., 1972, Parasexual interspecific plant hybridization, *Proc. Natl. Acad. Sci. U.S.A.* **69:**2292–2294.

Chapel, M., Montane, M. H., Ranty, B., Teissie, J., and Alibert, G., 1986, Viable somatic hybrids obtained by direct current electrofusion of chemically aggregated plant protoplasts, *FEBS Lett.* **196:**79–86.

Constabel, F., Dudits, D., Gamborg, O. L., and Kao, K. N., 1975*a*, Nuclear fusion in intergeneric heterokaryons, a note, *Can. J. Bot.* **53:**2092–2095.

Constabel, F., Kirkpatrick, J. W., Kao, K. N., and Kartha, K. K., 1975*b*, The effect of canavanine on the growth of cells from suspension cultures and on intergeneric heterokaryocytes of canavanine sensitive and tolerant plants, *Biochem. Physiol. Pflanz.* **168:**319–325.

Davey, M. R., and Kumar, A., 1983, Higher plant protoplasts—retrospect and prospect, in: *Plant Protoplasts,* International Review of Cytology, Supplement 16 (K. L. Giles, ed.), pp. 219–299, Academic Press, New York.

Evans, D. A., and Bravo, J. E., 1983, Plant protoplast isolation and culture, in: *Plant Protoplasts,* International Review of Cytology, Supplement 16 (K. L. Giles, ed.), pp. 33–53, Academic Press, New York.

Flores, H. E., Kaur-Sawhney, R., and Galston, A. W., 1981, Protoplasts as vehicles for plant propagation and improvement, *Adv. Cell Culture* **1:**241–279.

Fluhr, R., Aviv, D., Galun, E., and Edelman, M., 1984, Generation of heteroplastidic *Nicotiana* cybrids by protoplast fusion: Analysis for plastid recombinant types, *Theor. Appl. Genet.* **67**:491–497.

Fraley, R. T., Rogers, S. G., Horsch, R. B., Sanders, P. R., Flick, J. S., Adams, S. P., Bittner, M. L., Brand, L. A., Fink, C. L., Fry, J. S., Galluppi, G. R., Goldberg, S. B., Hoffmann, N. L., and Woo, S. C., 1983, Expression of bacterial genes in plant cells, *Proc. Natl. Acad. Sci. U.S.A.* **80**:4803–4807.

Fromm, M., Taylor, L. P., and Walbot, V., 1985, Expression of genes electroporated into monocot and dicot plant cells, *Proc. Natl. Acad. Sci. U.S.A.* **82**:5824–5828.

Gleba, Y. Y., and Sytnik, K. M., 1984, *Protoplast Fusion*, Springer-Verlag, New York.

Glimelius, K., Eriksson, T., Grafe, R., and Muller, A. J., 1978, Somatic hybridization of nitrate reductase deficient mutants of *Nicotiana tabacum* by protoplast fusion, *Physiol. Plant.* **44**:273–277.

Horsch, R. B., and Jones, G. E., 1980, A double filter paper technique for plating cultured plant cells, *In Vitro* **16**:103–108.

Horsch, R. B., Fraley, R. T., Rogers, S. G., Sanders, P. R., Lloyd, A., and Hoffmann, N., 1984, Inheritance of functional foreign genes in plants, *Science* **223**:496–498.

Horsch, R. B., Fry, J. E., Hoffmann, N. L., Eichholtz, D., Rogers, S. G., and Fraley, R. T., 1985, A simple and general method for transferring genes into plants, *Science* **227**:1129–1231.

Kao, K. N., 1975*a*, A nuclear staining method for plant protoplasts, in: *Plant Tissue Culture Methods* (O. L. Gamborg and L. R. Wetter, eds.), pp. 60–62, Prairie Research Laboratory, Saskatoon.

Kao, K. N., 1975*b*, A method for fusion of plant protoplasts with polyethylene glycol, in: *Plant Tissue Culture Methods* (O. L. Gamborg and L. R. Wetter, eds.), pp. 22–27, Prairie Research Laboratory, Saskatoon.

Kao, K. N., 1977, Chromosomal behavior in somatic hybrids of soybean—*Nicotiana glauca, Mol. Gen. Genet.* **150**:225–230.

Kao, K. N., 1981, Plant protoplast fusion and somatic hybrids, in: *Plant Tissue Culture* (H. Han, ed.), pp. 331–339, Proceedings of the Beijing Symposium, Pitman, London.

Kao, K. N., and Michayluk, M. R., 1974, A method for high frequency intergeneric fusion of plant protoplasts, *Planta* **115**:355–367.

Kartha, K. K., Gamborg, O. L., Constabel, F., and Kao, K. N., 1974, Fusion of rapeseed and soybean protoplasts and subsequent division of heterokaryocytes, *Can. J. Bot.* **52**:2435–2436.

Kohn, H., Schieder, R., and Schieder, O., 1985, Somatic hybrids in tobacco mediated by electrofusion, *Plant Sci.* **38**:121–128.

Koop, H.-U., 1985, Regeneration of plants after electrofusion of selected pairs of protoplasts, *Eur. J. Cell Biol.* **39**:46–49.

Maliga, P., Lazar, G., Joo, F., Nagy, A. H., and Menczel, L., 1977, Restoration of morphogenetic potential in *Nicotiana* by somatic hybridization, *Mol. Gen. Genet.* **157**:291–296.

Menczel, L., Galiba, G., Nagy, F., and Maliga, P., 1982, Effect of radiation dosage on efficiency of chloroplast transfer by protoplast fusion in *Nicotiana, Genetics* **100**:487–495.

Miller, R. A., Gamborg, O. L., Keller, W. A., and Kao, K. N., 1971, Fusion and division of nuclei in multinculeated soybean protoplasts, *Can. J. Genet. Cytol.* **13**:347–353.

Morikawa, H., Sugino, K., Hayashi, Y., Takeda, J., Senda, M., Hirai, A., and Yamada, Y., 1986, Interspecific plant hybridization by electrofusion in *Nicotiana, Biotechnology* **4**:57–60.

Nagy, J. I., and Maliga, P., 1976, Callus induction and plant regeneration from protoplasts of *Nicotiana sylvestris, Z. Pflanzenphysiol.* **78**:453–455.

Negrutiu, I., DeBrouwer, D., Watts, J. W., Sidorov, V. I., Dirks, R., and Jacobs, M., 1986, Fusion of plant protoplasts: a study using auxotrophic mutants of *Nicotiana plumbaginifolia* Viviani, *Theor. Appl. Genet.* **72**:279–286.

Pohl, H. A., 1978, *Dielectrophoresis,* Cambridge University Press, Cambridge.

Potrykus, I., Paszkowsi, J., Saul, M. W., Petruska, J., and Shillito, R. D., 1985*a*, Molecular and general genetics of a hybrid foreign gene introduced into tobacco by direct gene transfer, *Mol. Gen. Genet.* **199**:169–177.

Potrykus, I., Saul, M. W., Petruska, J., Paszkowski, J., and Shillito, R. D., 1985*b*, Direct gene transfer to cells of a graminaceous monocot, *Mol. Gen. Genet.* **199**:183–188.

Power, J. B., Berry, S. F., Frearson, E. M., and Cocking, E. C., 1977, Selection procedures for the production of inter-species somatic hybrids of *Petunia hybrida* and *Petunia parodii, Plant Sci. Lett.* **7**:51–55.

Puite, K. J., van Wikselaar, P., and Verhoeven, H., 1985, Electrofusion, a simple and reproducible technique in somatic hybridization of *Nicotiana plumbaginifolia* mutants, *Plant Cell Rep.* **4**:274–276.

Scheurich, P., and Zimmermann, U., 1981, Giant human erythrocytes by electric-field-induced cell-to-cell fusion, *Naturwissenschaften* **68**:45–47.

Shillito, R. D., Paszkowski, J., and Potrykus, I., 1983, Agarose plating and a bead type culture technique enable and stimulate development of protoplast-derived colonies in a number of plant species, *Plant Cell Rep.* **2**:244–247.

Sidorov, V. A., Menczel, L., Nagy, F., and Maliga, P., 1981, Chloroplast transfer in *Nicotiana* based on metabolic complementation between irradiated and iodoacetate treated protoplasts, *Planta* **152**:341–345.

Tempelaar, M. J., and Jones, M. G. K., 1985*a*, Directed electrofusion between protoplasts with different responses in a mass fusion system, *Plant Cell Rep.* **4**:92–95.

Tempelaar, M. J., and Jones, M. G. K., 1985*b*, Fusion characteristics of plant protoplasts in electric fields, *Planta* **165**:205–216.

Vasil, I. K., and Vasil, V., 1980, Isolation and culture of protoplasts, *Int. Rev. Cytol. (Suppl.)* **11A**:1–20.

Vienken, J., and Zimmermann, U., 1982, Electric field-induced fusion: Electro-hydraulic procedure for production of heterokaryon cells in high yield, *FEBS Lett.* **137**:11–13.

Watts, J. W., and King, J. M., 1984, A simple method for large-scale electrofusion and culture of plant protoplasts, *Biosci. Rep.* **4**:335–342.

Zimmermann, U., 1982, Electric field-mediated fusion and related electrical phenomena, *Biochim. Biophys. Acta* **694**:227–277.

Zimmermann, U., and Scheurich, P., 1981, High frequency fusion of plant protoplasts by electric fields, *Planta* **151**:26–32.

Zimmermann, U., Buchner, K. H., and Arnold, W. M., 1984, Electrofusion of cells: recent developments and relevance for evolution, in: *Charge and Field Effects in Biosystems* (M. J. Allen and P. N. R. Usherwood, eds.), pp. 293–318, Abacus Press, Normal, Ill.

Chapter 25

Chemically Induced Fusion of Plant Protoplasts

James A. Saunders and George W. Bates

1. INTRODUCTION AND OVERVIEW

In recent years, the fusion of plant protoplasts has become an important mechanism of transferring genetic information between sexually incompatible species. The potential totipotency of plant culture systems together with the development of protoplast isolation and fusion procedures has increased the use of somatic hybridization as a viable tool of molecular biologists.

This chapter reviews the procedures for the isolation, fusion, and culture of plant protoplasts. The fusogens discussed are limited to those derived from chemical sources, although successful somatic fusions have been induced by other physical phenomena including electrofusion (see Chapters 17 and 18, this volume) and centrifugation. The formation of cell lines and hybrid plants as a result of these somatic cell fusions have been evaluated in numerous reports (Constabel, 1984; Galun and Aviv, 1983; Gamborg *et al.*, 1981; Gleba and Sytnik, 1984). The criteria for the successful generation of a somatic hybrid is generally regarded to be the introduc-

James A. Saunders • Germplasm Quality and Enhancement Laboratory, USDA Plant Genetics and Germplasm Institute, Beltsville, Maryland 20705. George W. Bates • Department of Biological Science and Institute of Molecular Biophysics, Florida State University, Tallahassee, Florida 32306.

tion or transfer of genetic material, of either nuclear or cytoplasmic origin from parental lines to stable progeny. The transfer of cytoplasmic DNA, which includes either the chloroplast and/or mitochondria, is referred to as a cytoplasmic hybrid. A true hybrid is formed when two nuclei present in the same cell fuse to form a single stable nucleus.

The transfer of nuclear DNA between two somatic cells initially involves the successful cytoplasmic fusion of two or more parental cells, which in itself is a formidable task. Cell walls must be enzymatically broken down or removed and intact cell membranes must be perturbed in such a fashion as to induce a fusion susceptible state in both cell membranes. Whereas this temporary instability exists in the cell membrane, the cells must to be brought close enough together to permit the cell membranes to fuse. Obviously, the degree to which the cell membrane is perturbed is highly critical, not only to preserve the structural integrity of the cell, but in order to avoid a possible toxic response in cell viability.

The concept of totipotency or the development of an entire plant from a single somatic plant cell is one that has long intrigued plant biologists. This concept was suggested as early as 1838 by the cell theory of Schleiden and elaborated on by Schwann (1839) and Virchow (1858). In these classic studies, the authors indicated that each cell came from an existing previous cell and that cells in any living organism could be traced back to some common ancestry. The techniques for the production of fully regenerated plants from a single cell or protoplast is now common in cell culture laboratories and the availability of totipotent cells which can be easily manipulated suggests interesting possibilities for the introduction of desirable genetic traits through the mechanism of protoplast fusion.

2. GENETIC RECOMBINATION

Once cytoplasmic fusion is achieved, a cell exists which has two or more nuclei and this cell, if derived from different parental cells is called a heterokaryon. Homokaryons represent a somatic fusion between cells from the same parental cell line.

2.1. Nuclear Fusion

If nuclear encoded traits are to be transferred by somatic cell fusion with the goal of achieving stable progeny, then nuclear fusion must follow cell fusion before the first cell division. Nuclear fusion occurs naturally in the normal cycle of sexual reproduction with meiotic cells of sexually compatible species, however, nuclear fusion in mitotic cells can be a barrier to somatic hybridization. Thus, even successful cytoplasmic fusions often result in the segregation of nuclei at the first cell division if indeed the cell undergoes any division. Several reports indicate that the formation of nuclear hybrids with the fusion of the heterokaryon nuclei occurs in the majority of fusion products; however, the number of heterokaryons that eventually undergo nuclear fusion seems to be highly variable (Menczel, 1983; Medgyesy *et al.,* 1980). The expression of nuclear genomes of both parental cells, or even the partial expression, would require successful cytoplasmic fusion and successful nuclear fusion within the same pair of cells.

2.2. Cybrid Formation

One alternative to the incorporation of different somatic nuclei into the same cell is the incorporation of foreign cytoplasmic genomes of the mitochondria and/or chloroplasts into stable recombinants. These organelles, which have semiindependent expression of their own genetic material, have been successfully transferred through somatic cell fusion in a number of species. These genetic recombinants have been termed cybrids to distinguish them from true nuclear hybrids (Galun and Aviv, 1983). In nuclear somatic hybridization, a somatic cell is formed that contains the entire nuclear genome and the entire cytoplasmic contents of both parental cells. Although these cells are true hybrids they often divide to form cells which differ in the composition of their organellar components due to organelle segregation during subsequent cell divisions (Flick and Evans, 1982; Kumar *et al.,* 1982). By contrast, as shown in Figure 1, during cybrid formation the nucleus from one of the parental cells is either lost or experimentally eliminated to form a cell with a single

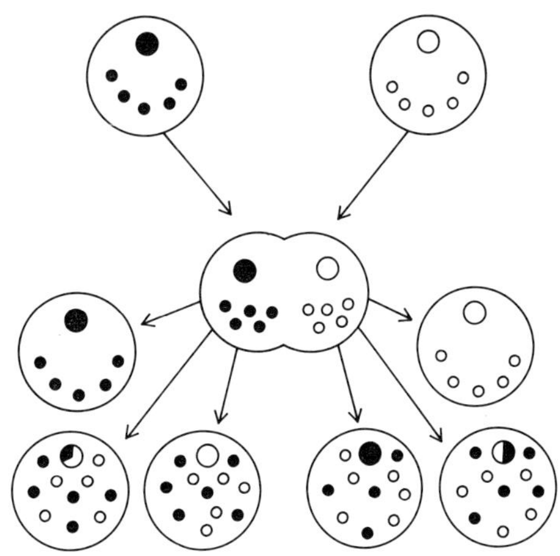

Figure 1. Schematic representation of somatic fusions between protoplasts with respect to nuclear and cytoplasmic DNA. Possible fusion products include separation of fused cells back to parental cells, complete fusion of nuclei and retention of cytoplasmic DNA, exclusion of one nucleus and retention of cytoplasmic DNA as in the donor–recipient approach to cell fusion, and fusion of both nuclei with subsequent partial loss of nuclear genes.

somatic nucleus and a mixed population of cytoplasmic organelles. After the period of organelle segregation during cell division, the cytoplasmic organelles will eventually stabilize in the newly formed cybrid and the expression of the DNA of these cytoplasmic recombinations can be followed through genetic or phenotypic traits.

The induction of these cybrids has been described in detail (Zelcer *et al.*, 1978; Aviv and Galun, 1980; Aviv and Galun, 1985) and involves the use of the donor–recipient approach to somatic fusion events. Prior to fusion, the donor cell, is mildly X-irradiated, while the recipient cell is left untreated as normal. The two populations of cells are mixed in equal numbers and the mixture is exposed to the somatic fusogen. In the cells that resulted from the fusion of a donor cell and a recipient cell the irradiated nucleus fails to divide and is quickly lost. After fusion, the expression of nuclear traits of the cybrid is solely contributed by the recipient cell, which contains a mixture of cytoplasmic organelles. This intriguing procedure has been used to incorporate several cytoplasmic traits into recipient cell lines which have resulted in stable expression in subsequent progeny. These progeny cells are then maintained in tissue culture and under appropriate conditions can regenerate entire new plants from the fusion event.

3. ISOLATION OF PROTOPLASTS

In plants the removal of the cell wall is a prerequisite to the fusion of somatic cells. The cell wall may be removed by either mechanical or enzymatic techniques to form a naked cell or plant protoplast.

3.1. History of Protoplast Isolation

Plant protoplasts were first isolated in 1892 using plasmolyzed cells of the sugar beet root (Klercker, 1892). This technique involved the mechanical maceration of specialized tissues with large vacuolated cells such as fruit and storage tissue. These tissues were plasmolyzed, sliced into thin strips, and then placed into hypotonic aqueous solutions. The protoplasts were released from the sectioned tissue by the rapid osmotic swelling in these solutions. Although this early technique did not produce large numbers of protoplasts, it clearly indicated the potential usefulness of the plant cell without the cell wall as a new tool for the cytologist. This procedure for protoplast isolation was later modified and mechanized by Leigh and Branton to produce larger numbers of intact isolated protoplasts and vacuoles (Leigh and Branton, 1976). Although the procedure has limited application because of its requirement for tuberous type tissue, it has provided a useful mechanism for comparative studies on protoplasts isolated by mechanical and enzymatic techniques.

Efficient large-scale production of protoplasts was implemented by the enzymatic digestion techniques of Cocking (1960). In this important study, which represented the first report for enzymatic protoplast isolation in plants, cellulase isolated from the fungus *Myrothecium* was used to isolate protoplasts from root tips of tomatoes. Since these protoplasts were stable in a buffered sucrose osmoticum for at least 6 hr, the path was opened for all subsequent progress in protoplast culture and plant somatic hybridization.

3.2. Enzymation Digestion

The enzymatic digestion of the plant cell wall is by far the most widely used method for the production of protoplasts. Protoplasts have been isolated from virtually every conceivable organ including the leaf, root, fruit, flower parts, and other specialized organs although leaves, cotyledons, and hypocotyls are the preferred sources. In addition, protoplasts are commonly made from cells in tissue culture including cell suspensions, callus, and differentiated organs. These culture systems have ultimately been derived from a variety of sources; however, stem pith, root and leaf meristems, and seedling callus culture are favorite tissues for initiating sterile cultures. Table I gives some representative examples of the variety of tissue types which have been successfully used for the production of protoplasts by enzymatic digestion techniques. Although by no

Table I. Representative Tissues Successfully Used for the Isolation of Protoplasts

Plant	Source of tissue	Reference
Atropa belladona	Mesophyll leaf	Lorz and Potrykus (1979)
Brassica napus	Mesophyll leaf	Kartha *et al.* (1974)
Datura innoxia	Mesophyll leaf	Schieder (1975)
Nicotiana tabacum	Mesophyll leaf	Takebe *et al.* (1971)
Beta vulgaris	Storage root	Schmidt and Poole (1980)
Zea mays	Root	Bawa and Torrey (1971)
Petunia hybrida	Flower petals	Wakasa (1973)
Lycopersicon esculentum	Fruit	Cocking and Gregory (1965)
Solanum nigrum	Berries	Raj and Herr (1970)
Avena sativa	Coleoptiles	Ruesink and Thimann (1965)
Linum usitatissimum	Hypocotyl	Gamborg and Shyluk (1976)
Hordeum vulgare	Seed Aleurone layer	Taiz and Jones (1971)
Glycine max	Root nodules	Davey *et al.* (1973)
Daucus carota	Cell-suspension culture	Grambow *et al.* (1972)
Vitus vinifera	Callus culture	Skene (1974)

means comprehensive, Table I serves to demonstrate the widespread utility and application of protoplast technology.

Three basic components are necessary for the enzymatic digestion of plant tissues to form protoplasts. These are the digestive enzymes, a buffer at an appropriate pH, and the osmoticum, which prevents the lysis of the protoplasts.

The various commercial preparations of digestive enzymes used in protoplast isolation differ somewhat in the enzymatic activities that are present and the purity of the extracts; however, the ultimate source of the digestive enzymes is usually derived from fungal isolates (i.e., *Rhizopus* or *Aspergillus* or crude extracts of *Trichoderma viride*). For the most part these preparations are not highly purified and contain several different types of enzymatic activities, as well as contaminants, and unknown components. The non-cellulytic enzymes which are present in the commerical digestive enzyme preparations include various types of hydrolytic and protolytic enzymes, including proteases, glycosidases, acid phosphotases, catalyases, lipases, ATPases, and others (Cocking, 1972; Butenko, 1979; Saunders, 1979; Vasil and Vasil, 1980; Gamborg *et al.,* 1981; and Saunders and Gillespie, 1984). The effects of these hydrolytic enzymes on the physiology and biochemistry of the protoplast have not been fully elucidated; however, there is direct evidence from several laboratories that numerous physiological factors may be affected. For example, Anderson *et al.* (1984, 1985) showed that commercial cell wall digesting enzymes promote the release of ethylene, a plant hormone which often triggers senescence, from protoplasts isolated from several different plant species. To overcome these potential deleterious effects of the digestive enzymes, several procedures have been used to further purify the commercial enzyme preprations. These purification procedures have included dialysis, column chromatography (Sephadex, cellulose, and DEAE), ammonia sulfate precipitation, differential centrifigation of insoluble residue, and other techniques (Schenk and Hildebrant, 1969; Kao *et al.,* 1971; Berghem and Pettersson, 1973).

In general commercial digestive protoplast enzyme preparations may be grouped into three different categories: (1) those that contain pectinase (i.e., Macerase,* Pectinase, Pectolyase, Macerozyme), (2) those that contain cellulase (i.e., Cellulase, Cellulysin, Driselase, Onozuka R-10), and (3) those that contain hemicellulase (i.e., Cellulysin, Hemicellulase).

The appropriate enzyme or combination of enzymes frequently depends on the source of the tissue which is being used for the isolation of

*Mention of a tradename or proprietary product does not constitute a guarantee of the product by these authors or their affiliated organizations nor does it imply approval to the exclusion of other products that may also be suitable.

the protoplasts, however, it is usually necessary to employ some combination of a pectinase (to separate the individual cells) and a cellulase (to actually digest the cell wall). Using a 1–4% solution of the digestive enzyme mixture protoplasts are obtained from most plant tissues within a 4–12 hr period. The time necessary for protoplast isolation may be shortened by raising the temperature and by procedures which facilitate the penetration of the digestive enzymes into the tissue, e.g., abrading or pulling away the epidermis or vacuum infiltration of the enzyme solution.

The buffer which is used for the digestion of protoplasts from plant tissues may be selected for an appropriate pH range of 5.5–5.8. Several different buffers have been successfully used for protoplast isolation including citrate–phosphate, MES–NaOH, and Tris–maleate usually at concentrations of 10–25 mM. The osmoticum which is used to stabilize the protoplasts is generally chosen for its compatability with the tissue type in use; however, plant protoplasts have been successfully isolated in isotonic or mildy hypertonic solutions of mannitol, sorbitol, or sucrose.

4. PROTOPLAST FUSION

The fusion of protoplasts has been documented for some time. In 1909 Kuster first observed the spontaneous fusion of mechanically isolated protoplasts, although the first report of induced fusion was not until 1970 when protoplasts were fused using sodium nitrate (Kuster, 1909; Power *et al.,* 1970). Since that time a variety of chemical fusogens have been evaluated for their effectiveness. Table II lists those fusogens that have been shown to yield positive somatic fusions of plant protoplasts and the organisms from which the protoplasts were isolated.

In order to achieve fusion, the plasma membrane of two or more protoplasts must be physically appressed for some time interval. Since the plasma membrane has a net negative surface charge, adjacent cells, with similar charges tend to repel each other. The surface charge of the cell membrane may be modified by the pH of the buffer, by polycations added to the media, or by the dehydrating effect of various chemical fusogens. In any case, the plant cell membrane must be considered a fluid, dynamic structure that can fuse with other membranes under appropriate conditions.

Several of the fusogens listed in Table II actually function only to induce cell agglutination are not true membrane fusogenic agents. These compounds override the repellant force of the membrane surface charges causing the cell to clump into tight clusters. In these cases, subsequent cell fusion is probably due to the intrinsic fluidity of the cell membrane rather than to the fusogen itself.

Table II. Chemical Fusogens Used with Plant Protoplasts

Fusogen	Plant	Mode of action	Reference
Gelatin	*Allium fistulosum*	Aggregation some fusion	Kameya (1973)
	Brassica chinensis		
Rabbit immune sera	*Viciahajastana*	Aggregation	Hartmann *et al.* (1973)
	Glycine max		
Salt solutions (NaCl, KCl, $NaNO_3$, KNO_3)	*Brassica oleracea*	Fusion	Kameya and Takahashi (1972)
	Brassica chinensis		
Dextran plus electrolytes (NaCl, KCl, $CaCl_2$)	*Brassica pekinensis*	Fusion	Kameya (1979)
High pH plus high Ca^{2+}	*Nicotiana tabacum*	Fusion	Keller and Melchers (1973)
	Nicotiana knightiana		Menczel *et al.* (1981)
PEG	*Hordeum vulgare*	Fusion and aggregation	Kao and Michayluk (1974)
	Glycine max		Kao and Constabel (1974)
	Zea mays		
	Pisum sativum		
	Vicia hajastana		
PEG plus high pH and high Ca^{2+}	*Nicotiana glaucia*	Fusion and aggregation	Kao *et al.* (1974)
	Nicotiana langsdorffii		Binding and Nehls (1978)
	Vicia fiaba		Constabel *et al.* (1976)
	Petunia hybrida		
	Glycine max		
	Pisum sativum		

Polyvinyl alcohol	*Nicotiana tabacum*	Fusion	Nagata (1978)
Sodium nitrate	*Zea mays*	Fusion	Power *et al.* (1970)
	Avena sativa		Potrykus (1971)
	Torenia baillonii		
	Torenia fournieri		
Phytolectins (phytohemagglutin, wheat germ agglutin, concanavalin A, poly-L-lysine)	*Hordeum vulgare*	Aggregation	Chin and Scott (1979)
	Triticum astivum		Glimelius *et al.* (1974)
	Nicotiana tabacum		Burgess and Fleming (1974)
	Daucus carota		
PEG plus DMSO	*Daucus carota*	Fusion	Haydu *et al.* (1977)
	Secale cereale		
Pronase E	*Brassica pekinensis*	Fusion	Kameya (1979)
Dextran sulfate	*Senecio vulgaris*	Fusion	Kameya (1975)
	Nicotiana tabacum		
	Brassica pekinensis		
	Daucus carota		
	Allium fistulosum		
	Physalis alkelengi		
Seawater	*Funaria hygrometricia*	Aggregation and fusion	Binding (1966)
	Physcomitrium eurystomum		Binding (1974)
	Petunia hybrida		
Lysolethicin	*Pisum sativum*	Aggregation and fusion	Constabel and Kao (1974)
	Vicia fiaba		

The fusion of plant protoplasts differs from fusions of animal or microbial cells in several important respects. One of these is the presence of a cellulosic cell wall in mature plant cells. Additionally, the plasmalemma or cell membrane of plant cells is interconnected through specialized perforations in the cell wall called the plasmodesmata. When the "cell" or protoplast is removed from the confines of the cell wall either through mechanical or enzymatic digestion techniques these plasmodesmatal connections are ordinarily ruptured releasing unconnected protoplasts. Occasionally protoplasts being released from the cell walls will spontaneously fuse, although such spontaneous fusions, which are a result of these plasmodesmatal connections, are rare. These spontaneous fusions are thought to be caused by the expansion of plasmodesmata-linking adjacent cells (Ruesink, 1979) and is probably not equivalent to fusions between unrelated cells.

The spontaneous fusion of protoplasts which are not linked through existing plasmodesmata has also been observed, both in protoplasts from meiotic cells such as microsporocytes (Ito, 1973) and pollen cells (Bhojwan and Cocking, 1972) and in certain somatic cell lines maintained in cell suspension cultures for long periods of time (Boss and Grimes, 1985). Protoplasts of these cell culture lines have developed a spontaneous fusogenic nature such that multinucleate cells can be induced with very simple procedures. Calcium appears to play an intimate role in these types of fusions and the yield of fusion products seems to be the result of interactions between the calcium concentrations in the media and modifications of the membrane biosynthesis in these cells. The selection of fusogenic cells or cell lines that have enhanced frequencies of spontaneous fusions and the role of calcium in fusogenic carrot cell lines are discussed in detail in Chapter 7 (Boss, 1986).

In general, the most widely used chemical fusion procedure involves pretreating plasmolyzed protoplasts with high concentrations of polyethylene glycol (PEG), followed by the dilution of the PEG with a buffer (pH 9–11) containing 10–50 mM calcium. This technique has been used successfully to produce both interspecific and intergeneric protoplast fusions (Carlson *et al.,* 1972; Fowke *et al.,* 1975; Kameya *et al.,*1981).

Polyethylene glycol is available as a polymeric compound in a range of molecular weights (200–20,000). The fusion of plant protoplasts using PEG requires the careful control of several important parameters, including the duration of exposure and the temperature of the fusion media as well as the molecular weight and concentration of the PEG itself. PEG with a molecular weight of less than 1000 does not cause a cohesive agglutination of the protoplasts and subsequent cell fusions are rare. Above a molecular weight of 6000, PEG becomes very viscous and severe technical difficulties are encountered in the manipulation of the cells.

Cells which have been tightly agglutinated with higher MW PEG often stick to the vessel or each other so aggressively that the recovery of the protoplasts is difficult even after the dilution of the fusion media. In addition, analysis of the results of the fusion procedure is severely hampered by the massive clumping of cells. A 0.33 M solution of PEG with a molecular weight of approximately 1500 seems to be an effective working range for mesophyll protoplasts, although PEGs at 4000 and 6000 MW are commonly used successfully (Ruesink, 1979; Mastrangelo, 1979).

5. MECHANISM OF ACTION OF CHEMICAL FUSOGENS

Several stages in the cell fusion process are necessary to achieve a successful cytoplasmic fusion of two or more protoplasts. First, the cell membranes of adjacent protoplasts must come into close physical contact. Secondly, the evidence suggests that lateral diffusion of membrane surface proteins results in the formation of lipid enriched regions of the plasma membrane, which may be subsequently involved in lipid–lipid interactions between adjacent membranes (Ahkong *et al.,* 1975; see Chapter 14, this volume). Fowke *et al.* (1975) suggested that there are several areas across the cell surface which are involved in these membrane interactions; however, their ultrastructural evidence suggests that the areas of first cytoplasmic continuity are established at the edges of the adhering surfaces of adjacent cells. As these regions expand, rearrangement of the lipid–protein membrane occurs with the eventual formation of pores or continuities between cells. These pores continue to enlarge and broaden into areas of cytoplasmic mixing.

Mixing of the cytoplasm seems to occur at differential rates according to the source of the protoplasts. Galbraith and Galbraith (1979) observed a rapid mixing of fluorescent labels using fused doubly stained soybean protoplasts. Following electrofusion, protoplasts isolated from *Nicotiana rustica* showed active cytoplasmic streaming within each hemisphere of fused protoplasts which had rounded up in culture media; however, very little intermixing of cytoplasm occurred between each hemisphere of the fused cell for as long as two hours after fusion (J. A. Saunders, unpublished data). In any case, mixing of cytoplasms in the fused cells does eventually occur to the extent that the source of individual cytoplasmic components can no longer be identified without morphological or biochemical markers.

Multinucleate protoplasts or heterokaryons that undergo cell division rapidly lose one of the nuclei unless nuclear fusion follows cytoplasmic fusion. Even if nuclear fusion occurs the cell may lose nuclear genes through chromosomal elimination or the loss of expression of the nuclear genome. Thus, even successful cytoplasmic and nuclear fusion does not ensure that a particular trait will be incorporated into stable hybrids formed during the fusion process.

Several reports have suggested that there is a relationship between the degree to which an individual cell is vacuolated and its propensity to undergo a fusion (Gleba and Sytnik, 1984; Boss and Grimes, 1985). Boss *et al.* (1983) have shown that cells that are densely cytoplasmic with small vacuoles, typical of meristemic cells, undergo fusion more readily than protoplasts with thin strips of cytoplasm between adjacent vacuolated cells. They have characterized and isolated fusion permissive or fusogenic cells lines from densely cytoplasmic embryogenic carrot cell cultures and have correlated the role of calcium with the fusogenic nature of these cell lines.

5.1. Cell Agglutination

In general chemical fusogens can stimulate the fusion of cells in one of two ways. They may induce cell agglutination or they may alter the membrane in such a way as to make it susceptible to fusion. PEG is one of the best known cell agglutinators; however, it probably also functions in a dual role as a membrane modifier as well.

In addition to PEG, lectins such as concanavalin A (Con A), and other phytohemagglutins as well as immune antisera and dextran have all been used to induce cell aggregation. These substances cause a rapid clumping of the protoplasts which taken to its extremes results in a mass of cells that precipitates in the bottom of the vessel or slide. During this period of tight adhesion, adjacent cells may begin to fuse, or an additional membrane modifying agent may be required to initiate fusion.

5.2. Membrane Modifiers

Several reports link the mechanism of fusion to modifications of plasma membrane surface charges. For example, Kao *et al.* (1974) suggested that the effect of sodium nitrate in the elimination of the negative protoplast surface charge is a key aspect of the fusogenic character of this compound. Similar statements concerning modifications of membrane charges have been attributed to other fusogens, including PEG (Ruesink, 1971; Constable and Kao, 1974), and calcium (Kao and Michayluk, 1974; Nagata and Melchers, 1978). The negative surface charge which is present on the protoplasts probably arises largely from phosphate groups associated with the cell membrane. When protoplasts are treated with acid phosphatase, the overall surface charge of the cell membrane is severely reduced (Nagata and Melchers, 1978).

5.3. PEG Fusion

The popular use of PEG for the fusion of plant protoplasts stems from its dual role both as a cell agglutinator and a membrane modifier.

Within this volume Boni and Hui have discussed the mechanisms of action for PEG using model membrane and human erythrocytes. PEG causes an extensive dehydration of cell membrane by binding and structuring available free water. Although other fusogens such as dextrans, seawater, and gelatin also cause dehydration, PEG has an ability to bind with phospholipids within the cell membrane. The formation of areas on the cell surface which are free of intramembranous particles (IMP) or protein subunits has been suggested as an important prerequisite for membrane fusion. These lipid-enriched IMP-free areas undergo changes in membrane fluidity and are thought to be the specific sites of membrane fusion.

When treated with PEG, the agglutination of protoplasts is highly dependent on the temperature of the media which the cells are suspended in, as well as the concentration and duration of the PEG treatment. Although several studies have documented that PEG treatment by itself is sufficient to cause the fusion of protoplasts from several species (Table II), there have been reports that contaminants present in commercial PEG preparations may be involved in the fusion process. Honda *et al.* (1981) showed that when commercial-grade PEG was purified, it lost its fusogenic character. A low-molecular-weight contaminant of the PEG, which was suggested to be *tert*-butylated pydroxyanisole, or a similar compound, when added to the purified PEG restored the fusogenic nature of the media. The authors indicate that the role of the PEG is simply to form tight complexes of polymeric chains of PEG between the membrane bound proteins of the aggregating cells. These hydrogen bound protein/ polysaccharide complexes enabled the low molecular weight fusogen to interact with either the membrane protein globules or the phospholipid bilayer to cause subsequent fusion. Both Smith *et al.* (1982) and Arnold *et al.* (1983) confirmed Honda's original observation that some of the fusogenic character of PEG is due to contaminating substances in commercial preparations; however, even purified PEG retained a fraction of its fusogenic nature.

5.3.1. Protocol for Protoplast Fusion

The most commonly used procedure for the fusion of plant protoplasts involves the use of PEG, high calcium concentrations, and high pH conditions. These factors, however, are not the only parameters which must be controlled to achieve successful fusion products. Figure 2 gives a

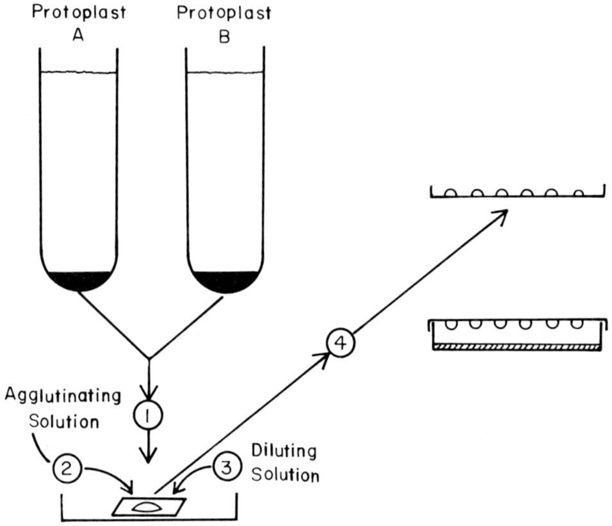

Figure 2. Protocol for PEG fusion of protoplasts. Step 1: Dilute protoplast mixture to 10^4 cells/ml in 0.5 M glucose, 3 mM $CaCl_2$, 0.7 mM KH_2PO_4, pH 5.7. Step 2: Add 3 vol of agglutinating solution of 0.09 mM PEG 6000, 10 mM $CaCl_2$, 0.7 mM KH_2PO_4, pH 5.5. Step 3: Add 5 vol of diluting solution of 50 mM $CaCl_2$, 0.5 M glucose, 50 mM glycine buffer, pH 10.0. Step 4: Wash cells with culture media and culture in hanging drop suspensions.

general outline for a fusion protocol that can be used for protoplasts isolated from several different sources with minor modifications.

Protoplasts must be isolated simultaneously from both parental cell types. The specific details for the isolation of protoplasts using this technique have been described in detail (Gamborg *et al.,* 1981; Dodds and Roberts, 1982; Reinert and Yeoman, 1982). In brief, the protoplasts are released from the tissue by enzymatic digestion, filtered through a 60-μm mesh to remove cell debris, and concentrated by centrifugation (100 × *g* 1–2 min). The protoplasts should be resuspended and rewashed with isotonic media to remove any residual contaminating digestive enzymes from the suspension. After mixing the two cell populations together and adjusting the concentration of protoplasts, 200 μl of the suspension is transferred to a slide or coverslip suspended in a petri plate. The concentration of the cells in this suspension may be adjusted for higher fusion yields by increasing the cell density; however, this often decreases the number of binuleate cells and favors the formation of multinucleate heterokaryons. After allowing the protoplasts to settle on the slide, 600 μl of the agglutinating solution is slowly added to the protoplasts causing them to clump into tight units and to adhere to the glass surface. After 15 min at 25°C, 500 μl of diluting solution is added carefully to the cells, and the suspension is incubated for an additional 10 min. This step is repeated while excess liquid is removed from the slide with a pipette. The cells should be washed repeatedly with culture media before being cultured. When cultured in small hanging drops as shown in Figure 3, the protoplast should begin to synthesize rudiments of a cell wall within a few hours. Complete regeneration of the cell wall requires 1–2 days; subsequent cell division will result in rapid growth of the cells within 2 weeks (Figs. 3A,B). Cell division will not occur until the cell wall has been regenerated.

Many modifications of this procedure for the fusion of protoplasts have been suggested, and it is clear that the protocol must be optimized for each particular cell type that is being examined. In addition, each step in the fusion process should be evaluated not only for yield of fusion products, but also for its overall effect on cell viability. For example, the use of higher concentrations of PEG or higher-molecular-weight PEG is known to cause tighter adhesion of the protoplasts; however, it also limits the recovery of viable fused cells. Similarly, the elution of PEG with high calcium alkaline solutions promotes fusion but it may do so at the expense of viable fusion products (Constabel, 1984).

A cytological analysis of the fused cells is usually the first method of evaluating a fusion sequence. Irreversible agglutination of the protoplasts due to the fusogen can be immediately assessed for the effects on cell viability. Typically, the number of nuclei per cell and thus the yield of fusion products, can be determined by selective staining techniques. Figure 4A,B shows carrot protoplasts fused using PEG procedures with and without 4′,6-Diamidino-2-phenyl indole (DAPI) stain. DAPI is a selective stain for DNA that clearly shows in Figure 4B the presence of multiple nuclei within a common cell, indicating the potential for nuclear fusion. Alternatively, the fusion of chlorophyll containing and achlorophyllous cells allows identification of live heterokaryons although the number of nuclei per cell cannot be determined.

One popular modification of chemical fusion protocols is the use of low-speed centrifugation during the aggregation and fusion stages. The increased centrifugal forces created by the centrifugation seems to promote tighter adhesion and promote higher yields of fused cells.

6. SELECTION OF HYBRIDS FOLLOWING FUSION

The selection of hybrids following protoplast fusion constitutes one of the most important and difficult aspects of somatic hybridization. Regardless of which fusogen is employed, most of the cells will be either unfused, fused within the same parental type, or the product of multiple

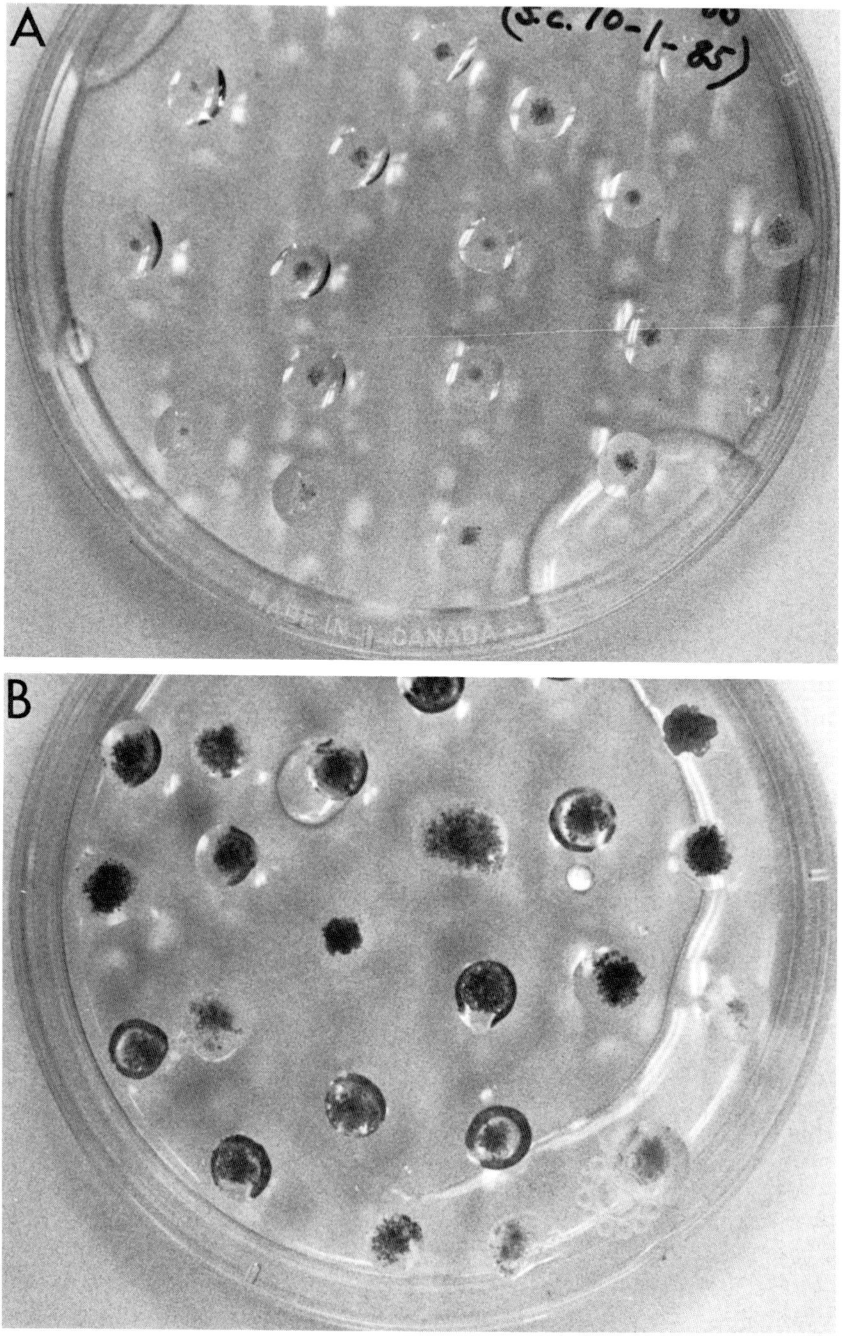

Figure 3. Protoplasts from a fusion between *Daucus carota* and *Daucus capillifolius* cultured by hanging drop suspension techniques in modified MS media. (A) Two days after initiation of the suspension culture. (B) Two weeks after the initiation of the culture.

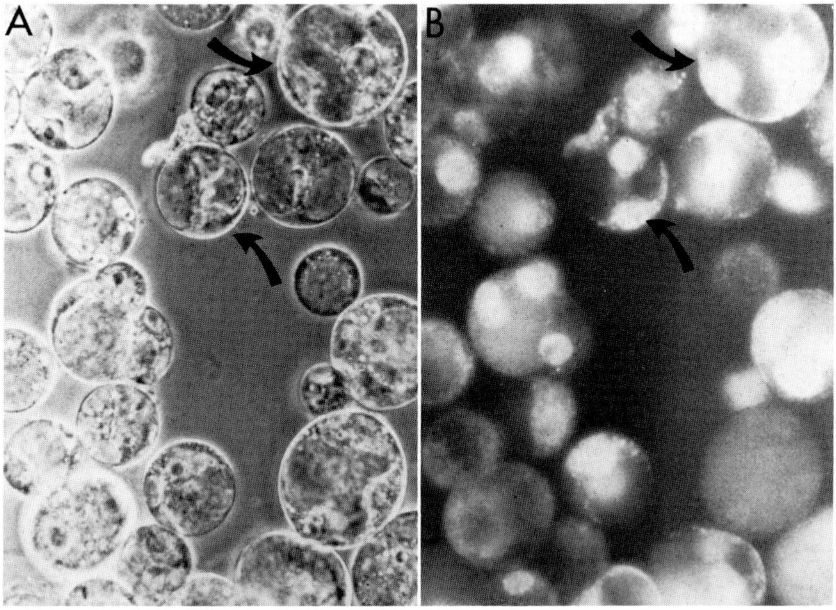

Figure 4. DAPI staining of protoplasts isolated from *Daucus carota* and *Daucus capillifolius* fused with PEG. Arrows show the presence of multiple nuclei stained in fused cells, however, under bright-field microscopy the number of nuclei are difficult to determine. (A) Bright-field microscopy. (B) Fluorescence microscopy. (Photo courtesy of Dr. B. F. Matthews, USDA.)

fusion events. The separation of these unwanted cells from the desired fusion products may be the determining factor in the success or failure of experiments designed to transfer a particular genetic trait through somatic fusion procedures.

6.1. Nuclear Selection Systems

In general several different types of selection mechanisms have been used with varying degrees of success in the selection of desired fusion products. The most direct procedure for the isolation of hybrids is the use of micropipettes to pick out fused heterkaryons and manually transfer them to cell culture media. The advantages of this technique is that only fused cells are incorporated into the culture media and the percent of hybrids resulting from these cells can be accurately determined. Aside from the obvious limitations with this technique in terms of the numbers of cells that can be processed and the length of time required to select each heterokaryon, there are additional problems, which are encountered. The heterokaryon must be morphologically distinct from each of the parental types. This can be accomplished through the use of achlorophyllous cell lines fused with protoplasts that contain green chloroplasts such as those from mesophyll leaves; however, PEG-fused cells are often difficult to manipulate when the cells have clumped together. Alternatively, differential staining of the parental protoplasts prior to fusion may subsequently enable heterkaryons to be identified and isolated by manual procedures. Another difficulty that may be encountered is that most plant cell culture systems are recalcitrant to vigorous growth in very low cell densities. The culture of protoplasts at densities of less than 10^4 cells/ml normally requires the use of conditioned media or nurse cultures and these cultural procedures are well established for only a few species. Conditioned media is complete media that has supported the growth of similar cell types grown at high densities; however, all previous cells have been filtered or centrifuged out of the solution. The culture of protoplasts in relatively low cell densities may be accomplished using hanging drop cultures, as shown in Figure 3, particularly when conditioned media is used. This system can be used with very low volumes of culture media, hence, very low numbers of cells.

Another useful selection scheme for recovering somatic hybrids involves the fusion of mutant cell lines which are resistant to different toxins or require specific growth factors in the culture media. Parental cell lines must be chosen such that only hybrids in which both parental phenotypes are expressed, will be able to survive the selective pressure. A clear example of this technique is the use of cell lines that can grow and divide in cell culture without essential supplements that are necessary for the growth of normal or wild-type cell lines. The fusion of amino acid analogue-resistant mutants in carrot cell hybridization studies demonstrates this procedure (Matthews and Widholm, 1985). Although the use of biochemically selectable traits in mutant cell lines has been successfully applied to the production of somatic hybrids, it has several limitations as well. The principal limitation of this procedure is the availability of specific mutant cell lines. Mutant cell lines are not as prevalent in plant research as in microbial and animal studies, although there are a few excellent exceptions. This requirement for specific mutant cell lines limits the widespread use of protoplasts isolated from differentiated tissue and restricts fusion to cell culture lines which have been previously isolated and shown that they are stable in tissue culture.

An alternative procedure to the use of biochemical selectable screens is the use of culture conditions or regeneration capabilities which are inherent in the parental cell lines. For example, Itoh and Futsuhara (1983) described the production of somatic hybrids between *Petunia hybrida* and *Petunia parodii* in which the *P. hybrida* cell suspension culture had lost the ability to regenerate shoot tissue. In turn, *P. parodii* could not grow beyond the small cell colony stage and thus was not able to regenerate differentiated tissue. The somatic hybrids which were produced by the use of PEG fusion techniques from these two species were able to grow and regenerate successfully.

6.2. Cytoplasmic Selection Systems

Maliga *et al.* (1982) described an excellent example of a selection system for cybrids following protoplast fusion which is based on cytoplasmic rather than nuclear genes. Cytoplasmically coded streptomycin-resistant *Nicotiana tabacum* protoplasts enucleated by centrifugation through gradients were fused with *Nicotiana plumbaginifolia* protoplasts that were sensitive to streptomycin. Since the enucleated *N. tabacum* could not divide and *N. plumbaginifolia* was streptomycin sensitive, only fused cybrids should be able to grow and form calli.

A final mechanism for the selection of somatic hybrids from the mixture of cell types produced by a typical fusion procedure, is to dilute the mixture of protoplasts to an appropriate volume and culture all the progeny to callus and subsequently cell differentiation stages. This mechanism is used in situations where a particular morphological trait is being transferred from parental cell lines to progeny and no other selection system is available. It has the obvious disadvantage of requiring large numbers of cell cultures to be grown to at least a callus stage requiring the use of extensive cell culture and/or greenhouse facilities.

7. GENETIC VARIABILITY IN TISSUE CULTURE AND SOMATIC HYBRIDS

Genetic variability associated with plant tissue culture and regeneration of plant cells can be a serious problem in the formation of stable cell lines and in the regeneration of somatic hybrid plants. Several problems are encountered during the culture of a selected cell line:

1. The culture of plant cells often results in cytogenetic abnormalities including polyploidy, aneuploidy, and translocations (Zosimovich and Kunakh, 1975; Gleba and Sytnik, 1984).
2. Plants obtained during the regeneration of cells from callus culture often have variable chromosome numbers due to losses during cell division or incomplete incorporation of the nuclear genome during the fusion process.
3. Further reductions of the nuclear chromosome number may result in the stabilization of hybrid plants after regeneration of plants resulting in genetically unstable systems (Sacristan and Melchers, 1969, 1977).

These problems seem to be inherent in the use of tissue culture as a manipulative tool; however, they do not preclude the use of somatic hybridization as an effective mechanism for the transfer of genetic traits from one species to another. The instability of the chromosome number of hybrids should caution the investigator to evaluate closely the experimental cell lines throughout the course of the study. To date very little progress has been achieved in the practical application of somatic hybrid production on economic crops outside the Solanaceae.

8. DIFFERENCES IN CELL FUSION BETWEEN PLANT AND ANIMAL CELLS

While research activity in plant protoplast fusion has been active since its inception in the 1970's somatic cell fusion using animal cells faces different problems. In many cases the fusogens that are effective in the fusion of plant cells are also used on animal cells. For example, PEG and calcium are both used routinely for fusion of animal cells although the procedures and techniques are usually modified slightly (Ahkong *et al.*, 1980). There are, however, several important differences between the cells derived from the animals vs. those from plants. One of the more apparent structural differences in the cell types is the predominance of a cellulosic cell wall in the plant cell fusion. In addition, after the removal of the cell wall plant cells will only grow and divide if the cell wall regenerates. In culture, many animal cells often grow as single cell isolates and the cell cycles can be conveniently synchronized to investigate a particular stage in the growth of the cells. Plant cells almost always grow as clumps or clusters of cells even in cell suspension cultures and cell division synchronization is very difficult to induce.

One major advantage to the use of plant cells for fusion studies is their capacity for totipotency in the regeneration of an entire organism from a single fusion event. This capability, more than any other, makes the use of somatic fusion of plant cells a valuable tool in the transfer of selected genetic traits.

Finally, one must remember that the plant systems contain two sources of extranuclear DNA, i.e., those of the mitochrondria and chloroplasts. These semiautonomous, self-replicating organelles have the potential for transferring selected traits to foreign cells and establishing a stable cybrid that will express the genetic code contained in the transformed organelles.

REFERENCES

Ahkong, Q. F., Fisher, D., Tampion, W., and Lucy, J. A., 1975, Mechanisms of cell fusion, *Nature (Lond.)* **253:**194–195.

Ahkong, Q. F., Botham, G. M., Woodward, A. W., and Lucy, J. A., 1980, Calcium activated thiol-proteinase activity in the fusion of rat erythrocytes induced by benzyl alcohol, *Biochem. J.* **192:**829–836.

Anderson, J. D., Chalutz, E., and Mattoo, A. K., 1984, Purification and properties of the ethylene-inducing factor from the cell wall digesting mixture, cellulysin, in: *Ethylene: Biochemical, Physiological and Applied Aspects* (Y. Fuchs and E. Chalutz, eds.), pp. 189–198, Nijhoff/Junk, The Hague.

Anderson, J. D., Chalutz, E., and Mattoo, A. K., 1985, Induction of ethylene biosynthesis by cell wall digesting enzymes, in: *Cellular and Molecular Biology of Plant Stress,* Vol. 22 (J. L. Key, T. Kosuge), pp. 263–273, Alan Liss, New York.

Arnold, K., Pratsch, L., and Gawrisch, K., 1983, Effect of poly(ethylene glycol) on phospholipid hydration and polarity of the external phase, *Biochim. Biophys. Acta* **728:**121–128.

Aviv, D., and Galun, E., 1980, Restoration of fertility in cytoplasmic male sterile (CMS) *Nicotiana sylvestris* by fusion with X-irradiated *N. tabacum* protoplasts, *Theor. Appl. Genet.* **58:**121–127.

Aviv, D., and Galun, E., 1985, An in vitro procedure to assign pigment mutations in *Nicotiana* to either the chloroplast or the nucleus, *J. Hered.* **76:**135–136.

Bawa, S. B., and Torrey, J. G., 1971, "Budding" and nuclear division in cultured protoplasts of corn, *Convolulus,* and onion, *Bot. Gaz.* **132:**240–245.

Berghem, L. E. R., and Pettersson, L. G., 1973, The mechanism of enzymatic cellulose degradation: Purification of a cellulolytic enzyme from *Trichoderma viride* active on highly ordered cellulose, *Eur. J. Biochem.* **37:**21–30.

Bhojwani, S. S., and Cocking, E. C., 1972, Isolation of protoplasts from pollen tetrads, *Nature (New. Biol.)* **239:**29–30.

Binding, H., 1966, Regeneration and fusion behavior of isolated protoplasts of mosses, *Z. Pflanzenphysiol.* **55:**305–321.

Binding, H., 1974, Fusion experiments with isolated protoplasts of *Petunia hybrida* L, *Z. Pflanzenphysiol.* **72:**422–426.

Binding, H., and Nehl, R., 1978, Somatic cell hybridization of *Vicia faba* + *Petunia hybrida, Molec. Gen. Genet.* **164:**137–143.

Boss, W. F., and Grimes, H. D., 1985, Dynamics of calcium-induced fusion of fusogenic carrot protoplasts, In: *Frontiers of Membrane Research in Agriculture,* Vol. 9 (J. B. St. John, P. C. Jackson, and E. Berlin, eds.), pp. 64–68, Rowman and Allanheld, Totowa, New Jersey.

Boss, W. F., Allen, N. S., and Grimes, H. D., 1983, Developmentally regulated fusion of carrot protoplasts, in: *Sixth International Protoplast Symposium,* Basel, Switzerland, pp. 96–97.

Burgess, J., and Fleming, E. N., 1974, Ultrastructural studies of the aggregation and fusion of plant protoplasts, *Planta* **118:**183–193.

Butenko, R. G., 1979, Cultivation of isolated protoplasts and hybridisation of somatic plant cells, *Int. Rev. Cytol.* **59:**323–392.

Carlson, P. S., Smith, H. H., and Dearing, R. D., 1972, Parasexual Interspecific Plant Hybridization. *Proc. Natl. Acad. Sci. U.S.A.* **69:**2292–2294.

Chin, J. C., and Scott, K. J., 1979, Effect of phytolectins on isolated protoplasts from plants, *Ann. Bot.* **43:**33–44.

Cocking, E. C., 1960, A method for the isolation of plant protoplasts and vacuoles, *Nature (Lond.)* **187:**962–963.

Cocking, E. C., 1972, Plant cell protoplasts—isolation and development, *Annu. Rev. Plant Physiol.* **23**:29–50.

Cocking, E. C., and Gregory, D. W., 1965, The large scale isolation of protoplasts from immature tomato fruit, *J. Cell Biol.* **24**:143–146.

Constabel, F., and Kao, K. N., 1974, Agglutination and fusion of plant protoplasts by polyethylene glycol, *Can. J. Bot.* **52**:1603–1606.

Constabel, F., Weber, G., Kirkpatrick, J. W., and Paul, K., 1976, Cell division of intergeneric protoplast fusion products, *Z. Pflanzenphysiol.* **79**:1–7.

Constabel, F., 1984, Fusion of protoplasts by polyethylene glycol (PEG), in: *Cell Culture and Somatic Cell Genetics of Plants,* Vol. I (I. K. Vasil, ed.), pp. 414–422, Academic Press, New York.

Davey, M. R., Cocking, E. C., and Bush, E., 1973, Isolation of legume root nodule protoplasts, *Nature (Lond.)* **24**:460–461.

Dodds, J. H., and Roberts, L. W., 1982, *Experiments in Plant Tissue Culture,* Cambridge University Press, New York.

Flick, C. E., and Evans, D. A., 1982, Evaluation of cytoplasmic segregation in somatic hybrids of *Nicotiana:* Tentoxin sensitivity, *J. Hered.* **73**:264–266.

Fowke, L. C., Rennie, P. J., Kirkpatrick, J. W., and Constabel, F., 1975, Ultrastructural characteristics of intergenetic protoplast fusion, *Can. J. Bot.* **53**:272–278.

Galbraith, D. W., and Galbraith, J. E. C., 1979, A method for identification of fusion of plant protoplasts derived from tissue cultures, *Z. Pflanzenphysiol.* **93**:149–158.

Galun, E., and Aviv, D., 1983, Cytoplasmic hybridization genetic and breeding applications, in: *Application of Plant Tissue Culture Methods for Crop Improvement,* Vol. I (D. A. Evans, W. R. Sharp, and P. V. Ammirato, eds.), pp. 358–392, Macmillan, New York.

Gamborg, O. L., and Shyluk, J. P., 1976, Tissue Culture, protoplasts and morphogenesis in flax, *Bot. Gaz.* **137**:301–306.

Gamborg, O. L., Shyluk, J. P., and Shanin, E. A., 1981, Isolation, fusion and culture of plant protoplasts, in: *Plant Tissue Culture Methods and Applications in Agriculture* (T. A. Thorpe, ed.), pp. 115–153, Academic Press, New York.

Gleba, Y. Y., and Sytnik, K. M., 1984, Protoplasts fusion, genetic engineering in higher plants, (R. Shoeman, ed.), Springer-Verlag, New York.

Glimelius, K., Wallin, A., and Eriksson, T., 1974, Agglutinating effects of concanavalin A on isolated protoplasts of *Daucus carota, Physiol. Plant* **31**:225–230.

Grambow, H. J., Kao, K. N., Miller, R. A., and Gamborg, O. L., 1972, Cell division and plant development from protoplasts of carrot cell suspension cultures, *Planta* **103**:348–355.

Hartman, J. X., Kao, K. N., Gamborg, O. L., and Miller, R. A., 1973, Immunological methods for the agglutination of protoplasts from cell suspension cultures of different genera, *Planta* **112**:45–56.

Haydu, Z., Lazar, G., and Dudits, D., 1977, Increased frequency of polyethylene glycol induced protoplast fusion by dimethylsulfoxide, *Plant Sci. Lett.* **10**:357–360.

Honda, K., Maeda, Y., Sasakawa, S., Ohno, H., and Tsuchida, E., 1981, The components contained in polyetheylene glycol of commercial grade (PEG-6,000) as a cell fusogen, *Biochem. Biophys. Res. Commun.* **101**(1):165–171.

Ito, M., 1973, Studies on the behavior of meitotic protoplasts. II. Induction of a high fusion frequency in protoplasts from liliaceous plants, *Plant Cell Physiol.* **14**:865–872.

Itoh, K., and Futsuhara, Y., 1983, Restoration of the ability to regenerate shoots by somatic hybridization between *Petunia hybrida* and *P. parodii, Jpn. J. Breed.* **33**(2):130–137.

Kameya, T., and Takahashi, N., 1972, The effects of inorganic salts on fusion of protoplasts from roots and leaves of *Brassica* species, *Jpn. J. Genet.* **47**:215–217.

Kameya, T., 1973, The effects of gelatin on aggregation of protoplasts from higher plants, *Planta* **115**:77–82.

Kameya, T., 1975, Induction of hybrids through somatic cell fusion with dextran sulfate and gelatin, *Jpn. J. Genet.* **50**:235–246.

Kameya, T., 1979, Studies on plant cell fusion: Effects of dextran and pronase E on fusion, *Cytologia* **44**:449–456.

Kameya, T., Horn, M. E., and Widholm, J. M., 1981, Hybrid shoot formation from fused *Daucus carota* and *D. capillifolius* protoplasts, *Z. Pflanzenphysiol.* **104**:459–466.

Kao, K. N., Gamborg, O. L., Miller, R. A., and Keller, W. A., 1971, Cell divisions in cells regenerated from protoplasts of soybean and *Haplopappus gracilis, Nature (New Biol.)* **232**:124.

Kao, K. N., and Constabel, F., 1974, Agglutination and fusion of plant protoplasts by polyethylene glycol, *Can. J. Bot.* **52**:1603–1606.

Kao, K. N., and Michayluk, M. R., 1974, A method for high-frequency intergeneric fusion of plant protoplasts, *Planta* **115**:355–367.

Kao, K. N., Constabel, F., Michayluk, M. R., and Gamborg, O. L., 1974, Plant protoplast fusion and growth of intergeneric hybrid cells, *Planta* **120**:215–227.

Kartha, K. K., Michayluk, M. R., Kao, K. N., Gamborg, O. L., and Constabel, F., 1974, Callus formation and plant regeneration from mesophyll protoplasts of rape plants, *(Brassic napus,* L. cv. Zephyr), *Plant Sci. Lett.* **3**:265–271.

Keller, W. A., and Melchers, G., 1973, The effects of high pH and calcium on tobacco leaf protoplast fusion, *Z. Naturforsch.* **28C**:737–741.

Klercker, J. von, 1892, Eine Methode zur Isolierung Lebender Protoplasten, *Ofvers. Vitens. Akad. Forhandl.* **49**:463–474.

Kumar, A., Cocking, E. C., Bovensberg, W. A., and Kool, A. J., 1982, Restriction endonuclease analysis of chloroplast DNA in interspecies somatic hybrids of petunia, *Theoret. Appl. Genet.* **62**:377–383.

Kuster, E., 1909, Uber die Verschmelzung Nackter Protoplasten, *Ber. Dtsch. Bot. Ges.* **27**:589–598.

Leigh, R. A., and Branton, D., 1976, Isolation of vacuoles from root storage tissue of *Beta vulgaris* L, *Plant Physiol.* **58**:656–662.

Lorz, H., and Potrykus, I., 1979, Regeneration of plants from mesophyll protoplasts of *Atropa belladonna, Experientia* **35**:313–314.

Maliga, P., Lorz, H., Lazar, G., and Nagy, F., 1982, Cytoplast-protoplast fusion for interspecific chloroplast transfer in *Nicotiana, Mol. Gen. Genet.* **185**:211–215.

Mastrangelo, I. A., 1979, Protoplast fusion and organelle transfer, in: *Nicotiana, Procedures for Experimental Use,* USDA Technical Bulletin #1586, pp. 65–73.

Matthews, B. F., and Widholm, J. M., 1985, Organelle DNA compositions and isoenzymes expression in an interspecific somatic hybrid of *Daucus, Mol. Gen. Genet.* **198**:371–376.

Medgyesy, P., Menczel, L., and Maliga, P., 1980, The use of cytoplasmic streptomycin resistance; chloroplast transfer from *Nicotiana tabacum* into *Nicotiana sylvestris* and isolation of their somatic hybrids, *Mol. Gen. Genet.* **179**:693–698.

Menczel, L., 1983, Protoplast fusion for in vitro genetic modification of plants, *Whats New Plant Physiol.* **14**(3):9–12.

Menczel, L., Nagy, F., Kiss, Z. R., and Maliga, P., 1981, Streptomycin resistant and sensitive hybrids of *Nicotiana tabacum + Nicotiana knightiana:* Correlation of resistance with *N. Tabacum* Plastids, *Theoret. Appl. Genet.* **59**:191–195.

Nagata, T., 1978, A novel cell-fusion method of protoplasts by polyvinyl alcohol, *Naturwissenschaften* **65**:263–264.

Nagata, T., and Melchers, G., 1978, Surface charge of protoplasts and their significance in cell–cell interaction, *Planta* **142**:235–238.

Potrykus, I., 1971, Intra and Interspecific fusion of protoplasts from petals of *Torenia bailloni* and *Torenia fournieri, Nature (New Biol.)* **231**:57–58.

Power, J. B., Cummins, S. E., and Cocking, E. C., 1970, Fusion of isolated plant protoplasts, *Nature (Lond.)* **255:**1016–1018.

Raj, B., and Herr, J. M., 1970, The isolation of protoplasts from the placental cells of *Solanum nigrum* L, *Protoplasma* **69:**291–300.

Reinert, J., and Yeoman, M. M., 1982, *Plant Cell and Tissue Culture, A Laboratory,* Springer-Verlag, New York.

Ruesink, A. W., 1971, The plasmamembrane of *Avena* coleoptile protoplasts, *Plant Physiol.* **47:**192–195.

Ruesink, A. W., 1979, Fusion of higher plant protoplasts, *Methods Enzymol.* **58:**359–367.

Ruesink, A. W., and Thimann, K. V., 1965, Protoplasts from the *Avena* coleoptile, *Proc. Natl. Acad. Sci. U.S.A.* **54:**56–64.

Sacristan, M. D., and Melchers, G., 1969, The caryological analysis of plants regenerated from tumorous and other callus cultures of tobacco, *Mol. Genet.* **105:**317–333.

Sacristan, M. D., and Melchers, G., 1977, Regeneration of plants from habituated and agrobacterium transformed single cell clones of tobacco, *Mol. Genet.* **152:**111–117.

Saunders, J. A., 1979, Investigations of vacuoles isolated from tobacco. I. Quantitation of nicotine, *Plant Physiol.* **64:**74–78.

Saunders, J. A., 1985, Electrically induced fusion of cells and protoplasts, in: *Frontiers of Membrane Research in Agriculture,* Vol. 9 (J. B. St. John, E. Berlin, and P. C. Jackson, eds.), pp. 147–156, Rowman and Allanheld, Totowa, New Jersey.

Saunders, J. A., and Gillespie, J. M., 1984, Localization and substrate specificity of glycosidases in vacuoles of *Nicotiana rustica, Plant Physiol.* **76:**885–888.

Schenk, R. U., and Hildebrant, A., 1969, Production of protoplasts from plant cells in liquid culture using purified commercial cellulases, *Crop Sci.* **9:**629–631.

Schieder, O., 1975, Regeneration of Haploid and Diploid *Datura innoxia* Mill. mesophyll protoplasts to plants, *Z. Pflanzenphysiol.* **76:**462–466.

Schleiden, M. J., 1838, Beitrage zur Phytogenesis, *Arch. Anat. Physiol. Wiss. Med.* 137–176.

Schmidt, R., and Poole, R. J., 1980, Isolation of protoplasts and vacuoles from storage tissue of red beet, *Plant Physiol.* **66:**25–28.

Schwann, T., 1839, Mikroskopische Untersuchungen über die Ubereinstimmung in der Struktur und dem Wachstume der Tiere und Pflanzen, #176 in: *Ostwalds Klassiker der Exakten Wissenschaften,* Engelmann, Leipzig, 1910.

Skene, K. G. M., 1974, Culture of protoplasts from grape vine pericarp callus, *Aust. J. Plant Physiol.* **1:**371–376.

Smith, C. L., Ahkong, Q. F., Fisher, F., and Lucy, J. A., 1982, Is purified poly(ethylene glycol) able to induce cell fusion?, *Biochim. Biophys. Acta.* **692:**109–114.

Taiz, L., and Jones, R. L., 1971, The isolation of barley-aleurone protoplasts, *Planta* **101:**95–100.

Takebe, I., Labib, G., and Melchers, G., 1971, Regeneration of whole plants from isolated mesophyll protoplasts of tobacco, *Naturwissenschaften* **58:**318–320.

Vasil, I. K., and Vasil, V., 1980, Isolation and culture of protoplasts in perspectives in plant cell and tissue culture (I. K. Vasil, ed.), *International Review of Cytology,* Suppl. 11B, pp. 1–19, Academic Press, New York.

Virchow, R., 1858, *Die Cellularpathologie in Ihrer Begrundung auf Physiologische und Pathologische Gewebelehre,* Hirschwald, Berlin.

Wakasa, K., 1973, Isolation of protoplasts from various plant organs, *Jpn. J. Genet.* **48:**279–289.

Zelcer, A., Aviv, D., and Galun, E., 1978, Interspecific transfer of cytoplasmic male sterility by fusion between protoplasts of normal *Nicotiana sylvestris* and X-ray irradiated protoplasts of male sterile *N. tabacum, Z. Pflanzenphysiol.* **90:**397–407.

Zosimovitch, V. P., and Kunach, V. A., 1975, Level, types and origin of chromosome aberration in culture of isolated plant tissues, *Sov. Genet.* **11:**37–46.

Chapter 26

Nongenetic Techniques for Isolating Fusion Products between Different Cell Types

Woodring E. Wright

1. INTRODUCTION

Until recently, virtually all investigations into the behavior of cell hybrids have employed genetic selection techniques, such as the hypoxanthine-aminopterin-thymidine (HAT) system (Littlefield, 1964), to isolate the cell fusion products. In this approach, one starts with two cell lines, one of which lacks enzyme A and the other of which lacks enzyme B. After fusing the two cell types together, they are placed in a specially designed culture medium in which only cells that have both enzymes are able to grow. Under these conditions, neither parental cell line is able to divide. However, since the cell line that lacks enzyme A has enzyme B and vice versa, fusion products between the two do contain both enzymes and the cell hybrids are able to multiply. Although systems based on this approach have been extremely useful, there are a variety of practical and theoretical problems associated with their use. Obviously, only special cell lines that lack the necessary enzymes can be used. Although in some cases this problem can be surmounted by mutagenizing and isolating cells carrying the required defect, this can only be done with immortal cell lines.

WOODRING E. WRIGHT • Department of Cell Biology, University of Texas Southwestern Medical School, Dallas, Texas 75235.

The cultured life span of normal diploid cells (Hayflick, 1965) is limited, so that by the time mutant clones are isolated they are essentially senescent. Some half-selective strategies have been developed in which an established immortal cell line is fused to a normal diploid cell type which does not divide or only divides a limited number of times, and where it is thus unnecessary to select against the normal diploid parental cell type. By and large, however, the inability to use genetic selection techniques with normal cells represents a major limitation for many types of studies.

A second limitation that has not been widely appreciated is that, since these selective systems depend on the ability of cells to divide in the selective medium, a substantial bias is introduced. In most reports fewer than 1% of the cells that actually fuse give rise to growing hybrid colonies, and in some cases it is as few as 0.001% (Miller and Ruddle, 1976). Although this may not matter in experimental systems in which cell division and the phenotype of interest are unlinked, many situations are being studied using somatic cell hybridization in which the two processes are intimately associated. For example, most normal differentiated cells are either slowly dividing or postmitotic. Limiting the analysis of cell hybrids to the rare fusion product capable of extensive cell division will obviously bias the results against the expression of differentiated products. Similar considerations apply to cell-fusion investigations of tumorigenesis.

Heterokaryons are the initial fusion product between two cell types, when the different nuclei reside within a common cytoplasm. Once the heterokaryons divide, their progeny are referred to as cell hybrids. Cell hybrids are generally mononucleated, so that both genomes then reside within a single nucleus. Because most genetic selection systems depend on the selective ability of the fusion products to divide, by definition they are incapable of isolating the heterokaryons. Because of this, the behavior of heterokaryons has remained largely unexplored. It is now becoming clear that the regulatory interactions that occur in the heterokaryons before the onset of DNA synthesis can be very different from those that are observed in dividing cell hybrids (Wright, 1984*a,b*; Blau *et al.,* 1983).

2. NONGENETIC SELECTION TECHNIQUES

A variety of approaches have been developed to circumvent the biases and limitations associated with the genetic selection techniques. Velocity sedimentation has been used to separate the larger polynucleated cell fusion products from the unfused mononucleated parental cells (Hohman and Shows, 1979; Eisenberg and Migeon, 1979). Although this method has the advantage of being used on very large numbers of cells, it only provides a relatively modest enrichment, since it cannot distinguish between homokaryons (self–self fusions) and true heterokaryons. The fluorescence activated cell sorter (FACS) has been used to isolate fusion products between cells carrying different fluorescent markers (Jonkind and Verkerk, 1982). For example, cells can be tagged with fluorescein- or rhodamine-labeled fatty acids (Keeler *et al.,* 1977) or with vital dyes such as the mitochondrial stain rhodamine 123. After the cells are fused, they are subjected to two-color fluorescent sorting, so that only the heterofusion products that exhibit both fluorescent patterns are isolated. This approach is elegant, but only relatively small numbers of cells can be manipulated, and it requires access to expensive and sophisticated equipment.

I have developed two nongenetic systems for isolating heterokaryons and cell hybrids. The first exploits the differing specificities of two irreversible biochemical inhibitors: iodoacetamide (directed primarily toward sulfhydryl groups) and diethylpyrocarbonate (which reacts mainly with histidine residues). One cell population is treated with a lethal concentration of iodoacetamide while the other is treated with diethylpyrocarbonate. Both treatments are carried out for a short time (30 min) in the cold; although a lethal lesion is introduced, the cells are temporarily prevented from expressing the metabolic consequences of the lesion and dying. The unreacted inhibitors are then washed away and the two cell types mixed together and fused with polyethylene glycol. Because the two inhibitors have different targets, a different spectrum of molecules are inactivated in each cell type. Iodoacetamide-treated cells contain active versions of molecules that had been inactivated in the diethylpyrocarbonate treated cells and vice versa. Complementation thus occurs, and the heterokaryons survive, whereas the unfused parental cells die as a consequence of their lethal biochemical treatments (Wright, 1978, 1982). Most of the results concerning cell differentiation discussed in Section 4 were obtained using this selective system. However, this technique also has several important drawbacks. Not surprisingly, iodoacetamide and diethylpyrocarbonate are sufficiently nonspecific that many molecules are affected by both agents. Consequently, if high doses are used, so much overlap in the effects of these agents is produced that complementation no longer occurs and the heterokaryons die. It is thus necessary to use a very narrow window of concentrations that is just sufficient to kill the parental cells while still permitting complementation in the heterokaryons. The technique is thus somewhat tricky to use. In addition, although a variety of controls can be performed to show that the biochemical treatments are not producing long-term changes in the parameter being studied, lingering doubts remain as to the consequences of these horrendous treatments. I thus developed a second selective system that is much simpler to use and that avoids these difficulties.

3. TOXIN-ANTITOXIN SELECTION

Okada and Rechsteiner (1982) devised a very elegant technique for introducing macromolecules into cells. Cells are first allowed to pinocytose a hypertonic sucrose solution containing the molecule. Normally, these pinosomes would be transported to the lysosomes, where their contents would be degraded. However, if the cells are exposed to a brief *hypotonic* shock after pinocytosing for 10 min, the hypertonic pinosomes swell and burst, releasing their contents into cytoplasm. This method permits macromolecules such as antibodies to be osmotically injected into mass populations of cells, and thus allows the strategy illustrated in Figure 1 to be performed. Ricin is a plant toxin that inactivates the 60 S ribosomal subunit, whereas diphtheria toxin inactivates elongation factor 2. Since both toxins act catalytically, it can take as few as one molecule per cell to inhibit protein synthesis and kill the cell. Antiricin antibodies are injected

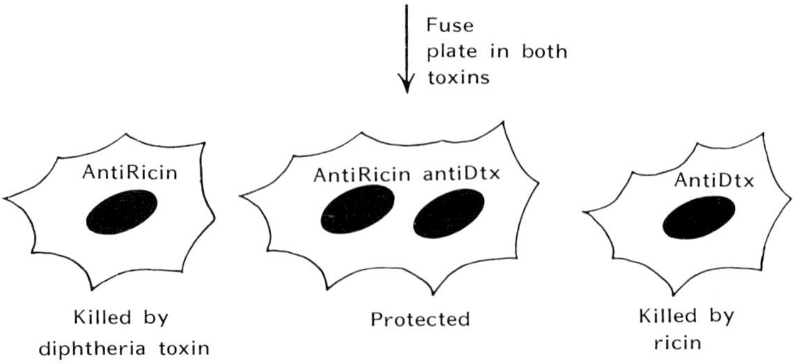

Figure 1. Strategy for toxin–antitoxin selection of heterokaryons. Cells osmotically injected with different antitoxin antibodies are fused together and plated in the presence of both toxins. Only heterokaryons containing both antitoxins survive.

into one cell type, and antidiphtheria toxin antibodies are osmotically lysed into another. The two cell types are now fused together and plated in a lethal concentration of both toxins. The antiricin-injected cells will be killed by the diphtheria toxin, the antidiphtheria toxin-injected cells will be killed by the ricin, but the heterokaryons will contain both antibodies and thus be protected from the effects of both toxins. Using this approach, we have been able to obtain populations of normal diploid human fibroblasts in which 99% of the nuclei are present in heterokaryons and where virtually all heterokaryons that were formed survived selection (Wright, 1984*c*).

The basic selection protocol is to osmotically inject each cell type, wash the cells thoroughly to remove all noninternalized antibody, fuse with polyethylene glycol (PEG), plate overnight in various toxin concentrations, and then feed toxin-free medium the following day. The killed parental cells detach from the dish by the third day following cell fusion, so that aliquots of the surviving cells can be fixed and stained to analyze for the purity of the population (percentage of nuclei in cells containing more than one nucleus). A detailed consideration of each step is presented in the following sections.

3.1. Osmotic Injection

The original injection procedure of Okada and Rechsteiner (1982) consisted of preparing a solution of 0.5 M sucrose and 10% (v/v) PEG (1000 M_r) in serum-free medium containing the macromolecule. Cells were then exposed to this solution for 10 min at 37°C, either as a cell monolayer or as a suspension of detached cells. The advantage of using suspended cells is that a much greater number of cells could be treated in a much smaller volume of hypertonic sucrose. The cells were then hypotonically shocked by diluting the 0.1–0.2 ml of hypertonic solution with 10 ml of medium–distilled water (60 : 40). Two minutes later, the cells were washed (or centrifuged and resuspended in the case of the detached cells) several times with isotonic medium and then fed maintenance medium. The important parameters in this protocol include the concentration of sucrose, the use of PEG, the temperature, the time of treatment, and the time and degree of the hypotonic treatment. In order to optimize each variable for a particular cell type, an adequate assay system is needed.

3.1.1. Assay Systems

3.1.1a. Horseradish Peroxidase. Horseradish peroxidase provides an extremely sensitive tracer for measuring the amount of bulk phase pinocytosis (Steinman *et al.*, 1974). Total uptake can easily be measured by lysing the cells in 2% Triton X-100 in distilled water, then using a soluble substrate such as ABTS (2′2′azino-di-(3-ethylbenzthiazoline sulfonic acid), 0.3 mg/ml in pH4 citrate buffer to which 0.01% hydrogen peroxide is added just before use) to measure peroxidase activity. The reaction can be stopped after 5–10 min by the addition of one-quarter volume of 1.25% sodium fluoride; the absorbance at 490 nm is then measured and compared with a standard curve. Total uptake does not distinguish between enzyme that is still contained within pinosomes and that which is free in the cytoplasm. Okada and Rechsteiner (1982) used cell fractionation techniques to separate soluble and particulate compartments and quantitate the free cytoplasmic peroxidase following osmotic injection. In our hands, this separation was not sufficiently clean to permit an adequate optimization of the different variables. Although we have found horseradish peroxidase to be very useful for quantitating total uptake (and thus the maximum amount that would be released if all of the pinosomes were lysed), it has not been satisfactory for quantitating the amount actually released. Horseradish peroxidase (HRP) histochemistry, however, is extremely useful in determining if one is near or far from optimal conditions. The cells are fixed (2% glutaraldehyde in saline for 3 min at room temperature) approximately 3 hr after osmotic injection. They are then stained for 20–90 min with a substrate that gives an insoluble product (diaminobenzidine, 0.5 mg/ml in 0.1 M Tris buffer pH7–7.5 to which

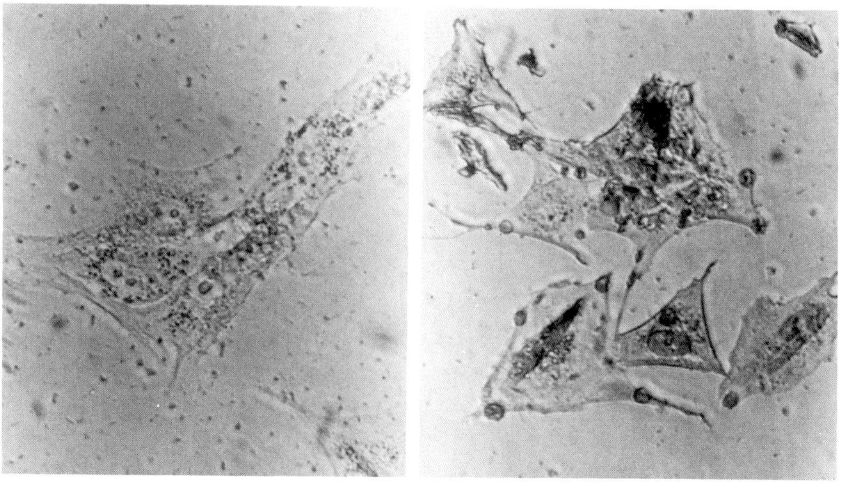

Figure 2. Horseradish peroxidase histochemistry of control (left) and osmotically injected cells (right). In the absence of hypertonic treatments, horseradish peroxidase activity is granular (left, presumably in lysozomes). Following osmotic lysis of the pinosomes, the peroxidase activity is diffusely distributed throughout the nucleus and cytoplasm (right).

0.03% hydrogen peroxide is added just before use). The amount of particulate versus diffuse staining (Fig. 2) then indicates whether one is lysing only a few or most of the pinosomes.

3.1.1b. Antibody Protection. A rough measure for quantitating either the effective titre of an antibody or the efficiency of osmotic injection is to inject an antitoxin antibody and then to measure the amount of protection obtained. This is done by plating control cells (noninjected, mock injected, or injected with an irrelevant antibody) and injected cells in serial threefold dilutions of the toxin. The next day, the cells are fed toxin-free medium. By the third day after injection, most of the cells that were killed by the exposure to the toxin will have died and detached from the dish, so the viability for each toxin dose can be easily determined by estimating the percentage of cells still attached with respect to a no-toxin control. Figure 3 shows the results of a typical experiment. The degree of protection is defined as the increase in the LD_{50} for the injected as compared with the control cells. The main disadvantage of this approach is its lack of sensitivity. Dilution experiments have shown that it takes about a 10-fold increase in antibody concentration in the hypertonic solution in order to produce a threefold increase in the amount of protection con-

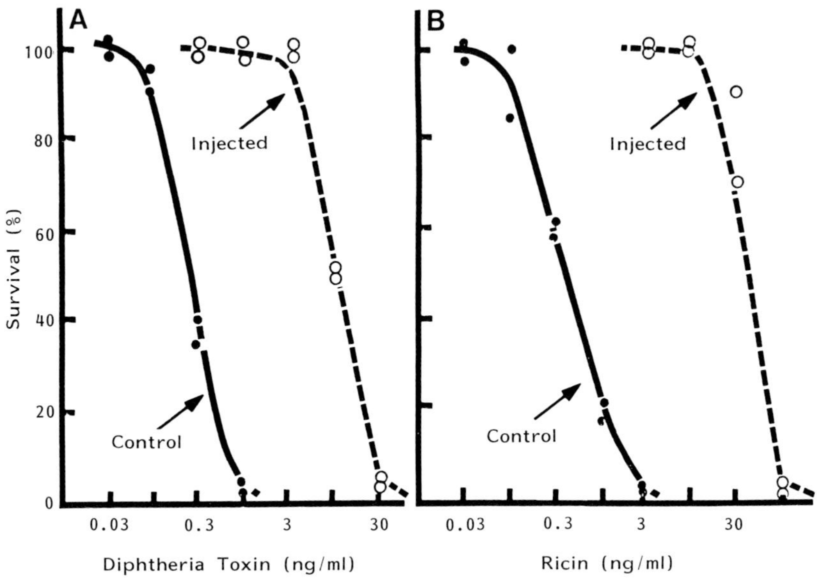

Figure 3. Antitoxin protection against toxin killing. The relative survival in each aliquot was estimated from the number of human fibroblast cells remaining attached 2 days after an overnight toxin treatment for (A) antidiphtheria toxin-injected cells and (B) antiricin-injected cells.

ferred. Since the estimation of the LD_{50} is only accurate within about a factor of two, this assay becomes totally inadequate for determining if one has actually optimized the injection protocol (e.g., it would not distinguish between methods in which 30% versus 60% of the pinosomes had ruptured).

3.1.1c. Ricin A Chain. Ricin is composed of two subunits, the catalytic A chain and the cell-binding component, the B chain. Since it is unable to bind to cells by itself, the purified A chain is nontoxic. It thus provides a very sensitive probe for the injection of macromolecules using the osmotic lysis technique. Ricin A chain that is released into the cytoplasm will inactivate the 60 S ribosomal subunit and inhibit protein synthesis, whereas ricin A chain contained within unbroken pinosomes travels to the lysosome, where it is degraded. As the rate of inhibition of protein synthesis is directly proportional to the amount of ricin A chain liberated in the cytoplasm, this approach provides a much more sensitive measure of the efficiency of the osmotic injection technique than the degree of protection afforded by antitoxin antibodies. In a typical experiment, we would inject cells with a constant amount of ricin A chain (e.g., 15 ng/ml) using different osmotic lysis conditions and then wash away the free ricin A chain and put the cells back in culture. Three hours later, the cells would be labeled for 30 min with [^{35}S]methionine; then the cells would be lysed, TCA precipitated, and scintillation counted to determine the rate of protein synthesis. The relative amount of inhibition produced under the different conditions would then indicate which condition gave the greatest amount of osmotic lysis.

In the process of optimizing the conditions for human diploid lung fibroblasts described below, the following general sequence was followed. Horseradish peroxidase was injected and histochemistry performed 3 hr later. This demonstrated that, although some horseradish peroxidase was being released into the cytoplasm, most of it was remaining in pinosomes and going to lysosomes. Various parameters were then investigated using the ricin A chain approach. Finally, the revised protocol was tested with HRP histochemistry, which demonstrated that the modifications were effective and now most of the peroxidase activity was being released into the cytoplasm.

3.1.2. Optimization of Conditions

3.1.2a. Sucrose Concentration. The efficiency of lysis of the pinosomes was directly proportional to the concentration of sucrose in the hypertonic medium; 1 M sucrose resulted in approximately three times as much ricin

A chain being released into the cytoplasm as 0.5 M sucrose. Unfortunately, higher concentrations of sucrose were extremely toxic; 1 M sucrose was chosen as the best compromise between efficiency and toxicity.

3.1.2b. Temperature. The rate of pinocytosis shows a high temperature dependence, with a tenfold increase in rate for each 2.7°C increase in temperature (Steinman *et al.,* 1974). This suggests that the highest possible temperature should be used. Unfortunately, the toxicity of hypertonic sucrose is directly proportional to temperature. For example, whereas 50% of the cells survived a 10-min exposure to 1 M sucrose at 37°C, only 15% survived a similar treatment at 40°C. The fact that a 3°C temperature change makes such a dramatic difference in cell viability demonstrates that great care should be exercised in maintaining a reproducible temperature. Although lower temperatures give greater viability, they also result in both lower rates of pinocytosis and lower efficiencies of lysis of the pinosomes. We decided to optimize the remaining conditions using the physiological temperature of 37°C.

3.1.2c. Time. Experiments in which cells were pulsed with HRP for various intervals during a 10-min hypertonic treatment showed that 50% of the total was taken up within the first 3 min, and 70% within the first 5 min. Since the toxicity of a 5-min treatment was less than that of a 10-min one, we decided to utilize 5 min as our standard treatment time.

3.1.2d. PEG Concentration. Okada and Rechsteiner (1982) reported that including 10% PEG 1000 in their hypertonic sucrose solution was essential for obtaining rupture of the pinosomes in mouse L cells. Using ricin A chain, we could detect no change in the efficiency of pinosome rupture in human diploid fibroblasts with or without PEG. Since 10% PEG decreased the viability resulting from a 10-min treatment with 0.5 M sucrose from 80% to 60%, we have eliminated it from our hypertonic protocol.

3.1.2e. Hypotonic Treatment. Initial experiments using the original hypertonic protocol (0.5 M sucrose with 10% PEG for 10 min) suggested that increased pinosome rupture was obtained when the cells were shocked with increasingly hypotonic solutions. However, the toxicity of the treatments paralleled their increased efficiency. Since HRP histochemistry showed that the above modifications (1 M sucrose without PEG for 5 min) already resulted in lysing most of the pinosomes and little would be gained from further modifications, we did not pursue the use of more hypotonic treatments. In addition, since medium–distilled water

(60 : 40) displayed little toxicity during at least a 15-min treatment, it seemed prudent to minimize the number of critical variables and maintain the hypotonic exposure within conditions that exhibited a large toleration for variation.

3.1.2f. Calcium Concentration. Because the efficiency of pinosome rupture and toxicity seemed to parallel each other in so many of the above variables, we hypothesized that some of the toxicity might in fact result from the lysis of pinosomes. One possible mechanism for this might be changes in intracellular calcium concentrations. Extracellular calcium concentrations are about 1000 times greater than intracellular ones. HRP experiments indicated that the cells were taking up about 10 fl of solution during a 10-min exposure at 37°C. This represents about 1/1000th of the cytoplasmic volume of a cell 30 μm in diameter in which one-third of the total volume is nuclear. The 10 fl of uptake might thus be enough to double the intracellular calcium concentration, wreak havoc on a multitude of calcium-regulated processes, and cause cell death. Reducing the calcium concentration in the hypertonic solution by adding EGTA or using Ham's F12 (which has about 17% as much calcium in it as Dulbecco's MEM) did result in a substantial decrease in toxicity. However, it also resulted in a 50% decrease in the amount of ricin A chain released into the cytoplasm. Although a tradeoff is involved, we believe that in most experiments the reduction in toxicity is the more important consideration and thus recommend using a calcium concentration of 0.3 mM during the hypertonic treatment. In any case, it is important to recognize calcium as a critical variable, because the concentration of calcium in serum (e.g., a rabbit antitoxin serum) is probably about 3–4 mM, as compared with only 0.3 mM in Ham's F12 medium. If one makes a hypertonic solution, for example, by mixing equal volumes of 2 M sucrose (dissolved in Ham's F12) and the antitoxin serum, the concentration of calcium in the actual experimental solution would be greater than that in test solutions made using Ham's F12 alone. Furthermore, if one then purified the antibodies and used them instead of the crude serum, the calcium concentration would be different from in the initial tests using the crude serum. Although we have not yet chosen a definitive calcium concentration, further studies into its role in osmotic lysis are clearly indicated.

3.2. PEG Fusion

Dimethylsulfoxide (DMSO) has been found to broaden the concentration range over which polyethylene glycol is effective in producing membrane fusion (Norwood *et al.,* 1976). Although many fusion protocols call for the use of 50% PEG, it is our experience that a solution of DMSO–PEG 1000–serum-free medium (10 : 35 : 55) provides for very effective fusion conditions with a dramatically reduced level of toxicity. Cells are centrifuged, aspirated completely to give a dry pellet, and then resuspended by trituration for 15 sec using approximately 0.1 ml of PEG solution per 5–10 million cells. The suspended cells are then rotated gently for an additional 45 sec and the reaction stopped by the addition of 2 ml of complete medium. The cells are then very gently distributed into their final plating medium. We have not found it necessary to remove the PEG from the plating medium as long as the PEG solution has been diluted more than 40-fold and the cells are going to be fed the following day.

3.3. Current Toxin–Antitoxin Selection Protocol

The following sample protocol calls for several centrifugations and washes. For reasons that are unclear to us, regardless of the speed and time of centrifugation, approximately 20% of the cells are lost during each spin. Consequently, after four spins, one is left with only about 40% of the initial cell number (0.8^4). It is thus essential to make each wash as efficient as possible. If cells are centrifuged and the supernatant is aspirated, 10–20 μl of liquid is left adhering to the walls of the tube. This substantially reduces the efficiency of the subsequent wash. The most effective procedure is to pellet the cells (e.g., 700 $\times g$ for 2 min, which is about 2000 rpm in a typical benchtop centrifuge), pour off the supernatant, recentrifuge briefly by accelerating until 1000 rpm is reached, and then aspirate off the remaining fluid. This procedure is used during each centrifugation in the following protocol.

Five million cells of each cell type are trypsinized, centrifuged, and resuspended in 0.2 ml of hypertonic antitoxin serum containing 1 M sucrose. After being incubated at 37°C for 5 min, 10 ml of hypotonic solution (40% distilled water in medium) is added. Two minutes later, the cells are spun, resuspended in 10 ml of isotonic medium, spun, and resuspended again in isotonic medium. At this point, aliquots can be removed, spun, and plated to determine the degree of protection against the toxin (as in Fig. 3). The remaining cells are mixed with the cells of the opposite type, spun, fused with PEG 1000–DMSO, diluted in complete medium (35 : 10 : 55), and plated overnight in the desired toxin concentrations. Initially, this might be a 3 $\times$ 3 matrix of toxin concentrations (e.g., 1, 3, and 10 ng/ml ricin and 0.1, 0.3, and 1 ng/ml abrin). Later, after the optimal conditions have been determined, a single combination can be used. The cells are fed toxin-free complete medium the next day and then analyzed as needed.

4. APPLICATIONS

The complementation analysis of human genetic syndromes permits one to determine the number of alleles, and thus presumably the number of different steps in a metabolic pathway that can lead to a given phenotype. It thus provides an important first step in defining the biochemical basis of a particular syndrome. Complementation analysis has been difficult to perform because of the inability to isolate fusion products between normal diploid cells. Most studies have been limited to syndromes in which single cells can be analyzed, so that the problem of contaminating parental cells is circumvented by only analyzing cells containing more than one nucleus. In order to demonstrate the usefulness of the toxin–antitoxin system for complementation analysis, we showed that the known complementation groups of the syndrome xeroderma pigmentosa produced the predicted complementation patterns in heterokaryons purified using antiricin and antidiphtheria toxin antibodies (Wright, 1984*c*). The toxin–antitoxin selective system should thus make complementation analysis a routine approach for analyzing human genetic diseases in which an abnormal phenotype has been identified in cultured cells.

The ability to isolate heterokaryons also provides an important tool for the study of cell differentiation. Cell hybrid studies have led to the general conclusion that differentiated functions are suppressed in hybrids between cells of different developmental lineages (Davidson, 1974; Ringertz and Savage, 1976). We (Wright, 1984*a,b*) and others (Blau *et al.,* 1983) have now shown that very different interactions occur at the level of the initial fusion product, the heterokaryon. For example, rat skeletal myosin light chain synthesis is induced following the fusion of a rat glial cell to a differentiated chick skeletal myocyte. By contrast, myogenesis is suppressed when the same rat glial cell line is used to form dividing hybrids with a line of rat myoblasts (Wright, 1984*a*). The simplest interpretation of these superficially contradictory results is that in the heterokaryon one is looking at the ability of regulatory molecules already present in a differentiated myocyte to act on the glial genome. On the other hand, in the cell hybrids, one is looking at the ability of the myogenic precursor component to influence the decision of the entire cell to terminally differentiate. This decision is subject to a complicated series of steps: The cells must have the appropriate receptors for and be able to respond to the environmental signals that normally stimulate myogenesis and must successfully complete all the intermediate steps leading to the final expression of the myogenic phenotype. In other words, in order to be myogenic, the myogenic phenotype must be dominant at every step in the entire sequence in cell hybrids, whereas in heterokaryons involving an already differentiated myocyte it is only the final regulatory stage(s) being examined. The ability to examine specific stages in development should make heterokaryons a powerful tool for studying the mechanism regulating cell differentiation.

The most straightforward use of the toxin–antitoxin selection technique is thus for the isolation of heterokaryons. We have found that it can routinely produce populations in which greater than 95% of the nuclei are present in heterokaryons, which are eminently suitable for a variety of biochemical analyses. Its use for the production of dividing cell hybrids is more problematic and requires very careful controls. Many of the heterokaryons contain more than two nuclei, so that if one restricts the analysis to mono- versus binucleates and counts cells rather than nuclei, the purity of the population can drop from 95% to 60%. Since it is likely that the cloning efficiency of a contaminating parental cell is much greater than that of a binucleated heterokaryon, most of the colonies that form may in fact contain parental cells rather than hybrids. It is thus essential to have an independent measure (e.g., karyotype, isozyme pattern) to verify that a particular colony is an actual hybrid, since the simple ability to form a colony following toxin–antitoxin selection does not provide sufficient evidence to conclude that it is a hybrid.

A good illustration for the necessity for verifying the hybrid nature of growing colonies is the use of the toxin–antitoxin system for producing hybridomas. In these experiments, normal spleen cells were injected with antiricin antibodies. One-half the cells were mixed with myeloma cells and fused with polyethylene glycol. As a control, the other half was first fused to itself and only then mixed with the myeloma cells. The cells were then plated overnight in 3 ng/ml ricin and washed and distributed at various dilutions among 96-well dishes the next day. A week later, the number of colonies in the two groups was determined. Although the control contained 5000 colonies, the actual spleen cell–myeloma fusion contained 50,000 colonies, suggesting that 90% of those colonies should represent true hybrids. When individual colonies were picked and karyotyped, however, only a rare colony exhibited more chromosomes than the parental myeloma. We have repeated this experiment in a variety of ways and consistently obtain the result that the amount of survival in the control does not reflect the amount of parental cell contamination of the hybridized cells. Our working hypothesis (for which we have no evidence) is that when cells are fused with polyethylene glycol, multiple points of nascent membrane fusion are established. If complete fusion is established, these many points of contact coalesce to provide permanent cytoplasmic continuity between the two cells. However, there may be many situations in which an insufficient number of points of membrane fusion are established, and the cells then separate back to being two individual cells. If some cytoplasmic exchange can occur during this process, limited amounts of antiricin antibodies in the spleen cell could be transferred to a myeloma cell without heterokaryon formation. This myeloma cell would then be slightly more resistant to ricin killing than the control myeloma cells, giving the observed results. It is important to note that 50,000 colonies represents 0.1% of the input myeloma cells and thus a very small fraction of the initial material. In the human fibroblast experiments described above, approximately 20% of the input cells were recovered as heterokaryons in the toxin–antitoxin selected populations. A 0.1% increase in contaminating parental cells as compared to that predicted by the controls would have been totally undetectable. However, if only 1 in 1000 of the heterokaryons is capable of successfully making the transition and becoming a dividing hybrid cell, such small changes become important.

It is possible to enucleate cells using the combination of high centrifugal forces and cytochalasin B (Wright, 1973; Veomett, 1982). Anucleate cytoplasms produced by this technique remain viable for up to a day or so and maintain the ability to perform a variety of cellular processes, including pinocytosis. It is thus possible to inject antitoxin antibodies osmotically into anucleate cytoplasts and then fuse them to whole cells, forming heteroplasmons that can be isolated following toxin treatment. The toxin–antitoxin system thus provides a flexible approach for isolating fusion products between a variety of cells or cell fragments and should provide a valuable addition to the repertoire of selection techniques. Although in many cases it can probably be used to isolate cell hybrids, its major use and importance should be in providing purified populations of heterokaryons for the detailed analysis of regulatory interactions.

REFERENCES

Blau, H. M., Chin, C. -P., and Webster, C., 1983, Cytoplasmic activation of human nuclear genes in stable heterocaryons, *Cell* **32:**1171–1180.

Davidson, R. C., 1974, Control of expression of differentiated functions in somatic cell hybrids, in: *Somatic Cell Hybridization* (R. L. Davidson and F. de la Cruz, eds.), pp. 131–150, Raven Press, New York.

Eisenberg, C. R., and Migeon, B. R., 1979, Enrichment of human heterokaryons by ficoll gradient for complementation analysis of iduronate sulfatase deficiency, *Somatic Cell Genet.* **5:**1079–1089.

Hayflick, L., 1965, The limited *in vitro* lifetime of human diploid cell strains, *Exp. Cell Res.* **37:**614–636.

Hohman, L. K., and Shows, T. B., 1979, Complementation of genetic disease: A velocity sedimentation procedure for the enrichment of heterokaryons, *Somat. Cell Genet.* **5:**1013–1029.

Jongkind, J. F., and Verkerk, A., 1982, Nonselective isolation of fibroblast heterokaryons, hybrids, and cybrids by flow sorting, in: *Techniques in Somatic Cell Genetics* (J. Shay, ed.), pp. 81–100, Plenum Press, New York.

Keeler, P. M., Person, S., and Snipes, S., 1977, A fluorescent enhancement assay of cell fusion, *J. Cell Sci.* **28**:167–177.

Littlefield, J., 1964, Selection of hybrids from mating of fibroblasts *in vitro* and their presumed recombinants, *Science* **145**:709–710.

Miller, R. A., and Ruddle, F. H., 1976, Pluripotent teratocarcinoma–thymus somatic cell hybrids, *Cell* **9**:45–55.

Norwood, T. H., Zeigler, C. J., and Martin, G. M., 1976, Dimethyl sulfoxide enhances polyethylene glycol-mediated somatic cell fusion, *Somat. Cell Genet.* **2**:263–270.

Okada, C. Y., and Rechsteiner, M., 1982, Introduction of macromolecules into cultured mammalian cells by osmotic lysis of pinocytic vesicles, *Cell* **29**:33–41.

Ringertz, N. R., and Savage, R. E., 1976, *Cell Hybrids,* Academic Press, New York.

Steinman, R. M., Silver, J. M., and Cohn, Z. A., 1974, Pinocytosis in fibroblasts. Quantitative studies *in vitro, J. Cell Biol.* **63**:949–969.

Veomett, G., 1982, Techniques for enucleation of mammalian cells, in: *Techniques in Somatic Cell Genetics* (J. W. Shay, ed.), pp. 67–79, Plenum Press, New York.

Wright, W. E., 1973, Production of mass populations of anucleate cytoplasms, in: *Methods in Cell Biology,* Vol. VII (D. M. Prescott, ed.), pp. 203–210, Academic Press, New York.

Wright, W. E., 1978, The isolation of heterokaryons and hybrids by a selective system using irreversible biochemical inhibitors, *Exp. Cell Res.* **112**:395–407.

Wright, W. E., 1982, The selection of heterokaryons and cell hybrids using the biochemical inhibitors iodoacetamide and diethylpyrocarbonate, in: *Techniques in Somatic Cell Genetics* (J. Shay, ed.), pp. 48–65, Plenum Press, New York.

Wright, W. E., 1984*a,* The induction of muscle genes in neural cells, *J. Cell Biol.* **98**:427–435.

Wright, W. E., 1984*b,* The expression of differentiated functions in heterokaryons between skeletal myocytes, adrenal cells, fibroblasts and glial cells, *Exp. Cell Res.* **151**:55–69.

Wright, W. E., 1984*c,* Toxin–antitoxin selection for isolating somatic cell fusion products between any cell types, *Proc. Natl. Acad. Sci. U.S.A.* **81**:7822–7826.

Index